T0178455

Lecture Notes in Computer Science 13261

More information about this series at https://link.springer.com/bookseries/558

Paul Groth · Maria-Esther Vidal ·
Fabian Suchanek · Pedro Szekley ·
Pavan Kapanipathi · Catia Pesquita ·
Hala Skaf-Molli · Minna Tamper (Eds.)

The Semantic Web

19th International Conference, ESWC 2022
Hersonissos, Crete, Greece, May 29 – June 2, 2022
Proceedings

Springer

Editors
Paul Groth (iD)
University of Amsterdam
Amsterdam, Noord-Holland, The Netherlands

Fabian Suchanek
Institut Polytechnique de Paris "DIG"
Télécom ParisTech
Palaiseau, France

Pavan Kapanipathi (iD)
IBM Research - Thomas J. Watson Research
Yorktown Heights, NY, USA

Hala Skaf-Molli (iD)
University of Nantes
Nantes, France

Maria-Esther Vidal (iD)
Universidad Simón Bolívar
Leibniz Information Centre for Science
and Technology
Hannover, Niedersachsen, Germany

Pedro Szekley (iD)
University of Southern California
Marina del Rey, CA, USA

Catia Pesquita (iD)
LaSIGE, Fac de Ciencias,Edif C6, Pis0 3
Universidade de Lisboa
Lisbon, Portugal

Minna Tamper (iD)
Aalto University
Espoo, Finland

ISSN 0302-9743 ISSN 1611-3349 (electronic)
Lecture Notes in Computer Science
ISBN 978-3-031-06980-2 ISBN 978-3-031-06981-9 (eBook)
https://doi.org/10.1007/978-3-031-06981-9

Preface

This volume contains the main proceedings of the 19th edition of the European Semantic Web Conference (ESWC 2022). ESWC is a major venue for discussing the latest in scientific results and innovations related to the semantic web, knowledge graphs, and web data. While the community has become excellent at virtual discussions over the past two years, we have missed the kind of spontaneous networking that comes with a chat over a coffee or an exchange at a poster session. This year, we were able to have that much needed in-person exchange again in Crete.

ESWC has always been a conference that experiments with both its format and the kinds of content featured. This year's research track is an example of such experimentation. In the past, ESWC's main research track has featured multiple sub-tracks focused on particular topical categories; this year a single research track was instituted, welcoming all relevant topic areas. Additionally, the research track experimented for the first time with double-blind reviewing. Lastly, to enable authors to cite the relevant material, references were no longer counted in page limits. The in-use track also added innovative elements by introducing three paper types, focused on end-user applications, technologies within real world applications, and experiences in large knowledge graphs. The research and in-use tracks were complemented by the resource track – focused on resources to support the research community.

The main scientific program of ESWC 2022 contained 28 papers selected out of 119 submissions (82 research, 7 in-use, 30 resource): 16 papers in the research track, 4 in the in-use track, and 8 in the resources track. The overall acceptance rate was 23% (20% research, 57% in-use, 27% resources). As you will see within these proceedings, the accepted contributions push forward the scientific conversation related to semantic technologies, ranging from deep learning and knowledge graphs to integration for software development and ontologies. ESWC continues to highlight the best research in this area. The quality of the accepted papers put the maturity level reached by semantic web technologies into perspective. They provide evidence of the impact of these technologies in areas like data management, question answering, reasoning, programming languages, and machine learning.

This program was complimented with invited keynotes from three world renowned speakers: Matthias Niepert (University of Stuttgart and NEC Labs Europe); Tova Milo (Tel Aviv University); and Axel Ngonga (Paderborn University).

The conference also offered other opportunities to discuss the latest research and innovation work, including a poster and demo session, workshops and tutorials, a PhD symposium, an EU project networking session, and an industry track. Eleven workshops and tutorials covered topics ranging from knowledge graph construction and natural language processing to linked data in construction and music. A separate volume contains proceedings from these events.

ESWC 2022 is a reflection of the work of a community. The general and program chairs would like to thank all of those involved. First, our thanks go to the 21 members of the organizing team. You did an amazing job. We would like to thank the 304 reviewers for providing their feedback on the scientific program and to their other community members. Here, the general chair extends his appreciation to the track chairs for ensuring not only a rigorous and efficient review process, but also an exciting program.

Conferences need outreach. Thanks go to Benno Kruit for managing the website and social media presence. Special thanks go to our proceedings chair, Minna Tamper, for all her work in preparing this volume with the help of Springer. We thank STI International for supporting the conference organization, and in particular Julia Weninger for her quick reactions. ESWC 2022 benefited from the support of sponsors. We thank them for the support of this community and our sponsorship chairs Albert Meroño and Joe Raad for securing them.

Lastly, even as we looked forward to coming together again as a community, as a conference based in Europe, we have been shocked by the Russian invasion of Ukraine. Our thoughts are with the colleagues and people impacted by the war.

April 2022
Paul Groth
Maria-Esther Vidal
Fabian Suchanek
Pedro Szekley
Pavan Kapanipathi
Catia Pesquita
Hala Skaf-Molli
Minna Tamper

Organization

General Chair

Paul Groth   University of Amsterdam, The Netherlands

Research Track Program Chairs

Maria-Esther Vidal   Leibniz University of Hannover and TIB - Leibniz
                             Information Centre for Science and
                             Technology, Germany
Fabian M. Suchanek   Télécom Paris, France

Resource Track Program Chairs

Catia Pesquita   LASIGE, Universidade de Lisboa, Portugal
Hala Skaf-Molli   University of Nantes, France

In-Use Track Program Chairs

Pedro Szekley   USC Information Sciences Institute, USA
Pavan Kapanipathi   IBM Research, USA

Workshops and Tutorials Chairs

Mehwish Alam   FIZ Karlsruhe – Leibniz Institute for Information
                             Infrastructure, Germany
Anastasia Dimou   KU Leuven, Belgium

Poster and Demo Chairs

Jodi Schneider   University of Illinois Urbana-Champaign, USA
Anisa Rula   University of Brescia, Italy

Symposium Chairs

Ilaria Tiddi   Vrije Universiteit Amsterdam, The Netherlands
Elena Simperl   King's College London, UK

Industry Track Program Chairs

Rinke Hoekstra   Elsevier, The Netherlands
Panos Alexopoulos   Textkernel, The Netherlands

Sponsorship

Albert Meroño   King's College London, UK
Joe Raad   University of Paris-Saclay, France

Project Networking

Valentina Presutti   University of Bologna, Italy
Marieke van Erp   KNAW Humanities Cluster, The Netherlands

Web and Publicity

Benno Kruit   VU Amsterdam, The Netherlands

Semantic Technologies

François Scharffe   Columbia University, USA

Proceedings

Minna Tamper   Aalto University, Finland

Program Committee

Ibrahim Abdelaziz   IBM, USA
Maribel Acosta   Ruhr University Bochum, Germany
Alessandro Adamou   Open University, UK
Nitish Aggarwal   Roku Inc., USA
Céline Alec   Université de Caen Normandie, France
Alsayed Algergawy   University of Jena, Germany
Andreea Iana   University of Mannheim, Germany
Grigoris Antoniou   University of Huddersfield, UK
Ghislain Auguste Atemezing   Mondeca, France
Maurizio Atzori   University of Cagliari, Italy
Sören Auer   TIB - Leibniz Information Center Science and
                                        Technology and University of Hannover,
                                        Germany

Nathalie Aussenac-Gilles	IRIT and CNRS, France
Payam Barnaghi	Imperial College London, UK
Pierpaolo Basile	University of Bari, Italy
Rafael Berlanga	Universitat Jaume I, Spain
Russa Biswas	Karlsruhe Institute of Technology, Germany
Eva Blomqvist	Linköping University, Sweden
Carlos Bobed	University of Zaragoza, Spain
Fernando Bobillo	University of Zaragoza, Spain
Katarina Boland	GESIS - Leibniz Institute for the Social Sciences, Germany
Loris Bozzato	Fondazione Bruno Kessler, Italy
Adrian M. P. Brasoveanu	MODUL Technology GmbH, Austria
Carlos Buil Aranda	Universidad Técnica Federico Santa María, Chile
Davide Buscaldi	LIPN, Université Paris 13, France
Elena Cabrio	I3S, CNRS, Inria, and Université Côte d'Azur, France
Jean-Paul Calbimonte	University of Applied Sciences and Arts Western Switzerland HES-SO, Switzerland
Valentina Anita Carriero	University of Bologna, Italy
Irene Celino	Cefriel, Italy
Yoan Chabot	Orange Labs, France
Pierre-Antoine Champin	LIRIS, Université Claude Bernard Lyon 1, France
Maria Chang	IBM, USA
Victor Charpenay	École des Mines de Saint-Étienne, France
David Chaves-Fraga	Universidad Politécnica de Madrid, Spain
Jiaoyan Chen	University of Oxford, UK
Gong Cheng	Nanjing University, China
Michael Cochez	Vrije Universiteit Amsterdam, The Netherlands
Pieter Colpaert	Ghent University and imec, Belgium
Simona Colucci	Politecnico di Bari, Italy
Olivier Corby	Inria, France
Oscar Corcho	Universidad Politécnica de Madrid, Spain
Francesco Corcoglioniti	Free University of Bozen-Bolzano, Italy
Julien Corman	Free University of Bozen-Bolzano, Italy
Marco Cremaschi	Università di Milano-Bicocca, Italy
Claudia d'Amato	University of Bari, Italy
Mathieu D'Aquin	Loria, University of Lorraine, France
Enrico Daga	Open University, UK
Marilena Daquino	University of Bologna, Italy
Jérôme David	Inria, France
Victor de Boer	Vrije Universiteit Amsterdam, The Netherlands
Daniele Dell'Aglio	Aalborg University, Denmark

Elena Demidova	University of Bonn, Germany
Ronald Denaux	Amazon, Spain
Kathrin Dentler	Triply, The Netherlands
Gayo Diallo	ISPED and LABRI, University of Bordeaux, France
Dennis Diefenbach	Université Jean Monet, France
Stefan Dietze	GESIS - Leibniz Institute for the Social Sciences, Germany
Christian Dirschl	Wolters Kluwer Germany, Germany
Milan Dojchinovski	Czech Technical University in Prague, Czech Republic
Ivan Donadello	Free University of Bozen-Bolzano, Italy
Mauro Dragoni	Fondazione Bruno Kessler, Italy
Aaron Eberhart	Kansas State University, USA
Fajar J. Ekaputra	Vienna University of Technology, Austria
Pavlos Fafalios	Institute of Computer Science, FORTH, Greece
Nicola Fanizzi	Università degli Studi di Bari "Aldo Moro", Italy
Alessandro Faraotti	IBM, Italy
Daniel Faria	LASIGE, Universidade de Lisboa, Portugal
Catherine Faron Zucker	Université Nice Sophia Antipolis, France
Anna Fensel	University of Innsbruck, Austria
Javier D. Fernández	F. Hoffmann-La Roche AG, Switzerland
Mariano Fernández López	Universidad CEU San Pablo, Spain
Jesualdo Tomás Fernández-Breis	Universidad de Murcia, Spain
Sebastián Ferrada	Linköping University, Sweden
Sebastien Ferre	Université de Rennes, CNRS, and IRISA, France
Agata Filipowska	Poznan University of Economics, Poland
Flavius Frasincar	Erasmus University Rotterdam, The Netherlands
Adam Funk	University of Sheffield, UK
Luis Galárraga	Inria, France
Fabien Gandon	Inria, France
Aldo Gangemi	Università di Bologna and CNR-ISTC, Italy
Raúl García-Castro	Universidad Politécnica de Madrid, Spain
Daniel Garijo	Universidad Politécnica de Madrid, Spain
Chiara Ghidini	Fondazione Bruno Kessler, Italy
Carole Goble	University of Manchester, UK
Jose Manuel Gomez-Perez	expert.ai, Spain
Simon Gottschalk	L3S Research Center, Germany
Alasdair Gray	Heriot-Watt University, Scotland
Kalpa Gunaratna	Samsung Research, USA
Christophe Guéret	Accenture Labs, Ireland
Peter Haase	Metaphacts, Germany

Torsten Hahmann	University of Maine, USA
Armin Haller	Australian National University, Australia
Harry Halpin	World Wide Web Consortium, Switzerland
Ismail Harrando	EURECOM, France
Olaf Hartig	Linköping University, Sweden
Maria M. Hedblom	Jönköping University, Sweden
Ivan Heibi	University of Bologna, Italy
Veronika Heimsbakk	Capgemini, Norway
Nicolas Heist	University of Mannheim, Germany
Nathalie Hernandez	IRIT, France
Sven Hertling	University of Mannheim, Germany
Aidan Hogan	Universidad de Chile, Chile
Katja Hose	Aalborg University, Denmark
Wei Hu	Nanjing University, China
Madelon Hulsebos	University of Amsterdam, The Netherlands
Ali Hurriyetoglu	Koc University, Turkey
Eero Hyvönen	Aalto University and University of Helsinki, Finland
Luis Ibanez-Gonzalez	University of Southampton, UK
Ryutaro Ichise	National Institute of Informatics, Japan
Filip Ilievski	USC Information Sciences Institute, USA
Antoine Isaac	Europeana and VU Amsterdam, The Netherlands
Prateek Jain	Nuance Communications Inc., India
Fuad Jamour	University of California, Riverside, USA
Krzysztof Janowicz	University of California, Santa Barbara, USA
Mustafa Jarrar	Birzeit University, Palestine
Julien Romero	SAMOVAR, Télécom SudParis, France
Simon Jupp	SciBite, UK
Maulik R. Kamdar	Elsevier Inc., USA
Naouel Karam	Fraunhofer FOKUS, Germany
Katariina Kari	Inter IKEA Systems, Finland
Tomi Kauppinen	Aalto University, Finland
C. Maria Keet	University of Cape Town, South Africa
Mayank Kejriwal	USC Information Sciences Institute, USA
Ali Khalili	Deloitte, The Netherlands
Haklae Kim	Samsung Electronics, South Korea
Sabrina Kirrane	Vienna University of Economics and Business, Austria
Tomas Kliegr	Prague University of Economics and Business, Czech Republic
Craig Knoblock	USC Information Sciences Institute, USA
Haridimos Kondylakis	Institute of Computer Science, FORTH, Greece

Stasinos Konstantopoulos	NCSR Demokritos, Greece
Roman Kontchakov	Birkbeck, University of London, UK
Manolis Koubarakis	National and Kapodistrian University of Athens, Greece
Maria Koutraki	L3S Research Center and Leibniz University Hannover, Germany
Kouji Kozaki	Osaka Electro-Communicatîon University, Japan
Ralf Krestel	ZBW - Leibniz Information Centre for Economics and Kiel University, Germany
Adila A. Krisnadhi	Universitas Indonesia, Indonesia
Tobias Käfer	Karlsruhe Institute of Technology, Germany
Jose Emilio Labra Gayo	Universidad de Oviedo, Spain
Frederique Laforest	LIRIS, INSA Lyon, France
Sarasi Lalithsena	IBM Watson, USA
Nelia Lasierra	F. Hoffmann-La Roche, Switzerland
Danh Le Phuoc	TU Berlin, Germany
Maxime Lefrançois	École des Mines de Saint-Etienne, France
Yuan-Fang Li	Monash University, Australia
Sven Lieber	Royal Library of Belgium (KBR) and Ghent University, Belgium
Pasquale Lisena	EURECOM, France
Ismini Lourentzou	Virginia Tech, USA
Jun Ma	Amazon, USA
Maria Maleshkova	University of Siegen, Germany
Beatrice Markhoff	LIFAT, Université de Tours, France
Miguel A. Martinez-Prieto	University of Valladolid, Spain
Jose L. Martinez-Rodriguez	CINVESTAV, Mexico
Franck Michel	I3S, CNRS and Université Côte d'Azur, France
Nandana Mihindukulasooriya	IBM Research, USA
Thomas Minier	Opendatasoft, France
Aditya Mogadala	Saarland University, Germany
Pascal Molli	LS2N, University of Nantes, France
Gabriela Montoya	Aalborg University, Denmark
Kody Moodley	Maastricht University, The Netherlands
Jose Mora	Universidad Politécnica de Madrid, Spain
Diego Moussallem	Paderborn University, Germany
Raghava Mutharaju	IIIT-Delhi, India
Lionel Médini	LIRIS, Université Claude Bernard Lyon 1, France
Sebastian Neumaier	St. Pölten University of Applied Sciences, Austria
Vinh Nguyen	National Library of Medicine, NIH, USA
Andriy Nikolov	AstraZeneca, UK
Natasha Noy	Google, USA

Cliff O'Reilly	City, University of London, UK
Oliver Lehmberg	Diffbot, USA
Inah Omoronyia	University of Glasgow, UK
Femke Ongenae	Ghent University and imec, Belgium
Francesco Osborne	Open University, UK
Matteo Palmonari	University of Milano-Bicocca, Italy
Harshvardhan J. Pandit	Trinity College Dublin, Ireland
George Papadakis	National Technical University of Athens, Greece
Heiko Paulheim	University of Mannheim, Germany
Tassilo Pellegrini	St. Pölten University of Applied Sciences, Austria
Maria Angela Pellegrino	Università degli Studi di Salerno, Italy
Sujan Perera	IBM Watson, USA
Nathalie Pernelle	LIPN, Université Sorbonne Paris Nord, France
Silvio Peroni	University of Bologna, Italy
Johann Petrak	University of Sheffield, UK
Guangyuan Piao	National University of Ireland Maynooth, Ireland
Pierre-Henri Paris	CNAM, France
Lydia Pintscher	Wikimedia Deutschland, Germany
Giuseppe Pirrò	Sapienza University of Rome, Italy
Dimitris Plexousakis	Institute of Computer Science, FORTH, Greece
André Pomp	University of Wuppertal, Germany
María Poveda-Villalón	Universidad Politécnica de Madrid, Spain
Nicoleta Preda	Université Paris Saclay, Versailles, France
Cédric Pruski	Luxembourg Institute of Science and Technology, Luxembourg
Tara Raafat	University of Surrey, UK
Srinivas Ravishankar	IBM Research, USA
Simon Razniewski	Max Planck Institute for Informatics, Germany
Diego Reforgiato	Università degli Studi di Cagliari, Italy
Blake Regalia	University of California, Santa Barbara, USA
Georg Rehm	DFKI, Germany
Achim Rettinger	Trier University, Germany
Artem Revenko	Semantic Web Company GmbH, Austria
Petar Ristoski	IBM Research - Almaden, USA
Giuseppe Rizzo	LINKS Foundation, Italy
Sergio José Rodríguez Méndez	Australian National University, Australia
Roghaiyeh Gachpaz Hamed	Trinity College Dublin, Ireland
Edelweis Rohrer	Universidad de la República, Uruguay
Julian Rojas	Ghent University, Belgium
Maria Del Mar Roldan-Garcia	Universidad de Málaga, Spain
Henry Rosales-Méndez	University of Chile, Chile
Catherine Roussey	INRAE, France

Edna Ruckhaus	Universidad Politécnica de Madrid, Spain
Sebastian Rudolph	TU Dresden, Germany
Harald Sack	FIZ Karlsruhe - Leibniz Institute for Information Infrastructure and KIT, Germany
Angelo Antonio Salatino	Open University, UK
Muhammad Saleem	University of Leipzig, Germany
Emanuel Sallinger	TU Wien, Austria
Felix Sasaki	Cornelsen Verlag GmbH and TH Brandenburg, Germany
Ulrike Sattler	University of Manchester, UK
Fatiha Saïs	LRI, Paris-Saclay University, France
Marco Luca Sbodio	IBM Research, Ireland
Stefan Schlobach	Vrije Universiteit Amsterdam, The Netherlands
Daniel Schwabe	PUC-Rio, Brazil
Gezim Sejdiu	University of Bonn, Germany
Juan F. Sequeda	data.world, USA
Barış Sertkaya	Frankfurt University of Applied Sciences, Germany
Dominic Seyler	Baidu Research, USA
Pavel Shvaiko	Informatica Trentina, Italy
Gerardo Simari	Universidad Nacional del Sur and CONICET, Argentina
Evren Sirin	Clark & Parsia, LLC, USA
Hala Skaf-Molli	University of Nantes, France
Xingyi Song	University of Sheffield, UK
Adrián Soto	Fintual, Chile
Blerina Spahiu	Università degli Studi di Milano Bicocca, Italy
Marc Spaniol	Université de Caen Normandie, France
Kavitha Srinivas	IBM, India
Steffen Staab	Universität Stuttgart, Germany, and University of Southampton, UK
Nadine Steinmetz	TU Ilmenau, Germany
Armando Stellato	University of Rome Tor Vergata, Italy
Simon Steyskal	Siemens AG Austria, Austria
Umberto Straccia	ISTI-CNR, Italy
Heiner Stuckenschmidt	University of Mannheim, Germany
Gerd Stumme	University of Kassel, Germany
Vojtěch Svátek	Prague University of Economics and Business, Czech Republic
Ruben Taelman	Ghent University and imec, Belgium
Hideaki Takeda	National Institute of Informatics, Japan
Valentina Tamma	University of Liverpool, UK

Andrea Tettamanzi   Université Nice Sophia Antipolis, France
Andreas Thalhammer   F. Hoffmann-La Roche AG, Switzerland
Tobias Weller   University of Mannheim, Germany
Konstantin Todorov   LIRMM, University of Montpellier, France
Riccardo Tommasini   INSA Lyon, France
Anna Tordai   Elsevier, The Netherlands
Sebastian Tramp   eccenca GmbH, Germany
Cassia Trojahn   UT2J and IRIT, France
Raphaël Troncy   EURECOM, France
Umair Ul Hassan   National University of Ireland Galway, Ireland
Jürgen Umbrich   Onlim GmbH, Austria
Ricardo Usbeck   Hamburg University, Germany
Sahar Vahdati   InfAI, Germany
Ludger Van Elst   DFKI, Germany
Frank Van Harmelen   Vrije Universiteit Amsterdam, The Netherlands
Miel Vander Sande   Meemoo, Belgium
Ruben Verborgh   Ghent University and imec, Belgium
Serena Villata   I3S, CNRS and Université Côte d'Azur, France
Boris Villazón-Terrazas   Majorel, Spain
Fabio Vitali   University of Bologna, Italy
Domagoj Vrgoc   Pontificia Universidad Católica de Chile, Chile
Andreas Wagner   Schaeffler AG, Germany
Kewen Wang   Griffith University, Australia
Ruijie Wang   University of Illinois Urbana-Champaign, USA
Rigo Wenning   W3C, France
Xander Wilcke   Vrije Universiteit Amsterdam, The Netherlands
Cord Wiljes   Nationale Forschungsdateninfrastruktur (NFDI)
                                    e.V., Germany
Gregory Todd Williams   Amazon Web Services, USA
Zhe Wu   eBay, USA
Josiane Xavier Parreira   Siemens AG Österreich, Austria
Fouad Zablith   American University of Beirut, Lebanon
Hamada Zahera   Paderborn University, Germany
Ondřej Zamazal   Prague University of Economics and Business,
                                    Czech Republic
Songmao Zhang   Chinese Academy of Sciences, China
Ziqi Zhang   Sheffield University, UK
Rui Zhu   University of California, Santa Barbara, USA
Antoine Zimmermann   École des Mines de Saint-Étienne, France
Matthäus Zloch   GESIS - Leibniz Institute for the Social Sciences,
                                    Germany
Amal Zouaq   Ecole Polytechnique de Montréal, Canada

Hanna Ćwiek-Kupczyńska	Institute of Plant Genetics, Polish Academy of Sciences, Poland
Umutcan Şimşek	Semantic Technology Institute Innsbruck, Austria

Additional Reviewers

Ayats, Hugo	Möller, Cedric
Braun, Christoph	Parvin, Parvaneh
Bruns, Oleksandra	Paulus, Alexander
Burgdorf, Andreas	Salman, Muhammad
Eckert, Kai	Santini, Cristian
Ettorre, Antonia	Schestakov, Stefan
Flouris, Giorgos	Scrocca, Mario
Gesese, Genet Asefa	Sha, Alyssa
He, Yuan	Siciliani, Lucia
Hoppe, Fabian	Sierra-Múnera, Alejandro
Hosseini Beghaeiraveri, Seyed Amir	Silvestre, Jorge
Hosseinzadeh Vahid, Ali	Simón Ramos, José Manuel
Jain, Nitisha	Singh, Gunjan
Kugler, Kai	Tietz, Tabea
König, Lukas	Vimercati, Manuel
Marcia, Diego	Werner, Simon
Marx, Edgard	Yan, Xi
Mohtashim, Mirza	Zhuang, Zhiqiang

Sponsors

Platinum Sponsors

VideoLectures.NET is an award-winning free and open access educational video lectures repository. The lectures are given by distinguished scholars and scientists at the most important and prominent events such as conferences, summer schools, workshops, and science promotional events in many fields of science. The portal is aimed at promoting science, exchanging ideas, and fostering knowledge sharing by providing high-quality didactic contents not only to the scientific community but also to the general public. All lectures, accompanying documents, information, and links are systematically selected and classified through the editorial process taking into account users' comments.

Gold Sponsors

SIEMENS

Siemens AG (Berlin and Munich) is a global technology powerhouse that has stood for engineering excellence, innovation, quality, reliability, and internationality for more than 170 years. The company is active around the globe, focusing on the areas of electrification, automation, and digitalization. One of the largest producers of energy-efficient, resource-saving technologies, Siemens is a leading supplier of efficient power generation and power transmission solutions and a pioneer in infrastructure solutions as well as automation, drive, and software solutions for industry. With its publicly listed subsidiary Siemens Healthineers AG, the company is also a leading provider of medical imaging equipment – such as computed tomography and magnetic resonance imaging systems – and a leader in laboratory diagnostics as well as clinical IT. In fiscal 2018, which ended on September 30, 2018, Siemens generated revenue of €83.0 billion and net income of €6.1 billion. At the end of September 2018, the company had around 379,000 employees worldwide. Further information is available at www.siemens.com.

Silver Sponsors

Elsevier is a global information analytics business that helps scientists and clinicians to find new answers, reshape human knowledge, and tackle the most urgent human crises. For 140 years, we have partnered with the research world to curate and verify scientific knowledge. Today, we're committed to bringing that rigor to a new generation of platforms. Elsevier provides digital solutions and tools in the areas of strategic research management, R&D performance, clinical decision support, and professional education, including ScienceDirect, Scopus, SciVal, ClinicalKey, and Sherpath. Elsevier publishes over 2,500 digitized journals, including *The Lancet* and *Cell*, 39,000 e-book titles, and many iconic reference works, including *Gray's Anatomy*. Elsevier is part of RELX Group, a global provider of information and analytics for professionals and business customers across industries.

Ontotext is a global leader in enterprise knowledge graph technology and semantic database engines. Ontotext employs big knowledge graphs to enable unified data access and cognitive analytics via text mining and integration of data across multiple sources. Ontotext GraphDBᴛᴍ engine and Ontotext Platform power business critical systems in the biggest banks, media companies, market intelligence agencies, and car and aerospace manufacturers. Ontotext technology and solutions are spread wide across the value chain of the most knowledge intensive enterprises in financial services, publishing, healthcare, pharma, manufacturing, and public sectors. Leveraging AI and cognitive technologies, Ontotext helps enterprises get competitive advantage by connecting the dots of their proprietary knowledge and putting in the context of global intelligence.

Springer

Springer is part of Springer Nature, a leading global research, educational, and professional publisher, home to an array of respected and trusted brands providing quality content through a range of innovative products and services. Springer Nature is the world's largest academic book publisher, publisher of the world's most influential journals, and a pioneer in the field of open research. The company numbers almost 13,000 staff in over 50 countries and has a turnover of approximately €1.5 billion. Springer Nature was formed in 2015 through the merger of Nature Publishing Group, Palgrave Macmillan, Macmillan Education, and Springer Science+Business Media. Find out more at www.springernature.com.

Bronze Sponsors

IOS Press is an independent, international STM publishing house established in 1987 in Amsterdam. One of our guiding principles is to embrace the benefits a lean organization offers. While our goal is to keep things simple, we strive to meet the highest professional standards. Our business practices are straightforward, transparent, and ethical. IOS Press serves the information needs of scientific and medical communities worldwide. IOS Press now publishes more than 100 international journals and approximately 75 book titles each year on subjects ranging from computer sciences and mathematics to medicine and the natural sciences. Please visit iospress.com to find out more.

metaphacts

metaphacts is a Germany-based company delivering metaphactory – a platform that empowers customers to accelerate their knowledge graph journey and drive knowledge democratization, improve data literacy, and reach smarter business decisions with data. The **metaphacts** team offers unmatched experience and know-how around enterprise knowledge graphs for our clients in areas such as pharma and life sciences, engineering and manufacturing, energy, finance, business, and cultural heritage.

Contents

Research

Enhancing Sequential Recommendation via Decoupled Knowledge Graphs 3
    Bingchao Wu, Chenglong Deng, Bei Guan, Yongji Wang,
    and Yuxuan Kangyang

An Analysis of Links in Wikidata 21
    Armin Haller, Axel Polleres, Daniil Dobriy, Nicolas Ferranti,
    and Sergio J. Rodríguez Méndez

Knowledge Graph Entity Type Prediction with Relational Aggregation
Graph Attention Network 39
    Changlong Zou, Jingmin An, and Guanyu Li

Union and Intersection of All Justifications 56
    Jieying Chen, Yue Ma, Rafael Peñaloza, and Hui Yang

Supervised Knowledge Aggregation for Knowledge Graph Completion 74
    Patrick Betz, Christian Meilicke, and Heiner Stuckenschmidt

Expressive Scene Graph Generation Using Commonsense Knowledge
Infusion for Visual Understanding and Reasoning 93
    Muhammad Jaleed Khan, John G. Breslin, and Edward Curry

Impact of the Characteristics of Multi-source Entity Matching Tasks
on the Performance of Active Learning Methods 113
    Anna Primpeli and Christian Bizer

Optimal ABox Repair w.r.t. Static $\mathcal{EL}$ TBoxes: From Quantified ABoxes
Back to ABoxes 130
    Franz Baader, Patrick Koopmann, Francesco Kriegel,
    and Adrian Nuradiansyah

Ensemble-Based Fact Classification with Knowledge Graph Embeddings 147
    Unmesh Joshi and Jacopo Urbani

The Problem with XSD Binary Floating Point Datatypes in RDF 165
    Jan Martin Keil and Merle Gänßinger

DCWEB-SOBA: Deep Contextual Word Embeddings-Based
Semi-automatic Ontology Building for Aspect-Based Sentiment
Classification 183
    Roos van Lookeren Campagne, David van Ommen, Mark Rademaker,
    Tom Teurlings, and Flavius Frasincar

Never Mind the Semantic Gap: Modular, Lazy and Safe Loading of RDF
Data 200
    Eduard Kamburjan, Vidar Norstein Klungre, and Martin Giese

Improving Question Answering Quality Through Language Feature-Based
SPARQL Query Candidate Validation 217
    Aleksandr Gashkov, Aleksandr Perevalov, Maria Eltsova,
    and Andreas Both

Learning Concept Lengths Accelerates Concept Learning in ALC 236
    N'Dah Jean Kouagou, Stefan Heindorf, Caglar Demir,
    and Axel-Cyrille Ngonga Ngomo

Dihedron Algebraic Embeddings for Spatio-Temporal Knowledge Graph
Completion 253
    Mojtaba Nayyeri, Sahar Vahdati, Md Tansen Khan,
    Mirza Mohtashim Alam, Lisa Wenige, Andreas Behrend,
    and Jens Lehmann

Hierarchical Topic Modelling for Knowledge Graphs 270
    Yujia Zhang, Marcin Pietrasik, Wenjie Xu, and Marek Reformat

Resources

Do Arduinos Dream of Efficient Reasoners? 289
    Alexandre Bento, Lionel Médini, Kamal Singh, and Frédérique Laforest

A Programming Interface for Creating Data According to the SPAR
Ontologies and the OpenCitations Data Model 305
    Simone Persiani, Marilena Daquino, and Silvio Peroni

LD Connect: A Linked Data Portal for IOS Press Scientometrics 323
    Zilong Liu, Meilin Shi, Krzysztof Janowicz, Blake Regalia,
    Stephanie Delbecque, Gengchen Mai, Rui Zhu, and Pascal Hitzler

Chowlk: from UML-Based Ontology Conceptualizations to OWL 338
    Serge Chávez-Feria, Raúl García-Castro, and María Poveda-Villalón

QuoteKG: A Multilingual Knowledge Graph of Quotes 353
  Tin Kuculo, Simon Gottschalk, and Elena Demidova

Stunning Doodle: A Tool for Joint Visualization and Analysis
of Knowledge Graphs and Graph Embeddings 370
  Antonia Ettorre, Anna Bobasheva, Franck Michel, and Catherine Faron

Capturing the Semantics of Smell: The Odeuropa Data Model for Olfactory
Heritage Information 387
  Pasquale Lisena, Daniel Schwabe, Marieke van Erp, Raphaël Troncy,
  William Tullett, Inger Leemans, Lizzie Marx, and Sofia Colette Ehrich

Stream Reasoning Playground 406
  Patrik Schneider, Daniel Alvarez-Coello, Anh Le-Tuan,
  Manh Nguyen-Duc, and Danh Le-Phuoc

In-Use Track

The Dow Jones Knowledge Graph 427
  Ian Horrocks, Jordi Olivares, Valerio Cocchi, Boris Motik, and Dylan Roy

CONSTRUCT Queries Performance on a Spark-Based Big RDF Triplestore ... 444
  Adam Sanchez-Ayte, Fabrice Jouanot, and Marie-Christine Rousset

Matching Multiple Ontologies to Build a Knowledge Graph
for Personalized Medicine 461
  Marta Contreiras Silva, Daniel Faria, and Catia Pesquita

FindSampo: A Linked Data Based Portal and Data Service for Analyzing
and Disseminating Archaeological Object Finds 478
  Heikki Rantala, Esko Ikkala, Ville Rohiola, Mikko Koho,
  Jouni Tuominen, Eljas Oksanen, Anna Wessman, and Eero Hyvönen

Author Index 495

Research

Enhancing Sequential Recommendation via Decoupled Knowledge Graphs

Bingchao Wu[1,3], Chenglong Deng[1,3], Bei Guan[1,3], Yongji Wang[1,2,3(✉)],
and Yuxuan Kangyang[2,3]

[1] Collaborative Innovation Center, Institute of Software,
Chinese Academy of Sciences, Beijing 100190, China
{bingchao2017,chenglong2018,guanbei}@iscas.ac.cn
ywang@itechs.iscas.ac.cn
[2] State Key Laboratory of Computer Science, Institute of Software,
Chinese Academy of Sciences, Beijing 100190, China
kyyx@ios.ac.cn
[3] University of Chinese Academy of Sciences, Beijing 100049, China

Abstract. Sequential recommendation can capture dynamic interest patterns of users based on user interaction sequences. Recently, there has been interest in integrating the knowledge graph (KG) into sequential recommendation. Existing works suffer from two main challenges: a) representing each entity in the KG as a single vector can confound heterogeneous information about the entity; b) triple-based facts are modeled independently, lacking the exploration of high-order connectivity between entities. To solve the above challenges, we decouple the KG into two subgraphs, namely CRoss-user Behavior-based graph and Intrinsic Attribute-based graph (Crbia), depending on the type of relation between entities. We further propose a CrbiaNet based on the two subgraphs. First, CrbiaNet obtains behavior-level and attribute-level semantic features from these two subgraphs independently by different graph neural networks, respectively. Then, CrbiaNet applies a sequential model incorporating these semantic features to capture dynamic preference of the users. Extensive experiments on three real-world datasets show that our proposed CrbiaNet outperforms previous state-of-the-art knowledge-enhanced sequential recommendation models by a large margin consistently.

Keywords: Sequential recommendation · Knowledge graph · Heterogeneous information · Graph neural network

1 Introduction

The recommendation system aims to suggest related items to users from a massive collection of items, thereby alleviating the problem of information overload. Sequential recommendation has been receiving increasing attention from researchers in the recommendation field. It is necessary to model dynamic user preference over time to provide accurate and high-quality recommendations. With the popularity and effectiveness of deep learning technologies in the fields

P. Groth et al. (Eds.): ESWC 2022, LNCS 13261, pp. 3–20, 2022.
https://doi.org/10.1007/978-3-031-06981-9_1

of computer vision and natural language processing, much of the literature on sequential recommendation has focused specifically on capturing sequential patterns from the historical interaction sequences sorted by time to predict future items for users via neural network models, such as GRU4Rec [6], Caser [22], SASRec [9], and BERT4Rec [21].

Although sequential recommendation has achieved great success in capturing dynamic user preference, it is limited by the fact that the vector of user preference is learned independently through each user's interaction sequence, and a large portion of the user interaction sequence is sparse [4]. Recently, many previous studies have focused on injecting the KG into sequential recommendation models through path-based methods (e.g., MASR [8] and KSRN [41]) and embedding-based methods (e.g., Chorus [27] and KERL [29]) to solve the aforementioned problems. The path-based approaches extract meta-paths that are relevant to user behavior sequences from the KG. However, these approaches rely heavily on expert knowledge to design reasonable meta-paths, and it is difficult to enumerate all potentially useful meta-paths [4]. The research in this paper is concerned with embedding-based approaches, which use the KG embedding methods to acquire the embedding of each entity in the KG. The existing embedding-based methods integrated into sequential recommendation models are divided into two categories, i.e., traditional distance-based models (e.g., TransE [1] and TransR [13]) and traditional semantic matching models (e.g., DistMult [35] and ComplEx [24]).

Fig. 1. (a) The heterogeneous information in the KG. (b) The high-order connectivity between items in the KG where the yellow dashed line indicates no directly connected edges between items. (Color figure online)

In the recommendation domain, there are two challenges in applying these two categories of embedding-based approaches to encode semantic features in the KG.

– Heterogeneous semantic information of items: the KG in the recommendation domain includes intrinsic attribute-level semantic information of items and behavior-level semantic information of items extracted from user logs [14]. A case is shown in Fig. 1-(a), the bottom two triples (*iPhone, brand, Apple*) and (*iPhone, category, Phone*) construct the attribute-level semantic information of *iPhone*, and the top two triples generate the behavior-level semantic information of *iPhone*. Existing embedding-based methods applied to sequential recommendation confound two types of heterogeneous information in a single vector.

- High-order connectivity between items: the embedding-based approaches mentioned above only model each fact consisting of a triplet individually, and ignore the high-order connectivity between items [28]. The high-order connectivityis a multi-hop relation path between items [30], which allows exploring deeper semantic information about items. A case is shown in Fig. 1-(b). Even though there are no directly connected edges between *iPhone* and *Apple Watch*, we can still capture the potential semantic relation through a multi-hop connection ($iPhone \longrightarrow MacBook \longrightarrow AirPods \longrightarrow Apple\ Watch$).

While the existing works (e.g., KSR [7], KERL [29] and GFE-SASRec [36]) utilize graph neural networks to model high-order connectivity, they only consider one type of KG or conflate heterogeneous information of items into a single vector. To overcome these challenges, we propose a sequential recommendation model CbiaNet[1] via merging decoupled knowledge graphs. First, we decouple the KG into two complementary subgraphs, named the cross-user behavior-based graph and the intrinsic attribute-based graph. Then, two knowledge sub-extractors encode the two subgraphs independently by graph neural networks to solve the problem of confounding heterogeneous semantics and to capture the higher-order connections between items. Next, a hierarchical knowledge aggregator combines the heterogeneous semantic information to generate high-level semantic features. Finally, a sequential model incorporating the high-level semantic features is developed to capture the dynamic preference of the users. We conduct experiments on three real-world datasets, and the experimental results show that our proposed CrbiaNet outperforms the existing state-of-the-art recommendation models. In addition, we extend the high-level semantic features to several sequential recommendation models, which also improves their performance.

2   Related Work

2.1   Sequential Recommendation

In order to model the dynamic interests of users, sequential recommendation methods utilize the user's historical interaction data. Markov chains are applied in traditional sequential recommendation methods by estimating the transition probability between items within the previous action sequence [19,20]. With the great success of deep learning methods in various fields, many efforts have been made to model users' historical interaction sequences by utilizing neural networks [6,9,12,21,22]. GRU4Rec [6] applies Gated Recurrent Units (GRU) to the session-based recommendation. NARM [12] further introduces attention-based GRU by assigning different weights to items of historical interaction sequences. Besides, Caser [22] and NextItNet [37] introduce Convolution Neural Network

---

[1] The codes are released at https://github.com/paulpig/sequentialRec.git..

(CNN) to learn sequential patterns as local features by using convolutional filters. Recently, various studies have validated that self-attention mechanisms effectively model dependencies between items [9, 21]. SASRec [9] utilizes left-to-right Transformer models [25] to predict the next item. BERT4Rec [21] uses bidirectional Transformer models (BERT [3]) to encode user interest vectors by optimizing a Cloze task [23]. Sequential recommendations focus only on the user's own interaction sequence, ignoring the similar co-occurrence across users between items and relationships between items at the attribute level.

2.2 Knowledge-Enhanced Recommendation

KGs have been applied in various recommendation models to improve the performance of the recommendation where KGs use triples to describe realistic facts, such as the user-item KG [41], the item-item KG [34], and the item-attribute KG [41]. Several graph-based recommendation models jointly encode behavior-level user-item relations and knowledge-level item-item relations to introduce semantic knowledge from KG into the recommender system, such as KHGT [32], UGRec [39], and SMIN [14]. However, the above graph-based models cannot capture the dynamic user preference, so more research is focused on how to utilize knowledge graphs to enhance sequential recommendation models. Existing studies on injecting knowledge graphs into sequential models are mainly divided into two categories: path-based and embedding-based methods. For path-based methods, MASR [8] introduces meta-paths from the knowledge graph to capture global contextual information and applies the sequential model to capture the local contextual information. KARN [41] combines users' historical behavior sequences and the path between the user and the target item for recommendation. For embedding-based methods, KERL [29] uses TransR to obtain semantic features from KG that are fused into the sequential models. Chorus [27], RCF [34], and KDA [26] use DistMult to extract semantic features of items from KG by bilinear objectives and use the semantic features as input to the sequential model. Despite these recent advancements, the above knowledge graph embeddings cannot capture the higher-order connections between items in KG. DHIMN [33] applies a GCN-based message-passing layer to capture the high-level semantic knowledge in the KG, but ignores heterogeneous information of item relations in KG.

3   Problem Definition and Notation

3.1   Cross-User Behavior-Based Graph (CRBGraph)

In the recommendation domain, item relations extracted from user logs naturally exist in the datasets [15, 26]. For example, the relation *also_buy* (*also_view*) between *iPhone* and *MacBook* means that users bought an *iPhone* and also bought (viewed) a *MacBook* afterwards. Here we represent these item relations with a cross-user behavior-based graph $\mathcal{G}_1$, defined as $\{(h, r, t)|h, t \in \mathcal{I}, r \in \mathcal{R}^b\}$

where $\mathcal{I}$ and $\mathcal{R}^b$ denote sets of item instances and item relations, respectively. The relations between item-item pairs are all positively correlated, so all types of item relations are reduced to a positive relation. This means that $r \in \{0, 1\}$ where $r = 1$ represents that there is a behavior-level link between the item-item pair.

3.2 Intrinsic Attribute-Based Graph (IAGraph)

In addition to behavior-level links between items, there are various types of item attributes, such as category and brand. Here we utilize the item-attribute pairs to generate an intrinsic attribute-based graph $\mathcal{G}_2$, defined as $\{(h, r', a)|h \in \mathcal{I}, a \in \mathcal{A}, r' \in \mathcal{R}^a\}$, where $\mathcal{I}$ and $\mathcal{A}$ denote sets of item instances and attribute values, and R^a is the set of attribute-level relations. For example, the triple (*iPhone*, *brand*, *Apple*) represents that the *brand* of the *iPhone* is *Apple*.

3.3 Task Description

Assume that there are M users and N items in the recommender system. Given the graphs $\mathcal{G}_1, \mathcal{G}_2$ and the interaction sequence $S^u = [i_1^u, i_2^u, \cdots, i_T^u]$ of user u where $i_1^u \in \mathcal{I}$ and T is the length of the interaction sequence, the knowledge-enhanced sequential recommendation task is denoted as follows:

$$i_u^* = argmax_{i_k \in \mathcal{I}} P(i_{T+1}^u = i_k | S^u, \mathcal{G}_1, \mathcal{G}_2)$$

where i_{T+1}^u is the predicted item at $T + 1$ time step, and P is the probability distribution over $\mathcal{I}$.

4 Method

The overview of CrbiaNet is shown in Fig. 2. The knowledge extractor is firstly employed to obtain heterogeneous item features from two distinct KGs, comprising a Behavior-level Knowledge Sub-extractor (BKS) and an Attribute-level Knowledge Sub-extractor (AKS). Then, the knowledge aggregator applies a hierarchical integration strategy to generate high-level semantic features by merging heterogeneous item features. Finally, a sequential interactions modeling layer merging high-level semantic features is employed to capture the dynamic user intention from the user's historical interaction sequence.

4.1 Knowledge Extractor

In this section, we design two types of graph neural networks to encode the behavior-level and attribute-level higher-order semantic features from the CRB-Graph and the IAGraph, respectively. To model the CRBGraph, we design a behavior-level knowledge sub-extractor that aggregates semantic features of neighbors based on the flow direction of message passing in the graph neural network. For IAGraph, we aggregate the neighborhood information to the central node through the relationship-aware attention mechanism of the attribute-level knowledge sub-extractor.

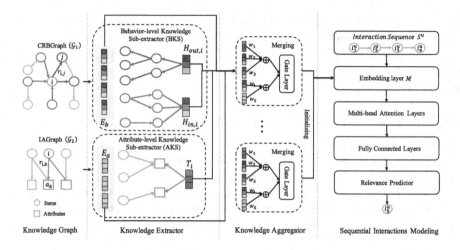

Fig. 2. The overall framework of our proposed model.

Behavior-Level Knowledge Sub-extractor (BKS). CRBGraph is a directed homogeneous graph in which each triple contains the time-series relation between the head and the tail entity. For example, the triple (*phone, also_buy, phone case*) means that users bought a *phone* and then also bought a *phone case*. Each node in the CRBGraph appears as a head entity in some related triples and as a tail entity in the rest of the related triples. This indicates that each node in the CRBGraph contains two types of time-series relations. To capture these time-series relations, we construct two-sided semantic features for each node, $\mathbf{H}_{in}$ and $\mathbf{H}_{out}$.

Specially, given the item i, one-hop neighbors of i are divided into out-degree neighbors $\mathcal{N}_i^{out}$ and in-degree neighbors $\mathcal{N}_i^{in}$. For example, $\mathcal{N}_5^{out}$ of the item v_5 is $\{v_1, v_4, v_6\}$ and $\mathcal{N}_5^{in}$ is $\{v_2, v_3\}$ in Fig. 3-(a). One-side semantic feature of item i at k^{th} layer, in-degree feature $\mathbf{H}_{in,i}^{(k)}$, aggregates the features $\mathbf{H}_{out,i}^{(k-1)}$ at $(k-1)^{th}$ layer in neighbors $\mathcal{N}_i^{in}$. The other-side feature $\mathbf{H}_{out,i}^{(k)}$ aggregates $\mathbf{H}_{in,i}^{(k-1)}$ in $\mathcal{N}_i^{out}$. The formula for the aggregation operation is:

$$\mathbf{H}_{in,i}^{(k)} = \sum_{j \in \mathcal{N}_i^{in}} \frac{1}{\sqrt{|\mathcal{N}_j^{in}|}\sqrt{|\mathcal{N}_i^{out}|}} \mathbf{H}_{out,j}^{(k-1)}, \tag{1}$$

$$\mathbf{H}_{out,i}^{(k)} = \sum_{j \in \mathcal{N}_i^{out}} \frac{1}{\sqrt{|\mathcal{N}_j^{out}|}\sqrt{|\mathcal{N}_i^{in}|}} \mathbf{H}_{in,j}^{(k-1)} \tag{2}$$

where $|\mathcal{N}_i^{out}|$ and $|\mathcal{N}_i^{in}|$ are the number of items in $\mathcal{N}_i^{out}$ and $\mathcal{N}_i^{in}$, respectively. A case is shown in Fig. 3-(b), the in-degree features $\{\mathbf{H}_{in,1}^{(k-1)}, \mathbf{H}_{in,4}^{(k-1)}, \mathbf{H}_{in,6}^{(k-1)}\}$ at $(k-1)^{th}$ layer of $\{v_1, v_4, v_6\}$ are propagated to the out-degree feature $\mathbf{H}_{out,5}^{(k)}$ at k^{th} layer of the item v_5 by Eq. 1. Note that $\mathbf{H}_{out}^{(0)} = \mathbf{H}_{in}^{(0)} = \mathbf{E}_b$, which means that $\mathbf{H}_{out}^{(0)}$ and $\mathbf{H}_{in}^{(0)}$ are from a shared embedding layer $\mathbf{E}_b$ to avoid overfitting.

Next, we stack more layers to capture higher-order item relations by Eq. 1 subject to k > 1 and obtain the final in-degree representation $\mathbf{H}_{in}$ by averaging

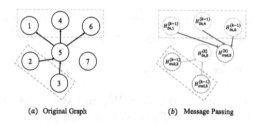

(a) Original Graph   (b) Message Passing

Fig. 3. (a) The items in the orange and blue dashed boxes are the out-degree and in-degree neighbors; (b) The red and green dashed circles indicate the out-degree and in-degree features, respectively. (Color figure online)

the in-degree item features at each layer. The final out-degree representation $\mathbf{H}_{out}$ is derived using the similar operation. Finally, we optimize the behavior-level knowledge sub-extractor using the BPR loss [18]:

$$L_{CB} = - \sum_{(i,j,j')\in \mathcal{D}_R} \ln \sigma(\hat{y}_{ij} - \hat{y}_{ij'}); \quad \hat{y}_{ij} = \mathbf{H}_{out,i}\mathbf{H}_{in,j}^T \tag{3}$$

where $\mathcal{D}_R$ is $\{(i,j,j')|(i,r,j)\in \mathcal{G}_1 \wedge r = 1, (i,r',j')\in \mathcal{G}_1 \wedge r' = 0\}$.

Attribute-Level Knowledge Sub-extractor (AKS). Another knowledge sub-extractor is applied to encode potential attribute-level knowledge of items via a graph neural network, which can explore the user's preference at the attribute level. The high correlation of the attribute information and the preference behavior has been verified in [11,31].

First, the translation-based method TransR [13] is applied to model the first-order connectivity of entities in the IAGraph. However, it lacks the encoding of high-order connectivity between entities. We further introduce a graph attention network consisting of message propagation layers and message aggregation layers. For the kth message propagation layer, we use the relation-aware attention mechanism to integrate neighbors of the central item i:

$$\mathbf{T}_{\mathcal{F}_i}^{(k)} = \sum_{(i,r,a)\in \mathcal{F}_i} \pi^{(k)}(i,r,a)\mathbf{T}_a^{(k)} \tag{4}$$

where $\mathcal{F}_i$ is the set of triples with the item i as the head entity in $\mathcal{G}_2$, and $\mathbf{T}_a^{(k)}$ is the feature of the entity a at k^{th} layer; $\pi^{(k)}(i,r,a)$ indicates the decay factor of the triple (i,r,a) in the message propagation [30]:

$$\pi^{(k)}(i,r,a) = \frac{\exp(f^{(k)}(i,r,a))}{\sum_{(i,r',a')\in \mathcal{F}_i} \exp(f^{(k)}(i,r',a'))} \tag{5}$$

$$f^{(k)}(i,r,a) = (\mathbf{W}_r \mathbf{T}_a^{(k)})^{\mathsf{T}} \tanh\left((\mathbf{W}_r \mathbf{T}_i^{(k)} + \mathbf{T}_r^{(k)})\right)$$

where $\mathbf{W}_r$ is the relation-aware trainable parameter, and $\mathbf{T}_i^{(k)}$ and $\mathbf{T}_r^{(k)}$ are the features of the entity i and the relation r. For the kth message aggregation layer,

$\mathbf{T}_{\mathcal{F}_i}^{(k)}$ and $\mathbf{T}_i^{(k)}$ are aggregated by two types of feature interactions and a nonlinear transformation and then passed to the $(k+1)^{th}$ layer:

$$\mathbf{T}_i^{(k+1)} = \sigma\Big(\mathbf{W}_1(\mathbf{T}_i^{(k)} + \mathbf{T}_{\mathcal{F}_i}^{(k)})\Big) + \sigma\Big(\mathbf{W}_2(\mathbf{T}_i^{(k)} \odot \mathbf{T}_{\mathcal{F}_i}^{(k)})\Big) \tag{6}$$

where σ is a LeakyReLU activation layer; $\mathbf{W}_1$ and $\mathbf{W}_2$ are the trainable parameters; Note that $\mathbf{T}^{(0)} = \mathbf{E}_a$ where $\mathbf{E}_a$ is an embedding table.

To model the higher-order connectivity in the IAGraph, we stack more layers and average the features of entities at each layer to generate $\mathbf{T}$. To optimize this sub-extractor, we introduce the BPR-based loss L_{AT}:

$$L_{AT} = -\sum_{(i,r,a,a') \in \mathcal{D}_A} \ln \sigma(\bar{y}_{i,r,a} - \bar{y}_{i,r,a'}); \quad \bar{y}_{i,r,a} = \mathbf{T}_i \mathbf{W}_r^a \mathbf{T}_a^T \tag{7}$$

where σ is a sigmoid activation layer, and $\mathcal{D}_A$ is $\{(i,r,a,a')|(i,r,a) \in \mathcal{G}_2, (i,r,a') \notin \mathcal{G}_2\}$; $\mathbf{W}_r^a$ is the trainable parameter.

4.2 Knowledge Aggregator

To merge the heterogeneous content information of each item into a fixed size embedding, we design a knowledge aggregator by integrating the semantic features of items extracted from the BKS and the AKS in a hierarchical manner. These item features consist of two parts: 1) high-order semantic features of items, including high-order out-degree features $\mathbf{H}_{out}$, high-order in-degree features $\mathbf{H}_{in}$ and high-order attribute-based features $\mathbf{T}$; 2) item embeddings, containing embeddings $\mathbf{E}_b$ of input to the AKS and embeddings $\mathbf{E}_a$ of input to the BKS. The fused high-order semantic features $\mathbf{M}_k^h$ are integrated by an attention mechanism that dynamically assigns attention weights to the three high-order semantic features mentioned above:

$$\mathbf{M}_k^h = \sum_{V \in \{\mathbf{H}_{out}, \mathbf{H}_{in}, \mathbf{T}\}} w_v^k * \mathbf{V}_k,$$

$$w_v^k = \frac{exp(\mathbf{W}_f^1 tanh(\mathbf{W}_f^2 \mathbf{V}_k^T))}{\sum_{Q \in \{\mathbf{H}_{out}, \mathbf{H}_{in}, \mathbf{T}\}} exp(\mathbf{W}_f^1 tanh(\mathbf{W}_f^2 \mathbf{Q}_k^T))} \tag{8}$$

where $\mathbf{W}_f^1$ and $\mathbf{W}_f^2$ are the parameters of the attention mechanism. The fused item embeddings $\mathbf{M}_k^l$ are merged using the same attention mechanism. Next, a learnable gate is introduced to balance the contributions of the fused high-order item features $\mathbf{M}_k^h$ and the fused item embeddings $\mathbf{M}_k^l$:

$$\mathbf{G}_k = \sigma(\mathbf{W}_g^1 \mathbf{M}_k^h + \mathbf{W}_g^2 \mathbf{M}_k^l)$$

$$\mathbf{M}_k = \mathbf{G}_k \cdot \mathbf{M}_k^h + (1 - \mathbf{G}_k) \cdot \mathbf{M}_k^l \tag{9}$$

where $\mathbf{W}_g^1$ and $\mathbf{W}_g^2$ are the learnable parameters and σ is a sigmoid function; $\mathbf{M}$ is the high-level semantic knowledge.

4.3   Sequential Interactions Modeling (SIM)

In sequential interactions modeling, sequential models (e.g., GRU4Rec [6], SAS-Rec [9], and BERT4Rec [21]) are widely used to capture the dynamic user preference based on historical interaction sequences. In this paper, we apply SASRec to encode the user interest representation, which consists of an embedding layer and self-attention blocks [25]. To inject the rich semantic knowledge extracted from the two KGs into SASRec, the embedding layer of SASRec is initialized by the high-level semantic knowledge $\mathbf{M}$ extracted from the knowledge aggregator. Specifically, given $\mathbf{M}$ and a user's interaction sequence $\mathcal{S}^u = [i_1, i_2, \cdots, i_T]$, the input embedding is:

$$\mathbf{E}_{\mathcal{S}^u} = [\mathbf{M}_0 + \mathbf{P}_0, \mathbf{M}_1 + \mathbf{P}_1, \cdots, \mathbf{M}_T + \mathbf{P}_T] \tag{10}$$

where $\mathbf{P}$ is a position embedding table. Then, we apply self-attentive blocks to establish dependencies between interactive items and capture the dynamic preference of the user through multi-head attention layers (MH) and fully connected feed-forward layers (FNN):

$$\mathbf{H}_{\mathcal{S}^u} = FFN(MH(\mathbf{E}_{\mathcal{S}^u})) \tag{11}$$

where $\mathbf{H}_{\mathcal{S}^u}$ is the hidden representation of the user interaction sequence $\mathcal{S}^u$. For MH and FNN, [25] has a detailed definition. To optimize the SIM, we adopt a binary cross entropy loss as the objective function:

$$L_{SQ} = -\sum_{\mathcal{S}^u \in \mathcal{S}} \sum_{t \in [1,2,\cdots,T]} \left(\ln \sigma(\widetilde{y}_{tj}) + \sum_{k \notin \mathcal{S}^u} \ln(1 - \sigma(\widetilde{y}_{tk})) \right); \quad \widetilde{y}_{tj} = \mathbf{H}_{\mathcal{S}^u,t} \mathbf{M}_j^{h^T} \tag{12}$$

Note that the fused high-order item features $\mathbf{M}^h$ are used as semantic features of the target items to avoid overfitting.

4.4   Model Learning and Prediction

We use the pre-training and fine-tuning paradigm to better incorporate the semantic information extracted from KGs into the sequential recommendation model. Specifically, the BKS and the AKS are first pre-trained according to the optimization objectives in Eq. 3 and Eq. 7, and then fine-tuned together with the knowledge aggregator and the SIM using the optimization objective in Eq. 12. The final objective function of CrbiaNet is:

$$L_{CrbiaNet} = L_{SQ} + \alpha L_{CB} + \beta L_{AT} + \gamma(||\theta||_2^2) \tag{13}$$

where L_2 regularization on θ with the weight γ is designed to prevent overfitting, and α and β are the weights of the loss functions for different knowledge sub-extractors. In the inference phase, we only use the SIM as an online service to ensure the efficiency of the service.

5   Experiment

5.1   Experimental Settings

Datasets. We conduct experiments on the Amazon dataset [5], which includes the interactions between users and items and metadata of items with natural item relations [27] (e.g., *also_view*, *also_buy*) and attributes of items (e.g., price, brand and category). CRBGraph ($\mathcal{G}_1$) and IAGraph ($\mathcal{G}_2$) are constructed from the natural item relations and the attributes of items, respectively. We use three representative sub-datasets in the Amazon dataset: *Beauty*(Beauty), *Sports and Outdoors*(Sports), and *Toys and Games*(Toys). The detailed statistics of Amazon datasets are consistent with [40]. To construct user interaction sequences, we group user interaction records, sort them according to the timestamps ascendingly. We filter out users and items with less than five interaction records following previous studies [9,21].

Parameter Settings and Evaluation Metrics. CrbiaNet is trained by the Adam optimizer [10] with a learning rate of 0.001, where the batch size of the knowledge sub-extractors (BKS and AKS) and SIM are set as 2048 and 256, respectively. Gradients are clipped when the gradient norm is greater than five. The number of layers and the embedding dimensions are set to 2 and 64 for BKS, AKS, and SIM. Following previous sequential recommendation models [9,21], the maximum length of the user interaction sequence is set as 50. The weights α, β and γ are set to 1.0. Besides, the leave-one-out strategy is used for training and evaluation, and top-k HIT Ratio(HR@k) and top-k Normalised Discounted Cumulative Gain (NDCG@k) are considered to be ranking metrics. Following previous studies [6,9], we evaluate the performance of the models by combining the ground-truth item and 99 randomly sampled non-interactive negative items.

Baseline Methods. To validate the effectiveness of our proposed CrbiaNet model, we select nine previous representative models as baseline methods.

- **BPR** [18] is a classical Bayesian personalized ranking algorithm with implicit feedback based on stochastic gradient descent.
- **FM** [17] considers the combined features based on linear regression.
- **GRU4Rec** [6] applies GRU [2] to model user interaction sequences for session-based recommendations with a ranking loss function.
- **SASRec** [9] is a sequential recommendation model based on deep unidirectional transformers that capture dynamic user interests.
- **BERT4Rec** [21] uses BERT [3] to encode user interaction sequences by deep bidirectional transformers.
- **FDSA** [38] captures the dynamic user preference by simultaneously modeling both item-level and feature-level(attribute-level) sequences.
- **S^3-Rec** [40] adopts the paradigm of pre-training and fine-tuning, where attributes are employed in the pre-training phase.

Table 1. The performance of our proposed model and previous existing recommendation models on three datasets, where the best results and the second best results are marked in bold and underlined, respectively.

Datasets	Metric	BPR	FM	GRU4Rec	SASRec	BERT4Rec	FDSA	S³-Rec	Chorus	KDA\T	CrbiaNet
Beauty	HR@5	0.3602	0.1461	0.3487	0.3754	0.4034	0.4010	0.4502	0.4575	<u>0.4846</u>	**0.5123***
	NDCG@5	0.2601	0.0934	0.2580	0.2832	0.3080	0.2974	0.3407	0.3402	<u>0.3654</u>	**0.3875***
	HR@10	0.4659	0.2311	0.4460	0.4795	0.5052	0.5096	0.5506	0.5694	<u>0.6008</u>	**0.6204***
	NDCG@10	0.2944	0.1207	0.2893	0.3168	0.3408	0.3324	0.3732	0.3766	<u>0.4031</u>	**0.4225***
Sports	HR@5	0.3629	0.1603	0.3208	0.3538	0.3922	0.3855	0.4267	<u>0.4540</u>	0.4504	**0.4860***
	NDCG@5	0.2624	0.1048	0.2257	0.2493	0.2852	0.2756	0.3104	<u>0.3346</u>	0.3273	**0.3554***
	HR@10	0.4851	0.2491	0.4389	0.4805	0.5203	0.5136	0.5614	0.5823	<u>0.5831</u>	**0.6200***
	NDCG@10	0.3018	0.1334	0.2638	0.2900	0.3264	0.3170	0.3538	<u>0.3761</u>	0.3701	**0.3988***
Toys	HR@5	0.3140	0.0978	0.3284	0.3684	0.3926	0.3994	0.4420	0.4290	<u>0.4961</u>	**0.5149***
	NDCG@5	0.2286	0.0614	0.2422	0.2712	0.2979	0.2903	0.3270	0.3306	<u>0.3806</u>	**0.3974***
	HR@10	0.4138	0.1715	0.4293	0.4751	0.4959	0.5129	0.5530	0.5291	<u>0.6015</u>	**0.6217***
	NDCG@10	0.2607	0.0850	0.2746	0.3057	0.3313	0.3271	0.3629	0.3631	<u>0.4147</u>	**0.4320***

- **Chorus** [27] is a sequential recommendation model with natural item relations and corresponding temporal dynamics.
- **KDA** [26] injects natural item relations between items, attributes of items, and temporal evolution information as additional knowledge into the sequence recommendation. For the sake of fairness, the temporal evolution information is removed in this paper and named **KDA\T**.

5.2   Performance Comparison

Table 1 shows the results of all baselines and our proposed CrbiaNet model on all datasets. First, sequential recommendation methods (e.g., GRU4Rec, SAS-Rec, and BERT4Rec) outperform collaborative filtering methods (e.g., BPR and FM) because the dynamic user preference can be captured by encoding the history of the user's interaction with the recommender system. The performance of sequential recommendation models can be further improved by merging the attributes of items (e.g., FDSA and S³-Rec), which indicates the attribute-based side information is helpful for recommender systems. Chorus obtains better performance due to incorporating behavior-based (natural) item relations. In addition, KDA\T achieves the previous state-of-the-art performance on the three datasets by integrating both attributed-based and behavior-based KGs. One possible reason for this is that the complex relations between the target items and the items in the user's historical interaction sequence are explicitly captured by the KGs.

Then, CrbiaNet consistently outperforms the pure and attribute-enhanced sequential recommendation models in the three datasets, thanks to the rich heterogeneous semantic features injected into the sequential interaction model. Compared with pure sequential recommendation methods (e.g., GRU4Rec, SAS-Rec, and BERT4Rec), CrbiaNet achieves better recommendation performance, demonstrating that the underlying semantic knowledge embedded in the KGs is helpful for capturing the dynamic user preference. CrbiaNet is superior to

Table 2. The effectiveness of each component of our proposed CrbiaNet on the three datasets.

Model	Metric	Beauty	Sports	Toys
CrbiaNet	HR@10	0.6204	0.6200	0.6217
	NDCG@10	0.4225	0.3988	0.4320
CrbiaNet-BKS	HR@10	0.6033	0.5964	0.5924
	NDCG@10	0.4076	0.3801	0.4086
CrbiaNet-AKS	HR@10	0.5134	0.5189	0.5203
	NDCG@10	0.3396	0.3178	0.3422
CrbiaNet-ADD	HR@10	0.6066	0.6145	0.6192
	NDCG@10	0.4156	0.3930	0.4274
CrbiaNet-RANDOM	HR@10	0.4795	0.4805	0.4751
	NDCG@10	0.3168	0.2900	0.3057

FDSA and S^3-Rec incorporating only attribute-based knowledge, suggesting that behavior-based(natural) item relations are helpful for the recommendation. This shows that co-occurrence patterns from item-item pairs of historical interaction sequences of similar users mitigate the disadvantage of sparse user interaction behaviors.

Finally, our proposed CrbiaNet achieves the state-of-the-art performance in three datasets compared with previous knowledge-enhanced sequential recommendation models (Chorus and KDA\T). The following facts can illustrate these results: 1) independent modeling of CRBGraph and IAGraph allows encoding the heterogeneous semantic information of items more efficiently (see the Subsect. 5.4 for more discussion); 2) high-order connections between items in CRB-Graph and IAGraph can be captured by message passing mechanism in the knowledge extractor; 3) the knowledge aggregator effectively aggregates the heterogeneous semantic information of items, which helps to dynamically assign attention weights to different semantic features based on user interest.

5.3   Ablation Study

To investigate the impact of components in CrbiaNet, we compare CrbiaNet with its four variants:

- CrbiaNet-BKS: This model incorporates only the semantic features extracted from CRBGraph by the behavior-level knowledge sub-extractor (BKS) into the sequential interactions modeling (SIM) to demonstrate the impact of cross-user item relations on recommendation performance.
- CrbiaNet-AKS: This model uses only the semantic features extracted from the IAGraph via attribute-level knowledge sub-extractor (AKS) to inject into the SIM.
- CrbiaNet-ADD: This model replaces the complex knowledge aggregator (KA) with the simple addition operation to fuse heterogeneous semantic features to validate the effectiveness of the integration strategy.

- CrbiaNet-RANDOM: This model replaces the high-level semantic knowledge **M** extracted from KGs with an embedding layer with 0 mean and 0.01 standard deviation.

Table 2 shows HR@10 and NDCG@10 for these models in all three datasets. We summarize the following findings. First, the BKS of modeling the cross-user item relations in the CRBGraph is the most critical component of CrbiaNet. CrbiaNet-BKS offers significant performance gains over the other three variants in all three datasets, indicating that co-occurrence patterns from item-item pairs can guide the extraction of more accurate user interests. Second, CrbiaNet-AKS outperforms CrbiaNet-RANDOM by utilizing attribute-based semantic features extracted from the IAGraph, demonstrating the need to incorporate the attributes of items. In addition, CrbiaNet-AKS outperforms FDSA [38] on both NDCG@10 and HR@10, which validates that our proposed AKS can effectively extract attribute-aware high-level semantic knowledge. Last, the difference in performance between CrbiaNet and CrbiaNet-ADD suggests that the hierarchical knowledge integration strategy can better integrate heterogeneous semantic features from the KGs by dynamically assigning attention weights to features.

5.4 Effectiveness of Knowledge Extractor

To validate the effectiveness of our proposed knowledge extractor, we compare CrbiaNet with three variants in terms of graph construction and graph encoding:

- DisMult: To explore the effectiveness of extracting heterogeneous semantic features from CRBGraph and IAGraph independently, this model first constructs a unified knowledge graph by merging CRBGraph and IAGraph, and then uses DisMult [35] instead of the knowledge extractor in this paper (for more details see [34]).
- TransR(IA): This model replaces AKS with TransR [13] to validate the necessity of potential attribute-aware high-order semantic features for recommendations.
- TransR(CB): This model uses TransR [13] instead of BKS to encode CRBGraph to validate the effectiveness of behavior-level high-order item relations.

The results of these variants and CrbiaNet are shown in Fig. 4. CrbiaNet achieves better performance than DisMult. Two reasons may cause this phenomenon: 1) CrbiaNet encodes different types of KGs independently to avoid confusion of heterogeneous semantic features; 2) the bilinear diagonal model (DisMult) cannot map attribute-level and behavior-level semantic features to the identical semantic space. Compared to TransR(IA) and TransR(CB), CrbiaNet achieves the best performance on all three datasets. This shows that high-level semantic features are practical for sequential recommendations. In addition, the most significant performance gap is observed between CrbiaNet and TransR(CB), indicating that behavior-level high-order item relations play a crucial role in encoding the dynamic user preference.

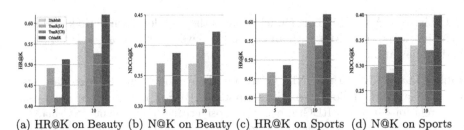

(a) HR@K on Beauty (b) N@K on Beauty (c) HR@K on Sports (d) N@K on Sports

Fig. 4. Performance of the knowledge extractor in CrbiaNet and other extractors on three datasets.

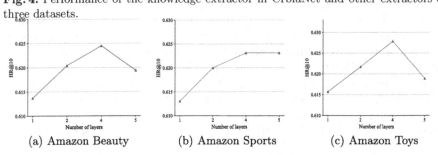

(a) Amazon Beauty   (b) Amazon Sports   (c) Amazon Toys

Fig. 5. Impact of knowledge extractor depth.

5.5   Impact of Knowledge Extractor Depth

This subsection considers the impact of the number of layers in the knowledge extractor to validate the necessity of high-order connections between items in the KGs. The results are summarized in Fig. 5. First, we can observe that recommendation performance is improved by stacking a certain number of layers in the knowledge extractor, indicating that stacking more layers can explore higher-order item relations in the KG and mine the potential preference of users. However, the recommendation performance of CribaNet on Amazon Beauty and Amazon Toys datasets decreases when more layers are stacked in the knowledge extractor. This shows that stacking too many layers in the knowledge extractor may lead to the problem of over-smoothing. This problem is prevalent in the graph neural networks [16], and we leave the exploration of solving this problem as future work. In addition, the over-smoothing problem does not affect CribiaNet on Amazon Sports dataset when the number of layers is stacked to five. The reason might be that there are more triples in the KG on the Sports dataset than the other two datasets, and longer-distance item relations are required to encode the heterogeneous semantic knowledge of items.

5.6   Compatibility of High-level Semantic Knowledge

To explore the validity and compatibility of High-level Semantic Knowledge **M** (HSK) mentioned in the Subsect. 4.2, we conduct an experiment employing the HSK and its three variants on four sequential models (GRU4Rec [6], NARM [12],

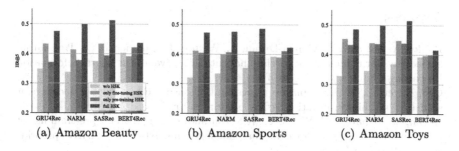

Fig. 6. The performance of CrbiaNet and its variants under different sequential models on three datasets.

SASRec [9], and BERT [21]): 1) **w/o HSK**: This method uses randomly initialized embeddings to replace the HSK; 2) **only fine-tuning HSK**: This method only uses the optimization objective in Eq. 12 to obtain the HSK through fine-tuning CrbiaNet; 3) **only pre-training HSK**: This approach only uses the optimization objectives in Eq. 3 and Eq. 7 to obtain the HSK through pre-training the knowledge extractor and keeps the HSK constant in the fine-tuning stage. 4) **full HSK**: This method first pre-trains the knowledge extractor to obtain the HSK, and then fine-tunes CrbiaNet to adapt the HSK to the sequential recommendation task.

The experimental results are shown in Fig. 6. First, we can observe that all sequential models achieve better performance than 'w/o HSK' when merging HSK, indicating that our proposed HSK is compatible and effective with the sequential recommendation models. In addition, the sequential models' performance decreases on both 'only fine-tuning HSK' and 'only pre-training HSK' compared to 'full HSK', which suggests that our proposed HSK can fully exploit the deeper underlying semantic features in the heterogeneous KGs.

6   Conclusion

In this paper, we propose a CrbiaNet for sequential recommendation by merging heterogeneous semantic features of entities extracted from decoupled KGs. In our approach, we decouple the original KG in the recommendation domain into two subgraphs, named the cross-user behavior-based graph and the intrinsic attribute-based graph. Then, we propose two knowledge sub-extractors to acquire higher-order features of entities with different semantics independently by graph neural networks. Finally, the high-order semantic features are combined and fed into the sequential recommendation model to enhance the representation of the user preference. We construct experiments on Amazon datasets, and the experimental results show that CrbiaNet outperforms the previous state-of-the-art recommendation models.

Acknowledgement. This research project was supported by the Foundation of Science and Technology Project of Hebei Education Department (Grants No. ZD2021063).

References

1. Bordes, A., Usunier, N., Garcia-Duran, A., Weston, J., Yakhnenko, O.: Translating embeddings for modeling multi-relational data. In: Advances in Neural Information Processing Systems, vol. 26 (2013)
2. Cho, K., et al.: Learning phrase representations using RNN encoder-decoder for statistical machine translation. arXiv preprint arXiv:1406.1078 (2014)
3. Devlin, J., Chang, M.-W., Lee, K., Toutanova, K.: BERT: pre-training of deep bidirectional transformers for language understanding. arXiv preprint arXiv:1810.04805 (2018)
4. Guo, W., et al.: Dual graph enhanced embedding neural network for ctrprediction. arXiv preprint arXiv:2106.00314 (2021)
5. He, R., McAuley, J.: Ups and downs: modeling the visual evolution of fashion trends with one-class collaborative filtering. In: Proceedings of the 25th International Conference on World Wide Web, pp. 507–517 (2016)
6. Hidasi, B., Karatzoglou, A., Baltrunas, L., Tikk, D.: Session-based recommendations with recurrent neural networks. arXiv preprint arXiv:1511.06939 (2015)
7. Huang, J., Zhao, W.X., Dou, H., Wen, J.-R., Chang, E.Y.: Improving sequential recommendation with knowledge-enhanced memory networks. In: The 41st International ACM SIGIR Conference on Research & Development in Information Retrieval, pp. 505–514 (2018)
8. Huang, X., Qian, S., Fang, Q., Sang, J., Xu, C.: Meta-path augmented sequential recommendation with contextual co-attention network. ACM Trans. Multimed. Comput. Commun. Appl. (TOMM) **16**(2), 1–24 (2020)
9. Kang, W.-C., McAuley, J.: Self-attentive sequential recommendation. In: 2018 IEEE International Conference on Data Mining (ICDM), pp. 197–206. IEEE (2018)
10. Kingma, D.P., Ba, J.: Adam: a method for stochastic optimization. arXiv preprint arXiv:1412.6980 (2014)
11. Kosinski, M., Stillwell, D., Graepel, T.: Private traits and attributes are predictable from digital records of human behavior. Proc. Natl. Acad. Sci. **110**(15), 5802–5805 (2013)
12. Li, J., Ren, P., Chen, Z., Ren, Z., Lian, T., Ma, J.: Neural attentive session-based recommendation. In: Proceedings of the 2017 ACM on Conference on Information and Knowledge Management, pp. 1419–1428 (2017)
13. Lin, Y., Liu, Z., Sun, M., Liu, Y., Zhu, X.: Learning entity and relation embeddings for knowledge graph completion. In: Twenty-Ninth AAAI Conference on Artificial Intelligence (2015)
14. Long, X., et al.: Social recommendation with self-supervised metagraph informax network. arXiv preprint arXiv:2110.03958 (2021)
15. Ma, W., et al.: Jointly learning explainable rules for recommendation with knowledge graph. In: The World Wide Web Conference, pp. 1210–1221 (2019)
16. Oono, K., Suzuki, T.: Graph neural networks exponentially lose expressive power for node classification. arXiv preprint arXiv:1905.10947 (2019)
17. Rendle, S.: Factorization machines. In: IEEE International Conference on Data Mining, pp. 995–1000. IEEE (2010)
18. Rendle, S., Freudenthaler, C., Gantner, Z., Schmidt-Thieme, L.: BPR: Bayesian personalized ranking from implicit feedback. arXiv preprint arXiv:1205.2618 (2012)
19. Rendle, S., Freudenthaler, C., Schmidt-Thieme, L.: Factorizing personalized Markov chains for next-basket recommendation. In: Proceedings of the 19th International Conference on World Wide Web, pp. 811–820 (2010)

20. Shani, G., Heckerman, D., Brafman, R.I., Boutilier, C.: An MDP-based recommender system. J. Mach. Learn. Res. **6**(9) (2005)
21. Sun, F., et al.: BERT4Rec: sequential recommendation with bidirectional encoder representations from transformer. In: Proceedings of the 28th ACM International Conference on Information and Knowledge Management, pp. 1441–1450 (2019)
22. Tang, J., Wang, K.: Personalized top-n sequential recommendation via convolutional sequence embedding. In: Proceedings of the Eleventh ACM International Conference on Web Search and Data Mining, pp. 565–573 (2018)
23. Taylor, W.L.: "Cloze procedure": a new tool for measuring readability. J. Q. **30**(4), 415–433 (1953)
24. Trouillon, T., Welbl, J., Riedel, S., Gaussier, É., Bouchard, G.: Complex embeddings for simple link prediction. In: International Conference on Machine Learning, pp. 2071–2080. PMLR (2016)
25. Vaswani, A., et al.: Attention is all you need. In: Advances in Neural Information Processing Systems, pp. 5998–6008 (2017)
26. Wang, C., Ma, W., Zhang, M., Chen, C., Liu, Y., Ma, S.: Toward dynamic user intention: temporal evolutionary effects of item relations in sequential recommendation. ACM Trans. Inf. Syst. (TOIS) **39**(2), 1–33 (2020)
27. Wang, C., Zhang, M., Ma, W., Liu, Y., Ma, S.: Make it a chorus: knowledge- and time-aware item modeling for sequential recommendation. In: Proceedings of the 43rd International ACM SIGIR Conference on Research and Development in Information Retrieval, pp. 109–118 (2020)
28. Wang, M., Qiu, L., Wang, X.: A survey on knowledge graph embeddings for link prediction. Symmetry **13**(3), 485 (2021)
29. Wang, P., Fan, Y., Xia, L., Zhao, W.X., Niu, S., Huang, J.: Kerl: a knowledge-guided reinforcement learning model for sequential recommendation. In: Proceedings of the 43rd International ACM SIGIR Conference on Research and Development in Information Retrieval, pp. 209–218 (2020)
30. Wang, X., He, X., Cao, Y., Liu, M., Chua, T.-S.: KGAT: knowledge graph attention network for recommendation. In: Proceedings of the 25th ACM SIGKDD International Conference on Knowledge Discovery & Data Mining, pp. 950–958 (2019)
31. Wu, L., Yang, Y., Zhang, K., Hong, R., Fu, Y., Wang, M.: Joint item recommendation and attribute inference: an adaptive graph convolutional network approach. In: Proceedings of the 43rd International ACM SIGIR Conference on Research and Development in Information Retrieval, pp. 679–688 (2020)
32. Xia, L., et al.: Knowledge-enhanced hierarchical graph transformer network for multi-behavior recommendation. In: Proceedings of the AAAI Conference on Artificial Intelligence, vol. 35, no. 5, pp. 4486–4493 (2021)
33. Xie, T., Xu, Y., Chen, L., Liu, Y., Zheng, Z.: Sequential recommendation on dynamic heterogeneous information network. In: 2021 IEEE 37th International Conference on Data Engineering (ICDE), pp. 2105–2110. IEEE (2021)
34. Xin, X., He, X., Zhang, Y., Zhang, Y., Jose, J.: Relational collaborative filtering: modeling multiple item relations for recommendation. In: Proceedings of the 42nd International ACM SIGIR Conference on Research and Development in Information Retrieval, pp. 125–134 (2019)
35. Yang, B., Yih, W.-T., He, X., Gao, J., Deng, L.: Embedding entities and relations for learning and inference in knowledge bases. arXiv preprint arXiv:1412.6575 (2014)
36. Yang, Z., Dong, S., Hu, J.: GFE: general knowledge enhanced framework for explainable sequential recommendation. Knowl.-Based Syst. **230**, 107375 (2021)

37. Yuan, F., Karatzoglou, A., Arapakis, I., Jose, J.M., He, X.: A simple convolutional generative network for next item recommendation. In: Proceedings of the Twelfth ACM International Conference on Web Search and Data Mining, pp. 582–590 (2019)
38. Zhang, T., et al.: Feature-level deeper self-attention network for sequential recommendation. In: IJCAI, pp. 4320–4326 (2019)
39. Zhao, X., Cheng, Z., Zhu, L., Zheng, J., Li, X.: UGRec: modeling directed and undirected relations for recommendation. arXiv preprint arXiv:2105.04183 (2021)
40. Zhou, K., et al.: S3-Rec: self-supervised learning for sequential recommendation with mutual information maximization. In: Proceedings of the 29th ACM International Conference on Information & Knowledge Management, pp. 1893–1902 (2020)
41. Zhu, Q., Zhou, X., Wu, J., Tan, J., Guo, L.: A knowledge-aware attentional reasoning network for recommendation. In: Proceedings of the AAAI Conference on Artificial Intelligence, vol. 34, no. 04, pp. 6999–7006 (2020)

An Analysis of Links in Wikidata

Armin Haller[1(✉)] , Axel Polleres[2] , Daniil Dobriy[2] , Nicolas Ferranti[2] ,
and Sergio J. Rodríguez Méndez[1]

[1] Australian National University, Canberra, ACT 2601, Australia
{armin.haller,sergio.rodriguezmendez}@anu.edu.au
[2] Vienna University of Economics and Business, Vienna, Austria
{axel.polleres,daniil.dobriy,nicolas.nicolas}@wu.ac.at

Abstract. Wikidata has become one of the most prominent open knowledge graphs (KGs) on the Web. Relying on a community of users with different expertise, this cross-domain KG is directly related to other data sources. This paper investigates how Wikidata is linked to other data sources in the Linked Data ecosystem. To this end, we adapt previous definitions of ontology links and instance links to the terminological part of the Wikidata vocabulary and perform an analysis of the links in Wikidata to external datasets and ontologies from the Linked Data ecosystem. As a side effect, this reveals insights on the ontological expressiveness of meta-properties used in Wikidata. The results of this analysis show that while Wikidata defines a large number of individuals, classes and properties within its own namespace, they are not (yet) extensively linked. We discuss reasons for this and conclude with some suggestions to increase the interconnectedness of Wikidata with other KGs.

1 Introduction

Wikidata, as a "multilingual Wikipedia for data" [25], has grown to a knowledge graph (KG) containing over 95M entities[1]. Since its beginning in 2012, Wikidata has been conceived as a KG that is built bottom-up by its many editors (plus, partially, automatic bots). As a backend, Wikidata uses Wikibase, an open-source software suite for creating collaborative knowledge bases, which allow its many editors to contribute to this KG. Being build bottom-up by domain experts who often also maintain the external original source of data that is being added, Wikidata already includes many links to other datasets, for example, through the reuse of external identifiers for entities (e.g., ORCID records for academics, DOIs for digital artefacts, or the Ensembl identifier for genes (e.g., Q14864292)). This allows the editors of Wikidata to (automatically) integrate data from external KGs that remain under the control of the original publisher. In fact, such automatic integration of external data through bots[2] already exists on Wikidata itself, e.g., a Citationgraph_bot that updates citation numbers of academic works. Consequently, Wikidata has become in practice a *data directory* that serves as entry point to external datasets, other knowledge graphs, or ID providers, respectively. These

---

[1] cf. https://www.wikidata.org/wiki/Special:Statistics.

[2] https://www.wikidata.org/wiki/Wikidata:Bots.

© The Author(s), under exclusive license to Springer Nature Switzerland AG 2022
P. Groth et al. (Eds.): ESWC 2022, LNCS 13261, pp. 21–38, 2022.
https://doi.org/10.1007/978-3-031-06981-9_2

observations motivate a more in-depth study on the linkage of Wikidata with other KGs and the types of links used for such linking.

Previous work has established link types definitions between datasets [13]. Broadly, this work defined two categories of links, ontology links and instance links. We aim to herein re-use and adapt these definitions and apply them to the Wikidata data model. To do so, we evaluate the HDT dump of the entire Wikidata KG from March 3rd, 2021[3].

For the analysis of ontology links, however, we can not directly use the established link types definitions in [13], since Wikidata does practically not rely on the RDFS/-OWL semantics and vocabularies. While – strictly speaking, in terms of its (RDFS and OWL) TBox constructs used – Wikibase and, as such, Wikidata, use a very simple ontology (i.e., wikiba.se/ontology), the *actual* ontology to describe entities in Wikidata is largely build bottom-up by the community itself, not using RDFS/OWL. Indeed, Wikidata partially tries to re-use and integrate external ontologies, but it does so by introducing its own meta-model, and only links to external ontologies through specific, again community-introduced, property relations, such as equivalent class (P1709). This flexibility allows the community to extend the knowledge graph rapidly by adding a rich set of statements about entities in the world without much concern for (logical) consistency expected in the stricter frameworks of RDFS and OWL. This liberty comes with drawbacks, though, with semantic errors or inconsistencies, such as incoherent meta-modeling of classes/instances [23] (i.e., using a taxonomy relation instance of (P31) or a subclass of (P279) relation for similar items[4]), being prevalent. However, many of these problems are eventually resolved through discussions among the editors. There have been some studies on such quality issues within Wikidata [18], but generally there is still little understanding of the quality and evolution of knowledge contained within Wikidata, particularly on the schema level and the schematic relations to other ontologies on the Web.

We therefore present an extension of the definitions of ontology links in [13] by mapping them to the informal, community-developed Wikidata meta-model. In the course of that, we also compare the available meta-properties in Wikidata to their respective corresponding properties in the OWL and RDFS vocabulary which allows us to draw some preliminary conclusions about the ontological expressivity used in Wikidata's meta-modeling. The mapping also allows us to analyse the extend of ontology links and instance links from Wikidata to other KGs. Specifically, we aim to investigate how central Wikidata is to the Linked Data ecosystem by testing the following hypotheses in our analysis.

First, for a KG to serve as a central hub for Linked Data, it should use classes and properties that are defined within its own namespace to represent entities in its KG. Not relying on external ontologies to provide semantics to entities within makes a KG robust to changes in the semantics or availability of external ontologies and as such, a reliable link target for other KGs. It has been observed in our previous study [13] that DBpedia, an existing central hub for Linked Data, exhibits this phenomena that we test in our first hypothesis.

---

[3] https://www.rdfhdt.org/datasets/.

[4] For instance, the pattern { [] wdt:P279 ?X; wdt:P31 ?X. } indicates ambiguous subclass vs. instance of usage on 2131 entities, run on 9 Dec 2021 at https://w.wiki/4XQw.

H1 Wikidata defines the vast majority of its terminological entities and properties in its authoritative namespace.

Our next set of hypotheses are concerned with the extend to which Wikidata is linked to other ontologies in the Linked Data ecosystem.

H2.1 As a central KG, the ratio of class links to classes defined within Wikidata is much larger than the same ratio for other datasets in the Linked Data ecosystem.

H2.2 As a central KG, the ratio of property links to properties defined within Wikidata is much larger than the same ratio for other datasets in the Linked Data ecosystem.

H3 As a central KG, Wikidata does not type entities using classes from external ontologies, i.e., classes using a namespace other than the authoritative namespace of Wikidata.

Our next two hypotheses are concerned with the extend of which Wikidata is linked to other KGs on an instance level and if the link targets are indeed RDF data.

H4.1 As a central KG, Wikidata includes links from entities defined in its authoritative namespace to entities defined in other KGs and the ratio of such instance links to entities defined in Wikidata is much larger than for other datasets in the Linked Data ecosystem.

H4.2 The amount of instance links to RDF resources is relatively higher than to other types of Web resources, i.e., the content type of the target URI in an instance link is a common RDF serialisation.

In our last hypothesis we test for how many of the entities defined within Wikidata, it is (claims to be) the only authoritative source. A central hub for Linked Data should **not** be the authoritative source for entities, but rather only provide a persistent identity for an entity, while linking to the authoritative external source.

H5 Wikidata establishes equivalence or some weaker forms of likeness relations for the majority of its unique individuals that are part of an instance link, i.e., between entities defined within the Wikidata authoritative namespace and entities defined in other authoritative namespaces.

The remainder of this paper is structured as follows. In Sect. 2 we discuss the ontology in Wikidata and provide our mapping semantics between the Wikidata meta-model and RDFS/OWL. In Sect. 3 we describe our methodology to analyse link types in Wikidata. Section 4 presents the results of this analysis and the hypotheses tested on the entire Wikidata RDF corpus. We discuss related work in Sect. 5 before we conclude in Sect. 6.

2   The Wikidata Ontology Schema

In terms of a formal backbone terminology, Wikidata relies on Wikibase's minimal pre-defined schema, i.e., wikiba.se/ontology that is used to describe the wiki *pages* of an entity on Wikidata, and, among other things, defines what constitutes a statement for an entity through the wikiba.se/ontology#Statement class. However, for our research, this ontology is somewhat irrelevant, as we are looking at internal and external links between entities and the schema (properties and classes used at the statement level) in Wikidata's itself, rather than the Wikibase meta-model. The actual vocabulary used to describe entities in Wikidata is collaboratively built, bottom-up, and indeed its own meta-modelling properties, similar to RDFS/OWL vocabulary properties, have been introduced to this end in the Wikidata namespace. That is, while Wikidata follows the RDF model, it does not use the RDFS or OWL semantics for its ontological meta-model: it rather conflates[5] what in the traditional Semantic Web stack is defined in RDFS and OWL, i.e., the knowledge about things, groups of things, and relations between things, with what would normally be defined in upper-level or domain ontologies. In this section we will therefore discuss the specific meta-modelling classes and relations that are introduced in Wikidata, their relations and – where possible – their mapping to RDFS/OWL. This mapping will form the basis of our link analysis. We *emphasize* that our proposed mapping is one possible interpretation of the (evolving) meta-model in Wikidata, with the specific purpose of providing formal semantics for our link analysis: we acknowledge that the community does not provide such a mapping by design, in order to avoid (too) strong formal ontological commitment.

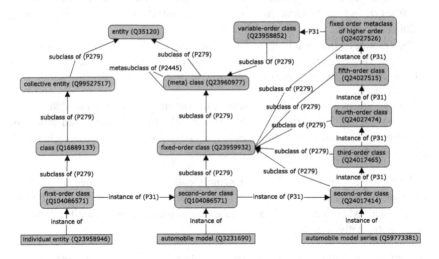

Fig. 1. Overview of the top-level class hierarchy of the Wikidata ontology

2.1   Classes in Wikidata

Figure 1 presents the top-level class hierarchy of Wikidata. Wikidata formally distinguishes between items that are classes, for example, person (Q215627), and items that are instances, for example, Barack Obama (Q76) who is an instance of (P31) human, which itself is a subclass of (P279) person (Q215627). These instances are related to its class via the instance of (P31) relation. Classes in Wikidata are items that are in

---

[5] See   https://www.wikidata.org/wiki/Wikidata:WikiProject_Ontology/Top-level_ontology_list for the top two layers of the ontology.

the object position of an instance of (P31) relation or in the subject or object position in a subclass of (P279) statement. There is also a is meta-subclass of (P2445) relation, but it is hardly used in Wikidata[6]. The Wikidata top-level class for class items is class (Q16889133), which itself is a subclass of entity (Q35120). Wikidata also distinguishes between first-order classes (Q104086571) and second-order classes (Q104086571)[7].

A metaclass (Q19361238) is defined that is the superclass of fixed-order classes. As such, a second-order class is a metaclass, the instances of which are classes of individuals, for example, the aircraft class (Q1875621) is a second-order class whose members (instances) are first-order classes, including for example wide-body quadjet (Q19394992) and aircraft functional class (Q20027953) which has a subclass wide-body twinjet (Q18683432). While the latter (i.e., the use of an aircraft functional class only for wide-body twinjet, but not for wide-body quadjets) is an example of a non-intuitive subclass hierarchy in Wikidata, it may also reflect different modelling choices by different users. We have argued previously [14] that such a bottom-up development may eventually lead to a more broadly accepted Web ontology. Also, while Wikidata does distinguish between classes and instances, it does not mandate that instances can not also be subclasses of (P279) a class or classes can not be defined as instances of (P31) an instance. While such meta-modeling is not per se forbidden in OWL2 (i.e., through "punning"[8]), in Wikidata it often appears with entities that should be either a class or an instance, but not both. For example, Wiener Schnitzel (Q6497852) is a subclass of schnitzel which is a subclass of meat dish, while at the same time Wiener Schnitzel is also an instance of veal dish which itself is a subclass of meat dish. The only assumption in Wikidata is that entity (Q35120) is the class of all items and therefore all items are an instance of entity (Q35120), as well as all classes are subclasses of entity (Q35120) which is not unlike the role of rdfs:Resource in the RDF meta-model.

2.2 Properties in Wikidata

Properties in Wikidata use the full generality of RDF properties in the sense that they represent both binary (object) relations and (atomic value) attributes. That is, properties are used to define arbitrary item-property-value triples where the value can either be an item or a literal. On top of that, reminding one of RDF's reification mechanism, each such statement can also be qualified, i.e., additional information can be added to the statement (e.g., contextual or provenance information). Indeed, the relation between Wikidata's qualified statements to reification and other potential meta-statement encodings in RDF has been discussed in detail by Hernández et al. [15].

Properties, like entities, have their own Wikidata page and use opaque identifiers starting with "P". Wikidata reuses some RDFS/OWL properties, e.g., rdf:type,

---

[6] i.e., there are 37 uses of P2445 in total in Wikidata as of August 2021.

[7] There are higher orders of second-order class, i.e., third-, fourth- and fifth-order classes, each of which is an instance of the higher ordered class, all of which are subclasses of the fixed-order class (Q23959932).

[8] cf. https://www.w3.org/TR/owl2-new-features/#Simple_metamodeling_capabilities.

rdfs:label, owl:complementOf, owl:someValuesFrom, owl:allValuesFrom etc. However, as discussed earlier, these are merely used to define statements about pages in Wikidata using the wikiba.se ontology, rather than for terminological statements about Wikidata classes and properties. Yet, in order to define such terminological properties within the Wikidata KG, "properties for properties" (73 as of November 2021)[9] are defined in the Wikidata "ontology".

As compared to other RDF KGs, where typically the RDFS/OWL namespaces are used for terminological properties and a separate ontology namespace is used for domain entities/relations, Wikidata's terminological properties use the same namespace as properties that describe entities/relations. For example, in KGs using the RDFS/OWL semantics, relations such as "subClassOf" use the RDFS namespace while a relation like "name" uses a domain-ontology namespace such as FOAF, whereas in Wikidata "subclass of" (P279) and "given name" (P735) share the same namespace.

In the following we discuss under which circumstances we treat which of these properties as equivalent to their related RDFS/OWL properties as per Table 1: additional to introducing its own already mentioned instance of (P31), and subclass of (P279) relations, to describe taxonomic relations and identify class memberships and hierarchies, respectively, Wikidata introduces a property to describe the hierarchical relation between properties, i.e., subproperty of (P1647), which we consider equivalent to the rdfs:subPropertyOf relation. However, note that this property is used in Wikidata exclusively to link properties within the Wikidata namespace, i.e., it is not used for linking to external vocabularies. For linking to properties external to the Wikidata namespace, external subproperty (P2236) and external superproperty (P2235) are introduced. We consider these equivalent to rdfs:subPropertyOf, or its inverse, respectively.

Domain and range properties are not directly defined in Wikidata. These property restrictions[10] can be stated in Wikidata using a qualified property constraint (P2302) on the property, where a skolemized IRI is assigned to the entity that is defined as either a type constraint (Q21503250) for the domain classes of the property, or, respectively, as value-type constraint (Q21510865) for the range classes of the property; the respective target class is referenced with the property (P2308). For example, the domain for the date of birth property (P569) is defined to be Human (Q5) (among others) with the following triples[11],

$$t_1 = \text{wdt:P569 p:P2302 \_:qe.}$$

$$t_2 = \text{\_:qe ps:P2302 wd:Q21503250.}$$

$$t_3 = \text{\_:qe pq:P2308 wd:Q5.}$$

---

[9] cf. https://www.wikidata.org/wiki/Wikidata:List_of_properties/Wikidata_property_for_proper ties.

[10] We note here again that subtle semantic differences such as constraining (i.e., CWA) vs implicit (i.e., OWA) semantics of certain properties are not relevant for the purpose of our link analysis.

[11] Prefixes are used as follows: wd: <http://www.wikidata.org/entity/>, wdt: <http://www. wikidata.org/prop/direct/>, pq: <http://www.wikidata.org/prop/qualifier/>, p: <http://www. wikidata.org/prop/>, ps: <http://www.wikidata.org/prop/statement/>.

Table 1. Mapping of Wikidata properties to RDFS/OWL properties

RDFS/OWL property	Equivalence established through	Wikidata property
rdf:type	equivalent property (P1628)	instance of (P31)
rdfs:subClassOf	equivalent property (P1628)	subclass of (P279)
rdfs:subPropertyOf	equivalent property (P1628)	subproperty of (P1647)
rdfs:subPropertyOf	equivalent property (P1628)	external subproperty (P2236)
Inverse rdfs:subPropertyOf	equivalent property (P1628)	external superproperty (P2235)
rdfs:range	equivalent property (P1628)	expressed via property constraint (P2302)
rdfs:domain	equivalent property (P1628)	expressed via property constraint (P2302)
rdfs:label	documented as matching[a]	rdfs:label
rdfs:comment	documented as matching[a]	schema:description
rdf:first	documented as matching[a]	expressed via series ordinal (P1545)
rdf:rest	documented as matching[a]	expressed via series ordinal (P1545)
rdfs:member	documented as matching[a]	part of (P361)
Inverse rdfs:member	inverse property (P1696) of part of (P361)	has part (P527)
owl:equivalentProperty	equivalent property (P1628)	equivalent property (P1628)
owl:equivalentClass	equivalent property (P1628)	equivalent class (P1709)
owl:inverseOf	equivalent property (P1628)	inverse property (P1696)
owl:di erentFrom	equivalent property (P1628)	di erent from (P1889)
owl:unionOf	equivalence intended[b]	union of (P2737)
owl:disjointUnionOf	equivalence intended[b]	disjoint union of (P2738)
owl:onProperty	no documented equivalence	possible candidates: property constraint (P2302)
owl:sameAs	no documented equivalence	possible candidates: exact match (P2888), said to be the same as (P460)
owl:disjointWith	no documented equivalence	N/A
owl:propertyDisjointWith	no documented equivalence	N/A
owl:propertyChainAxiom	no documented equivalence	N/A
owl:assertionProperty	no documented equivalence	N/A

[a] cf. https://www.wikidata.org/wiki/Wikidata:Relation_between_properties_in_RDF_and_in_Wikidata.
[b] cf. https://www.wikidata.org/wiki/Wikidata:Property_proposal/Archive/48#P2737.

where _:qe is actually a skolemized blank node with the IRI wd:statement/ P569-F9768BAA-6BB3-4710-A3E1-B6FB9432D372. Note that for our analysis of links, when only considering whether an external ontology is referenced on a property, we do not need to distinguish if the object that belongs to an external namespace is a domain or range class, i.e., we will not need to check in our SPARQL query below if the target object is a type constraint (Q21503250) or value-type constraint (Q21510865).

In order to state equivalence between two properties, Wikidata introduces the property equivalent property (P1628). Disjointness between properties cannot be stated in Wikidata: in fact, disjointness between properties was proposed by the community (and voted on)[12], but eventually not included. While the reason for its non-inclusion is unclear, it is challenging to maintain disjointness with other properties in a bottom-up created KG where properties can be added arbitrarily.

There is also no relation to define property chain axioms (i.e., owl:propertyChainAxiom), nor is there support for negative property assertions, i.e., a relation similar to owl:assertionProperty does not exist.

As shown above, Wikidata uses the relation property constraint (P2302) to define restrictions on a property. While such a restriction can have more than one triple, it is otherwise very similar to owl:onProperty; as such, for the purpose of our analysis, we consider it equivalent. Restrictions are linked to either a class or property using the OWL properties, owl:onClass, owl:onProperty.

3 Links in Wikidata

Based on the above correspondences of Wikidata's terminological properties with RDF-S/OWL, we are ready to define different link types in the Wikidata data model. Here we will rely on definitions of link types in the RDFS/OWL model defined in our earlier work [13] in terms of resp. SPARQL queries on the Wikidata model[13]. This enables us to directly provide a quantitative and qualitative analysis of the discovered links in terms of the resp. query results.

3.1 Dataset Corpus and Authoritative Namespaces

In order to analyse links to other datasets in the Wikidata KG, we first need to establish a list of authoritative namespace URIs that are defined by KGs other than Wikidata. For that, we are using the dataset corpus that was defined and published in [13], i.e., the LODCloud, augmented with historically available datasets that were cached in the LODLaundromat [4] and provided as a downloadable corpus in HDT [8]. The Wikidata HDT file (using the http://www.wikidata.org namespace URI) was added to that corpus. The resulting corpus consists of 431 Linked Datasets, each encoded in HDT for a total size of 104 GB (uncompressed 353 GB), with a total number of 17,841,499,814 (i.e., ≈17.8 billion) triples.

3.2 Ontology Corpus

As with the dataset corpus, we are reusing the ontology corpus published by Haller et al. [13], i.e., a crawl of the unique classes and properties in prefix.cc as well as the declared classes and properties in each dataset (the 431 from above). While not every ontology is registered in prefix.cc (a total of 2,794 ontologies are registered as of August 2021), our process also follows all import statements in those ontologies. Given

---

[12] https://www.wikidata.org/wiki/Wikidata:Property_proposal/disjoint_with.

[13] All code implemented in Python is available at: https://github.com/arminhaller/LinksInLOD.

that ontologies are supposed to be a shared conceptualisation of a domain, if no other ontology reuses the ontology, it is unlikely to be used in many datasets, nor is it likely been used in Wikidata either.

3.3 Link Type Analysis

As per the definitions in our previous work [13] we distinguish two general types of links, Ontology (TBox) Links and Instance (ABox) Links. Ontology Links are further classified into *class links*, *instance typing links*, *property links* and *instance role links*. Instead of providing re-definitions of those links to match the meta-model of Wikidata, we provide reformulations of the operationalised SPARQL queries that retrieve those link types and that implement the mapping relations between the Wikidata meta-model and the RDFS/OWL semantics as defined in Table 1.

For instance, let the dataset ds_{WD} be Wikidata with the set of its authoritative namespaces (i.e., the namespaces denoting Wikidata-defined URIs) being $NS_{ds_{WD}}$ = {http://www.wikidata.org/entity/, http://www.wikidata.org/prop/direct/, https://www.wikidata.org/wiki/Special:EntityData/}. Further, let ds_2 denote the Disease Ontology, with NS_{ds_2} = { http://identifiers.org/doid/} and let further ds_3 denote the schema.org vocabulary, with NS_{ds_3} = {http://schema.org/}. We shall denote the mentioned namespaces with the prefixes wd:, wdt:, data:, doid:, and schema:, respectively. If we consider now the triple,

$$t_1 = \text{wd:Q84263196 wdt:P2888 doid:0080600.}$$

in ds_{WD}, stating that COVID-19 (Q84263196) is an **exact match** (P288) to the class DOID:0080600 in the Human Disease Ontology, it shall be considered an *instance link*, from ds_{WD} to ds_2: while doid:0080600 is a class in the Human Disease Ontology, it is used in an instance position in the triple above from Wikidata, therefore it is an *instance link*. The next triple we consider

$$t_2 = \text{wd:Q84263196 wdt:P31 wd:Q609748.}$$

defines COVID-19 (Q84263196) as an **instance of** (P31) of the emerging communicable disease (Q609748) class. While this is not a link, but rather an internal ontological reference within ds_{WD}, however,

$$t_2' = \text{wd:Q84263196 rdf:type schema:Dataset.}$$

is indeed an *ontology link*, more specifically, an *instance typing link* from ds_{WD} to ds_3. In fact, every wiki page in Wikidata is defined as of type **schema:Dataset**. Next,

$$t_3 = \text{wdt:P569 wdt:P1628 schema:birthDate.}$$

is an example of a *property link* from ds_{WD} to ds_3. Finally,

$$t_4 = \text{wd:Q5 wdt:P1709 schema:Person}$$

is a *class link* from ds_{WD} to ds_3.

In order to define these link types more clearly, in the following we provide SPARQL queries on the Wikidata data model that correspond to the link types defined in [13], adapted to the correspondences in Table 1.

Ontology (TBox) Links. With the query shown in Listing 1.1 we retrieve all external classes, i.e., classes using a namespace other than the Wikidata namespace (using the FILTER statement) that are not explicitly declared as an RDFS/OWL class (which no class in Wikidata is) or as a type of class (Q16889133), but are used to *i)* define an instance (i.e., they are used in an assertional axiom), *ii)* define a terminological axiom that either extends or narrows a class through a subclass of relation (P279), *iii)* define a class' equivalence (P1709), union of (P2737), disjoint union of (P2738) or *iv)* define the domain or range of a property (pq:P2308).

Listing 1.1. SPARQL query used to retrieve all external classes.

```
SELECT DISTINCT ?C WHERE {
    {[] a ?C. } UNION
    {[] wdt:P279 ?C. } UNION {?C wdt:P279 []. } UNION
    {?C wdt:P2738 [].} UNION
    {?C wdt:P1709 [].} UNION {[] wdt:P1709 ?C.} UNION
    {?C wdt:P2737 [].} UNION
    {[] pq:P2308 ?C. }
    FILTER (!regex(str(?C), "http://www.wikidata.org","i"))
        .

}
```

For each class URI retrieved through this query, we check its occurrence in either the subject or object position in any triple in the KG. The number of resulting triples constitutes the number of *Class Links* in the Wikidata KG.

For *Property Links* we follow a similar process. With the query shown in Listing 1.2, we retrieve all external properties (i.e., properties using a namespace other than the authoritative Wikidata namespace) that are not explicitly declared as a property but are used: *i)* within a subproperty relation (P1647) or external sub/superproperty relation (P2236, P2235), *iii)* in a property restriction or to define the domain or range of a class (P2302), or *iv)* to define a properties' equivalence (P1628), inverseness with/to another property (P1696), different from (P1889), complement of (P8882).

Listing 1.2. SPARQL query used to retrieve external properties.

```
SELECT DISTINCT ?P WHERE {
    {?P wdt:P1647 []. } UNION {[] wdt:P1647 ?P. } UNION
    {[] wdt:P2236 ?P. } UNION
    {[] wdt:P2235 ?P. } UNION
    {?P wdt:P1628 []. } UNION {[] wdt:P1628 ?P. } UNION
    {?P wdt:P1696 []. } UNION {[] wdt:P1696 ?P. } UNION
    {?P wdt:P1889 []. } UNION {[] wdt:P1889 ?P. } UNION
    {?P wdt:P8882 []. } UNION {[] wdt:P8882 ?P. } UNION
    {?P wdt:P2302 []. }
    FILTER (!regex(str(?P), "http://www.wikidata.org","i"))
        .

}
```

For each property URI retrieved through this query, we check its occurrence in the predicate position in any triple in the dataset. The number of resulting property URIs constitutes the occurrence of *Property Links* in the dataset.

Instance Links (ABox Links). Before we can compute the number of *Instance Links* from an individual in the Wikidata namespace to any individual in an external namespace, we first need to find all unique individuals in the KG.

1. We find all individuals of classes/properties that are declared (i.e., individuals that are defined as a type of a class/property using (P31)). For each retrieved unique individual, we check if they are defined in the Wikidata namespace. If not, the triple they appear in is counted as an *Instance Typing Link*.
2. We then find all individuals that are reused from a non-authoritative namespace URI in the subject position without being explicitly declared as a type of a class or property. To retrieve those, we first query all triples in the dataset and then check for each unique subject URI that is not in the Wikidata namespace, if it is already in the set of declared instances (as of step 1), or if it is in the set of classes and properties (cf. Sect. 3.2). If it is neither, we count the triple as an *Instance Link*.
3. We then follow a similar process for each individual reused from a non-authoritative namespace URI in the object position. For each unique object URI, we check the following conditions: *i)* the subject URI does not contain the Wikidata namespace URI, *ii)* the predicate is not an instance of (P31) relation, and *iii)* the object URI is not already contained within the set of declared instances. If none of these conditions are satisfied, we record it as an *Instance Link*.

4 Evaluation of Links

In the following, we discuss the results of the link analysis of Wikidata. All tests were performed on a machine with 8vCPUs, 380 GB RAM and 5 TB hard disk space.

4.1 General Statistics of the Wikidata KG

Before we analyse the number and types of links in Wikidata, we present some general statistics of the Wikidata KG that we computed using its HDT file in Table 2. The first noteworthy observation we can make is that the ratio between unique subjects to unique predicates in Wikidata is 1/41810, whereas in the LOD dataset corpus [13] the ratio was 1/3900 if using the mean and 1/19 if using the median. Since the Wikidata KG with 1.69bn triples is much larger than the largest KG in the LOD corpus, i.e., the 2016 version of DBpedia with 1.04bn triples, which itself was much larger than the mean number of triples (i.e. 16.92 m) for all datasets in the corpus, suggests that the number of predicates in a KG grows following an asymptotic function. This seems natural, as while the number of entities in a general KG such as Wikidata is potentially infinite, the attributes that can be assigned to those entities are somehow limited.

Supporting our first hypothesis, Wikidata defines all of its 89 m unique individuals using its own ontology. The number of unique individuals is also larger than the number of claimed unique entities defined in Wikidata at the time the HDT file was generated (March 2021), i.e., 73 m, meaning that all entities (plus some more) are defined within the Wikidata namespace. While the number of unique subjects, with 1.62bn is a lot larger than the number of unique individuals (i.e., 89 m), this is due to the fact that Wikidata introduces skolemized IRIs for qualified statements on an entity, and not because of entities being defined using external ontologies, i.e., instance typing links (see below). As such, on average for each unique entity about 18 entities are created as part of the subgraph for that entity and that redirect in the Linked Data API to that target entity. For example, for the entity Barack Obama (Q76) there are 394 skolemized IRIs (as of November 2021), such as http://www.wikidata.org/entity/statement/q76-F23589FF-58A6-438B-BC7E-79F6B436AFD0 that describes a qualifier about Barack Obama's education at (P69) Harvard Law School which he completed with an academic degree (P512) of Juris Doctor with an end time (P582) of 1991. These qualified statements do not have their own page on Wikidata, but they do resolve to the page where they are defined through the Linked Data API.

Table 2. General statistics of the Wikidata KG

# Triples	1,693,668,039
# Unique Subjects	1,625,057,179
# Unique Predicates	38,867
# Unique Objects	2,538,585,808

Table 3. Class/property statistics in Wikidata

# Unique Individuals	89,120,227
# Unique Declared Classes	0
# Unique Undeclared Classes:	2,522,595
# Unique Declared Properties:	74,309
# Unique Undeclared Properties:	29,167

4.2   Ontology Links

Before we set out to test our hypotheses related to ontology links in Wikidata, we first present some general statistics on the use of classes and properties in Wikidata in Table 3. For our analysis we distinguish between declared and undeclared classes, i.e., class URIs that are defined within the authoritative namespace of the KG using a triple {[] rdf:type owl:Class.} or {[] rdf:type rdfs:Class.}, and class URIs that are merely reused from a different namespace URI. Since Wikidata does not use rdf:type relations for class definitions (as above), all 2,522,595 unique classes defined in Wikidata are undeclared according to the RDFS/OWL semantics.

In contrast to class URIs, properties in Wikidata are declared using the owl:DatatypeProperty and owl:ObjectProperty types. In fact, each property, denoted by an identifier starting with "P" includes up to nine datatype and object property definitions, each with a different URI of that property identifier (i.e., strictly speaking different properties) as defined by the wikiba.se ontology, e.g., for date of birth (P569) this includes http://www.wikidata.org/prop/direct/P569 which is defined as an owl:DatatypeProperty and http://www.wikidata.org/prop/statement/P569 as a shorthand property to find the statements this property is used in which is also defined as an owl:DatatypeProperty. Therefore the number of declared properties in March 2021 (i.e., 74,309) is more than six times larger than the claimed number of properties on Wikidata, i.e., 9,367 properties as of November 2021[14,15] (Table 4).

---

[14] https://www.wikidata.org/wiki/Wikidata:List_of_properties.

[15] No longitudional data is published on the Wikidata site, but the growth in the number of properties between July and November 2021 was 3.4%.

Table 4. Link Type statistics

# Class Links	3,955
# Property Links	835
# Instance Typing Links	0 (173,168,537)
# Instance Links	173,177,045

Class Links. There are only 3,955 class links defined in Wikidata. This is comparable to the other datasets in the LOV corpus that were analysed previously [13], where the number of class links is relatively constant around 100–10,000 per dataset. However, Wikidata uses 2.5 m classes (compared to an average of 6,379 classes per dataset), with a ratio of class links per class of only 0.0015, while the average ratio of class links per class for the datasets in the LOV cloud is 11.27. One of the reasons why the ratio is so low, is that many instances in Wikidata are also defined as classes (see above), contrary to many other KGs where there is a strict separation between TBox and ABox axioms. Also, the user interface's (Wikibase) autocomplete feature when creating links to classes only works for classes in the Wikidata namespace, but not for external URIs of classes. Still, as a central hub of the Linked Data ecosystem, one would expect Wikidata to have more such links, particularly given the bottom-up development of the Wikidata ontology. The 2016 version of DBpedia [2], for example, includes 8,258 class links for its 3,197 classes with a ratio of 2.58, even though its ontology is built by experts top-down. We therefore need to reject our hypothesis H2.1. To increase the number of class links, a relation similar to "external subproperty" should be introduced in Wikidata to define external subclass relations on a class. A lookup service (based for example on the LOV API [24]) could then guide Wikidata editors to the existence of external classes.

Property Links. There are only 835 property links for a total of 74,309 properties in Wikidata, i.e., there are on average only 0.01 property links per property, a ratio that is much lower than for the LOD corpus. We therefore must also reject our hypothesis H2.2, that Wikidata includes many more property links per property than other datasets. One of the reasons for this low property link ratio might be that while there exist several properties in the Wikidata ontology that are specifically designed to link to external ontologies or allow external URIs to be used, i.e., equivalent property (P1628), di er-ent from (P1889), external subproperty (P2236), external superproperty (P2235) these are only relatively recent additions. External subproperty and superproperty which are used 94 and 159 times, respectively, were only added in May 2017 and May 2018, after many of the 74k properties in Wikidata have been defined. There would need to be a concerted effort by the community to update existing properties with these relations.

Instance Typing Links. There are no instance typing links in Wikidata that use the instance of (P31) relation. Since, to the best of our knowledge, Wikibase does not allow users to add an external URI when using the instance of relation it is not unexpected that there are no such links. There are 173 m instance typing links using the rdf:type relation. However, since they are all (auto-generated) links to define a Wikidata page as a schema:Dataset and a schema:Article we excluded them. Therefore we can confirm our hypothesis H3 that as a KG with a general and sufficiently comprehensive ontology, Wikidata types all entities using its own ontology and therefore includes no instance typing links.

4.3   Instance Links

Wikidata includes many links from entities (unique individuals) defined in its authoritative namespace to entities defined in other KGs. With 173 m such links, it means that 10.22% of all triples in the Wikidata graph link to individuals that use a namespace other than the Wikidata namespace. However, the ratio of such links to entities (at 1.94) is much lower than with other datasets in the LOD ecosystem (8.6) and we therefore need to reject our hypothesis H4.1.

Even this number includes many links to Wikipedia. In fact, every entity in Wikidata that also has a Wikipedia entry includes hundreds of links to Wikipedia. However, while they are considered links according to our definition, none of the target resources are, in fact, RDF resources, but the Wikipedia entity is created in the Wikidata namespace (using the Wikipedia URL).

While for ontology links we are able to verify for all links if the target URI is an RDF resource, with the large number of instance links, we can not. However, to test our hypothesis H4.2 a sample of 1,924,940 target URIs from all instance links was randomly collected. We then built a simple crawler that checked for each URI if a document in RDF format can be retrieved at the target URI, i.e., classifying the links in two main groups: Web resources, and RDF entities. Table 5 shows that the majority of links point to resources other than RDF. While we therefore need to reject our hypothesis H4.2, the fact that a quarter of resources are, in fact, RDF resources is encouraging, given that for many entities there may not yet exist an RDF representation outside of Wikidata. To distinguish RDF from non-RDF resources in links, i.e., to distinguish 1-star linked data from higher-ordered linked data [5], Wikidata should automatically qualify links based on the target format of the linked resource (Table 6).

Table 5. Instance links content-types statistics

# URL Not Found	18,081 (0.9%)
# Other Errors	138,656 (7.2%)
# Timeout	218,542 (11.4%)
# RDF Entities	471,088 (24.5%)
# Web Resources	1,078,573 (56.0%)

Table 6. Instance Link types statistics

# owl:sameAs Links	0
# Exact Match (P2888) Links	3,268,021
# Said to be the Same (P460) Links	2
# Inverse Property (P1696) Links	0

We also checked how many of the instance links use an equivalence or some weaker forms of likeness relations to test our Hypothesis 5. Unsurprisingly, no instance link uses the owl:sameAs relation, as Wikibase does not allow its use and encourages the use of the exact match (P2888) relation. However, with 3,268,021 such links, at most[16]

---

[16] Some individuals might use more than one exact match relation.

only 3.7% of all unique individuals use the exact match relation to an individual defined in a namespace other than the Wikidata namespace. P460 and P1696 are not used. We therefore must also reject Hypothesis 5 (Table 7).

Table 7. Hypotheses testing

H1	Wikidata defines the vast majority of its terminological entities and properties in its authoritative namespace	Supported
H2.1	The ratio of class links to classes in Wikidata is higher than in the LOD ecosystem	Rejected
H2.2	The ratio of property links to properties in Wikidata is higher than in the LOD ecosystem	Rejected
H3	Wikidata does not type entities using classes from external ontologies	Supported
H4.1	Wikidata's ratio of instance links to entities is higher than for other datasets in the LOD ecosystem	Rejected
H4.2	Most instance links point to RDF Web resources	Rejected
H5	Wikidata includes similarity relations for a majority of its instance links	Rejected

5 Related Work

There are many works that analyse different quality aspects of Wikidata. Erxleben et al. [9] introduce RDF exports that connect Wikidata to the Linked Data Web. In [6] an axiomatic theory for multi-level modeling is used to analyse Wikidata content and to identify a significant number of problematic classification and taxonomic statements. Färber et al. [11] present an extensive survey of open KGs, including Wikidata. Freire & Isaac [10] present an assessment of Wikidata for high-quality machine interpretation of its alignment properties to RDF/S, OWL, SKOS, and schema.org.

Piscopo & Simperl [18] present a systematic literature review of 28 papers about data quality in Wikidata, categorised by quality dimensions addressed. The completeness aspect of Wikidata is analysed in [3], which cites some tools and services that address various quality aspects around the WikiMedia projects. Pillai et al. [16] compare Wikidata with other KGs from the perspectives of completeness of its relations, timeliness of the data, and accessibility as the data quality criteria. Abian et al. [1] present an approach based on cross-comparing date values (the concept of contemporary constraint) to discover inconsistent temporal data in Wikidata. Piscopo & Simperl [17] study the relationship between different Wikidata user roles and the quality of the Wikidata ontology by proposing a framework to evaluate the ontology as it evolves. Samuel [21] introduces the WDProp tool that provides to human users an overview and statistics of various multi-language aspects of Wikidata properties, such as labels, descriptions, and aliases. Shenoy et al. [23] present a quality analysis of Wikidata focusing on correctness.

Other work exists that analyse interlinking in linked data in general [20,26] or quality studies and approaches that considered interlinking of linked data as an assessment metric [7,12,19,22].

None of the above works, however, have analysed how interlinked and central Wikidata is to the LOD ecosystem, and more specifically, analysed the number and types of links defined within Wikidata as presented in this paper.

6 Conclusion

We have analysed the number and types of links in Wikidata to evaluate how central Wikidata is to the Linked Open Data ecosystem. While Wikidata is the largest, most comprehensive general knowledge KG on the Web using also a comprehensive, bottom-up developed ontology that is used to type its many entities, it is not (yet) serving as a central hub for linked data on the Web.

For its relative lack of instance links, this means that either the Wikidata editors deem such links as obsolete, or that these links are yet to be included or that they already exist, but rather as incoming links from the external dataset to Wikidata. However, as a bottom-up created KG, there is the possibility for anyone who owns a dataset to actually create an outgoing link in the Wikidata namespace to the dataset they own. Many bots (332, https://www.wikidata.org/wiki/Wikidata:Bots) have been created for exactly this reason (i.e., automatically creating outgoing links from Wikidata to other datasets), and they improve the discoverability, and as such the visibility, of the external dataset. Every dataset publisher, for their own benefit, should therefore consider creating those outgoing links in the Wikidata KG.

Comparatively for its size, Wikidata also includes less ontology links than other datasets in the LOD ecosystem. While this can be partially explained by the fact that many individuals defined in Wikidata are also classes, skewing the ratio between classes and class links, this does not apply to property links, where there is no such distinction. Most properties in Wikidata are not linked to external properties at all, even though specific properties exist in the Wikidata ontology (e.g., external subproperty, external superproperty) to do so. While we have suggested in this paper that some changes to the user interface of Wikibase may encourage editors to provide more such links, a fundamental rethink of ontology design may have to occur too. Specifically, common best-practise in ontology engineering is to include links from an ontology to other ontologies (i.e., through import statements or URI reuse). However, in the case of the Wikidata ontology, the developers of domain ontologies should consider to create those links to their ontologies in the Wikidata namespace, rather than in the other direction.

Wikidata also does not (yet) provide many equivalence or weaker forms of likeness relations from its entities to external entities. There is an onus on the Wikidata editor community to ensure that such links are increasingly provided, given that Wikidata should generally not be the authoritative source of entities, but link to an authoritative representation of an entity through, for example, the exact match relation in Wikidata. However, as above, the lack of such links may also be an indication that entities defined in Wikidata do not yet exist or never will exist in the LOD ecosystem.

As future works, we first would like to analyse the evolution of links on Wikidata over time using several historical snapshots of the published Wikidata HDT files. Also, a deeper analysis of the entities that are linked (e.g., what are the top-ranked instance and ontology namespaces referenced from Wikidata) is planned for a future work.

Acknowledgment. This research has received funding from the Teaming.AI project, which is part of the European Union's Horizon 2020 research and innovation program under grant agreement No 957402.

References

1. Abián, D., Bernad, J., Trillo, R.: Using contemporary constraints to ensure data consistency. In: Proceedings of the 34th ACM/SIGAPP Symposium on Applied Computing, pp. 2303–2310, April 2019
2. Auer, S., Bizer, C., Kobilarov, G., Lehmann, J., Cyganiak, R., Ives, Z.: DBpedia: a nucleus for a web of open data. In: Aberer, K., et al. (eds.) ASWC/ISWC -2007. LNCS, vol. 4825, pp. 722–735. Springer, Heidelberg (2007). https://doi.org/10.1007/978-3-540-76298-0_52
3. Balaraman, V., Razniewski, S., Nutt, W.: Recoin: relative completeness in Wikidata. In: Wiki Workshop 2018 co-located with the Web Conference 2018 in Lyon, France, 24 April 2018, April 2018
4. Beek, W., Rietveld, L., Bazoobandi, H.R., Wielemaker, J., Schlobach, S.: LOD laundromat: a uniform way of publishing other people's dirty data. In: Mika, P., et al. (eds.) ISWC 2014. LNCS, vol. 8796, pp. 213–228. Springer, Cham (2014). https://doi.org/10.1007/978-3-319-11964-9_14
5. Berners-Lee, T.: Linked Data. W3C Design Issues, July 2006. http://www.w3.org/DesignIssues/LinkedData.html
6. Brasileiro, F., Almeida, J.P.A., Carvalho, V.A., Guizzardi, G.: Applying a multi-level modeling theory to assess taxonomic hierarchies in Wikidata. In: Proceedings of the 25th International Conference Companion Volume on World Wide Web, pp. 975–980 (2016)
7. Debattista, J., Auer, S., Lange, C.: Luzzu - a methodology and framework for linked data quality assessment. J. Data Inf. Qual. **8**(1), 4:1–4:32 (2016)
8. Debattista, J., Lange, C., Auer, S., Cortis, D.: Evaluating the quality of the LOD cloud: an empirical investigation. Semant. Web **9**(6), 859–901 (2018)
9. Erxleben, F., Günther, M., Krötzsch, M., Mendez, J., Vrandečić, D.: Introducing Wikidata to the linked data web. In: Mika, P., et al. (eds.) ISWC 2014. LNCS, vol. 8796, pp. 50–65. Springer, Cham (2014). https://doi.org/10.1007/978-3-319-11964-9_4
10. Freire, N., Isaac, A.: Technical usability of Wikidata's linked data. In: Abramowicz, W., Corchuelo, R. (eds.) BIS 2019. LNBIP, vol. 373, pp. 556–567. Springer, Cham (2019). https://doi.org/10.1007/978-3-030-36691-9_47
11. Färber, M., Bartscherer, F., Menne, C., Rettinger, A.: Linked data quality of DBpedia, Freebase, OpenCyc, Wikidata, and YAGO. Semant. Web **9**(1), 77–129 (2018)
12. Guéret, C., Groth, P., Stadler, C., Lehmann, J.: Assessing linked data mappings using network measures. In: Simperl, E., Cimiano, P., Polleres, A., Corcho, O., Presutti, V. (eds.) ESWC 2012. LNCS, vol. 7295, pp. 87–102. Springer, Heidelberg (2012). https://doi.org/10.1007/978-3-642-30284-8_13
13. Haller, A., Fernández, J.D., Kamdar, M.R., Polleres, A.: What are links in linked open data? A characterization and evaluation of links between knowledge graphs on the web. J. Data Inf. Qual. **12**(1), 1–34 (2020)
14. Haller, A., Polleres, A.: Are we better off with just one ontology on the web? Semant. Web **11**(1), 87–99 (2020)
15. Hernández, D., Hogan, A., Krötzsch, M.: Reifying RDF: what works well with Wikidata? In: Proceedings of the 11th International Workshop on Scalable Semantic Web Knowledge Base Systems, vol. 1457, pp. 32–47. CEUR-WS.org (2015)

16. Pillai, S.G., Soon, L.-K., Haw, S.-C.: Comparing DBpedia, Wikidata, and YAGO for web information retrieval. In: Piuri, V., Balas, V.E., Borah, S., Syed Ahmad, S.S. (eds.) Intelligent and Interactive Computing. LNNS, vol. 67, pp. 525–535. Springer, Singapore (2019). https://doi.org/10.1007/978-981-13-6031-2_40

17. Piscopo, A., Simperl, E.: Who models the world?: collaborative ontology creation and user roles in Wikidata. Proc. ACM Hum.-Comput. Interact. 2(CSCW), 141:1–141:18 (2018)

18. Piscopo, A., Simperl, E.: What we talk about when we talk about Wikidata quality: a literature survey. In: Proceedings of the 15th International Symposium on Open Collaboration, New York, NY, USA (2019)

19. Raad, J., Beek, W., van Harmelen, F., Pernelle, N., Saïs, F.: Detecting erroneous identity links on the web using network metrics. In: Vrandečić, D., et al. (eds.) ISWC 2018. LNCS, vol. 11136, pp. 391–407. Springer, Cham (2018). https://doi.org/10.1007/978-3-030-00671-6_23

20. Radulovic, F., Mihindukulasooriya, N., García-Castro, R., Gómez-Pérez, A.: A comprehensive quality model for linked data. Semant. Web 9(1), 3–24 (2018)

21. Samuel, J.: Towards understanding and improving multilingual collaborative ontology development in Wikidata. In: Proceedings of Wiki Workshop 2018 co-located with the Web Conference 2018, Lyon, France, April 2018

22. Sarasua, C., Staab, S., Thimm, M.: Methods for intrinsic evaluation of links in the web of data. In: Blomqvist, E., Maynard, D., Gangemi, A., Hoekstra, R., Hitzler, P., Hartig, O. (eds.) ESWC 2017. LNCS, vol. 10249, pp. 68–84. Springer, Cham (2017). https://doi.org/10.1007/978-3-319-58068-5_5

23. Shenoy, K., Ilievski, F., Garijo, D., Schwabe, D., Szekely, P.: A study of the quality of Wikidata. arXiv preprint arXiv:2107.00156 (2021)

24. Vandenbussche, P., Atemezing, G., Poveda-Villalón, M., Vatant, B.: Linked open vocabularies (LOV): a gateway to reusable semantic vocabularies on the web. Semant. Web 8(3), 437–452 (2017)

25. Vrandecic, D., Krötzsch, M.: Wikidata: a free collaborative knowledgebase. Commun. ACM 57(10), 78–85 (2014)

26. Zaveri, A., Rula, A., Maurino, A., Pietrobon, R., Lehmann, J., Auer, S.: Quality assessment for linked data: a survey. Semant. Web 7(1), 63–93 (2016)

Knowledge Graph Entity Type Prediction with Relational Aggregation Graph Attention Network

Changlong Zou, Jingmin An, and Guanyu Li$^{(\boxtimes)}$

Faculty of Information Science and Technology, Dalian Maritime University,
Dalian, Liaoning, China
`liguanyu@dlmu.edu.cn`

Abstract. Most of the knowledge graph completion methods focus on inferring missing entities or relations between entities in the knowledge graphs. However, many knowledge graphs are missing entity types. The goal of entity type prediction in the knowledge graph is to infer the missing entity types that belong to entities in the knowledge graph, that is, (entity, entity type=?). At present, most knowledge graph entity type prediction models tend to model entities and entity types, which will cause the relations between entities to not be effectively used, and the relations often contain rich semantic information. To utilize the information contained in the relation when performing entity type prediction, we propose a method for entity type prediction based on relational aggregation graph attention network (RACE2T), which consists of an encoder relational aggregation graph attention network (FRGAT) and a decoder (CE2T). The encoder FRGAT uses the scoring function of the knowledge graph completion method to calculate the attention coefficient between entities. This attention coefficient will be used to aggregate the information of relations and entities in the neighborhood of the entity to utilize the information of the relations. The decoder CE2T is designed based on convolutional neural network, which models the entity embeddings output by FRGAT and entity type embeddings, and performs entity type prediction. The experimental results demonstrate that the method proposed in this paper outperforms existing methods. The source code and dataset for RACE2T can be downloaded from: https://github.com/GentlebreezeZ/RACE2T.

Keywords: Knowledge graph · Entity type · Relational aggregation · Attention · Scoring function · Convolutional neural network

1 Introduction

Knowledge graph stores information mainly in the triple [5], denoted as (e_i, r_k, e_j), where e_i is the head entity, e_j is the tail entity, and r_k is the relation between e_i and e_j. Besides the triples, knowledge graphs usually contain many entity type instances in the form of entity-entity type tuples [30] (denoted as (e, t)), indicating that an entity e is of a certain entity type t, for example, (*Chicago*, *Film*) and

© The Author(s), under exclusive license to Springer Nature Switzerland AG 2022
P. Groth et al. (Eds.): ESWC 2022, LNCS 13261, pp. 39–55, 2022.
https://doi.org/10.1007/978-3-031-06981-9_3

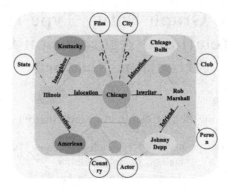

Fig. 1. Knowledge graph triples. The solid shape represents the entity; the hollow shape represents the entity type; the solid line represents the relation; the dotted line points to the entity type.

(*Chicago, City*). The entity type information in the knowledge graphs is widely used in various downstream tasks, such as entity alignment [32], entity linking [3], knowledge graph completion [25]. Missing entity types can undermine the effectiveness of algorithms in such tasks. However, knowledge graphs, especially cross-domain knowledge graphs, often face quality problems with incomplete entity type information. For example, 10% of the entities with type */music/artister* in FB15K [2], do not have type */people/person* in Freebase [13].

The missing entity type of knowledge graph can be solved by entity type prediction, which is a subproblem of knowledge graph completion. Early methods for predicting entity types in knowledge graphs are mainly based on probability distributions, such as SDtype [19]. Recently, representation learning [7] has gradually become the basis of knowledge graph-related research. For knowledge graphs, representation learning is the learning low-dimensional embedding vector representations of objects in the knowledge graph by methods such as machine learning. The learned embedding vectors preserve the semantic information of the objects in the knowledge graph and can be used for various downstream tasks based on the knowledge graph. In this paper, we mainly use representation learning to infer the missing entity types of entities in the knowledge graph.

At present, most entity type prediction models based on representation learning directly model entities and entity types, such as ETE [13], and ConnectE [31]. Thus these models have a common drawback of not effectively utilizing the knowledge graph triples. To be precise, the information of relations is not utilized, or these models deal with each entity-entity type tuple independently, without utilizing the semantically rich relations inherent in the neighborhood of entities in the knowledge graph. As shown in Fig. 1, the semantic information of the relations *Islocation* and *Iswriter* helps to infer that the type of entity *Chicago* may include *City* and *Film*.

GAT (Graph Attention Network) [22] is an effective tool for generating knowledge graph entity embeddings. It assigns different weights to entities in the neighborhood of the entity according to their importance. This weight will be used to aggregate the neighborhood information of the entity. However, GAT cannot utilize relational information.

Based on the above statements, in this work, we introduce a method for entity type prediction based on relational aggregation graph attention network (RACE2T), consisting of an encoder relational aggregation graph attention network (FRGAT) and a decoder CE2T. The encoder FRGAT is designed based on GAT and is mainly used to utilize relation information. FRGAT uses the scoring function of the knowledge graph completion method to calculate the attention between entities and entities in its neighborhood and uses that attention to aggregate information about entities and relations. Meanwhile, considering the limited ability of expressiveness of existing entity type prediction models, we propose a convolutional neural network-based entity type prediction model CE2T as a decoder. CE2T is composed of convolution, projection, and inner product layers. It models the entity embeddings output by FRGAT and entity type embeddings and performs entity type prediction. We demonstrated the effectiveness of our proposed model on the FB15K and FB15KET and YAGO43K and YAGO43KET datasets and obtained advanced results.

2   Related Works

2.1   Knowledge Graph Completion Models

Translation-based models treat relations as translations from head entities to tail entities and use energy-based scoring functions. The basic idea of the TransE [2] model is that if the triple (e_i, r_k, e_j) holds, then the sum of the head entity embedding and the relation embedding should be as close as possible to the tail entity embedding, i.e., $e_i + r_k \approx e_j$. TransH [23] and TransD [6] use a projection strategy and are extensions of TransE.

Semantic matching models use scoring functions based on the similarity of head and tail entities under a given relation. RESCAL [18] follows a relational learning approach based on a tensor factorization model that considers the inherent structure of relational data. DisMult [29] is a simplified version of RESCAL and uses a basic bilinear scoring function to match the underlying semantics of entities and relations in vector space. HOLE [17] combines the expressive ability of RESCAL with the simplicity of DisMult through the use of unique circular correlation operation. In addition to translation models and semantic matching models, there are ConvE [4], ConvKB[1] [15], and CapsE[2] [16], based on convolutional neural networks. Meanwhile, some models use additional information for knowledge graph completion [24].

In this paper, we will use the scoring functions of TransE [2], TransH [23], TransD [6], and DisMult [29] to calculate the attention between entities.

---

[1] https://github.com/daiquocnguyen/ConvKB.

[2] https://github.com/daiquocnguyen/CapsE.

2.2 Knowledge Graph Entity Type Prediction Models

CUTE [26] is a cross-language knowledge graph entity type prediction model, which mainly uses cross-language entity links between Chinese and English entities to construct training data. However, CUTE is based on non-representation learning. MuLR [28] learns multi-level embedding representations of entities through character, word, entity descriptions, and entity embeddings and then performs entity type prediction. HMGCN [9] is a knowledge graph entity type prediction model based on GCN [11], which considers relations, entity description information, and Wikipedia categories. Cat2Type [1] is similar to HMGCN in that both use Wikipedia categories for knowledge graph entity type prediction. APE [8] utilizes the attribute, structure, and type information of entities in the knowledge base for entity type prediction and learns entity embeddings through neural networks. FIGMENT [27] is mainly used to judge the types of entities in the corpora, including global models and context models. However, it additionally requires a large annotated corpora.

In short, the most significant difference between RACE2T and MuLR [28], FIGMENT [27], APE [8], HMGCN [9], and Cat2Type [1] is that RACE2T mainly uses entity-entity type tuples (e, t) and triples to learn embedding representations for objects. However, these models do not, and they usually require additional information (e.g., entity description information, corpus). Therefore, the following mainly introduces entity type prediction models that use entity-entity type tuples and triples for modeling.

Ref [14] proposed two knowledge graph entity type prediction models for encoding entity-entity type tuples (e, t), namely linear model and embedding model. The scoring function of the linear model is $\phi(e, t) = \boldsymbol{t}^{\mathrm{T}}\boldsymbol{e}$, where $\boldsymbol{e}$ is the entity embeddings and $\boldsymbol{t}$ is the entity type embeddings. The scoring function of the embedding model is $\phi(e, t) = \boldsymbol{t}^{\mathrm{T}}\boldsymbol{V}\boldsymbol{U}^{\mathrm{T}}\boldsymbol{e}$, where $\boldsymbol{U}$ and $\boldsymbol{V}$ are the projection matrices. However, both models do not use the knowledge graph triples, or they do not use the relations between entities.

The knowledge graph entity type prediction model proposed in the Ref [13] uses an asynchronous approach to learn the embedding representation of entities, relations and entity types. First, using knowledge graph completion methods, such as RESCAL [18], TransE [2], HOLE [17], and ContE [12], learn the entity embedding $\boldsymbol{e}$. Second, keep the entity embeddings $\boldsymbol{e}$ unchanged during training, and update the entity type embeddings by minimizing the distance between the entity embedding and the entity type embedding, namely, RESCAL-ET, HOLE-ET, TransE-ET, and ETE[3]. Their scoring function is $\phi(e, t) = \|\boldsymbol{e} - \boldsymbol{t}\|_{\ell 1}$, where $\|\boldsymbol{x}\|_{\ell 1}$ represents the $\ell 1$ norm of the vector $\boldsymbol{x}$. Although these methods use the relations between entities during training, the semantic information the relations is not utilized in making entity type predictions.

ConnectE[4] [31] uses entity-entity type tuples and entity type triples[5] for training and entity type prediction. However, entity type triples are created

---

[3] https://github.ncsu.edu/cmoon2/kg.

[4] https://github.com/Adam1679/ConnectE.

[5] For details of entity type triples, see Ref [31].

does not consider the semantics that entities represent when they correspond to different types. Meanwhile, entity type triples lead to data leakage in the test set. The scoring function of ConnectE is: $\phi(e,t) = \|M \cdot e - t\|_{\ell2}$, where M is projection matrix, $\|x\|_{\ell2}$ represents the $\ell2$ norm of the vector x. ConnectE uses an asynchronous approach to learn embeddings of entities, relations and entity types, and its training process is divided into three stages. Firstly, the model uses TransE [2] to train entity embeddings and relation embeddings. Secondly, the model trains the entity type embeddings and the projection matrix M by minimizing $\phi(e,t)$. Finally, the entity type triples are trained using TransE and only the embedding of the relations is changed, the entity type embedding remains unchanged.

It can be concluded that none of the above entity type prediction models do not effectively utilize the relations in the knowledge graph triple. Although ConnectE [31] utilizes relations through entity type triples, the entity type triples cause the leakage of the test set. In order to effectively use the relation information when predicting entity types, this paper uses the encoder FRGAT to aggregate the informations of entities and relations in the neighborhood of a given entity and uses the decoder CE2T to predict the entity type of the knowledge graph. Meanwhile, these methods mainly adopt an asynchronous way to learn the embeddings of objects (entity, relation and entity type). While the method in this work uses synchronous way to learn the embeddings of objects.

3   Methods

The method (RACE2T) in this paper adopts the form of encoder-decoder. The encoder is FRGAT. The decoder is CE2T, which is designed based on the convolutional neural network and is specially used to predict the entity type of knowledge graph. The overall framework is shown in Fig. 2.

3.1   Problem Definition and Symbol

The goal of knowledge graph entity type prediction is to infer the type t of a given entity e. Entity initial embedding matrix $E \in \mathbb{R}^{|\mathcal{E}| \times D}$, relation initial embedding matrix $R \in \mathbb{R}^{|\mathcal{R}| \times D}$, entity type embedding matrix $T \in \mathbb{R}^{|\mathcal{T}| \times \ell}$, where $\mathcal{E}$, $\mathcal{R}$ and $\mathcal{T}$ respectively represent the collection of all entities, relations, and entity types, $|\mathcal{E}|$, $|\mathcal{R}|$, and $|\mathcal{T}|$ respectively represent the number of entities, relations, and entity types, D is the dimension of the initial embedding vector of entities and relations, and ℓ is the dimension of the embedding vector of entity types. The output matrix of the last layer of FRGAT is $E_0 \in \mathbb{R}^{|\mathcal{E}| \times d}$, and d represents the dimension of FRGAT output embeddings.

3.2   Encoder: FRGAT

For a triple (e_i, r_k, e_j), its entity and relation embeddings are e_i, r_k and e_j, respectively. To make the model obtain sufficient expressive ability, the input entity and relation embeddings are converted into a higher-dimensional embedding using projection operation: $\boldsymbol{h}_i = \boldsymbol{e}_i \boldsymbol{W}_1$, $\boldsymbol{z}_k = \boldsymbol{r}_k \boldsymbol{W}_2$, and $\boldsymbol{h}_j = \boldsymbol{e}_j \boldsymbol{W}_1$, where $\boldsymbol{e}_i \in \boldsymbol{E}$, $\boldsymbol{r}_k \in \boldsymbol{R}$, $\boldsymbol{e}_j \in \boldsymbol{E}$, $\boldsymbol{W}_1 \in \mathbb{R}^{D \times D_1}$, $\boldsymbol{W}_2 \in \mathbb{R}^{D \times D_1}$, $D_1 > D$. At the same time, $\boldsymbol{W}_1 = \boldsymbol{W}_2$ will be restricted to ensure that entity embedding and relation embedding are in the same semantic space after projection. Then, in FRGAT, the calculation equation for the attention between the entity e_i and the entity e_j is as follows:

$$a_{ijk} = \sigma \left(f \left(\boldsymbol{h}_i, \boldsymbol{z}_k, \boldsymbol{h}_j \right) \right) \tag{1}$$

where $\sigma \left(\cdot \right)$ represents the activation function, a_{ijk} represents the importance (attention) of e_j to e_i, $f \left(\cdot, \cdot, \cdot \right)$ represents the scoring function in the knowledge graph completion method. Since there may be more than one other entities in the first-order neighborhood of entity e_i, we use *softmax* to normalize a_{ijk}:

$$\alpha_{ijk} = soft \max \left(a_{ijk} \right) = \frac{\exp \left(a_{ijk} \right)}{\sum_{n \in \mathcal{N}_i} \sum_{r \in \mathcal{R}_{in}} \exp \left(a_{inr} \right)} \tag{2}$$

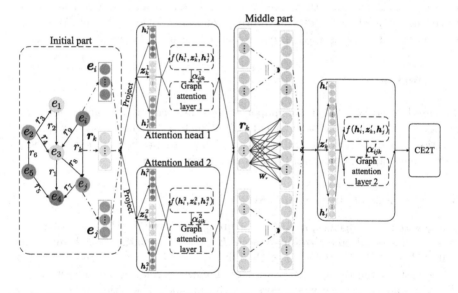

Fig. 2. The overall framework of RACE2T. The FRGAT in this figure includes two graph attention layers. The first layer includes two attention heads, and the second layer includes one.

where $\mathcal{N}_i$ is the entity set contained in the first-order neighborhood of entity e_i, $\mathcal{R}_{in}$ is the set of relations between linked entity e_i and the entities contained in $\mathcal{N}_i$, and α_{inr} represents the importance of entity e_n to entity e_i. To aggregate the information of entities and relations, we use the message propagation mechanism proposed in Ref [21] to perform a combination operation on the embedding vector of the entity and the relation in the form of $\mathbb{R}^D \times \mathbb{R}^D \rightarrow \mathbb{R}^D$. In this case, the new embedding of the entity e_i is expressed as:

$$h_i' = \sigma \left(\sum_{j \in \mathcal{N}_i} \sum_{k \in \mathcal{R}_{ij}} \alpha_{ijk} \psi \left(h_j, z_k \right) \right) \tag{3}$$

where $\psi(\cdot, \cdot)$ represents combination operation.

Similar to GAT [22], a connected and independent multi-head attention mechanism is used to stabilize the learning process of the model and enhance the generalization ability of the model. Under the multi-head attention mechanism: $h_i^m = e_i W_1^m$, $z_k^m = r_k W_2^m$, $h_j^m = e_j W_1^m$, and $\alpha_{ijk}^m = soft \max \left(\sigma \left(f \left(h_i^m, z_k^m, h_j^m \right) \right) \right)$, where $W_1^m \in \mathbb{R}^{D \times D_1}$, $W_2^m \in \mathbb{R}^{D \times D_1}$, m represents the m-th attention head, as shown in the *attention head 1 and attention head 2* of Fig. 2. The new embedding of entity e_i is expressed as:

$$h_i' = \overset{M}{\underset{m=1}{\Big\|}} \sigma \left(\sum_{j \in \mathcal{N}_i} \sum_{k \in \mathcal{R}_{ij}} \alpha_{ijk}^m \psi \left(h_j^m, z_k^m \right) \right) \tag{4}$$

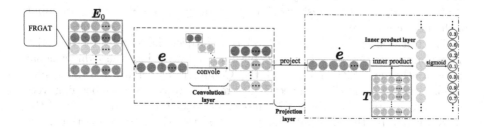

Fig. 3. The framework of CE2T.

where M represents the number of heads in the multi-head attention mechanism, and $\|$ represents the vector connect operation, as shown in the *middle part* of Fig. 2. At the same time, in order to use the embedding vector of the relation in the following aggregation process, the conversion matrix $W_r \in \mathbb{R}^{D \times M \cdot D_1}$ is used to perform a linear transformation on the embedding vector of the input relation. The relation embedding after transformation is expressed as:

$$z_k' = r_k W_r \tag{5}$$

To reduce the number of parameters, we use the average to obtain the final embedding vector of the entity in the last layer of the FRGAT model instead of linking embeddings from multiple attention heads. The final entity embedding vector is used as the input of the decoder (CE2T), which is expressed as follows:

$$h_i' = \sigma \left(\frac{1}{M} \sum_{m=1}^{M} \sum_{j \in \mathcal{N}_i} \sum_{k \in \mathcal{R}_{ij}} \alpha_{ijk}^m \psi \left(h_j^m, z_k^m \right) \right) \tag{6}$$

3.3   Decoder: CE2T

Inspired by the application of convolutional neural network in knowledge graph completion [4,15], we propose a convolutional neural network-based knowledge graph entity type prediction model (CE2T[6]) as the decoder of this paper. The framework of CE2T is shown in Fig. 3.

The scoring function of CE2T is defined as:

$$\phi(e,t) = \sigma\left(vec\left(\sigma\left(e * \boldsymbol{\Omega}\right)\right) \cdot \boldsymbol{M}\right) t \qquad (7)$$

where $e \in \mathbb{R}^d$ represents the final embedding of the entity output by FRGAT, $t \in \mathbb{R}^\ell$ represents the embedding vector of the entity type, $\boldsymbol{\Omega}$ represents the set of convolution kernels, $\boldsymbol{M}$ represents the projection matrix, $*$ represents convolution operation, and $vec\left(\cdot\right)$ represents vectorization operation.

As shown in Fig. 3, during the forward propagation process, the entity embedding e of the CE2T input needs to be found in the entity embedding matrix $\boldsymbol{E}_0$ of the FRGAT output. Suppose the convolution kernel of the convolution layer $\boldsymbol{\Omega} \in \mathbb{R}^{|\Omega| \times 1 \times f}$, the step size is c_s, where $|\boldsymbol{\Omega}|$ represents the number of convolution kernels, and $1 \times f$ is the size of the convolution kernel. Then the output of the convolutional layer is $\mathcal{F} = \sigma\left(e * \boldsymbol{\Omega}\right) \in \mathbb{R}^{|\Omega| \times a \times b}$, where $a = 1$, $b = \left(d - f\right)/c_s + 1$. Then reshape $\mathcal{F}$ into vector $vec\left(\mathcal{F}\right) \in \mathbb{R}^{|\Omega| \cdot a \cdot b}$ as the input of the projection layer.

The function of the projection layer is to project the features of the entity embedding vector extracted by the convolutional layer into the ℓ-dimensional space. Both the CE2T model and the ConnectE model use a similar projection strategy, but the ConnectE [31] model projects the embedding vector of the entity, while the CE2T model projects the feature vector of the entity output by the convolutional layer. The weight of the projection layer is $\boldsymbol{M} \in \mathbb{R}^{|\Omega| \cdot a \cdot b \times \ell}$. The entity projection vector projected to the ℓ-dimensional space is: $\dot{e} = \sigma\left(vec\left(\mathcal{F}\right) \cdot \boldsymbol{M}\right) \in \mathbb{R}^\ell$. Finally, entity e and entity type t similarity score is calculated by the inner product operation between the entity projection vector $\dot{e}$ and the entity type embedding t.

3.4   Training

To accelerate the training of the RACE2T model, we use the 1-N scoring proposed in the Ref [4], i.e., we score both the projection vector of an entity and the embedding vector of all entity types, as shown in the *inner product layer* of Fig. 3. At the same time, to minimize the cross-entropy loss to train the parameters of RACE2T, we use the sigmoid function to normalize $\phi\left(e,t\right)$ to between 0 and 1, which is $\boldsymbol{p} = sigmoid\left(\phi\left(e,t\right)\right)$. The loss function of RACE2T is defined as:

$$\mathcal{L} = -\frac{1}{|T|} \sum_{i=1}^{|T|} \boldsymbol{y}_i \log\left(\boldsymbol{p}_i\right) + \left(1 - \boldsymbol{y}_i\right) \log\left(1 - \boldsymbol{p}_i\right) \qquad (8)$$

---

[6] https://github.com/GentlebreezeZ/CE2T.

where $y \in \mathbb{R}^{|T|}$ is the binary label vector. If (e, t) is true, the corresponding position of y is 1. Otherwise, it is 0. The optimizer uses Adam [10].

3.5   Performing

For each entity that appears in the test set, the entity type predicted by the RACE2T model is:

$$\hat{t} = \arg \max_{t \in T} sigmoid \left(\phi \left(e, t \right) \right) \tag{9}$$

4   Experiments

4.1   Datasets

The datasets used for knowledge graph entity type prediction in this paper are FB15KET [13], and YAGO43KET [13] (the form is: (entity, entity type)), and the corresponding knowledge graph triple datasets are FB15K [2], and YAGO43K [13]. The entity types in FB15KET, and YAGO43KET are mapped to the entities in FB15K, and YAGO43K, respectively. The statistics of datasets are shown in Table 1. FRGAT uses triple datasets to utilize the information about the relation, and CE2T uses entity-entity type tuple datasets to learn the embedding of entity types and perform entity type prediction.

Table 1. Statistics of datasets.

Dataset	$|\mathcal{E}|$	$|T|$	#Train	#Valid	#Test
FB15KET	14951	3851	136618	15749	15780
YAGO43KET	41723	45182	375853	42739	42750
Dataset	$|\mathcal{E}|$	$|\mathcal{R}|$	#Train	#Valid	#Test
FB15K	14951	1345	483142	–	–
YAGO43K	42335	37	331687	–	–

4.2   Experimental Setup

Evaluation Metrics: Use ranking-based metric for evaluation [2]. First, for each tested entity-entity type tuple, we remove the entity type. Then, the ranking of the entity types predicted by RACE2T is calculated according to Eq. (9). Finally, its exact rank is obtained by the correct entity type. Two metrics are used for evaluation: the mean reciprocal rank (MRR) [2] and the proportion of correct entity types that predict the top k (HITS@k, k = 1, 3, 10) [2].

Model Parameters Setting: The optimal parameters of the RACE2T model are determined by grid search. Specifically: the embedding dimension of entity and relation is adjusted in $\{50, 100\}$, the embedding dimension of entity type is adjusted in $\{100, 200, 300\}$, the dimension of hidden layer of FRGAT is adjusted in $\{200, 300\}$, the output dimension of FRGAT is adjusted in $\{200, 400, 600\}$, the batch size is adjusted in $\{128, 256, 512\}$, the number of layer is adjusted in $\{1, 2, 3\}$, the number of attention head is adjusted in $\{1, 2, 3, 4\}$, the learning rate is adjusted in $\{0.0001, 0.0005, 0.00\ 1, 0.01\}$, the number of convolution kernels is adjusted in $\{10, 32, 64, 128\}$, the size of the convolution kernel is adjusted in $\{1 \times 2, 1 \times 4, 1 \times 8\}$, and the stride size of convolution operation is adjusted in $\{1, 2, 4, 8\}$. We use Xavier to initialize model parameters. Detailed parameter settings can be found on our GitHub[7].

4.3   Entity Type Prediction Experiments

The baselines choose RESCAL [18], RESCAL-ET [13], HOLE [17], HOLE-ET [13], TransE [2], TransE-ET [13], ConvKB [15], CapsE [16], ETE [13], ConnectE (E2T)[8] [31] and ConnectE(E2T+TRT)[9] [31]. The experimental results are shown in Table 2.

From Table 2, we can see that RACE2T outperforms existing baselines on FB15kET and YAGO43kET. We attribute these results to the fact that RACE2T reasonably aggregate the informations of entities and relations in a given entity neighborhood, and that information about entities and relations helps to infer the types to which entities belong. Without using the encoder FRGAT, the decoder CE2T proposed in this paper also achieved good performance, reflecting that the

Table 2. Entity type prediction results. The baseline results are taken from Ref [31].

Dataset	FB15KET				YAGO43KET			
Model	MRR	HITS@1	HITS@3	HITS@10	MRR	HITS@1	HITS@3	HITS@10
RESCAL [18]	0.19	0.0971	0.1958	0.3758	0.08	0.0424	0.0831	0.1531
RESCAL-ET [13]	0.24	0.1217	0.2792	0.5072	0.09	0.0432	0.0962	0.1940
HOLE [17]	0.22	0.1329	0.2335	0.3816	0.16	0.0902	0.1728	0.2925
HOLE-ET [13]	0.42	0.2940	0.4804	0.6673	0.18	0.1028	0.2013	0.3490
TransE [2]	0.45	0.3151	0.5145	0.7393	0.21	0.1263	0.2324	0.3893
TransE-ET [13]	0.46	0.3356	0.5296	0.7116	0.18	0.0919	0.1941	0.3558
ConvKB [15]	0.45	0.3365	0.5180	0.7462	0.19	0.1136	0.2481	0.3897
CapsE [16]	0.46	0.3461	0.5279	0.7320	0.21	0.1263	0.2498	0.3946
ETE [13]	0.50	0.3851	0.5533	0.7193	0.23	0.1373	0.2628	0.4218
ConnectE(E2T) [31]	0.57	0.4554	0.6231	0.7812	0.24	0.1354	0.2620	0.4451
ConnectE(E2T+TRT) [31]	0.59	0.4955	0.6432	0.7992	0.28	0.1601	0.3085	0.4792
CE2T	0.57	0.4681	0.6354	0.7834	0.29	0.2131	0.3225	0.4472
RACE2T	**0.64**	**0.5607**	**0.6884**	**0.8172**	**0.34**	**0.2482**	**0.3762**	**0.5230**

---

[7] https://github.com/GentlebreezeZ/RACE2T.

[8] ConnectE (E2T) represents that entity type triples are not used.

[9] ConnectE (E2T+TRT) represents that entity type triples are used.

Table 3. 1-1 and 1-N entity type prediction results on FB15KET and YAGO43KET.

	1-1							
Dataset	FB15KET				YAGO43KET			
Model	MRR	HITS@1	HITS@3	HITS@10	MRR	HITS@1	HITS@3	HITS@10
ETE	0.57	0.4655	0.6358	0.7596	0.27	0.1958	0.3289	0.4701
ConnectE(E2T)	0.63	0.5396	0.6959	0.8042	0.31	0.1950	0.3382	0.4965
ConnectE(E2T+TRT)	0.64	0.5502	0.7027	0.8146	0.32	0.2062	0.3560	0.5104
CE2T	0.64	0.5436	0.7039	0.8135	0.35	0.2706	0.3876	0.5053
RACE2T	**0.71**	**0.6347**	**0.7548**	**0.8620**	**0.39**	**0.3091**	**0.4322**	**0.5623**
	1-N							
Dataset	FB15KET				YAGO43KET			
Model	MRR	HITS@1	HITS@3	HITS@10	MRR	HITS@1	HITS@3	HITS@10
ETE	0.48	0.3652	0.5462	0.6971	0.19	0.1188	0.2129	0.3971
ConnectE(E2T)	0.53	0.4201	0.5982	0.7619	0.22	0.1292	0.2493	0.4185
ConnectE(E2T+TRT)	0.54	0.4521	0.6110	0.7712	0.25	0.1501	0.2802	0.4375
CE2T	0.54	0.4378	0.6078	0.7677	0.27	0.1902	0.3001	0.4267
RACE2T	**0.61**	**0.5312**	**0.6619**	**0.7996**	**0.32**	**0.2298**	**0.3533**	**0.4987**

improvement of model expression ability can improve the performance of entity type prediction.

In the knowledge graph, an entity often has multiple entity types (1-N). As shown in Fig. 1, the type of *Chicago* can be either *City* or *Film*. Since RACE2T aggregates information in entity neighborhoods, especially information about relations, hence RACE2T can be suitable for modeling the 1-N case. To verify that RACE2T can model the 1-N case better, we divided the test sets of FB15KET and YAGO43KET into two parts, one part is 1-1 and the other part is 1-N. Then, using ETE [13], ConnectE [31], CE2T, and RACE2T for entity type prediction, respectively. The experimental results are shown in Table 3.

From Table 3, it is not difficult to see that RACE2T perform better than ETE [13], ConnectE [31] and CE2T on 1-1 and 1-N. Since RACE2T reasonably aggregates entity and relation information in entity neighborhoods, and this information helps RACE2T capture the differences in entities when they correspond to different types. Therefore, RACE2T can model the 1-N case, while ETE, ConnectE, and CE2T cannot do well.

Table 4. Entity type prediction results on FB15KET and YAGO43KET.

Dataset	FB15KET				YAGO43KET			
Model	MRR	HITS@1	HITS@3	HITS@10	MRR	HITS@1	HITS@3	HITS@10
GAT+CE2T	0.6102	0.5151	0.6582	0.7939	0.3257	0.2352	0.3589	0.5051
RACE2T(TansE)	**0.6212**	**0.5291**	**0.6709**	**0.8030**	0.3301	0.2394	0.3651	0.5098
RACE2T(TansH)	0.6196	0.5269	0.6683	0.8020	0.3305	0.2387	0.3647	**0.5122**
RACE2T(TansD)	0.6195	0.5273	0.6687	0.8029	0.3266	0.2360	0.3603	0.5073
RACE2T(DisMult)	0.6202	0.5280	0.6685	0.8020	**0.3308**	**0.2397**	**0.3659**	0.5115

From Tables 2 and 3, we found that: the RACE2T did not perform as well on YAGO43KET relative to CE2T as they did on FB15KET relative to CE2T. This phenomenon may be related to the number of different relations connected to the entities. FB15K used to train RACE2T contains 1345 relations, while YAGO43K contains only 37 relations, which means that RACE2T has less relation information available on YAGO43K, leading to a decrease in RACE2T performance. More detailed experiments about the above statements will be given in Sect. 4.6.

4.4 Attention Calculation Function Analysis

For a triple (e_i, r_k, e_j), its entity and relation embedding are e_i, r_k and e_j, respectively. We choose the scoring function of the following knowledge graph completion method to calculate the attention between entities:

1) TransE [2]: For $e_i \in \mathbb{R}^l$, $r_k \in \mathbb{R}^l$, $e_j \in \mathbb{R}^l$. Scoring function: $f(e_i, r_k, e_j) = -\|e_i + r_k - e_j\|_{\ell 2}$.

2) TransH [23]: For $e_i \in \mathbb{R}^l$, $r_k \in \mathbb{R}^l$, $e_j \in \mathbb{R}^l$, $W_{r_k} \in \mathbb{R}^l$. Scoring function: $f(e_i, r_k, e_j) = -\|\hat{e}_i + r_k - \hat{e}_j\|_{\ell 2}$, where $\hat{e}_i = e_i - W_{r_k}^{\mathrm{T}} e_i W_{r_k}$, $\hat{e}_j = e_j - W_{r_k}^{\mathrm{T}} e_j W_{r_k}$.

3) TransD [6]: For $e_i \in \mathbb{R}^l$, $r_k \in \mathbb{R}^l$, $e_j \in \mathbb{R}^l$, $e_i^p \in \mathbb{R}^l$, $e_j^p \in \mathbb{R}^l$, $r_k^p \in \mathbb{R}^l$. Scoring function: $f(e_i, r_k, e_j) = -\|\hat{e}_i + r_k - \hat{e}_j\|_{\ell 2}$, where $\hat{e}_i = e_i + (e_i^p)^{\mathrm{T}} e_i r_k^p$, $\hat{e}_j = e_j + (e_j^p)^{\mathrm{T}} e_j r_k^p$.

4) DisMult [29]: For $e_i \in \mathbb{R}^l$, $r_k \in \mathbb{R}^l$, $e_j \in \mathbb{R}^l$. Scoring function: $f(e_i, r_k, e_j) = e_i^{\mathrm{T}} \mathrm{diag}(r_k) e_j$.

Take GAT[10] [22] as the comparison object, and CE2T as the decoder. For a more objective comparison, $\psi(u, v) = u$ (the mode used by GAT) is selected for the combination mode of FRGAT entity and relation embedding vectors, which means that the information of relations is not aggregated in aggregating information. The results of entity type prediction are shown in Table 4.

As can be seen from Table 4, the results of entity type prediction for RACE2T improve about 1% over GAT+CE2T for all evaluation metrics. It indicates that the way to calculate the attention between entities using the scoring function of the knowledge graph completion method is slightly better than the traditional way and illustrates the feasibility of FRGAT. At the same time, using the scoring function of the knowledge graph completion method to calculate the attention coefficient will not generate additional space overhead. They directly use the embedding of entities and relations for calculation, such as TransE [2] and DisMult [29].

4.5 Combination Mode Analysis

Inspired by the Ref [2,20,29], we choose the following ways to combine the embedding vectors of entities and relations:

---

[10] https://github.com/Diego999/pyGAT.

- void: $\psi\left(\boldsymbol{u}, \boldsymbol{v}\right) = \boldsymbol{u}$
- sub: $\psi\left(\boldsymbol{u}, \boldsymbol{v}\right) = \boldsymbol{u} - \boldsymbol{v}$
- mult: $\psi\left(\boldsymbol{u}, \boldsymbol{v}\right) = \boldsymbol{u} \circ \boldsymbol{v}$
- rotate: $\psi\left(\boldsymbol{u}, \boldsymbol{v}\right) = [\boldsymbol{u}_1 \circ \boldsymbol{v}_1 - \boldsymbol{u}_2 \circ \boldsymbol{v}_2; \boldsymbol{u}_1 \circ \boldsymbol{v}_2 + \boldsymbol{u}_2 \circ \boldsymbol{v}_1]$

where $\circ$ represents Hadamard product, $\boldsymbol{x}_1$ represents the first half of vector $\boldsymbol{x}$, $\boldsymbol{x}_2$ represents the second half of vector $\boldsymbol{x}$, $[\cdot; \cdot]$ represents the vector connect operation. The experiment is performed on FB15K and FB15KET, and results are shown in Table 5.

Table 5. The results of entity type prediction on FB15KET.

$f\left(\cdot, \cdot, \cdot\right) \rightarrow$	TransE		TransH		TransD		DisMult	
Model $\downarrow$	MRR	HITS@1	MRR	HITS@1	MRR	HITS@1	MRR	HITS@1
f+RACE2T(void)	0.6212	0.5291	0.6196	0.5269	0.6195	0.5237	0.6202	0.5280
f+RACE2T(sub)	0.6426	0.5579	0.6402	0.5524	0.6367	0.5484	**0.6456**	**0.5607**
f+RACE2T(mult)	0.6328	0.5448	0.6312	0.5432	0.6304	0.5427	0.6339	0.5451
f+RACE2T(rotate)	0.6412	0.5546	0.6388	0.5547	0.6352	0.5436	0.6427	0.5462

From Table 5, for the RACE2T model, when the combination of entity embedding and relation embedding is: $\psi\left(\boldsymbol{u}, \boldsymbol{v}\right) = \boldsymbol{u} - \boldsymbol{v}$, RACE2T achieve the best performance. For different $\psi\left(\cdot, \cdot\right)$ except for $\psi\left(\boldsymbol{u}, \boldsymbol{v}\right) = \boldsymbol{u}$, the performance gap of RACE2T is not apparent. From the experimental results of entity type prediction in Table 2 (CE2T), Table 4 (GAT+CE2T), and Table 5, we found that using graph networks (for example, GAT) to utilize the neighborhood information of the entity can provide the performance of the model substantial improvement. Meanwhile, from the experimental results of RACE2T (sub/mult/rotate) and RACE2T (void), it can be concluded that integrating the embeddings of relations into the process of aggregating the neighborhood information of the entity can further improve the performance of the model, and it is also verified that the information of relations is helpful to predict the missing entity types of entities.

4.6   the Number of Different Relations Analysis

Figure 4 shows the distribution of the number of different relations connected by entities in the FB15K and YAGO43K datasets. Since the relation distribution in the FB15K dataset is more even than the relation distribution in YAGO43K, this experiment is carried out on FB15K and FB15KET. The experimental results are shown in Table 6.

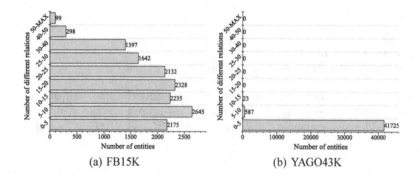

(a) FB15K   (b) YAGO43K

Fig. 4. Distribution of the number of different relations connected by entities.

As shown in Table 6, the performance of RACE2T is always better than that of CE2T, which benefits from the utilization of relations between entities by RACE2T or the information aggregation by RACE2T. Meanwhile, from the results of CE2T and RACE2T in Table 6, we found that more information aggregated is not the better. When an entity connects more different relations, RACE2T aggregates more information about different types of entities and relations in aggregating information. Aggregating more different types of entity and relation information may cause the entity to lose the original information and decrease the performance of RACE2T.

Table 6. The results of entity type prediction on FB15KET.

Distribution	CE2T			RACE2T		
	MRR	HITS@1	HITS@3	MRR	HITS@1	HITS@3
$[0, 5]$	0.4786	0.3836	0.5193	0.5189	0.4241	0.5673
$[5, 10]$	0.5745	0.4777	0.6248	0.6503	0.5691	0.6834
$[10, 15]$	0.5987	0.4912	0.6550	0.6858	0.6025	0.7196
$[15, 20]$	0.5899	0.4753	0.6580	0.6744	0.5909	0.7196
$[20, 25]$	0.5869	0.4703	0.6576	0.6465	0.5525	0.6944
$[25, 30]$	0.5803	0.4671	0.6437	0.6540	0.5624	0.7013
$[30, 40]$	0.5782	0.4746	0.6276	0.6323	0.5447	0.6713
$[40, 50]$	0.5405	0.4437	0.6026	0.5647	0.4794	0.6039
$[50, Max]$	0.4718	0.3596	0.5369	0.5109	0.4137	0.5640

From Table 6, we can observe that when the number of different relations connected by the entity is between 5–40, RACE2T improves about 8% higher than CE2T on MRR, HITS@1, and HITS@3. When the number of different relations connected by the entity is between 0–5, RACE2T improves about 4% higher than CE2T on MRR, HITS@1, and HITS@3. As shown in Fig. 4b, the number of different relations connected by entities in YAGO43K is mainly concentrated between 0–5, which indirectly leads to RACE2T did not perform as well on YAGO43KET relative to CE2T as they did on FB15KET relative to CE2T.

5 Conclusion and Future Work

This work proposes a method for entity type prediction in knowledge graphs with relational aggregation graph attention network called RACE2T. It includes an encoder and a decoder. The focus of RACE2T is on utilizing relational information when predicting the type of an entity. Therefore, we introduce relational aggregation graph attention network (FRGAT) as an encoder to aggregate the information of entities and relations in the entity neighborhood to utilize the information of relations. Meanwhile, we design a convolutional neural network-based knowledge graph entity type prediction model as the decoder of this paper, called: CE2T. Its role is to measure the similarity between entity embedding and entity type embedding. In addition, we provide various experiments to verify the validity of our model.

Our future research interest is to apply disentangled representation learning (learning disentangled representations of entities) to RACE2T, specifically: using disentangled representation learning for knowledge graph entity type prediction.

Acknowledgements. This work was supported by the National Natural Science Foundation of China (No. 61976032).

References

1. Biswas, R., Sofronova, R., Sack, H., Alam, M.: Cat2type: Wikipedia category embeddings for entity typing in knowledge graphs. In: K-CAP, pp. 81–88 (2021). https://doi.org/10.1145/3460210.3493575
2. Bordes, A., Usunier, N., García-Durán, A., Weston, J., Yakhnenko, O.: Translating embeddings for modeling multi-relational data. In: NIPS, pp. 2787–2795 (2013)
3. Chen, S., Wang, J., Jiang, F., Lin, C.: Improving entity linking by modeling latent entity type information. In: AAAI, pp. 7529–7537 (2020)
4. Dettmers, T., Minervini, P., Stenetorp, P., Riedel, S.: Convolutional 2D knowledge graph embeddings. In: AAAI, pp. 1811–1818 (2018)
5. Jain, N., Kalo, J.-C., Balke, W.-T., Krestel, R.: Do embeddings actually capture knowledge graph semantics? In: Verborgh, R., et al. (eds.) ESWC 2021. LNCS, vol. 12731, pp. 143–159. Springer, Cham (2021). https://doi.org/10.1007/978-3-030-77385-4_9
6. Ji, G., He, S., Xu, L., Liu, K., Zhao, J.: Knowledge graph embedding via dynamic mapping matrix. In: ACL, pp. 687–696 (2015). https://doi.org/10.3115/v1/p15-1067
7. Ji, S., Pan, S., Cambria, E., Marttinen, P., Yu, P.S.: A survey on knowledge graphs: representation, acquisition, and applications. IEEE Trans. Neural Netw. Learn. Syst. 1–21 (2021). https://doi.org/10.1109/TNNLS.2021.3070843

8. Jin, H., Hou, L., Li, J., Dong, T.: Attributed and predictive entity embedding for fine-grained entity typing in knowledge bases. In: Proceedings of the 27th International Conference on Computational Linguistics, pp. 282–292 (2018)

9. Jin, H., Hou, L., Li, J., Dong, T.: Fine-grained entity typing via hierarchical multi graph convolutional networks. In: EMNLP, pp. 4968–4977 (2019). https://doi.org/10.18653/v1/D19-1502

10. Kingma, D.P., Ba, J.: Adam: a method for stochastic optimization. In: ICLR, pp. 1–15 (2015)

11. Kipf, T.N., Welling, M.: Semi-supervised classification with graph convolutional networks. In: ICLR, pp. 1–14 (2017)

12. Moon, C., Harenberg, S., Slankas, J., Samatova, N.F.: Learning contextual embeddings for knowledge graph completion. In: PACIS, pp. 248–253 (2017)

13. Moon, C., Jones, P., Samatova, N.F.: Learning entity type embeddings for knowledge graph completion. In: CIKM, pp. 2215–2218. ACM (2017)

14. Neelakantan, A., Chang, M.: Inferring missing entity type instances for knowledge base completion: new dataset and methods. In: NAACL, pp. 515–525 (2015). https://doi.org/10.3115/v1/n15-1054

15. Nguyen, D.Q., Nguyen, T.D., Nguyen, D.Q., Phung, D.Q.: A novel embedding model for knowledge base completion based on convolutional neural network. In: NAACL, pp. 327–333 (2018). https://doi.org/10.18653/v1/n18-2053

16. Nguyen, D.Q., Vu, T., Nguyen, T.D., Nguyen, D.Q., Phung, D.Q.: A capsule network-based embedding model for knowledge graph completion and search personalization. In: NAACL, pp. 2180–2189 (2019). https://doi.org/10.18653/v1/n19-1226

17. Nickel, M., Rosasco, L., Poggio, T.A.: Holographic embeddings of knowledge graphs. In: AAAI, pp. 1955–1961 (2016)

18. Nickel, M., Tresp, V., Kriegel, H.: A three-way model for collective learning on multi-relational data. In: ICML, pp. 809–816 (2011)

19. Paulheim, H., Bizer, C.: Type inference on noisy RDF data. In: Alani, H., et al. (eds.) ISWC 2013. LNCS, vol. 8218, pp. 510–525. Springer, Heidelberg (2013). https://doi.org/10.1007/978-3-642-41335-3_32

20. Sun, Z., Deng, Z., Nie, J., Tang, J.: Rotate: knowledge graph embedding by relational rotation in complex space. In: ICLR, pp. 1–18 (2019)

21. Vashishth, S., Sanyal, S., Nitin, V., Talukdar, P.P.: Composition-based multi-relational graph convolutional networks. In: ICLR, pp. 1–15 (2020)

22. Velickovic, P., Cucurull, G., Casanova, A., Romero, A., Liò, P., Bengio, Y.: Graph attention networks. In: ICLR, pp. 1–12 (2018)

23. Wang, Z., Zhang, J., Feng, J., Chen, Z.: Knowledge graph embedding by translating on hyperplanes. In: AAAI, pp. 1112–1119 (2014)

24. Xie, R., Liu, Z., Jia, J., Luan, H., Sun, M.: Representation learning of knowledge graphs with entity descriptions. In: AAAI, pp. 2659–2665 (2016)

25. Xie, R., Liu, Z., Sun, M.: Representation learning of knowledge graphs with hierarchical types. In: IJCAI, pp. 2965–2971 (2016)

26. Xu, B., Zhang, Y., Liang, J., Xiao, Y., Hwang, S., Wang, W.: Cross-lingual type inference. In: Navathe, S.B., Wu, W., Shekhar, S., Du, X., Wang, X.S., Xiong, H. (eds.) DASFAA 2016. LNCS, vol. 9642, pp. 447–462. Springer, Cham (2016). https://doi.org/10.1007/978-3-319-32025-0_28

27. Yaghoobzadeh, Y., Adel, H., Schütze, H.: Corpus-level fine-grained entity typing. J. Artif. Intell. Res. 61, 835–862 (2018). https://doi.org/10.1613/jair.5601

28. Yaghoobzadeh, Y., Schütze, H.: Multi-level representations for fine-grained typing of knowledge base entities. In: EACL, pp. 578–589 (2017). https://doi.org/10.18653/v1/e17-1055
29. Yang, B., Yih, W., He, X., Gao, J., Deng, L.: Embedding entities and relations for learning and inference in knowledge bases. In: ICLR, pp. 1–12 (2015)
30. Zhao, Y., Li, Z., Deng, W., Xie, R., Li, Q.: Learning entity type structured embeddings with trustworthiness on noisy knowledge graphs. Knowl. Based Syst. **215**, 106630 (2021). https://doi.org/10.1016/j.knosys.2020.106630
31. Zhao, Y., Zhang, A., Xie, R., Liu, K., Wang, X.: Connecting embeddings for knowledge graph entity typing. In: ACL, pp. 6419–6428 (2020). https://doi.org/10.18653/v1/2020.acl-main.572
32. Zhu, Q., et al.: Collective multi-type entity alignment between knowledge graphs. In: WWW, pp. 2241–2252 (2020)

Union and Intersection of All Justifications

Jieying Chen[1]([✉]), Yue Ma[2], Rafael Peñaloza[3], and Hui Yang[2]

[1] SIRIUS, Department of Information, University of Oslo, Oslo, Norway
jieyingc@ifi.uio.no
[2] LISN, Univ. Paris-Sud, CNRS, Université Paris-Saclay, Orsay, France
{ma,yang}@lri.fr
[3] University of Milano-Bicocca, Milan, Italy
rafael.penaloza@unimib.it

Abstract. We present new algorithms for computing the union and intersection of all justifications for a given ontological consequence without first computing the set of all justifications. Our goal is to use these sets to explain the consequences and, if needed, repair them. Through an empirical evaluation, we show that our approach behaves well in practice for expressive description logics. In particular, the union of all justifications can be computed much faster than with existing justification-enumeration approaches.

Keywords: Justifications · Axiom pinpointing · Ontology Repairs

1 Introduction

It is well known that ontology engineering is a delicate and error-prone task, which requires automated tools to avoid introducing unexpected or unwanted consequences. Indeed, as an ontology grows in size it becomes difficult to predict *a priori* what effect would introducing or modifying an axiom have over the represented notions. In these settings, it is not rare for a knowledge engineer to encounter unexpected consequences from the explicitly stated knowledge. In this case, the knowledge engineer should try to understand why is this a consequence, and perhaps how to get rid of it. To achieve this, it is helpful to focus exclusively on the axioms that are relevant for this consequence.

Axiom pinpointing [35] is the task of identifying the axioms in an ontology that are required for a consequence to follow. Primarily, its focus is on computing the class of all *justifications*: subset-minimal subontologies that entail the consequence. A dual notion is that of a *repair*: a subset-maximal subontology which does not entail the consequence. Justifications provide a way to understand the causes for a consequence, while repairs suggest a way to get rid of it.

This work is partially funded by the Norwegian Research Council via the SIRIUS centre (Grant Nr.237898) and the BPI-France (PSPC AIDA: 2019-PSPC-09).

P. Groth et al. (Eds.): ESWC 2022, LNCS 13261, pp. 56–73, 2022.
https://doi.org/10.1007/978-3-031-06981-9_4

Axiom pinpointing methods can be separated into two main classes, commonly known as *black-box* and *glass-box*.

Black-box approaches [24,25,34] use existing reasoners as an oracle, and require no further modification of the reasoning method. Therefore, these approaches work for ontologies written in any monotonic logical language (including expressive DLs such as $\mathcal{SROIQ}$ [21]), as long as a reasoner supporting it exists. In their most naïve form, black-box methods check all possible subsets of the ontology for the desired entailment and compute the justifications from these results. In reality, many optimisations have been developed to reduce the number of calls needed, and avoid irrelevant work. Glass-box approaches, on the other hand, modify a reasoning algorithm to output one or all justifications directly, from only one call. While the theory for developing glass-box methods has been developed for different reasoners [6–9,32], in practice not many of these methods have been implemented, as they require new coding efforts and (often) deactivating the optimisation techniques that make reasoners practical. A promising approach, first proposed in [42] is to reduce, through a reasoning simulation, the axiom pinpointing problem to a related enumeration problem from a propositional formula, and use state-of-the-art SAT-solving methods to enumerate all the justifications. This idea has led to effective axiom pinpointing systems developed primarily for the lightweight DL $\mathcal{EL}$ [4]; see [1–3,26,30,31].

The interest of axiom pinpointing goes beyond enumerating justifications. When the scope is to get rid of an unwanted consequence, one may be interested in the repairs, or their complement, commonly called *diagnoses*. Diagnoses can be derived from justifications via a hitting set computation, and viceversa [8]. Moreover, as there might exist exponentially many justifications (or repairs, or diagnoses) for a given entailment w.r.t. an ontology, even for ontologies in an inexpressive language such as the DL $\mathcal{EL}$, finding all justifications is not feasible in general. To alleviate this issue, one may approximate the information through the union and the intersection of all justifications. Every element in the intersection is a diagnosis by itself. From the union, a knowledge engineer has a more precise view on the problematic instances, and can make a detailed analysis.

Although much work has focused on methods for computing one or all justifications efficiently, to the best of our knowledge there is little work on computing their intersection or union without enumerating them first, beyond the approximations presented in [36,37]. Here, we first propose an algorithm for computing the intersection of all justifications. This algorithm has the same worst-case behaviour as the black-box algorithm of computing one justification, avoiding the worst-case exponential enumeration. Additionally, we present two approaches to compute the union of all justifications: one is based on the black-box algorithm for finding all justifications and the other approach uses the SAT-based tool *cmMUS*. An extended abstract of our paper was published on [16].

The paper is structured as follows. In Sect. 2 we recall relevant definitions of description logics and propositional logic. Section 3 presents the algorithm for computing the intersection of all justifications without computing any single justification, followed by two methods of computing the union of all justifications in Sect. 4. Afterwards, we explain how to use the union and intersection of all justifications to repair ontologies. Before concluding, an evaluation of our methods on real-world ontologies is presented in Sect. 6.

2   Preliminaries

We briefly recall the description logic $\mathcal{ALCH}$ [5] and the notions of justifications, repairs and ontology modules.

Let N_C and N_R disjoint sets of *concept-*, and *role names* respectively. The set of $\mathcal{ALC}$-concepts is built through the grammar rule

$$C ::= \top \mid \bot \mid A \mid C \sqcap C \mid C \sqcup C \mid \neg C \mid \exists r.C \mid \forall r.C,$$

where $A \in N_C$ and $r \in N_R$. An $\mathcal{ALCH}$ *TBox* $\mathcal{T}$ is a finite set of general concept inclusions (*GCIs*) of the form $C \sqsubseteq D$ and *role inclusions* $r \sqsubseteq s$, where C and D are $\mathcal{ALC}$ concepts and $r, s \in N_R$. From now on, we will call the TBox an *ontology*, and its elements (GCIs and role inclusions) will be called *axioms* in general. The DL $\mathcal{EL}$ is the restriction of $\mathcal{ALC}$ that does not allow for bottom $\bot$, negations $\neg$, nor value restrictions $\forall$.

The semantics of this logic is defined in terms of interpretations. An *interpretation* is a pair $\mathcal{I} = (\Delta^{\mathcal{I}}, \cdot^{\mathcal{I}})$ where $\Delta^{\mathcal{I}}$ is a non-empty *domain*, and $\cdot^{\mathcal{I}}$ is the *interpretation function*, which maps each concept name $A \in N_C$ to a subset $A^{\mathcal{I}} \subseteq \Delta^{\mathcal{I}}$, and each role name $r \in N_R$ to a binary relation $r^{\mathcal{I}} \subseteq \Delta^{\mathcal{I}} \times \Delta^{\mathcal{I}}$. The interpretation function is extended to $\mathcal{ALC}$-concepts as usual: $(\top)^{\mathcal{I}} := \Delta^{\mathcal{I}}$, $(\bot)^{\mathcal{I}} := \emptyset$, $(\neg C)^{\mathcal{I}} := \Delta^{\mathcal{I}} \backslash C^{\mathcal{I}}$, $(C \sqcap D)^{\mathcal{I}} := C^{\mathcal{I}} \cap D^{\mathcal{I}}$, $(C \sqcup D)^{\mathcal{I}} := C^{\mathcal{I}} \cup D^{\mathcal{I}}$, $(\exists r.C)^{\mathcal{I}} := \{x \in \Delta^{\mathcal{I}} \mid \exists y \in C^{\mathcal{I}} : (x, y) \in r^{\mathcal{I}}\}$, and $(\forall r.C)^{\mathcal{I}} := \{x \in \Delta^{\mathcal{I}} \mid \forall y \in \Delta^{\mathcal{I}} : (x, y) \in r^{\mathcal{I}} \Rightarrow y \in C^{\mathcal{I}}\}$.

The interpretation $\mathcal{I}$ *satisfies* $C \sqsubseteq D$ iff $C^{\mathcal{I}} \subseteq D^{\mathcal{I}}$ and it *satisfies* $r \sqsubseteq s$ iff $r^{\mathcal{I}} \subseteq s^{\mathcal{I}}$. We write $\mathcal{I} \models \alpha$ if $\mathcal{I}$ satisfies the axiom α. The interpretation $\mathcal{I}$ is a *model* of an ontology $\mathcal{O}$ if $\mathcal{I}$ satisfies all axioms in $\mathcal{O}$. An axiom α *is entailed by* $\mathcal{O}$, denoted as $\mathcal{O} \models \alpha$, if $\mathcal{I} \models \alpha$ for all models $\mathcal{I}$ of $\mathcal{O}$. We use $|\mathcal{O}|$ to denote the size of $\mathcal{O}$, i.e., the number of axioms in $\mathcal{O}$. $\text{sig}(\alpha)$ is a function that extracts a set of concept and role names that occur in the GCI α.

We are interested in the notions of *justification* and *repair*.

Definition 1 (Justification, repair). *Let $\mathcal{O}$ be an ontology and α a GCI.*

- *A justification for $\mathcal{O} \models \alpha$ is a subset $\mathcal{M} \subseteq \mathcal{O}$ such that $\mathcal{M} \models \alpha$ and for any $\mathcal{M}' \subsetneq \mathcal{M}$, $\mathcal{M}' \not\models \alpha$. $\text{Just}(\mathcal{O}, \alpha)$ denotes the set of all justifications of α w.r.t. $\mathcal{O}$.*
- *A repair for $\mathcal{O} \models \alpha$ is a subontology $\mathcal{R} \subseteq \mathcal{O}$ such that $\mathcal{R} \not\models \alpha$, but $\mathcal{O}' \models \alpha$ for any $\mathcal{R} \subsetneq \mathcal{O}' \subseteq \mathcal{O}$. We denote the set of all repairs as $\text{Rep}(\mathcal{O}, \alpha)$.*

Briefly, a justification is a minimal subset of an ontology that preserves the conclusion, while a repair is a maximal sub-ontology that does not entail the consequence.

In the context of error-tolerant reasoning, where the goal is to derive meaningful consequences while avoiding known errors from the ontology, three main entailment relations have been considered.

Definition 2 (Brave, cautious and IAR entailments). *Let α be a consequence of $\mathcal{O}$ and $\mathrm{Rep}(\mathcal{O}, \alpha)$ be the set of all repairs. A consequence β is:*

- bravely *entailed by $\mathcal{O}$ w.r.t. α if $\mathcal{R} \models \beta$ for some $\mathcal{R} \in \mathrm{Rep}(\mathcal{O}, \alpha)$;*
- cautiously *entailed by $\mathcal{O}$ w.r.t. α if $\mathcal{R} \models \beta$ for all $\mathcal{R} \in \mathrm{Rep}(\mathcal{O}, \alpha)$;*
- IAR *entailed by by $\mathcal{O}$ w.r.t. α if $\bigcap_{R \in \mathrm{Rep}(\mathcal{O}, \alpha)} \mathcal{R} \models \beta$.*

In other words, a brave entailment is one which could still hold after repairing the ontology. Cautious entailment is a stronger notion, requiring that the consequence holds regardless of how the ontology is repaired; thus every cautious entailment is also a brave one, but the converse is not true. IAR entailments are those that follow from the *intersection* of all repairs. Importantly, note that the intersection of all repairs and the union of all justifications complement each other. Hence, studying one yields the results for the other.

When dealing with large ontologies, it is useful to consider only a subset of axioms, which preserves all information about the entailments under consideration. A *module* is a sub-ontology that preserves some syntactic or semantic properties w.r.t. a restricted signature Σ. In general, it is hard to compute minimal modules for expressive ontologies. Still, there exists the notion of a *syntactic locality module* which can be computed efficiently, even for very expressive DLs [40]. *Lean kernels* [27,37] for the DL $\mathcal{ALC}$, and *minimal deductive modules* for the DL $\mathcal{EL}$ [13,14] usually define a smaller module. An important property of those modules (syntactic locality modules, lean kernels and minimal deductive modules) is that they are *justification-preserving*; i.e., all the justifications for a given consequence are contained in them. Due to this property, when computing justifications, their union, and their intersection, we will first compute such a module. As it is better understood, and well defined for expressive DLs, in the following we consider computing only locality-based modules. However, it should be considered that any justification-preserving module would suffice.

We now consider a propositional language with a finite set of propositional *variables* $L = \{p_1, p_2, \cdots, p_n\}$. A *literal* is a variable p_i or its negation $\neg p_i$. A *clause* $l_1 \vee l_2 \vee \cdots \vee l_k$ is a disjunction of literals, denoted by ω [11]. A Boolean formula in conjunctive normal form (CNF) is a conjunction of clauses. A CNF formula ϕ is *satisfiable* iff there exists a *truth assignment* $\mu_L : L \to \{0, 1\}$ such that μ_L satisfies all clauses in ϕ. We can also consider a CNF formula as a set of clauses. A subformula $\phi' \subseteq \phi$ is a *minimally unsatisfiable subformula (MUS)* iff ϕ' is unsatisfiable, but for every $\phi'_1 \subsetneq \phi'$ is satisfiable. Note that the notion of a MUS corresponds to that of a justification where the ontology language is propositional logic and the consequence under consideration is unsatisfiability.

3   Computing the Intersection of All Justifications

We first study the problem of computing the intersection of all justifications, which we call the *core*. Algorithm 1 provides a method for finding this core. The algorithm is inspired by the known black-box approach for finding justifications [9,23]. Starting from a justification-preserving module $\mathcal{M}$ (in this case, the locality-based module, Line 2), we try to remove one axiom β (Line 4). If the removal of the axiom β removes the entailment (Line 5), then β belongs to all justifications (β is *sine qua non* required for the entailment within $\mathcal{M}$), and is thus added to the core $\mathcal{C}$ (Line 6). It can be shown that Algorithm 1 correctly computes the intersection of all justifications.

Theorem 3. *Let $\mathcal{O}$ be an ontology and α a GCI. Algorithm 1 computes the intersection of all justifications (core) of $\mathcal{O}$ w.r.t. α.*

Proof. Let $\mathcal{S}$ be the set computed by Algorithm 1, so $\mathcal{S} = \{\beta \mid \mathcal{O} \setminus \{\beta\} \not\models \alpha\}$ (Line 5–6). Additionally, let $\mathcal{C}$ be the core. First, we prove that $\mathcal{S} \subseteq \mathcal{C}$. We assume towards a contradiction that there exists an axiom β such that $\beta \in \mathcal{S}$, but $\beta \notin \mathcal{C}$. As $\beta \notin \mathcal{C}$, there exists a justification $\mathcal{J}$ of $\mathcal{O}$ w.r.t. α such that $\beta \notin \mathcal{J}$. According to Definition 1, $\mathcal{O} \setminus \{\beta\} \models \alpha$, which contradicts to our assumption that $\beta \in \mathcal{S}$. The other direction $\mathcal{C} \subseteq \mathcal{S}$ is analogous. Therefore, $\mathcal{C} = \mathcal{S}$, that is, Algorithm 1 computes the core of all justifications of $\mathcal{O}$ w.r.t. α.

Algorithm 2, on the other hand, generalises the known algorithm for computing a single justification, by considering a (fixed) set $\mathcal{C}$ that is known to be contained in all justifications. If $\mathcal{C} = \emptyset$, the approach works as usual; otherwise, the algorithm avoids trying to remove any axiom from $\mathcal{C}$. This reduces the number of calls to the black-box reasoner, potentially decreasing the overall execution time.

The choice for a locality-based module in these algorithms is arbitrary, and any justification-preserving module would suffice. In particular, we could compute lean kernel [27,37] for $\mathcal{ALC}$ ontologies, and minimal subsumption modules [12,15] for $\mathcal{EL}$ ontologies instead, which is typically smaller thus reducing the number of iterations within the algorithms. However, as it could be quite expensive to compute such modules, it might not be worthwhile in some cases.

Algorithm 1, like all black-box methods for computing justifications, calls a standard reasoner $|\mathcal{M}|$ times. In terms of computational complexity, computing the core requires as many computational resources as computing a single justification. However, computing one justification might be faster in practice, as the size of $\mathcal{M}$ decreases throughout the execution of Algorithm 2. Clearly, if the core coincides with one justification $\mathcal{M}$, then $\mathcal{M}$ is the *only* justification.

Corollary 4. *Let $\mathcal{O}$ be an ontology, α a GCI; and let $\mathcal{C}$ be the core and $\mathcal{J}$ a justification for $\mathcal{O} \models \alpha$. If $\mathcal{C} = \mathcal{J}$, $\mathcal{J}$ is the only justification for $\mathcal{O} \models \alpha$.*

Algorithm 1. Computing the intersection of all justifications of $\mathcal{O}$ w.r.t. α

INPUT: an Ontology $\mathcal{O}$, a conclusion α
1: **function** COMPUTE-JUSTIFICATION-CORE($\mathcal{O}, \alpha$)
2:     $\mathcal{C} := \emptyset$
3:     $\mathcal{M} :=$ COMPUTE-LOCALITY-BASED-MODULE($\mathcal{O}, \mathsf{sig}(\alpha)$)
4:     **for** every axiom $\beta \in \mathcal{M}$ **do**
5:         **if** $\mathcal{M} \setminus \{\beta\} \not\models \alpha$ **then**
6:             $\mathcal{C} := \mathcal{C} \cup \{\beta\}$
7:     **return** $\mathcal{C}$

Algorithm 2. Using core to compute a single justification of $\mathcal{O}$ *w.r.t.* an conclusion

---

INPUT: an ontology $\mathcal{O}$, a conclusion α, the core $\mathcal{C}$
1: **function** SINGLE-JUSTIFICATION($\mathcal{O}, \alpha, \mathcal{C}$)
2:     $\mathcal{M} :=$ COMPUTE-LOCALITY-BASED-MODULE($\mathcal{O}, \mathsf{sig}(\alpha)$)
3:     **for** every axiom $\beta \in \mathcal{M}$ and $\beta \notin \mathcal{C}$ **do**
4:         **if** $\mathcal{M} \setminus \{\beta\} \models \alpha$ **then**
5:             $\mathcal{M} := \mathcal{M} \setminus \{\beta\}$
6:     **return** $\mathcal{M}$

---

4 Computing the Union of All Justifications

We now present two algorithms of computing the union of all justifications. The first algorithm follows a black-box approach that calls a standard reasoner as oracle using the core of justifications computed in the previous section. Importantly, it is known that no black-box method for computing the union of all justifications can call an oracle only a polynomial number of times, unless P = NP [38]. Our method is inspired by Reiter's Hitting Set Tree algorithm [39] and partially follows the approach originally developed in [23,45] for enumerating all justifications. For the second algorithm, we reduce the problem of computing the union of all justifications to the problem of computing the union of MUSes of a propositional formula. Note that the second algorithm works only for $\mathcal{ALCH}$ ontologies, while the first algorithm can be applied to ontologies with any expressivity, as long as a reasoner is available.

4.1 Black-Box Algorithm

The black-box algorithm for computing all justifications from [45] was inspired by the algorithm of computing all minimal hitting sets [39]. Some of the improvements to prune the search space were already proposed in [39]. Our method for computing the union of all justifications (Algorithm 3) works in a similar manner, but with a few key differences. To avoid computing all justifications, we prune the search space when all remaining justifications are fully contained in the union computed so far. In addition, we use the core to speed the search. As the axioms in the core must appear in every justification, we can reduce the number of calls made to the reasoner, and optimise the single justification computation. Finally, when we organise our search space, we do not need to consider the axioms in the core.

We now explain Algorithm 3 in detail. Given an ontology $\mathcal{O}$, a conclusion α, and the core $\mathcal{C} \subseteq \mathcal{O}$ of $\mathcal{O}$ w.r.t. α as input, a justification-preserving module $\mathcal{M}$ of $\mathcal{O}$ w.r.t. α is extracted from $\mathcal{O}$ (Line 2). The justification search tree Ψ is a four-tuple $(\mathcal{V}, \mathcal{E}, \mathcal{L}, \rho)$, where $\mathcal{V}$ is a finite set of nodes, $\mathcal{E} \subseteq \mathcal{V} \times \mathcal{V}$ is a set of edges, $\mathcal{L}$ is an edge labelling function, mapping every edge to an axiom $\alpha \in \mathcal{M}$, and $\rho \in \mathcal{V}$ is the root node. We initialise the variable Ψ to represent a justification search tree for $\mathcal{O}$ having only root node ρ. Besides, the variables $\mathbb{M} \subseteq 2^{\mathcal{M}}$, containing the justifications that have been computed so far, and $\mathbb{P} \subseteq \mathcal{V}$, containing the already

---

Algorithm 3. Computing the Union of All Justifications w.r.t. a Conclusion α

INPUT: an Ontology $\mathcal{O}$, a conclusion α, the intersection of all Justifications $\mathcal{C} \subseteq \mathcal{O}$

 1: **function** UNION-OF-ALL-JUSTIFICATIONS($\mathcal{O}, \alpha, \mathcal{C}$)
 2:      $\mathcal{M} :=$ COMPUTE-LOCALITY-BASED-MODULE($\mathcal{O}, \text{sig}(\alpha)$)
 3:      $\mathcal{U} := \mathcal{C}; \Psi := (\{\rho\}, \emptyset, \emptyset, \rho); \mathbb{Q} := [\rho]; \mathbb{P} := \emptyset; \mathbb{M} := \{\emptyset\}$
 4:      **while** $\mathbb{Q} \neq [\,]$ **do**
 5:          $v := \text{HEAD}(\mathbb{Q}), \mathbb{Q} := \text{REMOVEFIRSTELEMENT}(\mathbb{Q}), \mathbb{P} := \mathbb{P} \cup \{v\}$
 6:          $\mathcal{M}_{\text{ex}} := \text{LABELS}(\text{PATH}(\Psi, \rho, v))$
 7:          **if** IS-PATH-REDUNDANT($\Psi, \rho, \mathcal{M}_{\text{ex}}, \mathbb{P}$) **then**
 8:              **continue**
 9:          **if** $\mathcal{M} \setminus \mathcal{M}_{\text{ex}} \not\models \alpha$ **then**
10:              **continue**
11:          **if** $\mathcal{M} \setminus \mathcal{M}_{\text{ex}} \subseteq \mathcal{U}$ **then**
12:              **continue**
13:          $\mathcal{M} := \emptyset$
14:          **if** there exists $\mathcal{M}' \in \mathbb{M}$ such that $\mathcal{M}_{\text{ex}} \cap \mathcal{M}' = \emptyset$ **then**
15:              $\mathcal{M} := \mathcal{M}'$
16:          **else**
17:              $\mathcal{M} := \text{SINGLE-JUSTIFICATION}(\mathcal{M} \setminus \mathcal{M}_{\text{ex}}, \alpha, \mathcal{C})$
18:              **if** $\mathcal{M} = \mathcal{C}$ **then**
19:                  **return** $\{\mathcal{C}\}$
20:          $\mathbb{M} := \mathbb{M} \cup \{\mathcal{M}\}$
21:          $\mathcal{U} := \mathcal{U} \cup \mathcal{M}$
22:          **for** every $\beta \in \mathcal{M} \setminus \mathcal{C}$ **do**
23:              $v_\beta := \text{ADDCHILD}(\Psi, v, \beta)$
24:              $\mathbb{Q} := v_\beta :: \mathbb{Q}$
25:      **return** $\mathcal{U}$

---

explored nodes of Ψ, are both initialised with the empty set. The queue $\mathbb{Q}$ of nodes in Ψ that still has to be explored is also set to contain the node ρ as its only element.

The algorithm then enters a loop (Lines 4–24) that runs while $\mathbb{Q}$ is not empty. The loop extracts the first element v from $\mathbb{Q}$ and adds it to $\mathbb{P}$ (Line 5). The axioms that label the edges of the path π_v from ρ to v in Ψ are collected in the set $\mathcal{M}_{\text{ex}}$ (Line 6). After that, the algorithm checks whether π_v is redundant via the function IS-PATH-REDUNDANT($\Psi, \rho, \mathcal{O}_{\text{ex}}, \mathbb{P}$). The path π_v is *redundant* iff there exists an explored node $w \in \mathbb{P}$ such that (a) the axioms in $\mathcal{O}_{\text{ex}}$ are exactly the axioms labelling the edges of the path π_w from ρ to w in Ψ, or (b) w is a leaf node of Ψ and the edges of π_w are only labelled with axioms from $\mathcal{O}_{\text{ex}}$. Case (a) corresponds to *early path termination* in [23,39]: the existence of π_w implies that all possible extensions of π_v have already been considered. Case (b) implies that the axioms labelling the edges of π_w lead to the fact that α can not be entailed be the remaining TBox when removed from $\mathcal{M}$. Therefore, by monotonicity of $\models$, we infer that removing $\mathcal{O}_{\text{ex}}$ from $\mathcal{M}$ also has the same consequence implying that we do not need to explore π_v and all its extensions.

If $\mathcal{M} \setminus \mathcal{M}_{ex} \not\models \alpha$ (Lines 9–10), the current iteration can be terminated immediately as no subset of $\mathcal{M} \setminus \mathcal{M}_{ex}$ can be a justification of $\mathcal{M}$ w.r.t. α. In contrast to other black-box algorithms for computing justifications, we additionally check whether $\mathcal{M} \setminus \mathcal{M}_{ex}$ is a subset of $\mathcal{U}$. If it is the case, no new axioms belonging to the union of all justifications can appear in this sub-tree. Hence, the algorithm does not need to explore it any further. Subsequently, the variable $\mathcal{M}$ that will hold a justification of $\mathcal{M} \setminus \mathcal{M}_{ex}$ is initialised with $\emptyset$. At this point we can check if a justification $\mathcal{M}' \in \mathbb{M}$ has already been computed for which $\mathcal{O}_{ex} \cap \mathcal{M}' = \emptyset$ (Lines 14–15) holds, in which case we set $\mathcal{M}$ to $\mathcal{M}'$. This optimisation step can also be found in [23,39] and it allows us to avoid a costly call to the SINGLE-JUSTIFICATION procedure. Otherwise, in Line 17 we call SINGLE-JUSTIFICATION on $\mathcal{M} \setminus \mathcal{O}_{ex}$ to obtain a justification of α w.r.t. $\mathcal{M} \setminus \mathcal{O}_{ex}$. We then check whether $\mathcal{M}$ is equal to $\mathcal{C}$ (Lines 18–19), in which case the search for additional justifications can be terminated (recall Corollary 4). Otherwise, the justification $\mathcal{M}$ is added to $\mathbb{M}$ in Line 20 and the union of all justifications is updated in Line 21. Finally, for every $\beta \in \mathcal{M} \setminus \mathcal{C}$, the algorithm extends the tree Ψ in Lines 22–24 by adding a child v_α to v, connected by an edge labelled with β. Note that it is sufficient to take $\beta \notin \mathcal{C}$ as a set $\mathcal{M}$ with $\mathcal{C} \not\subseteq \mathcal{M}$ cannot be a justification of $\mathcal{O}$ w.r.t. α. The procedure finishes by returning the set $\mathcal{U}$.

Note that this algorithm only adds justifications to $\mathbb{M}$. For completeness, one can show that the locality-based module $\mathcal{M}$ of $\mathcal{O}$ w.r.t. $\mathsf{sig}(\alpha)$ contains all justifications of $\mathcal{O}$ w.r.t. α. Moreover, it is easy to see that the proposed optimisations do not lead to a justification not being computed. Overall, we obtain the following result.

Theorem 5. *Let $\mathcal{O}$ be an ontology, α a GCI, and $\mathcal{C} \subseteq \mathcal{O}$ the core of α w.r.t. $\mathcal{O}$. The procedure* UNION-OF-ALL-JUSTIFICATIONS *computes the union of all justifications of $\mathcal{O}$ w.r.t. α.*

Algorithm 3 terminates on any input as the paths in the module search tree Ψ for $\mathcal{O}$ constructed during the execution represent all the permutations of the axioms in $\mathcal{O}$ that are relevant for finding all minimal modules. It is easy to see that the procedure UNION-OF-ALL-JUSTIFICATIONS runs in exponential time in size of $\mathcal{O}$ in the worst case.

4.2   MUS Membership Algorithm (MUS-MEM)

It has been well-investigated that one can encode the problem of computing justifications to the problem of computing MUSes of CNF formula. One first needs to transfer all axioms and a given conclusion to CNF formulae and then uses a SAT-solver to compute a MUS. Finally, the corresponding axioms of MUS is a justification for a given conclusion. For a general overview on how this process works see [30].

Table 1. Inference rules of *condor*

$$\mathbf{R_A^+} \; \frac{}{H \sqsubseteq A} : A \sqsubseteq H \qquad\qquad \mathbf{R_A^-} \; \frac{H \sqsubseteq N \sqcup A}{H \sqsubseteq N} : \neg A \sqsubseteq H$$

$$\mathbf{R_\sqcap^n} \; \frac{\{H \sqsubseteq N_i \sqcup A_i\}_{i=1}^n}{H \sqsubseteq \sqcup_{i=1}^n N_i \sqcup M} : \sqcap_{i=1}^n A_i \sqsubseteq M \in \mathcal{O}$$

$$\mathbf{R_\exists^+} \; \frac{H \sqsubseteq N \sqcup A}{H \sqsubseteq N \sqcup \exists r.B} : A \sqsubseteq \exists r.B \in \mathcal{O}$$

$$\mathbf{R_\exists^-} \; \frac{H \sqsubseteq M \sqcup \exists r.K, K \sqsubseteq N \sqcup A}{H \sqsubseteq M \sqcup B \sqcup \exists r.(K \sqcap \neg A)} : \exists s.A \sqsubseteq B \in \mathcal{O}, \mathcal{O} \models r \sqsubseteq s$$

$$\mathbf{R_\exists^\perp} \; \frac{H \sqsubseteq M \sqcup \exists r.K, K \sqsubseteq \perp}{H \sqsubseteq M}$$

$$\mathbf{R_\forall} \; \frac{H \sqsubseteq M \sqcup \exists r.K, H \sqsubseteq N \sqcup A}{H \sqsubseteq M \sqcup B \sqcup \exists r.(K \sqcap B)} : A \sqsubseteq \forall s.B \in \mathcal{O}, \mathcal{O} \models r \sqsubseteq s$$

We now show how to compute the union of all justifications of a GCI by a membership approach. The idea is to check the membership of each axiom, i.e., whether it is a member of some justification. We further encode it to the problem of checking each CNF-formula whether it is a member of some MUS.

The MUS-MEM approach first, as a pre-processing step, computes a CNF formula ϕ using the consequence-based reasoner *condor* [43].[1] Afterwards, it computes the union of all justifications of $\alpha \in \mathcal{O}$ by checking the membership for each axiom using the SAT-tool *cmMUS* [22] over ϕ. The two steps of our method is detailed below:

1. **Compute CNF formula ϕ.** Let H, K denote (possibly empty) conjunctions of concepts, and M, N (possibly empty) disjunctions of concepts; *condor* classifies the TBox through the inference rules in Table 1. Each inference rule can be rewritten as a clause. For example, the $\mathbf{R_\exists^\perp}$ rule can be rewritten to the clause $\neg p_1 \vee \neg p_2 \vee p_3$ if we denote the GCIs $H \sqsubseteq M \sqcup \exists r.K, K \sqsubseteq \perp$, and $H \sqsubseteq M$ as literals p_1, p_2, p_3, respectively. The CNF formula ϕ is the conjunction of all the clauses corresponding to all the applied inference rules during the classification process [37,42].

2. **Check membership of each axiom using *cmMUS*.** Given a CNF formula ϕ and a subformula $\phi' \subseteq \phi$, the algorithm *cmMUS* is used to decide whether there is a MUS $\phi'' \subseteq \phi$ with $\phi' \cap \phi'' \neq \emptyset$. We set $cmMUS(\phi, \phi') = 1$ if there exists such MUS ϕ'' and 0 otherwise. To check membership, we need to define two objects:
   (a) a CNF-formula $\phi_\mathcal{O} = \wedge_{\beta \in \mathcal{O}} p_\beta$, where each literal p_β corresponds to an axiom $\beta \in \mathcal{O}$, and $\phi_\alpha = \neg p_\alpha$, where α is the given conclusion;
   (b) $\psi_\alpha = \phi \wedge \phi_\mathcal{O} \wedge \phi_\alpha$.

---

[1] We restrict to $\mathcal{ALCH}$ in this section as *condor* only accepts $\mathcal{ALCH}$-TBoxes.

Then we can see the following facts hold: Firstly, ψ_α is unsatisfiable. Moreover, each MUS $\psi' \subseteq \psi_\alpha$ corresponds to a justification of α. Finally, $\forall \beta \in \mathcal{O}$, $cmMUS(\psi_\alpha, p_\beta) = 1$ iff β belongs to some justifications of α.

An important optimization is based on the fact that only a small number of clauses in ϕ are related to the derivation of α. In practice, (i) $\phi' \subseteq \phi$ is the subformula contributing to the derivation of α obtained by tracing back from α, (ii) $\phi'_\mathcal{O} \subseteq \phi_\mathcal{O}$ is the subformula including only $\beta \in \mathcal{O}$ that appears in ϕ'. Using $\psi'_\alpha = \phi' \wedge \phi'_\mathcal{O} \wedge \phi_\alpha$ instead of ψ_α as the input of algorithm $cmMUS$ can significantly accelerate the $cmMUS$ algorithm.

Theorem 6. *Let $\mathcal{O}$ be an ontology and α a GCI. The procedure* MUS-MEM *algorithm computes the union of all justifications of $\mathcal{O}$ w.r.t. α.*

Regarding to the computational complexity of the MUS-MEM algorithm, in general, the classification of an $\mathcal{ALC}$ TBox requires exponential time. Since the MUS-membership problem is Σ_2^P-complete [28], it follows that this method runs in exponential time overall.

5   Repairing Ontologies

Similar to justifications, it is common to have multiple repairs for an unwanted consequence. Instead of treating all the repairs equally, in this section we propose a notion of optimal repair and provide a method for computing all such optimal repairs. Therefore, knowledge engineers can be better guided while repairing erroneous conclusions.

Definition 7 (Optimal Repair). *Let $\mathcal{O}$ be an ontology, α be a GCI, and* $\mathrm{Rep}(\mathcal{O}, \alpha)$ *be the set of all repairs for $\mathcal{O} \models \alpha$. $\mathcal{R} \in \mathrm{Rep}(\mathcal{O}, \alpha)$ is an optimal repair for $\mathcal{O} \models \alpha$, if $|\mathcal{R}| \geq |\mathcal{R}'|$ holds for every $\mathcal{R}' \in \mathrm{Rep}(\mathcal{O}, \alpha)$.*

That is, an optimal repair is a repair with the largest cardinality. It is also important to recall the notion of a hitting set. Given a set of sets $\mathbb{S}$, $\mathcal{S}$ is a minimal *hitting set* for $\mathbb{S}$ if $\mathcal{S} \cap s \neq \emptyset$ for every $s \in \mathbb{S}$. $\mathcal{S}$ is a smallest minimal hitting set if it is of minimal cardinality among all hitting sets. We can compute the set of all optimal repairs through a hitting set computation [8,29,41].

Proposition 8. *Let $\mathrm{Just}(\mathcal{O}, \alpha)$ be the set of all justifications for the GCI α w.r.t. the ontology $\mathcal{O}$. If $\mathbb{S}$ is the set of all smallest minimal hitting sets for $\mathrm{Just}(\mathcal{O}, \alpha)$, then $\{\mathcal{O} \setminus \mathcal{S} \mid \mathcal{S} \in \mathbb{S}\}$ is the set of all optimal repairs for $\mathcal{O} \models \alpha$.*

When the core is not empty, a set consisting of single axiom from this core is a smallest hitting set for all justifications. We get the following corollary of Proposition 8, stating how to compute all optimal repairs faster in this case.

Corollary 9. *Let $\mathcal{O}$ be an ontology, α be a GCI and $\mathcal{C}$ be the core for $\mathcal{O} \models \alpha$. If $\mathcal{C} \neq \emptyset$, then $\{\mathcal{O} \backslash \{\beta\} \mid \beta \in \mathcal{C}\}$ is the set of all optimal repairs for $\mathcal{O} \models \alpha$.*

It is easy to see that removing the union of all justifications from the given ontology results in the intersection of all repairs. Therefore, the union of all justifications can be used as a step towards deducing IAR entailments [36].

6 Evaluation

To evaluate the performance of our algorithms in real-world ontologies, we built a prototypical implementation.[2] The black-box algorithm is implemented in Java and uses the OWLAPI [20] to access ontologies and HermiT [19] as a standard reasoner. The MUS-membership algorithm (MUS-MEM) is implemented in Python and calls cmMUS [22] to detect whether a clause is a member of MUSes. The ontologies used in the evaluation come from the classification task at the 2014 ORE competition [33]. We selected the ontologies that have less than 10,000 axioms, for a total of 95 ontologies. In the experiments, we computed a single justification, the core, and the union of all justifications for all atomic concept inclusions that are entailed by the ontologies. An atomic concept inclusion is the inclusion of the form of $A \sqsubseteq B$, with $A, B \in \mathsf{N_C}$. All experiments ran on two processors Intel® Xeon® E5-2609v2 2.5 GHz, 8 cores, 64Go, Ubuntu 18.04.

Computation Time of the Core vs. a Single Justification. In terms of computational complexity, as discussed in Sect. 3, computing the core and a single justification are equally hard problems. But the size of the remaining ontology reduces during the latter process. Intuitively, if $\mathcal{O}' \subseteq \mathcal{O}$, checking whether a subsumption is satisfied by $\mathcal{O}'$ would be faster than checking it on $\mathcal{O}$. Therefore, theoretically, the computation time of the core and a single justification should be comparable and computing the core should be slightly easier than computing a single justification. In practice, our evaluation justifies it. Table 2 provides some basic statistics for comparing the time to compute the core against computing a single justification. We can see from Table 2 that the mean computation time of the core and a single justification are very similar, around 0.4s. Generally speaking, computing the core is usually faster than computing one justification as expected.

Table 2. Statistics of computation time (s) of core ($\mathcal{C}$) vs. single justification ($\mathcal{J}$)

	$\mathcal{J}$	$\mathcal{C}$
mean	0.400	0.456
std	3.649	4.273
min	0.001	0.001
25%	0.004	0.001
50%	0.009	0.002
75%	0.023	0.005
max	226.608	341.560

To best of our knowledge, there is no existing tools to compute the intersection of all justifications directly. A naïve algorithm is to compute the intersection of finding all justifications. Thus, the naïve algorithm is as hard as computing all justifications, i.e., runs in exponential time in size of $\mathcal{O}$ in the worst case. However, our algorithm runs only in polynomial time in $|\mathcal{O}|$ in all cases. Therefore, we do not compare our algorithm with the naïve algorithm in the evaluation.

Computation Time of the Union of All Justifications. As a benchmark, we compute all justifications and their union via the OWL API. As the MUS-MEM algorithm can compute the union of all justifications only for $\mathcal{ALCH}$ ontologies, we divide the ontologies into two categories: (i) $\mathcal{ALCH}$ ontologies

---

[2] The implementation is vailable at https://github.com/JieyingChenChen/Intersec tionAndUnionOfAllJust.

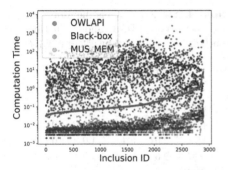

Fig. 1. Computation time (s) of the union for $\mathcal{ALCH}$-ontologies when there exist several justifications

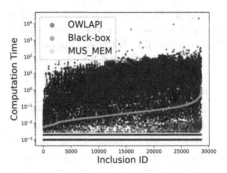

Fig. 2. Computation time (s) of the union for $\mathcal{ALCH}$-ontologies when there exists one justification

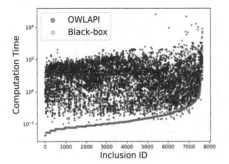

Fig. 3. Computation time (s) of the union for more expressive ontologies when there exist several justifications

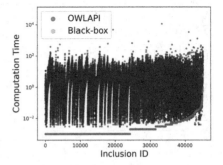

Fig. 4. Computation time (s) of the union for more expressive ontologies when there exists one justification

and (ii) with expressivity beyond $\mathcal{ALCH}$. The computation times for the union of all justifications for $\mathcal{ALCH}$ ontologies are shown in Figs. 1 and 2 for $\mathcal{ALCH}$ ontologies with several, or one justification, respectively. Figures 3 and 4 show the same information for the class of more expressive ontologies. Figures 1, 2, 3 and 4 plot the logarithmic computation time (in the Y-axis) of each test instance (in the X-axis). Each dot corresponds to computation time of the union by one of the methods tested. Note that the dots are much denser in Fig. 2 and 4 than Fig. 1 and 3 due to the fact that there exist more cases that have only one justification.

We order the conclusions along the X-axis by increasing order of computation time of MUS-MEM algorithms in Figs. 1 and Fig. 2, and by the black-box performance in the latter two figures. We observe from these plots that, generally, the green spots are located lower than the red and blue spots, which indicates that it took less time for black-box algorithm to compute the union of all justifications. When there exist several justifications, for more than 78% cases, MUS-MEM algorithm is faster than the OWLAPI. Detailed statistics of

Table 3. Statistics of computation time (s) of the union when there exist several justifications for $\mathcal{ALCH}$ ontologies

	Black-box	MUS-MEM	OWL API
mean	0.322	0.261	2.781
std	13.707	0.597	26.312
min	0.002	0.017	0.009
25%	0.004	0.069	0.123
50%	0.007	0.113	0.442
75%	0.016	0.259	1.726
max	970.834	10.255	1628.930

Table 4. Statistics of justification number, size of core ($\mathcal{C}$), a random justification ($\mathcal{J}$), and union of justifications ($\mathcal{U}$)

	#JUST	$|\mathcal{C}|$	$|\mathcal{J}|$	$|\mathcal{U}|$
mean	3.0	2.3	4.0	4.1
std	3.2	1.7	1.9	3.2
min	2.0	0.0	1.0	1.0
25%	2.0	1.0	1.0	1.0
50%	2.0	2.0	2.0	3.0
75%	3.0	3.0	5.0	6.0
max	92.0	20.0	28.0	32.0

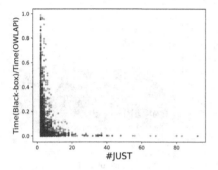

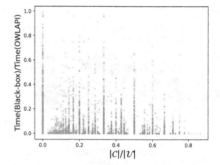

Fig. 5. Relation between #JUST and computation time of black-box/OWLAPI

Fig. 6. Relation between $|\mathcal{C}|/|\mathcal{U}|$ and computation time of black-box/OWLAPI

computation time of the union of all justifications when there exist several justifications for $\mathcal{ALCH}$ ontologies can be found in Table 3. We noticed that relatively large size of CNF-formulae (compared with the size of axioms) were generated in the cases that MUS-MEM algorithm is slower than the OWLAPI. Note that various numbers of CNF-formulae will be generated when we transform an axiom to CNF-formulae. Only less than 2% cases that MUS-MEM algorithm is faster than black-box algorithm.

Interestingly, we can see from Table 3 that the maximum computation time of the MUS-MEM algorithm is only 10.255 s; much lower than the black-box algorithm and OWLAPI. Additionally, according to the Table 3, the standard deviation of computation time of the MUS-MEM algorithm for all cases in Fig. 2 is only 0.597, which is much lower than the Black-box algorithm and OWLAPI. When we plot the data and visualize the relationship between the number of all justifications (#JUST) and computation time of the union, we found that it tends to take longer for black-box algorithm, especially OWLAPI to compute the union of all justifications when #JUST increases. But the MUS-MEM algorithm seems to be less sensitive to this factor compared with other approaches.

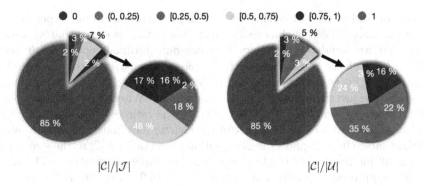

Fig. 7. Ratio of $|\mathcal{C}|$ to a random $|\mathcal{J}|$ (left) and ratio of $|\mathcal{C}|$ to $|\mathcal{U}|$ (right).

We consider two factors that influence the performance difference of the black-box algorithm and OWLAPI: the number of justifications and the ratio of $|\mathcal{C}|/|\mathcal{U}|$. The main difference between these two approaches is that the black-box algorithm uses several strategies to reduce search space. If there exists a large number of justifications, the search space of OWLAPI approach becomes very large and the black-box approach may have to traverse several uninformative branches. Our black-box algorithm only considers the axioms that are not included in the union and terminates earlier if there are no axioms that will be included in it, which significantly prunes the search space. Additionally, the black-box algorithm also uses core to optimise the algorithm. If the ratio of $|\mathcal{C}|/|\mathcal{U}|$ is larger, there are less axioms left for checking. In order to further investigate it, we plot the relations between these two factors and the ratio of computation time of the black-box algorithm and OWLAPI in Fig. 5 and Fig. 6. Y-axis represents the ratio of computation time of the black-box algorithm and OWLAPI in both figures. X-axis represents the number of justifications in Fig. 5 and, in Fig. 6, it represents the ratio of $|\mathcal{C}|/|\mathcal{U}|$. We can see that when the number of justifications increases, in Fig. 5, the ratio of computation time of the black-box algorithm and OWLAPI decreases, which means that black-box algorithm become much faster than OWLAPI. Similarly, in Fig. 6, the ratio of computation time of the two approaches also reduced when the ratio of $|\mathcal{C}|/|\mathcal{U}|$ increases, which means that the black-box algorithm has better performance than OWLAPI when the intersection of the core and the union is larger.

In general, we can conclude that, when available, MUS-MEM tends to perform better than a direct use of the OWLAPI. Although in most cases in our experiments, the MUS-MEM algorithm is slower than black-box algorithm, its time difference when computing justifications various less. The black-box algorithm can be used for more expressive ontologies and outperforms OWLAPI, especially when there exists a large number of justifications and the size of the core is relatively small compared with the union.

Size Comparisons for Justifications, Cores, and Unions of Justifications. Figure 7 illustrates the ratio of the size of the core to the size of a random justification and to the size of the union of all justifications. In our experiments, the core for only 2.35% conclusions is empty (magenta part of the charts), which means that we could use Corollary 9 to compute optimal repairs for 97.65% (100%-2.35%) of the cases. Moreover, for more than 85% cases, the size of a justification ($|\mathcal{J}|$) equals to the size of the core ($|\mathcal{C}|$), which indicates that there exists only one justification (blue part of the charts). When several justifications exist (the second chart from the left of Fig. 7), the ratio of $|\mathcal{C}|$ to a random $|\mathcal{J}|$ falls between 0.5 to 0.75 (yellow part of the charts) for almost half of the cases. The right-most chart displays the distribution of the ratio of $|\mathcal{C}|$ to the size of the union of all justifications $|\mathcal{U}|$ when there exist multiple justifications. The ratio distributes quite evenly between 0 (not including) to 0.75. Additionally, the core is empty for only 16% subsumptions even when several justifications exist. See to Table 4 for the statistics information of the size of the core, the union and a single justification when multiple justifications exist.

7   Conclusions

We presented algorithms for computing the core (that is, the intersection of all justifications) and the union of all justifications for a given DL consequence. Our black-box algorithm is based on repeated calls to a reasoner, and hence apply for ontologies and consequences of any expressivity, as long as a reasoner exists. Whilst our MUS-based approach for computing the union of all justifications depends on the properties of the $\mathcal{ALCH}$ consequence-based method implemented by *condor*. Still, the approach should be generalisable without major problems to any language for which consequence-based reasoning methods exists like, for instance, $\mathcal{SROIQ}$ [17,18]. As an application of our work, we study how to find optimal repairs effectively, through the information provided by the core and the union of all justifications.

Through an empirical analysis, run over more than 100,000 consequences from almost a hundred ontologies from the 2014 ORE competition we observe that our methods behave better in practice than the usual approach through the OWLAPI. Our experiments also confirm the observation that has already been made for light-weight ontologies [44], and to a smaller degree in the ontologies from the BioPortal corpus [10]; namely, that consequences tend to have one, or only a few, overlapping justifications. We also explored the fact that, in our experiments, the efficient core computation algorithm could find the optimal repairs in more than 97% of the test instances: those with non-empty core, where removing any axiom from it leads to an optimal repair.

It remains to be seen how these results change in the presence of larger ontologies. In particular, the instances considered had a limited number of justifications. We expect that the improvements observed would increase as more justifications are encountered.

References

1. Arif, M.F., Mencía, C., Ignatiev, A., Manthey, N., Peñaloza, R., Marques-Silva, J.: BEACON: an efficient SAT-based tool for debugging $\mathcal{EL}^+$ ontologies. In: Creignou, N., Le Berre, D. (eds.) SAT 2016. LNCS, vol. 9710, pp. 521–530. Springer, Cham (2016). https://doi.org/10.1007/978-3-319-40970-2_32
2. Arif, M.F., Mencía, C., Marques-Silva, J.: Efficient axiom pinpointing with EL2MCS. In: Hölldobler, S., Krötzsch, M., Peñaloza, R., Rudolph, S. (eds.) KI 2015. LNCS (LNAI), vol. 9324, pp. 225–233. Springer, Cham (2015). https://doi.org/10.1007/978-3-319-24489-1_17
3. Arif, M.F., Mencía, C., Marques-Silva, J.: Efficient MUS enumeration of horn formulae with applications to axiom pinpointing. In: Heule, M., Weaver, S. (eds.) SAT 2015. LNCS, vol. 9340, pp. 324–342. Springer, Cham (2015). https://doi.org/10.1007/978-3-319-24318-4_24
4. Baader, F., Brandt, S., Lutz, C.: Pushing the EL envelope further. In: Proceedings of the OWLED 2008 DC Workshop on OWL: Experiences and Directions (2008)
5. Baader, F., Calvanese, D., McGuinness, D.L., Nardi, D., Patel-Schneider, P.F. (eds.): The Description Logic Handbook: Theory, Implementation, and Applications, 2nd edn. Cambridge University Press, Cambridge (2010)
6. Baader, F., Hollunder, B.: Embedding defaults into terminological knowledge representation formalisms. J. Autom. Reason. **14**(1), 149–180 (1995)
7. Baader, F., Peñaloza, R.: Automata-based axiom pinpointing. J. Autom. Reason. **45**(2), 91–129 (2010)
8. Baader, F., Peñaloza, R.: Axiom pinpointing in general tableaux. J. Log. Comput. **20**(1), 5–34 (2010)
9. Baader, F., Peñaloza, R., Suntisrivaraporn, B.: Pinpointing in the description logic $\mathcal{EL}^+$. In: Hertzberg, J., Beetz, M., Englert, R. (eds.) KI 2007. LNCS (LNAI), vol. 4667, pp. 52–67. Springer, Heidelberg (2007). https://doi.org/10.1007/978-3-540-74565-5_7
10. Bail, S.P.: The justificatory structure of OWL ontologies. Ph.D. thesis, University of Manchester, UK (2013)
11. Chang, C.-L., Lee, R.C.-T.: Symbolic Logic and Mechanical Theorem Proving. Academic Press (2014)
12. Chen, J., Ludwig, M., Ma, Y., Walther, D.: Zooming in on ontologies: minimal modules and best excerpts. In: d'Amato, C., et al. (eds.) ISWC 2017. LNCS, vol. 10587, pp. 173–189. Springer, Cham (2017). https://doi.org/10.1007/978-3-319-68288-4_11
13. Chen, J., Ludwig, M., Ma, Y., Walther, D.: Computing minimal projection modules for $\mathcal{ELH}^r$-terminologies. In: Calimeri, F., Leone, N., Manna, M. (eds.) JELIA 2019. LNCS (LNAI), vol. 11468, pp. 355–370. Springer, Cham (2019). https://doi.org/10.1007/978-3-030-19570-0_23
14. Chen, J., Ludwig, M., Walther, D.: On computing minimal EL-subsumption modules. In: Proceedings of the Joint Ontology Workshops 2016 Episode 2: The French Summer of Ontology co-located with the 9th International Conference on Formal Ontology in Information Systems (FOIS 2016). CEUR Workshop Proceedings, Annecy, France, 6–9 July 2016, vol. 1660. CEUR-WS.org (2016)
15. Chen, J., Ludwig, M., Walther, D.: Computing minimal subsumption modules of ontologies. In: Proceedings of GCAI 2018, pp. 41–53 (2018)

16. Chen, J., Ma, Y., Peñaloza, R., Yang, H.: Union and intersection of all justifications (extended abstract). In: Proceedings of the 34th International Workshop on Description Logics, DL 2021. CEUR Workshop Proceedings, vol. 2954. CEUR-WS.org (2021)

17. Cucala, D.T., Grau, B.C., Horrocks, I.: Consequence-based reasoning for description logics with disjunction, inverse roles, number restrictions, and nominals. In: Lang, J. (ed.) Proceedings of the Twenty-Seventh International Joint Conference on Artificial Intelligence, IJCAI 2018, pp. 1970–1976. ijcai.org (2018)

18. Cucala, D.T., Grau, B.C., Horrocks, I.: Sequoia: a consequence based reasoner for SROIQ. In: Simkus, M., Weddell, G.E. (eds.) Proceedings of the 32nd International Workshop on Description Logics. CEUR Workshop Proceedings, vol. 2373. CEUR-WS.org (2019)

19. Glimm, B., Horrocks, I., Motik, B., Stoilos, G., Wang, Z.: HermiT: an OWL 2 reasoner. J. Autom. Reason. **53**(3), 245–269 (2014)

20. Horridge, M., Bechhofer, S.: The OWL API: a Java API for OWL ontologies. Semant. Web **2**(1), 11–21 (2011)

21. Horrocks, I., Kutz, O., Sattler, U.: The even more irresistible SROIQ. In: Doherty, P., Mylopoulos, J., Welty, C.A. (eds.) Proceedings of the Tenth International Conference on Principles of Knowledge Representation and Reasoning, pp. 57–67. AAAI Press (2006)

22. Janota, M., Marques-Silva, J.: cmMUS: a tool for circumscription-based MUS membership testing. In: Delgrande, J.P., Faber, W. (eds.) LPNMR 2011. LNCS (LNAI), vol. 6645, pp. 266–271. Springer, Heidelberg (2011). https://doi.org/10.1007/978-3-642-20895-9_30

23. Kalyanpur, A., Parsia, B., Horridge, M., Sirin, E.: Finding all justifications of OWL DL entailments. In: Aberer, K., et al. (eds.) ASWC/ISWC -2007. LNCS, vol. 4825, pp. 267–280. Springer, Heidelberg (2007). https://doi.org/10.1007/978-3-540-76298-0_20

24. Kalyanpur, A., Parsia, B., Sirin, E., Hendler, J.: Debugging unsatisfiable classes in OWL ontologies. J. Web Semant. **3**(4), 268–293 (2005)

25. Kalyanpur, A.A.: Debugging and repair of OWL ontologies. Ph.D. thesis (2006)

26. Kazakov, Y., Skočovský, P.: Enumerating justifications using resolution. In: Galmiche, D., Schulz, S., Sebastiani, R. (eds.) IJCAR 2018. LNCS (LNAI), vol. 10900, pp. 609–626. Springer, Cham (2018). https://doi.org/10.1007/978-3-319-94205-6_40

27. Koopmann, P., Chen, J.: Deductive module extraction for expressive description logics. In: Bessiere, C. (ed.) Proceedings of IJCAI 2020, pp. 1636–1643. ijcai.org (2020)

28. Liberatore, P.: Redundancy in logic I: CNF propositional formulae. Artif. Intell. **163**(2), 203–232 (2005)

29. Liffiton, M.H., Sakallah, K.A.: On finding all minimally unsatisfiable subformulas. In: Bacchus, F., Walsh, T. (eds.) SAT 2005. LNCS, vol. 3569, pp. 173–186. Springer, Heidelberg (2005). https://doi.org/10.1007/11499107_13

30. Manthey, N., Peñaloza, R., Rudolph, S.: Efficient axiom pinpointing in EL using sat technology. In: Description Logics (2016)

31. Manthey, N., Peñaloza, R., Rudolph, S.: SATPin: axiom pinpointing for lightweight description logics through incremental SAT. Künstliche Intell. **34**(3), 389–394 (2020)

32. Ozaki, A., Peñaloza, R.: Consequence-based axiom pinpointing. In: Ciucci, D., Pasi, G., Vantaggi, B. (eds.) SUM 2018. LNCS (LNAI), vol. 11142, pp. 181–195. Springer, Cham (2018). https://doi.org/10.1007/978-3-030-00461-3_13

33. Parsia, B., Matentzoglu, N., Gonçalves, R.S., Glimm, B., Steigmiller, A.: The OWL reasoner evaluation (ORE) 2015 competition report. J. Autom. Reason. **59**, 455–482 (2015)

34. Parsia, B., Sirin, E., Kalyanpur, A.: Debugging owl ontologies. In: Proceedings of the 14th International Conference on World Wide Web, pp. 633–640 (2005)

35. Peñaloza, R.: Axiom pinpointing. In: Cota, G., Daquino, M., Pozzato, G.L. (eds.) Applications and Practices in Ontology Design, Extraction, and Reasoning. Studies on the Semantic Web, vol. 49, pp. 162–177. IOS Press (2020)

36. Peñaloza, R.: Error-tolerance and error management in lightweight description logics. Künstliche Intell. **34**(4), 491–500 (2020)

37. Peñaloza, R., Mencía, C., Ignatiev, A., Marques-Silva, J.: Lean kernels in description logics. In: Blomqvist, E., Maynard, D., Gangemi, A., Hoekstra, R., Hitzler, P., Hartig, O. (eds.) ESWC 2017. LNCS, vol. 10249, pp. 518–533. Springer, Cham (2017). https://doi.org/10.1007/978-3-319-58068-5_32

38. Penaloza, R., Sertkaya, B.: Understanding the complexity of axiom pinpointing in lightweight description logics. Artif. Intell. **250**, 80–104 (2017)

39. Reiter, R.: A theory of diagnosis from first principles. Artif. Intell. **32**(1), 57–95 (1987)

40. Sattler, U., Schneider, T., Zakharyaschev, M.: Which kind of module should I extract? In: Proceedings of DL 2009. CEUR Workshop Proceedings, vol. 477. CEUR-WS.org (2009)

41. Schlobach, S., Cornet, R.: Non-standard reasoning services for the debugging of description logic terminologies. In: Gottlob, G., Walsh, T. (eds.) Proceedings of the Eighteenth International Joint Conference on Artificial Intelligence, pp. 355–362. Morgan Kaufmann (2003)

42. Sebastiani, R., Vescovi, M.: Axiom pinpointing in lightweight description logics via horn-SAT encoding and conflict analysis. In: Schmidt, R.A. (ed.) CADE 2009. LNCS (LNAI), vol. 5663, pp. 84–99. Springer, Heidelberg (2009). https://doi.org/10.1007/978-3-642-02959-2_6

43. Simancik, F., Kazakov, Y., Horrocks, I.: Consequence-based reasoning beyond horn ontologies. In: Walsh, T. (ed.) Proceedings of the 22nd International Joint Conference on Artificial Intelligence, IJCAI 2011, Barcelona, Catalonia, Spain, 16–22 July 2011, pp. 1093–1098. IJCAI/AAAI (2011)

44. Suntisrivaraporn, B.: Polynomial time reasoning support for design and maintenance of large-scale biomedical ontologies. Ph.D. thesis, Dresden University of Technology, Germany (2009)

45. Suntisrivaraporn, B., Qi, G., Ji, Q., Haase, P.: A modularization-based approach to finding all justifications for OWL DL entailments. In: Domingue, J., Anutariya, C. (eds.) ASWC 2008. LNCS, vol. 5367, pp. 1–15. Springer, Heidelberg (2008). https://doi.org/10.1007/978-3-540-89704-0_1

Supervised Knowledge Aggregation
for Knowledge Graph Completion

Patrick Betz[✉], Christian Meilicke, and Heiner Stuckenschmidt

University of Mannheim, Mannheim, Germany
{patrick,christian,heiner}@informatik.uni-mannheim.de

Abstract. We explore data-driven rule aggregation based on latent feature representations in the context of knowledge graph completion. For a given query and a collection of rules obtained by a symbolic rule learning system, we propose end-to-end trainable aggregation functions for combining the rules into a confidence score answering the query. Despite using latent feature representations for rules, the proposed models remain fully interpretable in terms of the underlying symbolic approach. While our models improve the base learner constantly and achieve competitive results on various benchmark knowledge graphs, we outperform current state-of-the-art with respect to a biomedical knowledge graph by a significant margin. We argue that our approach is in particular well suited for link prediction tasks dealing with a large multi-relational knowledge graph with several million triples, while the queries of interest focus on only one specific target relation.

1 Introduction

Knowledge graphs (KGs) represent structured knowledge in form of (subject, relation, object) facts. As KGs quite frequently suffer from missing data, the goal in link prediction or knowledge graph completion (KGC) is to predict new facts given an incomplete KG. While a vast amount of research is centered around knowledge graph embedding (KGE) models (e.g., [1,34]), rule-based approaches remain competitive in terms of performance [33]. They are fully interpretable as predictions are made by human-understandable symbolic rules and they easily allow for encoding domain knowledge into the overall model pipeline (e.g., [37]).

KGs are heavily used in the biomedical domain [11,18,25] in conjunction with general semantic web technologies to provide large linked data sources and ontologies such as Bio2RDF [2] and Hetionet[16]. These sources can be utilized for downstream tasks such as predictive diagnosis and processing KGs in the biomedical domain can differ from general KGC benchmark datasets. While the KGs may contain a substantial amount of relations, only one particular target relation might be of interest. The challenge is then to exploit the remaining graph context effectively, for instance, guiding models to exploit long-range dependencies while only providing supervision for the target relationships. For example, in the drug repurposing problem on the Hetionet dataset, we seek to

P. Groth et al. (Eds.): ESWC 2022, LNCS 13261, pp. 74–92, 2022.
https://doi.org/10.1007/978-3-031-06981-9_5

find treatable diseases for given compounds by answering queries with respect to the target relationship *Compound-treats-Disease*. The remaining relations in the KG such as *Compound-binds-Gene* or *Disease-associates-Gene* affect the evaluation scheme of the task only by their usefulness in regard to predicting the target relationship correctly.

KGE models are not interpretable, which is an important aspect in general and has even higher relevance in the biomedical domain. They have shown to perform worse when long-range dependencies need to be utilized [19]. Path-based methods, on the other hand, are specifically tailored towards utilizing graph context that exceeds one-hop neighbourhoods and indeed, the neuro-symbolic model PoLo [18] is shown to be effective on the Hetionet KG. By using reinforcement learning agents are trained to perform policy-based walks and are additionally guided by logical rules. However, the model can only process cyclical rules and it has been shown empirically that more specific types, i.e., rules containing constants, are necessary for achieving results on-par with state-of-the-art models [20,21] in the context of KGC.

In this work, we first employ the simple rule learner AnyBURL [20,21]. We mine knowledge in the form of logical rules from the biomedical KG and find that a simple aggregation baseline already outperforms the current state-of-the-art. We then seek to improve the performance further by improving the aggregation of the mined rules. We formulate the problem as data-driven and aim to learn the aggregation from the training data. We propose a novel aggregator with a strong inductive bias, which we call the sparse aggregator, that generalizes the standard aggregation functions by using latent feature representations. We suggest to train the sparse aggregator by using black-box optimization [27,29,32] and directly optimize the mean-rank on the training KG. Furthermore, we propose a simple scaling scheme to reduce the variance of the gradients which improves the performance. Finally, we experiment with a more complex model based on a modified self-attention encoder [10], which we call the dense aggregator.

We find that the sparse aggregator improves the base performance of Any-BURL in all our experimental settings and outperforms current state-of-the-art [18] on the biomedical KG. We also present results on three standard KG benchmarks, where we achieve competitive results. Finally, we demonstrate that the sparse aggregator remains fully interpretable even though it uses latent feature representations.[1]

2   Related Work

A KGE model is specified by a scoring function which outputs raw triple confidences and the models are trained by using likelihood based loss functions such as cross-entropy. A seminal model in the family of translation based models is TransE [4]. RESCAL [26] is based on tensor products and is extended by ComplEx [40] towards expressing asymmetric relations. Many alternative KGE models exist, for instance based on convolutional neural networks or graph-convolutions [9,41].

---

[1] The project repository and code can be found at this URL.

We refer to the studies in [33,46] for a more comprehensive overview. KGE models achieve strong performances in the field of KGC and are efficient to use, however, their predictions are in general hard to interpret which can be of specific interest depending on the respective target domain. This gives rise to symbolic learning where predictions can be pinned to human-understandable rules.

Symbolic rule learning for larger sized KGs is introduced in [13,14] who propose to learn closed connected rules and improved over inductive logic programming systems in terms of scalability and performance. This type of rule learning is outperformed by RuleN [22], the predecessor of the AnyBURL system [20,21], which achieves results that are competitive to the state-of-the-art in the context of KGC. More recent approaches focus on differentiable rule learning [7,45] by representing rules as chained TensorLog [7] operators. An unsupervised approach based on the rulesets from AnyBURL is proposed in [28]. Rules are clustered by calculating the Jaccard index for every possible rule pair. Subsequently, the rules are aggregated using noisy-or aggregation. In the context of multi-modal KGE models, rules are used as features in [15] where feature weights are trained jointly in a product-of-experts scoring function consisting of different modalities.

Symbolic representations in form of rules are also used in the context of neuro-symbolic learning where the logical inference procedure is relaxed into fuzzy-logic formulations and not only the learning but also the application of rules is made differentiable. In [12], forward chaining is formulated to be differentiable and [31] relax backward chaining in Prolog by introducing Neural Theorem Provers which are further improved towards efficiency and flexibility in [23,24]. Please note that these models have not yet shown to be scalable towards KGs with a comparable size as used in this work.

Path-based methods traverse the graph starting from an entity and with sequential transitions to neighbouring nodes. Transitions are made by agents guided by stochastic policies which are learned within a reinforcement learning framework. This is applied to perform triple classification in [44] and extended towards query answering in MINERVA [8]. Finally, PoLo extends these approaches by enabling to inject rules learned by an external system [18] or given from domain knowledge and applies the approach to the drug repurposing problem.

3   Rule-Based Knowledge Graph Completion

3.1   Preliminaries

A KG is a collection of (s, p, o) triples where s and o are entities while p is a relation. The KG forms a graph with relations represented as directed edges from s to o. In the context of KGC s is also called the head of the triple and o is called its tail. Models for KGC are concerned with predicting missing information in the KG, typically by answering queries of the form $(s, p, ?)$ and $(?, p, o)$. These queries are answered by scoring a set of candidates which allows to derive a ranking. Training a model takes place on a training set $\mathcal{K}^{train}$ while a validation set and a test set are used for the evaluation. A common form of evaluating KG models is by calculating ranking based metrics on the respective evaluation set.

The joint mean-rank is the average rank a model assigns to the true candidates when forming queries *(s, p, ?)* and *(?, p, o)* from all triples in the respective evaluation set. The mean reciprocal rank (MRR) takes the inverse $\frac{1}{rank_i}$ of the ranks and therefore lies in $[0, 1]$. The standard procedure is to filter the rankings with known triples from the remaining sets before calculating final ranks [4].

3.2 AnyBURL

AnyBURL [21] is a rule learning approach based on sampling paths from a given knowledge graph. These paths are generalized into rules by replacing constants with variables. The first edge in the path results into the head of the rule and the remaining edges yield the body of the rule. As edges in the graph represent relations a rule is composed of a set of non-grounded or partly grounded triples. AnyBURL is applicable to large datasets and can mine a large number of rules within a short period of time. We abstain from a more detailed description of the learning algorithm and refer to the respective publication. In the following, we briefly review the most relevant rule types AnyBURL learns and demonstrate how they are aggregated to create predictions.

Similar to related approaches, AnyBURL learns closed connected rules [14] also termed cyclic rules [21]. Here cyclic means the head variables X and Y which are connected by the target relation in the head of the rule are also connected via an alternative path represented by the body of the rule. These rules exclusively contain variables and we will give an example from the Hetionet KG in the following. Consider the rule $CtD(X, Y) \leftarrow \_PCiC(X, A) \wedge PCiC(A, B) \wedge CtD(B, Y)$. The rule is mined by AnyBURL and expresses with a certain confidence that a disease Y might be treated (CtD) with compounds which belong to the same pharmacological class (PCiC) as other compounds that are known to treat Y. We define the *confidence* of a rule as the number of body groundings that lead to a true prediction divided by the number of all body groundings estimated on the training data. Further examples for cyclical rules are given by Rule (R1) and (R2), discussed in Sect. 4.1.

Another rule type is given by acyclic rules with only one variable and constants, i.e., entities in the KG, at the remaining slots. AnyBURL is restricted in its default setting, which we used in our experiments, to learn rules of this type with only one body atom. The rule $citizen(X, UK) \leftarrow bornIn(X, London)$ is an example. It expresses that someone born in London is (probably) a UK citizen.

3.3 Knowledge Aggregation

Let us assume we are given a query $(s, p, ?)$,[2] a possible candidate answer c, and the set of rules $R_c := \{r_1, ..., r_k\}$ generating c, hence, all the rules that *fired* for the candidate. Note that R_c depends on the particular query but we do not reference this separately for brevity.

The chosen type of rule aggregation defines how the associated rule confidences $conf(r_1), ..., conf(r_k)$ are aggregated into a score ψ. This score can be

---

[2] All derivations throughout the work are equivalent for the head/subject direction.

interpreted as the standalone confidence for the candidate representing a true answer or it can be used to generate a ranking in regard to multiple candidate answers to $(s, p, ?)$. The default aggregation in AnyBURL is max-aggregation:

$$\psi_{Max}(c) := \max \{conf(r_1), ..., conf(r_k)\}. \tag{1}$$

Potential ties in a ranking are resolved by comparing the respective candidates by their second strongest rule. Another common aggregation technique, which usually performs worse, is noisy-or aggregation:

$$\psi_{NO}(c) := 1 - \prod_{i=1}^{k} \left(1 - conf(r_i)\right). \tag{2}$$

In terms of restrictiveness, noisy-or and max-aggregation are placed on opposite ends of the spectrum which can be seen easily by comparing Eqs. (1) and (2). Whereas only one rule determines the final score in max-aggregation, every single rule contributes when choosing noisy-or aggregation. The former is based on the underlying assumption that all rules are fully dependent, while the latter assumes rules to be mutually independent. Both assumptions are clearly wrong. Moreover, the score of noisy-or increases in the number of rules fired, *e.g.*, it also increases when many rules with low confidences are added to the input rule set. These observation inspire the sparse aggregator presented in Sect. 5.

4   Supervised Knowledge Aggregation

4.1   Challenges

The full version of AnyBURL mines a large number of rules. For example, on the Hetionet KG it learns more than 40k rules for the target relation. Additionally, $|R_c|$ can be large, that is, a single candidate for a query might be generated by hundreds of rules which need to be aggregated. Two main additional characteristics cause the difficulty of the aggregation problem: varying rule set cardinalities and rule redundancy.

Different Rule Set Cardinalities. As mentioned above the number of rules that generate a candidate answer, i.e., $|R_{c_i}|$ for some candidate c_i, can be large. However, this may depend strongly on the given candidate and it is also possible that only a few rules fired. This leads to the natural question of how to compare, for instance, a candidate for which only one strong rule fired to another candidate which was generated by multiple rules with lower confidences.

Rule Redundancies. Many of the mined rules are dependent in the sense that they fire for similar reasons. When an aggregation method does not take into account these redundancies, it will overestimate the final score whenever multiple similar rules generated a candidate. Let us consider the following two rules:

$$speaks(A, B) \leftarrow lives(A, C) \wedge cityOf(C, D) \wedge hasLanguage(D, B) \tag{R1}$$
$$speaks(A, B) \leftarrow lives(A, C) \wedge locatedIn(C, D) \wedge hasLanguage(D, B) \tag{R2}$$

While the rules are not identical, they are partly redundant. The second rule does not provide much more additional evidence if we knew already that the first rule generated a candidate. Noisy-or aggregation would simply treat both rules as independent and overestimate the final score, while max-aggregation would ignore one of the rules which might be the better choice in this example. However, ignoring all rules except of the strongest rule will underestimate the final score whenever a candidate is generated by different rules which represent independent knowledge.

Note that often there is no schema available that contains the knowledge that *cityOf* is more specific than *locatedIn*, nor can this knowledge be derived from the given knowledge graph which is usually incomplete and noisy. Thus, it is not possible to filter out redundant rules. Moreover, in many cases the dependencies are less clear-cut. Think, for example, about replacing the *lives* relation by the *diedIn* relation in Rule (R1) or Rule (R2).

4.2 Supervised Rule Aggregation

We emphasized the challenges of rule aggregation in the last section. The goal of this work is to investigate if a data-driven view that utilizes supervision on the training KG can be beneficial for making a step towards solving the problem. We will introduce a formal perspective in the following paragraph and in the next section the respective aggregation models will be presented.

Given a KG, $\mathcal{K}^{train} = \{(s_i, p_i, o_i)\}_{i=1}^{N}$, let $\mathcal{S}^p$ be the set of rules that were mined for relation p, i.e., all rules mined with relation p in the rule head. Let $\mathcal{P}(\mathcal{S}^p)$ be the power set. Using the definitions from above, for a query $(s, p, ?)$, one possible candidate answer c, and the set of rules R_c that generated the candidate, we have that $R_c \in \mathcal{P}(\mathcal{S}^p)$. For supervised rule aggregation, we seek to learn the parameters Θ of an aggregation function f_Θ with signature $f_\Theta : \mathcal{P}(\mathcal{S}^p) \to \mathbb{R}$ by minimizing a loss criterion $L(\Theta) := \sum L_{q_j}(\Theta)$ which is the sum of all individual losses for the queries q_j that can be formed from $\mathcal{K}^{train}$.

The aggregation function takes as input the subset of rules R_c that generated c and outputs a real-valued score which we interpret as the confidence for the candidate being a true answer. In the definition f_Θ is parameterized globally over relations, however, the question of parameter sharing between relations is a modeling decision and we will argue that it might be beneficial to not share parameters. Furthermore, we will in the following sections inject a strong inductive bias into f by using the rule confidences as an additional input which we omitted for brevity in the formal treatment above.

5 Latent-Based Aggregation

When viewing the problem as data-driven, a possibility would be to proposition-alize the data, i.e., define a binary feature for every rule which is one if the rule fired for a particular candidate and zero otherwise. We could then train a simple discriminative model on $\mathcal{K}^{train}$. However, this model would not take into account rule dependencies and, more importantly, for many relations such as the target relation of the Hetionet KG, the number of rules exceeds the number of triples. Such a model is therefore not applicable as the candidates could be perfectly separated in the feature space without learning anything useful. Therefore, we choose an implicit feature representation over an explicit one by encoding rules with latent representations while assigning a strong inductive bias to our model in form of the rule confidences.

5.1 Sparse Aggregation

The two aggregation techniques introduced in Sect. 3.3 make two opposing assumptions which are both permanently violated in the data as we discussed. In the next paragrahs, we present the *sparse* aggregator which uses more rules for the final confidence calculation than max-aggregation but it uses less rules than noisy-or aggregation while also being able to represent rule redundancies. In fact, when we set the latent input dimensionality to one, we recover max-aggregation wheres a larger dimensionality leads to a behavior closer to noisy-or. Furthermore, our formulation enables to learn the parameters on the KG, i.e., we can utilize the whole training set.

We encode rules with latent feature vectors $\mathbf{x} \in \mathbb{R}^d$ where d is the dimensionality. For a query $(s, p, ?)$, a candidate answer c, and the k rules R_c that generated the candidate, let $\mathbf{x}_j \in \mathbb{R}^d$ be the latent representation of rule r_j and let $\mathbf{v} \in [0, 1]^k$ be the confidence vector of the k rules, i.e., $\mathbf{v}_j = conf(r_j)$. We first normalize the latent vectors with the $\texttt{SoftMax}$ function such that each vector sums to one. Subsequently, we multiply the normalized vectors with the respective rule confidence. Define the normalized vector multiplied with its confidence as

$$\mathbf{x}_j^* := \texttt{SoftMax}(\mathbf{x}_j) \cdot \mathbf{v}_j, \tag{3}$$

where $\mathbf{v}_j \in [0, 1]$ and $\mathbf{x}_j^* \in [0, 1]^d$. Subsequently, we apply the $\texttt{Max}$ operator over the rules, that is, row-wise over the columns of the stacked matrix $\mathbf{X}^* \in [0, 1]^{(d,k)}$

$$\mathbf{s} := \texttt{Max}(\mathbf{X}^*), \tag{4}$$

where $\mathbf{s} \in [0, 1]^d$, i.e., $\mathbf{s}$ has the same dimensionality as each of the latent input vectors. It is important to note that only one single rule can contribute to a single entry of $\mathbf{s}$ due to the $\texttt{Max}$ operator. Finally, we calculate the final score ψ for the candidate c by using the noisy-or product,

$$\psi(c) := 1 - \prod_{i=1}^{d}(1 - \mathbf{s}_i). \tag{5}$$

The derived expression for ψ can be differentiated approximately with respect to the latent features by using automatic differentiation frameworks.[3] We will now discuss some important properties of this aggregation function.

---

[3] The common gradient approximation for the $\texttt{Max}$ function is, e.g., $\nabla max(y_1, y_2) = [1, 0]$ for $y_1 > y_2$ and $[0, 1]$ otherwise.

At Most d Rules Can Contribute to the Final Score. This follows from the fact that the Max function is taken over rules, i.e., for every row of $\mathbf{X}^*$ only one rule is considered and there exist d rows. We mentioned in the previous sections that hundreds of rules can fire for one candidate, with max-aggregation only regarding one rule and noisy-or aggregation using all the rules for the final score. With the sparse aggregator, we can balance this value by setting d accordingly, for instance, for the Hetionet dataset we set $d = 10$.

When We Set d = 1, We Recover Simple Max-Aggregation. To see this, for $d = 1$ we have that $x_j^* = 1 \cdot \mathbf{v_j} = conf(r_j)$, therefore, $s = max\{conf(r_1), ..., conf(r_k)\}$ with $\psi(c) = 1 - 1 - conf(r_{max}) = conf(r_{max})$ where r_{max} is the rule with the highest confidence. This property is ensured by the use of the SoftMax function as it normalizes a single value to 1 which would not be the case for alternative functions such as Sigmoid.

Potentially we can recover noisy-or aggregation by setting $d = |\mathcal{S}^p|$, and enforcing the matrix of all latent vectors to be the identity matrix, however, the goal is to learn to select a small strong subset of rules for a particular query. In the training stage, we aim to learn the latent representations by optimizing a loss criterion which will be discussed in the next section.

5.2   Optimizing Mean-Rank Using Black-Box Optimization

With the sparse aggregator we can easily represent rule redundancies by assigning similar latent features to redundant rules, reducing their joint influence on the score. In general, our model is closer to a discrete model than many machine learning or deep learning models. For instance, the SoftMax normalization restricts the updates of any learning algorithm to be a re-distribution of the rule confidences instead of a clear fitting of the data.

Our choice of supervised machine learning is not for the sake of it. Framing the model as purely discrete would result in an unfeasible search. For instance, only restricting the values to lie within $\{0, 1\}$ (without the SoftMax) would result in a search space with $2^{|\mathcal{S}^p| \cdot d}$ possible configurations for one relation p.

Fortunately, there exists recent work in machine learning that bridges the gap between continuous and discrete problems such as combinatorial optimization [3,27,29]. It has been demonstrated that these approaches can be applied to ranking-based metrics as they can be written as combinatorial optimization problems [32]. Using such a metric suits our problem because the model is not designed to fit the data tightly. Note that also rules with high confidences can fire for candidates which do not exist in the KG. A ranking metric can ignore these possible distortions as soon as the true candidate is ranked relatively high.

To that end, we define the loss criterion $L(\Theta)$ on the training data to be the mean-rank where Θ represents the latent features. For a given query $q = (s, p, ?)$ on the training data, let ψ be the vector of scores calculated according to Eq. (5) with $\psi(c_*)$ being the score for the true candidate c_*. The remaining scores belong to query candidates c_i that were generated by at least one rule and (s, p, c_i) does not appear in $\mathcal{K}^{train}$, that is, we exclusively filter with the training set. Note that conceptually these candidates can also be viewed as *negative* examples. We treat the maximum number of these negative candidates, sorted according to their ranking in max-aggregation, as a hyperparameter denoted by *top-n*. Moreover, let $\boldsymbol{rk}(\boldsymbol{\psi})$ be the vector of ranks and define rk_{c*} to be the rank of the true candidate. The mean-rank of the training data is simply the mean of ranks of the true candidates for the individual queries, therefore, the individual query loss is $L_q(\boldsymbol{rk}; \Theta) = rk_{c*}$. For applying gradient-based optimization, we need to calculate $\frac{dL_q}{d\Theta}$. From the chain rule, we obtain

$$\frac{dL_q}{d\Theta} = \frac{d\psi}{d\Theta} \frac{dL_q}{d\psi}, \tag{6}$$

where $\frac{d\psi}{d\Theta}$ can directly be calculated using auto-differentiation libraries as mentioned in the previous section. For $\frac{dL_q}{d\psi}$ we use black-box differentiation according to [32] and calculate

$$\frac{dL_q}{d\psi} = -\frac{1}{\lambda}\left[\boldsymbol{rk}(\boldsymbol{\psi}) - \boldsymbol{rk}\left(\boldsymbol{\psi} + \lambda \cdot \frac{dL_q}{d\boldsymbol{rk}}\right)\right], \tag{7}$$

where $\frac{dL_q}{d\boldsymbol{rk}}$ is a vector which is one at the entry of the true candidate and zero otherwise which follows from the definition of the query loss above. In practice during the forward pass, we obtain the scores ψ for the query and calculate the ranks. Subsequently, in the backward pass, the scores are perturbed to $\psi' = \psi + \lambda \cdot \frac{dL_q}{d\boldsymbol{rk}}$ and the new ranks are calculated leading to (7).

Consider the inner expression $\boldsymbol{rk}(\boldsymbol{\psi}) - \boldsymbol{rk}(\boldsymbol{\psi}')$ in Eq. (7) and let us only focus on the entry of the true candidate. Let's assume its rank is #50 before and #1 after the perturbation. Then we obtain $-\frac{49}{\lambda}$ which is in absolute terms magnitudes stronger compared to a case where the true candidate only improved by a few ranks. We seek to reduce this variance by scaling the gradient in accordance to the true candidate, in particular we compute

$$\frac{dL_q}{d\psi} = -\frac{1}{\lambda}\left[\boldsymbol{rk}(\boldsymbol{\psi}) - \boldsymbol{rk}(\boldsymbol{\psi}')\right]\frac{1}{rk_{c*} - 1}, \tag{8}$$

that is, we track the rank of the true candiate and scale the gradient with a proportional factor. This ensures a constant signal strength independent of the original position, for instance, in the example above we obtain $-\frac{49}{\lambda}\frac{1}{50-1} = -\frac{1}{\lambda}$.

5.3 Dense Aggregation

In the previous sections, we presented a data-driven perspective on knowledge aggregation and proposed a model based on latent features which has a strong inductive bias. We can relax this restriction and exploit the fact that language model architectures (e.g. [10,17,42]) are well calibrated for processing latent representations. In particular, we use a self-attention based architecture [10] that processes the input representations and outputs real-valued query confidences.

When dropping positional encodings, self-attention based architectures can be applied off-the-shelf on item set problems with varying input lengths.

Precisely, we use a slightly modified version of the PyTorch implementation of the BERT encoder [10]. The latent inputs $\{x_1, ..., x_k\}$ are fed into the encoder receiving the hidden representations $\{h_1, ..., h_k\}$ with same dimensionality. The modification that we implement affects how the self-attention is applied. We sort the inputs according to the rule confidences and we let the j'th rule only attend to the $j - 1$ rules with higher confidence. This reduces the possible distraction that can be caused from many weak input rules towards the high-confident rules. For the aggregation, we use simple average pooling on the hidden representations and feed the resulting vector into a fully connected linear layer which outputs one final score. The aggregator is termed *dense* aggregator as every rule in the input set contributes to the final score.

The model is expensive to train and, on average, it is inferior to the sparse aggregator in terms of performance although we were not able to explore a large part of the hyperparameter space due to runtime considerations. The model is also not practical in more general terms, however, we are merely interested in the question if this model can learn specific aspects of the data which are hidden from the other models. To investigate this, we evaluate a joint model of the sparse and dense aggregators. We tune weights β_{sparse} and $\beta_{dense} = 1 - \beta_{sparse}$ per relation and direction on the validation set for discovering potential differences between the two models. To make sure that these differences are significant, we restrict β_{sparse} to lie in $\{0,1\}$. This setting will be denoted by D+S in the experimental section. It is not applicable to Hetionet where only one target relation exists.

6 Experiments

In the following we describe experiments for a specific biomedical KG and three benchmark KGs commonly used in the field of KGC. We are mainly interested in the question how our learnable aggregation compares to baseline aggregation functions and to other models that have a similar degree of explainability.

6.1 Datasets

The Hetionet network combines knowledge from a large amount of biomedical studies into a KG containing 47k typed entities and 2.24 million triples with 24 possible relations [16]. We focus on the task of drug repurposing, i.e., finding new use cases for existing compounds. In the Hetionet KG this is expressed by answering queries of the form $CtD(X, ?)$ where the relation CtD means *Compound-treats-Disease*. It is discussed in [18] that Hetionet significantly differs from other benchmark KG's, for instance exhibiting a higher average node degree. Moreover, while the dataset is of considerable size, the target relationship appears in only 755 facts which highlights the challenge of exploiting graph context without having strong supervision. We use the train, valid, and test splits according to the public documentation of [18].

Table 1. Summary statistics.

Dataset	Entities	Relations	Triples
Hetionet	47,031	24	2,250,197
FB15k-237	14,505	237	310,116
WNRR	40,559	11	93,003
CoDEx-M	17,050	51	206,205

We further evaluate our approach on three general benchmark KGs. FB15k-237 [39] and WNRR [9] are frequently used in the field of KGC and Codex-M was designed with the goal to be more challenging than previous benchmarks [36]. For these datasets we use the common train, valid, and test splits. Table 1 shows summary statistics for the KGs.

6.2 Experimental Settings

For the Hetionet KG we mostly focus on the comparisons in [18]. That is, we compare to the interpretable models PoLo [18], MINERVA [8], and pLogicNet [30] and we also include the KGE results presented in [18] for the models TransE [4], ComplEx [40], ConvE [9] and RESCAL [26]. Furthermore, we include the HittER [6] no-context model implementation of the libKGE library for which we run the hyperparameter search provided by the library with 15 trials. We also include the RotatE [38] results from the TorchDrug library.

For the remaining KGs we additionally add the rule-based methods DRUM [35] and Neural-LP [45] if results are available. We abstain from a comprehensive comparison against KGE but we include the results from the official libKGE library and the underlying work about training KGE models [5,34] to put our approach in the context of strong KGE models trained under a well-tested and unified codebase. Moreover, the results for HittER$_{nc}$ are based on [19] and we also add the results of the recent model M^2GNN proposed in [43].

For all the datasets, we mine rules on the training splits of the KGs using AnyBURL and subsequently we learn the aggregation functions on the same training sets while utilizing the valid sets for hyperparameter search and early stopping. We train the sparse aggregator on the mean-rank on the training data using the scaled gradients. The dense aggregator is trained on a standard cross-entropy loss, i.e., maximizing the likelihood of the training data while using negative examples. The sparse aggregator does not share parameters between relations, that is, queries for different relations can be treated independently and hyperparameters could be searched relation-wise. For the sake of simplicity, we train the model globally, however, we save checkpoints per epoch and pick for every relation in head and tail direction the checkpoint that results in the highest MRR on the valid set. For the Hetionet KG this is not necessary as only one target relation exists and only tails are predicted. Hyperparameter configurations and further training details can be found in Appendix A.

Table 2. Filtered MRR in tail direction for Hetionet. The results for the learnable aggregators are averages over 5 runs. The first, middle and last part represents embedding-based models, interpretable models, and the results of this work, respectively.

Approach	h@1	h@3	h@10	MRR
TransE [4]	0.099	0.199	0.444	0.205
ComplEx [40]	0.152	0.285	0.470	0.250
ConvE [9]	0.100	0.225	0.318	0.180
RESCAL [26]	0.106	0.166	0.377	0.187
HittER$_{nc}$	0.316	0.517	0.740	0.453
RotatE	0.185	0.282	0.403	0.257
CompGCN [41]	0.172	0.318	0.543	0.292
pLogicNet [30]	0.225	0.364	0.523	0.333
MINERVA[8]	0.264	0.409	0.593	0.370
PoLo [18]	0.314	0.428	0.609	0.402
PoLo (pruned)	0.337	0.470	0.641	0.430
Max	0.272	0.444	0.642	0.398
Noisy-or	0.377	0.509	0.642	0.472
Dense	0.306	0.514	0.701	0.441
Sparse	0.380	0.525	0.694	0.490

6.3   Results

Table 2 shows results on the test set for the filtered MRR and filtered hits@k in tail direction for the drug repurposing problem and Table 3 shows results for the test sets of the joint MRR and joint hits@k on the remaining datasets.

On the Hetionet dataset, the sparse aggregator improves the recent state-of-the-art [18] of the interpretable models by 4.3, 5.5, 5.3, and 6% points for the metrics hits@1, hits@3, hits@10, and MRR, respectively. Interestingly, the noisy-or baseline already outperforms the state-of-the-art by a significant margin. Noteworthy, the sparse aggregator improves AnyBURL's max-aggregation by 9.2% points for the MRR and shows improvements of 1.8% points over noisy-or. The dense aggregator, on the other hand, performs significantly worse except for the hits@10 metric. Despite outperforming the previous models, it does not improve over the noisy-or baseline.

For the remaining datasets the sparse aggregator outperforms the interpretable methods and improves over the best AnyBURL baseline (note that noisy-or performs rather poor on these datasets) in all settings although the improvement is only marginal for the WNRR dataset. However, improvements of 2.3% points on FB15k-237 and 2% points on Codex-m are considered to be quite substantial in the KGC literature. Interestingly, the setting D+S which is explained in Sect. 5.3 achieves strong results on WNRR and Codex-M. This suggests that, despite being inferior overall, the dense aggregator captured some aspects of the data which are hidden from the sparse aggregator. We investigated this further and found that the dense aggregator is sensitive to *negative* signals.

Table 3. Results for FB15k-237, WNRR and Codex-M for the joint filtered MRR. The first, middle and last part of the table represents embedding-based models, interpretable models, and the results of this work, respectively. The results for *Dense* and *Sparse* are averages over 3 runs.

Approach	FB15k-237			WNRR			Codex-M		
	h@1	h@10	MRR	h@1	h@10	MRR	h@1	h@10	MRR
TransE	0.221	0.497	0.312	0.053	0.520	0.228	0.223	0.454	0.303
ComplEx	0.253	0.536	0.347	0.438	0.547	0.475	0.262	0.476	0.337
ConvE	0.248	0.521	0.338	0.411	0.505	0.442	0.239	0.464	0.318
RESCAL	0.263	0.541	0.355	0.439	0.517	0.467	0.244	0.456	0.317
HittER$_{nc}$	0.268	0.549	0.361	0.437	0.531	0.469	0.262	0.486	0.339
RotatE	0.240	0.522	0.333	0.439	0.553	0.478	–	–	–
M^2GNN	0.275	0.565	0.362	0.444	0.572	0.485	–	–	–
pLogicNet	0.237	0.524	0.332	0.398	0.537	0.441	–	–	–
MINERVA	0.217	0.456	0.293	0.413	0.513	0.448	–	–	–
DRUM [35]	0.255	0.516	0.343	0.425	0.586	0.486	–	–	–
Neural-LP [45]	–	0.362	0.240	0.371	0.566	0.435	–	–	–
Max	0.246	0.506	0.331	0.457	0.572	0.497	0.247	0.450	0.316
Noisy-or	0.247	0.494	0.329	0.391	0.559	0.446	0.218	0.427	0.289
Dense	0.245	0.510	0.335	0.466	0.587	0.507	0.261	0.465	0.331
Sparse	0.266	0.526	0.352	0.459	0.574	0.499	0.266	0.467	0.335
D+S	0.267	0.527	0.354	0.469	0.593	0.511	0.273	0.476	0.342

For instance, when adding rules with low confidences to an input set, the score of the dense aggregator might decrease. This behavior cannot be expressed by the sparse aggregator or the presented baselines which might open up interesting further directions.

Finally, Figs. 1 and 2 compare the sparse aggregator under different training settings. Figure 1 shows test results for five runs on Hetionet when using the scaled gradients proposed in Sect. 5 and the default formulation. Training with the scaled gradients leads to lower variance and higher average performance. Figure 2 compares mean-rank training with using a standard cross-entropy loss on Codex-M.

7   Interpretability

In the following we discuss an example for a query where the sparse aggregator generates a ranking that differs significantly from the ranking suggested by the rule with the highest confidence. Moreover, in contrast to noisy-or where all rules that fired contribute to the score, by setting $d = 10$ our model selects a compact subset of 10 rules of which we can further pick the ones with the highest impact. This procedure also resembles how a potential user can interact with the aggregation system. We use an example with short rules for the sake of simplicity.

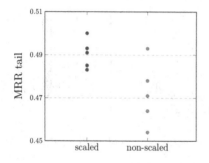

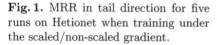

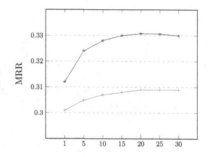

Fig. 1. MRR in tail direction for five runs on Hetionet when training under the scaled/non-scaled gradient.

Fig. 2. Valid MRR per epoch on Codex-M for Mean-Rank training ($\star$) and ordinary Cross-Entropy loss ($+$).

The target query asks for new diseases to which the compound *Methotrexate* (DB00563) can be applied. Our method ranks the correct answer *systemic lupus erythematosus* first, an autoimmune disease in which the body's immune system mistakenly attacks healthy tissue in many parts of the body.

To illustrate this example, we have chosen the eight rules with the highest confidences and depicted their normalized values for each of the ten dimensions side by side on the left part of Fig. 3. Different rules can be distinguished by their color and position within the group of bars that reflects the value of a specific dimension. Note that the most confident rule that created a candidate for the query had a confidence of 0.357, while the #1 candidate proposed by our method received a score of 0.806. This is caused by the fact that the rules that recited this candidate differ significantly with respect to their latent features. There are several dimension (in particular 5,6, 8 and 9), where we learned a significantly higher value for that specific rule compared to all other rules. This means at the same time that a large fraction of the overall power of this rule is assumed to be independent from the other rules and increases the overall score. If the rules had similar values in most of the dimensions, the resulting score would be close to the score of the maximum strategy.

A domain expert, who wants to understand why a certain candidate is ranked at #1, is probably interested in a small set of rules that had the highest influence on the prediction. These are obviously the rules that dominate in some of the dimensions. To quantify their contribution we computed for each rule the sum of values over all dimensions taking only those dimension into account where the rule received the maximal value compared to the other rules. The resulting scores are depicted on the right part of Fig. 3. This means that rules r_{229} and r_{264} are the most important two rules. Let us take a closer look at these rules and their meaning:

r_{229} `CtD(X,Disease::DOID:9074) <= CtD(X,Disease::DOID:7148)` - If a compound can be used as treatment for *rheumatoid arthritis* (DOID:7148), it might also be applied to treat the disease *lupus erythematosus* (DOID:9074).

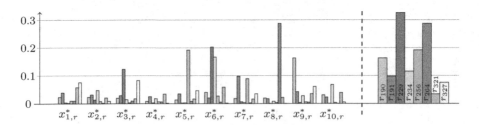

Fig. 3. Rule features when searching new treatments for *Methotrexate*.

Since *Methotrexate* is known to been used for *rheumatoid arthritis, lupus erythematosus* is predicted by the rule.

r_{264} CtD(Compound::DB00563,Y) <= D1A(Y,Anatomy::UBERON:0000043) - If a disease is diagnosed to affect a tendon (UBERON:0000043), that disease might be treated with *Methotrexate* (DB0056). Since *Methotrexate* is known to been used in such a context, it is predicted by the rule.

These two rules fire for different reasons. The first rule is based on observing that two different ailments which both correspond to autoimmune diseases are likely treatable by the same compound. The second rule focuses on a specific body part, the tendon, which – if afflicted by a disease – can often be treated with *Methotrexate*. In almost all cases tendons are either ruptured (no drug required) or inflamed which makes them treatable by *Methotrexate*. Figure 3 shows that our approach is capable to learn that these two rules are not redundant. It increases the score of a candidate that is proposed by both rules, which is in our case *lupus erythematosus*, a disease which can affect tendons and for which compounds have been applied that have also been applied to *rheumatoid arthritis*.

8   Conclusion

We showed empirically that simple rule learning approaches achieve strong results on the task of drug repurposing. We presented learnable knowledge aggregation in form of latent rule aggregation. To our best knowledge, this is a novel approach that differs fundamentally from symbolic and latent approaches proposed so far for KGC. In particular, we presented an aggregation function, the sparse aggregator, which can be learned on the training set and we proposed to employ black-box optimization. We found empirically that the sparse aggregator improves over baseline aggregation techniques. It is on-par with standard KGE methods and is state-of-the-art on the Hetionet dataset, while still maintaining interpretability. The sparse model learns how to aggregate rules as positive evidence, however, it is not capable to learn that a rule or a combination of rules makes a prediction less likely. It might thus be beneficial to incorporate negative evidence into the model which opens up directions for future research.

A   Experimental Details

A.1   Model Input

On the highest abstraction level, our models take as input a list of rules and output a real-valued score. More precisely, for a query $q = (s, p, ?)$ (same in head direction) we collect the top_n answer candidates c_i proposed by AnyBURL, that is, the candidates that were generated by at least one rule. For each of these candidates the respective list of rules defines the model input. Then, the descriptions of the main text apply. Finally, we obtain a vector of scores and likewise a ranking in regard to all candidate answers c_i. At test time, this ranking can directly be used for the evaluation. At training time, we distinguish the true answer/candidate c^* and the remaining candidates c' which we filter with the training set, i.e., we exclude a $c' \neq c^*$ if a triple (s, p, c') exists in train. For a ranking loss we can now calculate the query loss of q as explained in Sect. 5.2. For some arbitrary loss function such as cross-entropy, c^* defines the true candidate and the remaining candidates c' define the reference candidates or pseudo negative candidates.

A.2   Hyperparameters

For all the experiments, we use a max $top\text{-}n = 100$, the Adagrad optimizer, and a batch-size of 256. Training is performed by using early stopping based on the validation set. LibKGE based configuration files for the experiments are provided in the supplementary materials.

Sparse Aggregator. The hyperparameters that we are concerned with are dropout on the latent features, the latent dimension d, and the learning rate lr. For Hetionet we set $d = 10$, dropout $= 0.15$ and $lr = 0.9$. On Fb15k-237 we set $d = 40$, dropout $= 0.4$, and $lr = 0.02$. For WNRR we set $d = 50$, dropout $= 0.4$ and $lr = 0.03$. For Codex-m we set $d = 40$, dropout $= 0.4$ and $lr = 0.02$. For all the experiments we use a value of 5 for lambda when training on the mean-rank loss.

Dense Aggregator. The dense aggregator follows in its architecture the PyTorch BERT encoder with the modification as explained in Sect. 5.3. We use 4 heads and 4 layers throughout all the experiments. The feed-forward dimensionality within the encoder is 256. For Hetionet we use $d = 20$, dropout $= 0.15$ and $lr = 0.01$. For the remaining datasets we use $d = 56$, dropout $= 0.15$, and $lr = 0.005$. We set the maximal number of rules per input list to 50 for the dense aggregator for all the experiments.

A.3   Rule Sets

The base data for our experiments are the rules learned with AnyBURL. These are processed in a pre-processing pipeline to generate the inputs for the aggregators as explained above. For all the datasets we exclude AC2 rules and rules with an empty body. This leaves the AnyBURL performance mostly unchanged but we report newest AnyBURL results reported by the authors.

For WNRR rules are mined for 3600 s and we set the maximum length for cyclical rules equal to 5 as suggested in the AnyBURL documentation. All the learned rules are processed for training the models on this dataset. For the remaining datasets the default AnyBURL parameters are used. Here, we prune the learned rulesets slightly and only process rules that had at least 5 (10) true predictions for sparse (dense). On Hetionet rules are learned for 1000 s for dense and sparse. On Fb15k-237 rules are learned for 3600 (500) s for sparse (dense). Finally, on Codex-M rules are learend for 1000 (500) s for sparse (dense).

References

1. Ali, M., et al.: Bringing light into the dark: a large-scale evaluation of knowledge graph embedding models under a unified framework. IEEE Trans. Pattern Anal. Mach. Intell., 1–1 (2021). https://doi.org/10.1109/TPAMI.2021.3124805
2. Belleau, F., Nolin, M.A., Tourigny, N., Rigault, P., Morissette, J.: Bio2RDF: towards a mashup to build bioinformatics knowledge systems. J. Biomed. Inform. **41**(5), 706–716 (2008)
3. Betz, P., Niepert, M., Minervini, P., Stuckenschmidt, H.: Backpropagating through Markov logic networks. In: Proceedings of 15th International Workshop on Neural-Symbolic Learning and Reasoning, vol. 2986, pp. 67–81. CEUR (2021)
4. Bordes, A., Usunier, N., Garcia-Duran, A., Weston, J., Yakhnenko, O.: Translating embeddings for modeling multi-relational data. In: Neural Information Processing Systems (NIPS), pp. 1–9 (2013)
5. Broscheit, S., Ruffinelli, D., Kochsiek, A., Betz, P., Gemulla, R.: LibKGE-a knowledge graph embedding library for reproducible research. In: Proceedings of the 2020 Conference on Empirical Methods in Natural Language Processing: System Demonstrations, pp. 165–174 (2020)
6. Chen, S., Liu, X., Gao, J., Jiao, J., Zhang, R., Ji, Y.: Hitter: hierarchical transformers for knowledge graph embeddings. In: Proceedings of the 2021 Conference on Empirical Methods in Natural Language Processing (EMNLP) (2020)
7. Cohen, W., Yang, F., Mazaitis, K.R.: TensorLog: a probabilistic database implemented using deep-learning infrastructure. J. Artif. Intell. Res. **67**, 285–325 (2020)
8. Das, R., et al.: Go for a walk and arrive at the answer: Reasoning over paths in knowledge bases using reinforcement learning. arXiv preprint arXiv:1711.05851 (2017)
9. Dettmers, T., Minervini, P., Stenetorp, P., Riedel, S.: Convolutional 2D knowledge graph embeddings. In: Proceedings of the AAAI Conference on Artificial Intelligence, vol. 32, pp. 1811–1818 (2018)
10. Devlin, J., Chang, M.W., Lee, K., Toutanova, K.: BERT: pre-training of deep bidirectional transformers for language understanding. In: Proceedings of the 2019 Conference of the North American Chapter of the Association for Computational Linguistics: Human Language Technologies (Long and Short Papers), vol. 1, June 2019
11. Dörpinghaus, J., Jacobs, M.: Semantic knowledge graph embeddings for biomedical research: data integration using linked open data. In: SEMANTICS Posters&Demos (2019)

12. Evans, R., Grefenstette, E.: Learning explanatory rules from noisy data. J. Artif. Intell. Res. **61**, 1–64 (2018)
13. Galárraga, L., Teflioudi, C., Hose, K., Suchanek, F.M.: Fast rule mining in ontological knowledge bases with AMIE. VLDB J. **24**(6), 707–730 (2015)
14. Galárraga, L.A., Teflioudi, C., Hose, K., Suchanek, F.: AMIE: association rule mining under incomplete evidence in ontological knowledge bases. In: Proceedings of the 22nd International Conference on World Wide Web, pp. 413–422 (2013)
15. García-Durán, A., Niepert, M.: KBLRN: end-to-end learning of knowledge base representations with latent, relational, and numerical features. UAI (2018)
16. Himmelstein, D.S., et al.: Systematic integration of biomedical knowledge prioritizes drugs for repurposing. Elife **6**, e26726 (2017)
17. Hochreiter, S., Schmidhuber, J.: Long short-term memory. In: Neural Computation, vol. 9, pp. 1735–1780. MIT Press (1997)
18. Liu, Y., Hildebrandt, M., Joblin, M., Ringsquandl, M., Raissouni, R., Tresp, V.: Neural multi-hop reasoning with logical rules on biomedical knowledge graphs. In: Verborgh, R., et al. (eds.) ESWC 2021. LNCS, vol. 12731, pp. 375–391. Springer, Cham (2021). https://doi.org/10.1007/978-3-030-77385-4_22
19. Meilicke, C., Betz, P., Stuckenschmidt, H.: Why a naive way to combine symbolic and latent knowledge base completion works surprisingly well. In: 3rd Conference on Automated Knowledge Base Construction (2021)
20. Meilicke, C., Chekol, M.W., Fink, M., Stuckenschmidt, H.: Reinforced anytime bottom up rule learning for knowledge graph completion (2020)
21. Meilicke, C., Chekol, M.W., Ruffinelli, D., Stuckenschmidt, H.: Anytime bottom-up rule learning for knowledge graph completion. In: Proceedings of the 28th International Joint Conference on Artificial Intelligence (IJCAI). IJCAI/AAAI Press (2019)
22. Meilicke, C., Fink, M., Wang, Y., Ruffinelli, D., Gemulla, R., Stuckenschmidt, H.: Fine-grained evaluation of rule- and embedding-based systems for knowledge graph completion. In: Vrandečić, D., et al. (eds.) ISWC 2018. LNCS, vol. 11136, pp. 3–20. Springer, Cham (2018). https://doi.org/10.1007/978-3-030-00671-6_1
23. Minervini, P., Bošnjak, M., Rocktäschel, T., Riedel, S., Grefenstette, E.: Differentiable reasoning on large knowledge bases and natural language. In: Proceedings of the AAAI Conference on Artificial Intelligence, vol. 34, pp. 5182–5190 (2020)
24. Minervini, P., Riedel, S., Stenetorp, P., Grefenstette, E., Rocktäschel, T.: Learning reasoning strategies in end-to-end differentiable proving. In: International Conference on Machine Learning, pp. 6938–6949. PMLR (2020)
25. Mohamed, S.K., Nounu, A., Nováček, V.: Drug target discovery using knowledge graph embeddings. In: Proceedings of the 34th ACM/SIGAPP Symposium on Applied Computing, pp. 11–18 (2019)
26. Nickel, M., Tresp, V., Kriegel, H.: A three-way model for collective learning on multi-relational data. In: Getoor, L., Scheffer, T. (eds.) Proceedings of the 28th International Conference on Machine Learning, pp. 809–816. Omnipress (2011)
27. Niepert, M., Minervini, P., Franceschi, L.: Implicit MLE: backpropagating through discrete exponential family distributions. In: NeurIPS (2021)
28. Ott, S., Graf, L., Agibetov, A., Meilicke, C., Samwald, M.: Scalable and interpretable rule-based link prediction for large heterogeneous knowledge graphs (2020)
29. Pogančić, M.V., Paulus, A., Musil, V., Martius, G., Rolinek, M.: Differentiation of blackbox combinatorial solvers. In: International Conference on Learning Representations (2020)
30. Qu, M., Tang, J.: Probabilistic logic neural networks for reasoning. In: International Conference on Learning Representations (2020)

31. Rocktäschel, T., Riedel, S.: End-to-end differentiable proving. In: Advances in Neural Information Processing Systems 30: Annual Conference on Neural Information Processing Systems 2017, pp. 3788–3800 (2017)
32. Rolínek, M., Musil, V., Paulus, A., Vlastelica, M., Michaelis, C., Martius, G.: Optimizing rank-based metrics with blackbox differentiation. In: Proceedings of the IEEE/CVF Conference on Computer Vision and Pattern Recognition, pp. 7620–7630 (2020)
33. Rossi, A., Barbosa, D., Firmani, D., Matinata, A., Merialdo, P.: Knowledge graph embedding for link prediction: a comparative analysis. ACM Trans. Knowl. Discov. Data (TKDD) **15**(2), 1–49 (2021)
34. Ruffinelli, D., Broscheit, S., Gemulla, R.: You CAN teach an old dog new tricks! on training knowledge graph embeddings. In: International Conference on Learning Representations (2020)
35. Sadeghian, A., Armandpour, M., Ding, P., Wang, D.Z.: DRUM: end-to-end differentiable rule mining on knowledge graphs. In: Advances in Neural Information Processing Systems 32: Annual Conference on Neural Information Processing Systems, NeurIPS 2019, Vancouver, BC, Canada, pp. 15321–15331 (2019)
36. Safavi, T., Koutra, D.: CoDEx: a comprehensive knowledge graph completion benchmark. In: Proceedings of the 2020 Conference on Empirical Methods in Natural Language Processing (EMNLP), pp. 8328–8350. Association for Computational Linguistics, November 2020
37. Sola, D., Meilicke, C., van der Aa, H., Stuckenschmidt, H.: A rule-based recommendation approach for business process modeling. In: La Rosa, M., Sadiq, S., Teniente, E. (eds.) CAiSE 2021. LNCS, vol. 12751, pp. 328–343. Springer, Cham (2021). https://doi.org/10.1007/978-3-030-79382-1_20
38. Sun, Z., Deng, Z.H., Nie, J.Y., Tang, J.: Rotate: knowledge graph embedding by relational rotation in complex space. In: International Conference on Learning Representations (2019)
39. Toutanova, K., Chen, D.: Observed versus latent features for knowledge base and text inference. In: Proceedings of the 3rd Workshop on Continuous Vector Space Models and Their Compositionality, pp. 57–66 (2015)
40. Trouillon, T., Welbl, J., Riedel, S., Gaussier, É., Bouchard, G.: Complex embeddings for simple link prediction. In: Balcan, M., Weinberger, K.Q. (eds.) Proceedings of the 33nd International Conference on Machine Learning. JMLR Workshop and Conference Proceedings, vol. 48, pp. 2071–2080. JMLR.org (2016)
41. Vashishth, S., Sanyal, S., Nitin, V., Talukdar, P.: Composition-based multi-relational graph convolutional networks. In: International Conference on Learning Representations (2020)
42. Vaswani, A., et al.: Attention is all you need. In: Advances in Neural Information Processing Systems, pp. 6000–6010 (2017)
43. Wang, S., et al.: Mixed-curvature multi-relational graph neural network for knowledge graph completion. In: Proceedings of the Web Conference 2021, pp. 1761–1771 (2021)
44. Xiong, W., Hoang, T., Wang, W.Y.: DeepPath: a reinforcement learning method for knowledge graph reasoning. arXiv preprint arXiv:1707.06690 (2017)
45. Yang, F., Yang, Z., Cohen, W.W.: Differentiable learning of logical rules for knowledge base reasoning. In: Advances in Neural Information Processing Systems 30: Annual Conference on Neural Information Processing Systems, NeurIPS 2017, Long Beach, US (2017)
46. Zhang, J., Chen, B., Zhang, L., Ke, X., Ding, H.: Neural, symbolic and neural-symbolic reasoning on knowledge graphs. AI Open **2**, 14–35 (2021)

Expressive Scene Graph Generation Using Commonsense Knowledge Infusion for Visual Understanding and Reasoning

Muhammad Jaleed Khan[1]([⊠]) [iD], John G. Breslin[1,2] [iD], and Edward Curry[1,2] [iD]

[1] SFI Centre for Research Training in Artificial Intelligence, Data Science Institute, National University of Ireland, Galway, Galway, Ireland
{m.khan12,john.breslin,edward.curry}@nuigalway.ie
[2] Insight SFI Research Centre for Data Analytics, Data Science Institute, National University of Ireland, Galway, Galway, Ireland

Abstract. Scene graph generation aims to capture the semantic elements in images by modelling objects and their relationships in a structured manner, which are essential for visual understanding and reasoning tasks including image captioning, visual question answering, multimedia event processing, visual storytelling and image retrieval. The existing scene graph generation approaches provide limited performance and expressiveness for higher-level visual understanding and reasoning. This challenge can be mitigated by leveraging commonsense knowledge, such as related facts and background knowledge, about the semantic elements in scene graphs. In this paper, we propose the infusion of diverse commonsense knowledge about the semantic elements in scene graphs to generate rich and expressive scene graphs using a heterogeneous knowledge source that contains commonsense knowledge consolidated from seven different knowledge bases. The graph embeddings of the object nodes are used to leverage their structural patterns in the knowledge source to compute similarity metrics for graph refinement and enrichment. We performed experimental and comparative analysis on the benchmark Visual Genome dataset, in which the proposed method achieved a higher recall rate ($R@K = 29.89, 35.4, 39.12$ for $K = 20, 50, 100$) as compared to the existing state-of-the-art technique ($R@K = 25.8, 33.3, 37.8$ for $K = 20, 50, 100$). The qualitative results of the proposed method in a downstream task of image generation showed that more realistic images are generated using the commonsense knowledge-based scene graphs. These results depict the effectiveness of commonsense knowledge infusion in improving the performance and expressiveness of scene graph generation for visual understanding and reasoning tasks.

Keywords: scene graph · image representation · commonsense knowledge · visual reasoning · image generation

This publication has emanated from research conducted with the financial support of Science Foundation Ireland under Grant number 18/CRT/6223 and 12/RC/2289_P2. For the purpose of Open Access, the author has applied a CC BY public copyright licence to any Author Accepted Manuscript version arising from this submission.

1 Introduction

During the past few years, recent advances in deep learning techniques and multi-modal approaches have helped in solving several challenging problems in visual understanding tasks including object detection [57] and visual relationship detection [14,32,35]. Numerous efforts have been made to effectively capture and describe the image features and object relationships in a structured and explicit way. In this direction, Scene Graph Generation (SGG) [3,46,48] has attracted significant attention due to its capability to capture the detailed semantics of visual scenes by modelling objects and their relationships in a structured manner. Graph-based structured image representations like scene graphs are used in a wide range of visual understanding tasks including image reconstruction [11], image captioning [61], Visual Question Answering (VQA) [22,25], image retrieval [55], visual storytelling [54] and multimedia event processing [5,20]. The performance of SGG is compromised by challenges including bias and annotation issues in crowd-sourced datasets [7,23]. Several efforts have been made by researchers in this field to address these challenges by making use of state-of-the-art approaches, such as counterfactual analysis [48], self-supervised learning [40] and linguistic supervision [62]. However, there is still a need for significant improvement in the expressiveness, accuracy and robustness of SGG methods.

In addition to the objects and their relationships in scene graphs, higher-level visual reasoning for the downstream tasks mentioned in the last paragraph requires background information about the scene and its constituents to mimic the cognitive ability of humans to use commonsense reasoning. Leveraging and reasoning with commonsense knowledge is quite challenging because of its implicit nature; it is universally accepted and used by humans in everyday situations but generally disregarded when we speak or write. Most of the existing SGG methods use datasets that contain large collections of images along with annotations of objects, attributes, relationships, scene graphs, etc., such as, Visual Genome (VG) [23] and VRD [31]. These datasets have limited or no explicit commonsense knowledge, which limits the expressiveness of scene graphs and the higher-level reasoning capabilities in the downstream tasks unless commonsense knowledge is infused from external sources. There are several publicly available sources [21,43,44,50] that include different forms and notions of commonsense knowledge. Some consolidation efforts [9,17] have been made to unify the different sources into a global commonsense knowledge source to jointly exploit their diverse knowledge and coverage. These consolidated sources have been integrated and used in language processing methods [33,58] for improving their robustness and expressiveness. The consolidated commonsense knowledge sources have not been leveraged for visual understanding and reasoning yet, however, their capability to provide rich and diverse background information and relevant facts about the concepts in a scene can help in improving the performance of SGG and providing rich and expressive scene representations for downstream reasoning.

Figure 1 shows a motivating example of an image and its commonsense knowledge-based scene graph representation. The scene graph of the image contains the relationship triplets *(woman, holding, racket)* and *(woman, on,*

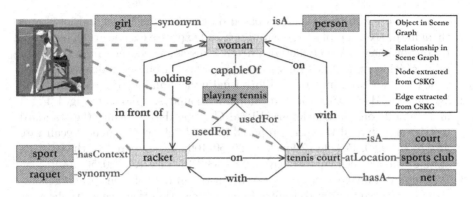

Fig. 1. A motivating example of a scene graph of an image with commonsense knowledge infusion using CommonSense Knowledge Graph (CSKG). The scene graph (blue) provides information about the objects and their pairwise relationships in the scene. The relevant nodes and edges extracted from CSKG (green) complement and enrich the scene graph by providing the necessary information about the possible spatial proximity of objects relative to each other and any possible interactions between objects, i.e. *(woman, at, tennis court)* and *(woman, holding, racket)*, and more importantly the background information and related facts, i.e. *(woman, capableOf, playing_tennis)* and *(racket, usedFor, playing_tennis)*, which allows higher-level reasoning to deduce *"the woman is playing tennis"*. (Color figure online)

tennis_court) representing the objects and their pairwise interactions. Though it is easy and straightforward for us to infer that the woman is playing tennis, it is challenging for machines to infer that without some external commonsense knowledge. The relevant nodes and edges extracted from the CommonSense Knowledge Graph (CSKG) [17] including *(woman, capableOf, playing_tennis)* and *(racket, usedFor, playing_tennis)* provide the necessary background information and facts for higher level reasoning. In this paper, we propose a commonsense knowledge-based SGG method that generates scene graph of an image and infuses the background knowledge and relevant facts about the concepts in the scene graph from CSKG [17], which is a large consolidated commonsense knowledge source. Graph embeddings were leveraged to compute the similarity of object nodes in the graph refinement and enrichment steps because similar entities tend to have similar vector representations in the embedding space [38]. The commonsense knowledge complements and enriches the scene graph relationships, which improves the performance of SGG and the expressiveness of scene graph representations. We evaluated the proposed method on the benchmark VG dataset and noted improvement of relationship prediction results for SGG. The encouraging experimental results depict the potential of commonsense knowledge in scene graph generation and its promising applications in visual understanding and reasoning. The main contributions of this paper include:

1. We propose a commonsense knowledge-based scene graph generation approach, which extracts background knowledge and relevant facts from

commonsense knowledge sources based on graph embeddings and integrates them in the scene graphs to generate rich and expressive scene graph representations of images. We employed a heterogeneous knowledge graph [17], containing rich commonsense knowledge consolidated from seven diverse sources, which has not been investigated for visual understanding and reasoning yet.

2. We performed experimental and comparative analysis (shown in Fig. 4, Fig. 5 and Table 2) on the benchmark Visual Genome dataset using the standard metric, and showed that the proposed method achieved a higher recall rate ($R@K = 29.89, 35.4, 39.12$ for $K = 20, 50, 100$) as compared to the existing state-of-the-art technique ($R@K = 25.8, 33.3, 37.8$ for $K = 20, 50, 100$).

3. We employed image generation as a downstream task of scene graph generation and showed improved results of image generation from scene graphs after commonsense knowledge infusion as shown in Fig. 6.

2 Related Work

2.1 Scene Graph Generation

Scene graph generation (SGG) is a challenging research problem and is actively investigated by researchers in computer vision. In the compositional methods, the subject, predicate and object are separately detected and aggregated later. Li et al. [26] used detected objects in an image to generate separate region proposals for subject, predicate and object; these region proposals are aggregated with features from a deep neural network (DNN) to reach a triplet prediction. Such methods are scalable, but they have very limited performance in the case of rare or unseen relations. The visual phrase models for visual relation detection treat relation triplets as a single entity. Sadeghi et al. [42] employed DNNs to predict objects as well as visual phrase or triplets and then refined those predictions by comparing them to other predictions in the image. Deep relational networks are also used for visual relation detection, in which the DNN also leverages the statistical dependency among objects and predicates [6]. The visual phrase models are less sensitive to the diversity of visual relations as compared to the compositional models, but they require a greater number of training examples in datasets with a large vocabulary of objects and predicates.

The more recent scene graph generation and visual relationship detection methods fuse visual and semantic embeddings in DNNs to detect visual relations on a large scale. Zhang et al. [67] extract visual features in three branches each for the subject, predicate and object, with the predicate branch fusing its features with the subject and object features at a later stage to leverage the interactions between subject and object for relation detection. During learning, features extracted from the text space are also embedded as labelling for the visual features. In a similar approach with improved precision, Peyre et al. [39] add a visual phrase embedding space during learning to enable analogical reasoning for predicting unseen relations and to improve robustness to appearance variations of visual relations. Tang et al. [48] attempted to address the problem of bias in SGG models due to the unbalanced distribution of relationships in datasets by leveraging causal inference and total direct effect.

Most of the existing works focus on visual and linguistic patterns in images while neglecting the background information and related facts about concepts in images and the structural patterns of scene graph elements in commonsense knowledge graphs, which have significant potential in understanding and interpretation of visual concepts. Only a few recent works mentioned in the next subsection explicitly leverage commonsense knowledge graphs for visual understanding and reasoning.

Table 1. Commonsense knowledge sources

Knowledge source	Knowledge type	Size	Example
ConceptNet [44]	Text-based knowledge about everyday objects, activities, relations, etc.	8M nodes, 36 relations & 21M edges	(chair, used for, sitting)
Wikidata [50]	General taxonomic knowledge about instances, concepts, relations etc.	75M objects, 1200 relations & 900M edges	(eating, subclass of, ingestion)
ATOMIC [43]	Procedural knowledge about pre/post conditions of events	0.3M nodes, 9 relations & 0.877M edges	(PersonX eating dinner, xEffect, satisfies hunger)
Roget [21]	Lexical knowledge about words, relations, etc.	72k words, 2 relations & 1.4M edges	(motorcycle, synonym, bike)
FrameNet [2]	Lexical knowledge about frames, roles, relations, etc.	1.2k frames, 12k roles, 1.9k edges & 13k lexical units	(cooking creation, has frame element, produced food)
Wordnet [36]	Lexical knowledge about words, concepts, relations, etc.	0.155M words, 10 relations & 0.176M synsets	(car, has part, air bag)
Visual Genome [23]	Visual knowledge about objects, relations and attributes in images	108k images, 3.8M nodes, 42k relations, 2.3M edges & 2.8M attributes	(food, on, plate), (woman, looking at, sandwich)
CSKG [17]	Consolidated commonsense knowledge from the above seven sources	2.16M nodes, 58 relations, 6M edges	(racket, used for, playing tennis)

2.2   Commonsense Knowledge Sources and Infusion

The acquisition and representation of commonsense knowledge and reasoning with it have been one of the major challenges in artificial intelligence since the 1960s s [34], which has led the research community to develop and curate several knowledge sources containing commonsense knowledge in different forms and contexts [16]. Some of the popular sources of commonsense knowledge along with their details are presented in Table 1. Some of these sources, especially ConceptNet [44], have been used in a few visual understanding and reasoning techniques. These techniques either extract relevant facts from a source and embed them in the model at a certain stage [11,37,45,66], or use graph-based message passing to embed the structural information from the source in the representations of the model [4,24,56,64]. Chen et al. [4] and Zellers et al. [66] incorporated commonsense knowledge from dataset statistics by employing pre-computed frequency priors in their predicate classification models to improve the performance of SGG. Wan et al. [51] proposed the use of a commonsense knowledge graph along with the visual features to enhance predicate detection for detected objects in visual relation detection. Gu et al. [11] retrieve relevant facts from a single source, i.e. ConceptNet [44] for each object, encode the facts into its features using recurrent neural networks and an attention mechanism in SGG. Kan et al. [19] infused commonsense knowledge from ConceptNet for zero-shot relationship prediction in SGG. The existing approaches mostly infuse triplets from the knowledge sources and ignore the rich structural information beyond individual triplets.

The knowledge sources are rich and diverse and cover different domains and contexts of commonsense knowledge, which can be consolidated to provide a rich and heterogeneous source of commonsense knowledge and to increase its impact in the downstream reasoning tasks. Zareian et al. [63] proposed GB-Net, which links the entities and edges in a scene graph to the corresponding entities and edges in a commonsense graph extracted from VG, WordNet and ConceptNet, and iteratively refine the scene graph using graph neural network-based message passing. Guo et al. [12] employed an instance relation transformer to extract relational and commonsense knowledge from VG and ConceptNet for SGG. These are the only SGG approaches that leverage multiple knowledge sources, while a subset [53] of DBpedia, ConceptNet and WebChild containing knowledge about visual concepts has been used in VQA [30,56]. The CommonSense Knowledge Graph (CSKG) [17] is currently the latest and largest consolidated source that integrates commonsense knowledge from the seven diverse and disjoint sources, including ConceptNet [44], Wikidata [50], ATOMIC [43], VG [23], Wordnet [36], Roget [21] and FrameNet [2]. Ma et al. [33] employed CSKG in language models and achieved the best performance in commonsense question answering by utilizing the diverse relevant knowledge from CSKG and aligning the knowledge with the task. To the best of our knowledge, the use and potential of CSKG have not yet been explored for visual understanding and reasoning tasks.

The knowledge-infusion methods also leverage knowledge graph embeddings, which are widely adopted in the vector representation of entities and relationships in knowledge graphs [38]. The knowledge graph embeddings capture the latent properties of the semantics in the KG, due to which similar entities are represented with similar vectors. The similarity of entities in the vector space is interpreted using vector similarity measures, such as cosine similarity. Knowledge graph embeddings have been used in several link prediction tasks including visual relationship detection [1] and recommender systems [52].

3   Proposed Method

The proposed commonsense knowledge-based scene graph generation method employs a DNN-based approach for detecting objects and their pairwise relationships in an image to generate its scene graph, which is followed by commonsense knowledge infusion using CSKG [17] for the enrichment of scene graph with background knowledge and relevant facts in the form of triplets. Figure 2 provides a detailed overview of the proposed method. The proposed method is built on the SGG toolkit [47].

Following the trend in recent SGG methods [48,49,59,66], we use Faster RCNN [41] for detecting objects in the images. We use ResNeXt-101-FPN architecture [29] as the backbone CNN for Faster RCNN. The Faster RCNN takes an image I as input and provides the object bounding boxes b and object class labels l of the n detected objects. The feature maps F are also extracted from the underlying CNN in the Faster RCNN.

$$\{b, l, F\} = FasterRCNN(I) \tag{1}$$

After detecting the objects and extracting the feature maps, the relationships between object pairs are predicted. RoIAlign [13] is applied to the image regions $I[b]$, which provides the region features a of each detected object.

$$a = RoIAlign(I[b]) \tag{2}$$

For all n objects, Bi-directional Long Short Term Memory (Bi-LSTM) layers [66] are used to encode a, $I[b]$ and l as the individual visual context features v_i.

$$v = BiLSTM(a, I[b], l) \tag{3}$$

The individual visual context features of objects are encoded by another set of Bi-LSTM layers and concatenated into combined pairwise object features $v_{ij} | i \neq j; i, j = 1, ..., n$.

$$v_{ij} = concat(BiLSTM(v_i), BiLSTM(v_j)) \tag{4}$$

In the same way, the pairwise object labels (l_i, l_j) are encoded through an embedding layer to compute the language prior p_{ij}. The contextual union features u_{ij} are extracted by applying RoIAlign to the union regions of pairwise objects in F.

$$u_{ij} = conv(RoIAlign(F[b_i \cup b_j])) \tag{5}$$

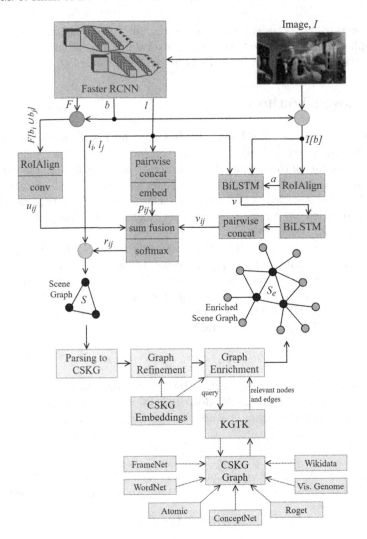

Fig. 2. The proposed commonsense knowledge-based scene graph generation method

Finally, all the three types of features representing the object pairs are fused using a summation feature fusion function [8] followed by a softmax function to predict the relationship class labels r_{ij} and the relationship class probabilities c_{ij}.

$$\{r_{ij}, c_{ij}\} = softmax(SUM(v_{ij}, u_{ij}, p_{ij})) \qquad (6)$$

The scene graph S is formed by linking the pairwise objects and relationships into a graph structure.

$$S = \{l_i, r_{ij}, l_j\} \qquad (7)$$

Algorithm 1: Graph refinement

Input: S, b
Output: S_r

1   $S_r = []$
2   **for** *each triplet* $\in S$ **do**
3     $e_1 = cskg\_emb(triplet[node1])$
4     $e_2 = cskg\_emb(triplet[node2])$
5     $b_1 = b[triplet[node1]]$
6     $b_2 = b[triplet[node2]]$
7     $metric_{sim} = cosine\_sim(e_1, e_2)$
8     $metric_{IoU} = compute\_IoU(b_1, b_2)$
9   **if** $metric_{sim} \leqslant \tau_{sim} \wedge metric_{IoU} \leqslant \tau_{iou}$ **then**
10      $S_r.append(triplet)$

Algorithm 2: Graph enrichment

Input: S, G_{cskg}
Output: S_e

1   $S_e = S$
2   **for** *each node* $\in S$ **do**
3     $e_1 = cskg\_emb(node)$
4     $triplets_{cskg} = query(G_{cskg}, node)$
5     $triplets_{cskg} = preprocess(triplets_{cskg})$
6   **for** *each triplet* $\in triplets_{cskg}$ **do**
7   **if** $node == triplet[node1]$ **then**
8        $e_2 = cskg\_emb(triplet[node2])$
9   **else**
10        $e_2 = cskg\_emb(triplet[node1])$
11       $s = cosine\_sim(e_1, e_2)$
12   **if** $s \geqslant \tau \wedge triplet \notin S_e$ **then**
13        $S_e.append(triplet)$
14     $S_e = postprocess(S_e)$

In order to infuse relevant triplets representing background knowledge and related facts from the CSKG [17], we parse the scene graph to a format compatible with the CSKG data model. Since similar entities tend to have similar vector representations in the embedding space [38], we leverage the graph embeddings to compute the similarity of nodes for various operations in the graph refinement and enrichment steps. The scene graph predictions are first refined using Algorithm 1 to discard any redundant or irrelevant predictions. The predicted objects with highly overlapping bounding boxes, similar names, or the same structural pattern in CSKG indicate the possibility of multiple redundant predictions of the same object. Such prediction errors are minimized at this stage by discarding the object nodes that have a high intersection over union (IoU) of its bounding box or a high similarity score of CSKG embedding with another object node.

We use the Knowledge Graph Toolkit (KGTK) [15] to query CSKG and extract triplets from CSKG that include a subject or object node in the predicted scene graph. After extraction, any duplicate triplets and the triplets with both nodes similar (e.g. (person, synonym, person) and (chair, similarTo, chair)) are discarded in the preprocessing step because they do not provide any useful information. Based on the embedding similarity of the object nodes and the extracted nodes, the extracted nodes with reasonable structural similarity with the corresponding object nodes are linked via extracted edges in the scene graph. If an extracted node is already present in the scene graph, the new edge is linked to the existing node, otherwise, the new node is created and linked in the scene graph. In postprocessing, the format of the enriched scene graph is adjusted according to the original scene graph representation so that the enriched scene graphs can be evaluated for performance comparison or can be used in a downstream reasoning task. Since the predicates integrated from VG are expressed as "LocatedNear" edge type in the CSKG, we replaced the predicates in triplets extracted from the VG source in CSKG with the most frequent predicate type between the nodes in the original VG dataset. This post-processing step uses statistical prior knowledge from VG about the possible predicates between a pair of objects (nodes) in relationships to further interpret the relationship predicate. Algorithm 2 gives an overview of the steps in extracting commonsense knowledge from CSKG and integrating it into the scene graph. The thresholds in both algorithms were set to 0.5 for the experimental evaluation. These thresholds determine the trade-off between the number and the accuracy of detected and infused relationships.

4 Experiments and Results

4.1 Experimental Setup

Dataset. We used the commonly used subset [59] of the Visual Genome dataset containing the most frequent 50 predicate classes and 150 object classes for training Faster RCNN, SGG model and image generation network. 70% of the training samples were used for training, out of which 5000 samples were used for validation during training. The remaining 30% samples were used for evaluation. The longer dimension of each image was resized to 1024 pixels and the shorter dimension is adjusted accordingly. We use the pre-trained CSKG embeddings [17] for computing the similarity of nodes in the graph refinement and enrichment steps of the proposed approach.

Evaluation Protocol. We used the cross-entropy loss to evaluate the training performance of the Faster RCNN and SGG models. Mean average precision (mAP) [10] was used to evaluate the object detection performance of Faster RCNN. For evaluating the performance of SGG before and after commonsense knowledge infusion, we used the most widely used metric, Recall@K ($R@K$) [31], which is defined as the fraction of times the correct relationship is predicted in the top K confident relationship predictions. We compared the performance of the proposed method and recent SGG methods using the standard metric and benchmark dataset. We also analysed some qualitative results of the proposed method. Additionally, we employed an existing image generation method [18] as a downstream task of scene graph generation to further evaluate the proposed method by comparing the results of image generation from scene graphs before and after commonsense knowledge infusion.

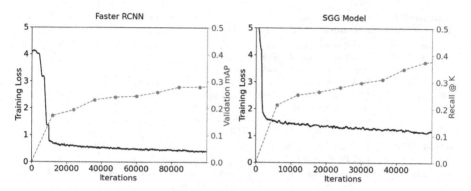

Fig. 3. Training progress plots along with periodic validation checks of the Faster RCNN and SGG models.

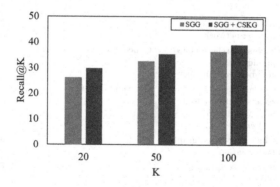

Fig. 4. Comparison of Recall@K of SGG before and after commonsense knowledge infusion.

Table 2. Comparison of the proposed method with the existing state-of-the-art SGG approaches in terms of Recall@K (R@K) on the Visual Genome dataset

SGG method	Approach	Commonsense knowledge source	R@20 (%)	R@50 (%)	R@100 (%)
Proposed Method (SGG+CSKG)	Scene graph enrichment via commonsense knowledge infusion from different sources	CSKG [17]	**29.89**	**35.4**	**39.12**
GLAT [65]	Transformer-based GNN for visual commonsense reasoning	–	–	–	38.8
Unbiased SGG [48]	Causal inference and total direct effect	–	25.8	33.3	37.8
Proposed Method (SGG Only)	Scene graph generation based on fusion of visual (region and object) and text features	–	26.1	32.7	36.5
GB-Net [63]	Message passing between scene graphs and commonsense graph	ConceptNet [44], WordNet [36], Visual Genome [23]	–	29.4	35.1
VCTree [49]	Dynamic tree structures and Bi-dir TreeLSTM	–	22	27.9	31.3
IRT-MSK [12]	Instance Relation Transformer with Multiple Structured Knowledge	ConceptNet [44], Visual Genome [23]	22.2	27.2	31.2
Neural Motifs [66]	Stacked Motif Networks	–	21.7	27.3	30.5
KERN [4]	Knowledge-embedded routing network	–	–	27.1	29.8
COACHER [19]	Zero-shot relationship prediction via commonsense infusion	ConceptNet [44]	13.42	19.31	22.22
KB-GAN [11]	Commonsense and reconstruction-based object and phrase refinement	ConceptNet [44]	–	13.65	17.57
FactorizableNet [27]	Clustering-based graph factorization	–	–	13.06	16.47
MSDN [28]	Scene description at object, phrase and caption levels	–	–	10.72	14.22
Graph RCNN [60]	RPN followed by Attention GCN	–	–	11.4	13.7
IMP [59]	Object and relationship feature refinement via message passing	–	–	3.44	4.24

4.2   Results

Training and Evaluation of Models. We trained the Faster RCNN model on the images and groundtruth annotations of objects in the Visual Genome dataset with Stochastic Gradient Descent (SGD) as an optimizer, batch size of 2 and initial learning rate of 0.002 which was decayed by a factor of 10 after 60k and 80k iterations. We froze the trained Faster RCNN and trained the whole SGG model on the images and groundtruth annotations of objects and relationships in the Visual Genome dataset using SGD as an optimizer, batch size of 4 and initial learning rate of 0.04 which was decayed by a factor of 10 twice during training when the validation performance stops improving noticeably. The plots of training loss and validation mAP for object detection and training loss and R@K for scene graph detection are shown in Fig. 3, which show a smooth convergence of the models during the training process. The Faster RCNN model achieved $29.19mAP$ (using 0.5 IoU threshold), while the SGG model achieved $R@K = 26.1, 32.7, 36.5$ for $K = 20, 50, 100$ on the test set. The training and evaluation of the SGG model was performed in the Scene Graph Detection (SGDet) setting.

Evaluation After Commonsense Knowledge Infusion. We repeated testing of the scene graph generation method after adding the proposed commonsense knowledge infusion steps and achieved $R@K = 29.89, 35.4, 39.12$ for $K = 20, 50, 100$ on the test set, which is considerably higher than the R@K values achieved for the scene graph generation without commonsense knowledge infusion steps, as shown in Fig. 4. The diverse commonsense knowledge integrated into the scene graphs from CSKG includes visual cues about the spatial proximity of objects in the scene relative to each other and physical interactions between the objects from the knowledge base of Visual Genome. This helps in mitigating some missed or wrong predictions made during scene graph generation and improves the recall rate for relationship prediction.

Comparative Analysis. A detailed comparative analysis of the proposed approach with the existing scene graph generation methods is presented in Table 2. The proposed method incorporates the latest, largest and most diverse commonsense knowledge source from a consolidation of 7 distinct sources, and thus achieves higher recall score ($R@K = 29.89, 35.4, 39.12$ for $K = 20, 50, 100$) for SGG on the benchmark Visual Genome dataset as compared to the state-of-the-art technique ($R@K = 25.8, 33.3, 37.8$ for $K = 20, 50, 100$).

Qualitative Results. Some qualitative results of the proposed method on Visual Genome images are shown in Fig. 5. In addition to the objects and their pairwise visual relationships, the commonsense knowledge-based scene graphs contain the background facts about the underlying concepts, additional knowledge about the spatial proximity of objects in the scene relative to each other, and possible physical interactions between the objects. The useful background facts include *(person, requires, eating)* and *(food, usedFor, eating)* in Fig. 5(a). The commonsense relationships about spatial proximity such as *(tree, on, street)* in Fig. 5(b) and the commonsense relationships about object interactions such as *(person, holding, surfboard)* in Fig. 5(c) complement the scene graph representations.

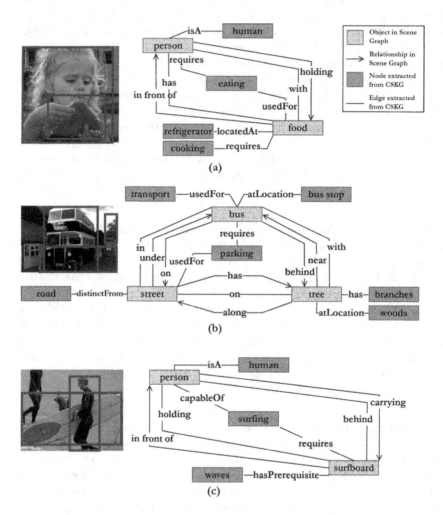

Fig. 5. Some qualitative results of the proposed commonsense knowledge-based scene graph generation method.

Downstream Task. The rich and heterogeneous scene representations generated by the proposed method can significantly improve the downstream visual reasoning tasks including image captioning, image generation, VQA, image retrieval, visual storytelling and multimedia event processing.

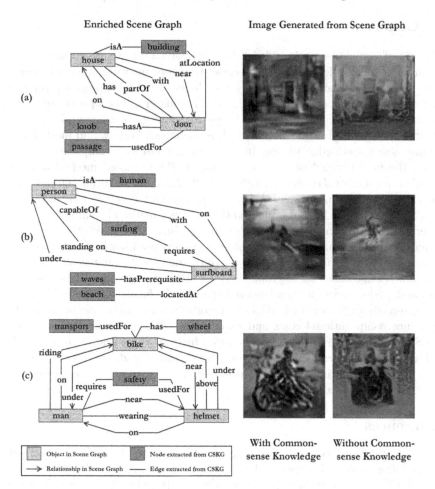

Fig. 6. Results of image generation using scene graphs generated by the proposed method.

We employed an existing image generation method [18] as a downstream task of scene graph generation to further evaluate the proposed method. We trained the image generation network on the Visual Genome subset that was used to train the scene graph generation model. The trained network was used to generate images from scene graphs before and after commonsense knowledge infusion. The results of image generation from scene graphs are presented in Fig. 6, which shows that the commonsense knowledge-based scene graphs generate more realistic images in which the semantic concepts in the input scene graph can be more clearly observed.

5   Conclusion

The use of commonsense knowledge for expressive and accurate visual understanding is inevitable due to its potential in complementing scene representations by providing necessary information for higher-level reasoning. In this paper, we propose a commonsense knowledge-based scene graph generation approach, which enriches the scene graph of an image with background knowledge and relevant facts extracted from CSKG, which is the latest, largest, and most diverse commonsense knowledge source. In the experimental and comparative analysis on the benchmark Visual Genome dataset, the proposed method achieved a higher recall rate ($R@K = 29.89, 35.4, 39.12$ for $K = 20, 50, 100$) as compared to the existing state-of-the-art technique ($R@K = 25.8, 33.3, 37.8$ for $K = 20, 50, 100$). We further evaluated the proposed method by employing image generation as a downstream task and showed improved qualitative results of image generation from scene graphs after commonsense knowledge infusion. The promising results depict the effectiveness of the rich and heterogeneous commonsense knowledge-based scene graph representations in improving the expressiveness and performance of visual reasoning tasks. In future work, we will investigate zero-shot and few-shot SGG using consolidated commonsense knowledge to reduce computational costs and requirement of training data and to allow the SGG model to predict unseen or rare object and predicate categories. We will also evaluate the efficacy of the proposed method in downstream reasoning tasks including multimedia event processing, image captioning, visual question answering and image retrieval.

References

1. Baier, S., Ma, Y., Tresp, V.: Improving visual relationship detection using semantic modeling of scene descriptions. In: d'Amato, C., et al. (eds.) ISWC 2017. LNCS, vol. 10587, pp. 53–68. Springer, Cham (2017). https://doi.org/10.1007/978-3-319-68288-4_4
2. Baker, C.F., Fillmore, C.J., Lowe, J.B.: The Berkeley framenet project. In: 36th Annual Meeting of the Association for Computational Linguistics and 17th International Conference on Computational Linguistics, vol. 1, pp. 86–90 (1998)
3. Chang, X., Ren, P., Xu, P., Li, Z., Chen, X., Hauptmann, A.: Scene graphs: a survey of generations and applications. arXiv preprint arXiv:2104.01111 (2021)
4. Chen, T., Yu, W., Chen, R., Lin, L.: Knowledge-embedded routing network for scene graph generation. In: Proceedings of the IEEE/CVF Conference on Computer Vision and Pattern Recognition, pp. 6163–6171 (2019)
5. Curry, E., Salwala, D., Dhingra, P., Pontes, F.A., Yadav, P.: Multimodal event processing: a neural-symbolic paradigm for the internet of multimedia things. IEEE Internet of Things J. https://doi.org/10.1109/JIOT.2022.3143171
6. Dai, B., Zhang, Y., Lin, D.: Detecting visual relationships with deep relational networks. In: Proceedings of the IEEE Conference on Computer Vision and Pattern Recognition, pp. 3076–3086 (2017)
7. Deng, J., Dong, W., Socher, R., Li, L.J., Li, K., Fei-Fei, L.: ImageNet: a large-scale hierarchical image database. In: 2009 IEEE Conference on Computer Vision and Pattern Recognition, pp. 248–255. IEEE (2009)

8. Feichtenhofer, C., Pinz, A., Zisserman, A.: Convolutional two-stream network fusion for video action recognition. In: Proceedings of the IEEE Conference on Computer Vision and Pattern Recognition, pp. 1933–1941 (2016)
9. Gangemi, A., Alam, M., Asprino, L., Presutti, V., Recupero, D.R.: Framester: a wide coverage linguistic linked data hub. In: Blomqvist, E., Ciancarini, P., Poggi, F., Vitali, F. (eds.) EKAW 2016. LNCS (LNAI), vol. 10024, pp. 239–254. Springer, Cham (2016). https://doi.org/10.1007/978-3-319-49004-5_16
10. Girshick, R., Donahue, J., Darrell, T., Malik, J.: Rich feature hierarchies for accurate object detection and semantic segmentation. In: Proceedings of the IEEE Conference on Computer Vision and Pattern Recognition, pp. 580–587 (2014)
11. Gu, J., Zhao, H., Lin, Z., Li, S., Cai, J., Ling, M.: Scene graph generation with external knowledge and image reconstruction. In: Proceedings of the IEEE/CVF Conference on Computer Vision and Pattern Recognition, pp. 1969–1978 (2019)
12. Guo, Y., Song, J., Gao, L., Shen, H.T.: One-shot scene graph generation. In: Proceedings of the 28th ACM International Conference on Multimedia, pp. 3090–3098 (2020)
13. He, K., Gkioxari, G., Dollár, P., Girshick, R.: Mask R-CNN. In: Proceedings of the IEEE International Conference on Computer Vision, pp. 2961–2969 (2017)
14. Hung, Z.S., Mallya, A., Lazebnik, S.: Contextual translation embedding for visual relationship detection and scene graph generation. IEEE Trans. Pattern Anal. Mach. Intell. **43**, 3820–3832 (2020)
15. Ilievski, F., et al.: KGTK: a toolkit for large knowledge graph manipulation and analysis. In: Pan, J.Z., et al. (eds.) ISWC 2020. LNCS, vol. 12507, pp. 278–293. Springer, Cham (2020). https://doi.org/10.1007/978-3-030-62466-8_18
16. Ilievski, F., Oltramari, A., Ma, K., Zhang, B., McGuinness, D.L., Szekely, P.: Dimensions of commonsense knowledge. arXiv preprint arXiv:2101.04640 (2021)
17. Ilievski, F., Szekely, P., Zhang, B.: CSKG: the commonsense knowledge graph. In: Verborgh, R., et al. (eds.) ESWC 2021. LNCS, vol. 12731, pp. 680–696. Springer, Cham (2021). https://doi.org/10.1007/978-3-030-77385-4_41
18. Johnson, J., Gupta, A., Fei-Fei, L.: Image generation from scene graphs. In: Proceedings of the IEEE Conference on Computer Vision and Pattern Recognition, pp. 1219–1228 (2018)
19. Kan, X., Cui, H., Yang, C.: Zero-shot scene graph relation prediction through commonsense knowledge integration. In: Oliver, N., Pérez-Cruz, F., Kramer, S., Read, J., Lozano, J.A. (eds.) ECML PKDD 2021. LNCS (LNAI), vol. 12976, pp. 466–482. Springer, Cham (2021). https://doi.org/10.1007/978-3-030-86520-7_29
20. Khan, M.J., Curry, E.: Neuro-symbolic visual reasoning for multimedia event processing: overview, prospects and challenges. In: Proceedings of the 29th ACM International Conference on Information and Knowledge Management (CIKM 2020) Workshops (2020)
21. Kipfer, B.: Roget's 21st Century Thesaurus in Dictionary form, 3rd edn. The Philip Lief Group, New York (2005)
22. Koner, R., Li, H., Hildebrandt, M., Das, D., Tresp, V., Günnemann, S.: Graphhopper: multi-hop scene graph reasoning for visual question answering. In: Hotho, A., et al. (eds.) ISWC 2021. LNCS, vol. 12922, pp. 111–127. Springer, Cham (2021). https://doi.org/10.1007/978-3-030-88361-4_7
23. Krishna, R., et al.: Visual genome: connecting language and vision using crowdsourced dense image annotations. Int. J. Comput. Vis. **123**(1), 32–73 (2017)
24. Lee, C.W., Fang, W., Yeh, C.K., Wang, Y.C.F.: Multi-label zero-shot learning with structured knowledge graphs. In: Proceedings of the IEEE Conference on Computer Vision and Pattern Recognition, pp. 1576–1585 (2018)

25. Lee, S., Kim, J.W., Oh, Y., Jeon, J.H.: Visual question answering over scene graph. In: 2019 First International Conference on Graph Computing (GC), pp. 45–50. IEEE (2019)

26. Li, Y., Ouyang, W., Wang, X., Tang, X.: VIP-CNN: visual phrase guided convolutional neural network. In: Proceedings of the IEEE Conference on Computer Vision and Pattern Recognition, pp. 1347–1356 (2017)

27. Li, Y., Ouyang, W., Zhou, B., Shi, J., Zhang, C., Wang, X.: Factorizable net: an efficient subgraph-based framework for scene graph generation. In: Proceedings of the European Conference on Computer Vision (ECCV), pp. 335–351 (2018)

28. Li, Y., Ouyang, W., Zhou, B., Wang, K., Wang, X.: Scene graph generation from objects, phrases and region captions. In: Proceedings of the IEEE International Conference on Computer Vision, pp. 1261–1270 (2017)

29. Lin, T.Y., Dollár, P., Girshick, R., He, K., Hariharan, B., Belongie, S.: Feature pyramid networks for object detection. In: Proceedings of the IEEE Conference on Computer Vision and Pattern Recognition (CVPR), pp. 2117–2125 (2017)

30. Liu, L., Wang, M., He, X., Qing, L., Chen, H.: Fact-based visual question answering via dual-process system. Knowl.-Based Syst. 107650 (2021)

31. Lu, C., Krishna, R., Bernstein, M., Fei-Fei, L.: Visual relationship detection with language priors. In: Leibe, B., Matas, J., Sebe, N., Welling, M. (eds.) ECCV 2016. LNCS, vol. 9905, pp. 852–869. Springer, Cham (2016). https://doi.org/10.1007/978-3-319-46448-0_51

32. Ma, C., Sun, L., Zhong, Z., Huo, Q.: ReLaText: exploiting visual relationships for arbitrary-shaped scene text detection with graph convolutional networks. Pattern Recogn. **111**, 107684 (2021)

33. Ma, K., Ilievski, F., Francis, J., Bisk, Y., Nyberg, E., Oltramari, A.: Knowledge-driven data construction for zero-shot evaluation in commonsense question answering. In: 35th AAAI Conference on Artificial Intelligence (2021)

34. McCarthy, J., et al.: Programs with Common Sense. RLE and MIT Computation Center (1960)

35. Mi, L., Chen, Z.: Hierarchical graph attention network for visual relationship detection. In: Proceedings of the IEEE/CVF Conference on Computer Vision and Pattern Recognition, pp. 13886–13895 (2020)

36. Miller, G.A.: WordNet: a lexical database for English. Commun. ACM **38**(11), 39–41 (1995)

37. Narasimhan, M., Schwing, A.G.: Straight to the facts: learning knowledge base retrieval for factual visual question answering. In: Proceedings of the European Conference on Computer Vision (ECCV), pp. 451–468 (2018)

38. Palmonari, M., Minervini, P.: Knowledge graph embeddings and explainable AI. In: Knowledge Graphs for Explainable Artificial Intelligence: Foundations, Applications and Challenges, pp. 49–72. IOS Press, Amsterdam (2020)

39. Peyre, J., Laptev, I., Schmid, C., Sivic, J.: Detecting unseen visual relations using analogies. In: Proceedings of the IEEE International Conference on Computer Vision, pp. 1981–1990 (2019)

40. Prakash, A., et al.: Self-supervised real-to-sim scene generation. In: Proceedings of the IEEE/CVF International Conference on Computer Vision, pp. 16044–16054 (2021)

41. Ren, S., He, K., Girshick, R., Sun, J.: Faster R-CNN: towards real-time object detection with region proposal networks. IEEE Trans. Pattern Anal. Mach. Intell. **39**(6), 1137–1149 (2016)

42. Sadeghi, M.A., Farhadi, A.: Recognition using visual phrases. In: CVPR 2011, pp. 1745–1752. IEEE (2011)

43. Sap, M., et al.: Atomic: an atlas of machine commonsense for if-then reasoning. In: Proceedings of the AAAI Conference on Artificial Intelligence, vol. 33, pp. 3027–3035 (2019)
44. Speer, R., Chin, J., Havasi, C.: ConceptNet 5.5: an open multilingual graph of general knowledge. In: Thirty-First AAAI Conference on Artificial Intelligence, pp. 4444–4451 (2017)
45. Su, Z., Zhu, C., Dong, Y., Cai, D., Chen, Y., Li, J.: Learning visual knowledge memory networks for visual question answering. In: Proceedings of the IEEE Conference on Computer Vision and Pattern Recognition, pp. 7736–7745 (2018)
46. Suhail, M., et al.: Energy-based learning for scene graph generation. In: Proceedings of the IEEE/CVF Conference on Computer Vision and Pattern Recognition, pp. 13936–13945 (2021)
47. Tang, K.: A scene graph generation codebase in pytorch (2020). https://github.com/KaihuaTang/Scene-Graph-Benchmark.pytorch
48. Tang, K., Niu, Y., Huang, J., Shi, J., Zhang, H.: Unbiased scene graph generation from biased training. In: Proceedings of the IEEE/CVF Conference on Computer Vision and Pattern Recognition, pp. 3716–3725 (2020)
49. Tang, K., Zhang, H., Wu, B., Luo, W., Liu, W.: Learning to compose dynamic tree structures for visual contexts. In: Proceedings of the IEEE/CVF Conference on Computer Vision and Pattern Recognition, pp. 6619–6628 (2019)
50. Vrandečić, D., Krötzsch, M.: Wikidata: a free collaborative knowledgebase. Commun. ACM **57**(10), 78–85 (2014)
51. Wan, H., Ou, J., Wang, B., Du, J., Pan, J.Z., Zeng, J.: Iterative visual relationship detection via commonsense knowledge graph. In: Wang, X., Lisi, F.A., Xiao, G., Botoeva, E. (eds.) JIST 2019. LNCS, vol. 12032, pp. 210–225. Springer, Cham (2020). https://doi.org/10.1007/978-3-030-41407-8_14
52. Wang, H., Zhang, F., Xie, X., Guo, M.: DKN: deep knowledge-aware network for news recommendation. In: Proceedings of the 2018 World Wide Web Conference, pp. 1835–1844 (2018)
53. Wang, P., Wu, Q., Shen, C., Dick, A., Van Den Hengel, A.: FVQA: fact-based visual question answering. IEEE Trans. Pattern Anal. Mach. Intell. **40**(10), 2413–2427 (2017)
54. Wang, R., Wei, Z., Li, P., Zhang, Q., Huang, X.: Storytelling from an image stream using scene graphs. In: Proceedings of the AAAI Conference on Artificial Intelligence, vol. 34, pp. 9185–9192 (2020)
55. Wang, S., Wang, R., Yao, Z., Shan, S., Chen, X.: Cross-modal scene graph matching for relationship-aware image-text retrieval. In: Proceedings of the IEEE/CVF Winter Conference on Applications of Computer Vision, pp. 1508–1517 (2020)
56. Wang, X., Ye, Y., Gupta, A.: Zero-shot recognition via semantic embeddings and knowledge graphs. In: Proceedings of the IEEE Conference on Computer Vision and Pattern Recognition, pp. 6857–6866 (2018)
57. Wu, X., Sahoo, D., Hoi, S.C.: Recent advances in deep learning for object detection. Neurocomputing (2020)
58. Xie, Y., Pu, P.: How commonsense knowledge helps with natural language tasks: a survey of recent resources and methodologies. arXiv preprint arXiv:2108.04674 (2021)
59. Xu, D., Zhu, Y., Choy, C.B., Fei-Fei, L.: Scene graph generation by iterative message passing. In: Proceedings of the IEEE Conference on Computer Vision and Pattern Recognition, pp. 5410–5419 (2017)

60. Yang, J., Lu, J., Lee, S., Batra, D., Parikh, D.: Graph R-CNN for scene graph generation. In: Proceedings of the European Conference on Computer Vision (ECCV), pp. 670–685 (2018)
61. Yang, X., Zhang, H., Cai, J.: Auto-encoding and distilling scene graphs for image captioning. IEEE Trans. Pattern Anal. Mach. Intell. **44**(5), 2313–2327 (2022). https://doi.org/10.1109/TPAMI.2020.3042192
62. Ye, K., Kovashka, A.: Linguistic structures as weak supervision for visual scene graph generation. In: Proceedings of the IEEE/CVF Conference on Computer Vision and Pattern Recognition (CVPR), pp. 8289–8299, June 2021
63. Zareian, A., Karaman, S., Chang, S.-F.: Bridging knowledge graphs to generate scene graphs. In: Vedaldi, A., Bischof, H., Brox, T., Frahm, J.-M. (eds.) ECCV 2020. LNCS, vol. 12368, pp. 606–623. Springer, Cham (2020). https://doi.org/10.1007/978-3-030-58592-1_36
64. Zareian, A., Karaman, S., Chang, S.F.: Weakly supervised visual semantic parsing. In: Proceedings of the IEEE/CVF Conference on Computer Vision and Pattern Recognition, pp. 3736–3745 (2020)
65. Zareian, A., Wang, Z., You, H., Chang, S.-F.: Learning visual commonsense for robust scene graph generation. In: Vedaldi, A., Bischof, H., Brox, T., Frahm, J.-M. (eds.) ECCV 2020. LNCS, vol. 12368, pp. 642–657. Springer, Cham (2020). https://doi.org/10.1007/978-3-030-58592-1_38
66. Zellers, R., Yatskar, M., Thomson, S., Choi, Y.: Neural motifs: scene graph parsing with global context. In: Proceedings of the IEEE Conference on Computer Vision and Pattern Recognition, pp. 5831–5840 (2018)
67. Zhang, J., Kalantidis, Y., Rohrbach, M., Paluri, M., Elgammal, A., Elhoseiny, M.: Large-scale visual relationship understanding. In: Proceedings of the AAAI Conference on Artificial Intelligence, vol. 33, pp. 9185–9194 (2019)

Impact of the Characteristics of Multi-source Entity Matching Tasks on the Performance of Active Learning Methods

Anna Primpeli$^{(\boxtimes)}$ ⓘ and Christian Bizer ⓘ

Data and Web Science Group, University of Mannheim, Mannheim, Germany
{anna,chris}@informatik.uni-mannheim.de

Abstract. Entity matching aims at identifying records in different data sources that describe the same real-world entity. Entity matching is the foundational technique for setting RDF links in the context of the Web of Data. By applying active learning methods for training entity matchers, it is possible to reduce the human labeling effort by selecting informative record pairs for labeling. Although active learning has been extensively studied for the two-data source matching case, it was only recently applied for the task of matching records in multi-source settings, such as the Web of Data. A multi-source matching task has certain inherent characteristics which do not apply for two-source matching tasks and which can be exploited by the active learning query strategy to further reduce the labeling effort. In this paper, we propose a set of profiling dimensions which capture these inherent characteristics of multi-source matching tasks and study their impact on the performance of different active learning methods for training entity matchers. To enable our analysis, we develop ALMSERgen, a multi-source matching task generator and curate a continuum of 252 matching tasks along the suggested profiling dimensions. We use the generated as well as five benchmark tasks to compare the performance of three query strategies: a committee-based strategy, a graph-based strategy, and a strategy that exploits grouping signals. Our results show that graph signals are relevant for multi-source matching tasks involving a large amount of records describing the same-real world entities with heterogeneous attribute values while using grouping signals is beneficial if there exists a small number of groups of matching tasks sharing the same underlying patterns.

Keywords: Entity Resolution · Active Learning · Multi-Source Entity Matching · Matching Task Profiling

1 Introduction

Entity matching (EM), also known as entity resolution, record linkage, and data deduplication, is the task of identifying records in one or more sources that refer to the same real-world entity [4,5]. EM is often treated as a supervised binary

© The Author(s), under exclusive license to Springer Nature Switzerland AG 2022
P. Groth et al. (Eds.): ESWC 2022, LNCS 13261, pp. 113–129, 2022.
https://doi.org/10.1007/978-3-031-06981-9_7

classification problem for which a labeled set of matching and non-matching record pairs is used for training [5,6,8]. Manually labeling training sets is expensive. Active learning is a supervised learning paradigm that aims at reducing the labeling effort by including the human annotator into the learning loop and iteratively selecting a small informative subset for labeling [29]. The informative labeled subset is used for training a classification model, to which we will refer to as *learner* in the rest of the paper.

Active learning has been extensively researched for matching records between two sources [3,22,26] while it has been barely applied for the task of matching records between multiple data sources [11,25]. Multi-source matching scenarios frequently appear in the context of link discovery [21] for the Web of Data [9]. Multi-source EM tasks have certain inherent characteristics which are different from the two-source EM tasks and can be exploited as signals by active learning methods to further reduce the labeling effort [25].

To demonstrate this, we use the example of Fig. 1. The example multi-source EM task comprises four data sources which contain records describing mobile phones (Fig. 1a). Combining pairwise the four data sources results in six two-source EM tasks (Fig. 1b). Given the overlap of entities among the data sources, the multi-source EM task can be viewed as a correspondence graph in which edges denote matches (Fig. 1c). Exploiting graph signals, such as graph transitivity, has already been shown to improve the performance of active learning methods by discovering potentially false negative and false positive record pairs among the predictions of the learner [25]. For example, if the learner's predictions for the record pairs in Fig. 1c are A1-B1:match, A1-D1:match, and D1-B1:non-match, considering graph transitivity and selecting the pair D1-B1 for annotation leads to the discovery of the pair D1-B1 as a false negative prediction.

Given the different attribute values of the phone records, different groups of two-source EM tasks with similar matching patterns arise (Fig. 1d). We consider a matching pattern as a disjunction of conjunctions of similarity-based features and threshold values. Exploiting the grouping signals during active learning can lead to the selection of more informative record pairs for labeling by, for example, annotating only representative pairs from each group.

However, the degree of graph and grouping signals may vary across different multi-source tasks and is highly dependent on the profile of the data sources to be matched. In our work, we explore the impact of the profiling characteristics of multi-source EM tasks on the performance of active learning methods which exploit different signals for selecting informative record pairs for labeling. To do so, we first propose a set of profiling dimensions for describing multi-source EM tasks. To enable our analysis, we develop ALMSERgen, a multi-source EM task generator, and generate a continuum of 252 multi-source EM tasks along the suggested dimensions. We evaluate the following three active learning query strategies on the generated tasks: 1.HeALER: a state-of-the-art committee-based query strategy [3], 2. ALMSER: a graph-based query strategy [25], and 3. ALMSERgroup: a newly introduced variation of the ALMSER query strategy which exploits grouping signals. By analyzing our evaluation results, we identify the best performing active learning query strategies for groups of multi-source EM

Data source A

ID	name	brand
A1	iphone 4s	applle
A2	one m9	htcc

Data source B

ID	name	brand
B1	iphone 4s	applee
B2	one m9	httc

Data source C

ID	name	brand
C1	iphone 4s smartphone	apple
C2	one m9 smartphone	htc

Data source D

ID	name	brand
D1	iphone 4s mobile	apple
D2	one m9 mobile	htc

(a) Data sources

Two-Source Task	Matches	Non-Matches
A-B	A1-B1	A1-B2
	A2-B2	A2-B1
A-C	A1-C1	A1-C2
	A2-C2	A2-C1
A-D	A1-D1	A1-D2
	A2-D2	A2-D1
B-C	B1-C1	B1-C2
	B2-C2	B2-C1
B-D	B1-D1	B1-D2
	B2-D2	B2-D1
C-D	C1-D1	C1-D2
	C2-D2	C2-D1

(b) Two-source tasks

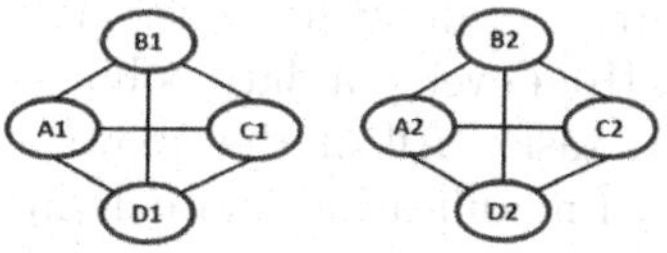

(c) Correspondence graph

Two-Source Task	Matching Pattern
A-B	ExactSim(name) > 0.0 ∧ LevenshteinSim(brand) > 0.75
A-C	
A-D	JaccardSim(name) > 0.66 ∧ LevenshteinSim(brand) > 0.75
B-C	
B-D	
C-D	JaccardSim(name) > 0.5 ∧ ExactSim(brand) > 0.0

(d) Groups of tasks

Fig. 1. Example of a multi-source EM task.

tasks sharing the same characteristics. Finally, we confirm the findings of our experimental analysis using five benchmark tasks from the related work.

The remainder of the paper is organized as follows: Sect. 2 discusses related work on active learning for two-source and multi-source matching, as well as on matching task generators. Section 3 introduces the set of dimensions for profiling multi-source matching tasks. Section 4 presents the multi-source task generator ALMSERgen which we use for generating a continuum of multi-source matching tasks. In Sect. 5, we present the experimental setup and results of our analysis on both the generated and the benchmark tasks. Finally, Sect. 6 concludes our paper and summarizes our findings.

2   Related Work

Entity matching (EM) is a central prerequisite for integrating data from multiple sources [4,5,23] as well as for setting RDF links in the context of the Web of Data [9,21]. There exists a large body of research on supervised and unsupervised multi-source EM [2,27,30], while active learning has been hardly used in this context [11,25]. Profiling EM tasks [24] and comparing the performance of different matchers in passive [1,5,15,17] and active learning settings [18] has been thoroughly studied for the two-source matching scenario. To allow a fair comparison, a large number of either benchmark [17,18,24] or generated EM tasks [12,28,33] are used for evaluation. However, to the best of our knowledge, there exists no work on studying the impact of the profile of multi-source EM tasks on the performance of different active learning methods.

Data Generators for EM. There exist several data generators for curating EM tasks for Linked Data [7,12,28] and which have been used for evaluating link discovery frameworks [1]. Such data generators produce EM tasks with varying degrees of difficulty considering a set of pre-defined dimensions. Hildebrandt et al. [10] develop a data pollution framework for modifying large-scale two-source EM tasks. All of the above frameworks inject value errors, such as token or word modification and deletion as part of the generation or pollution pipeline. However, existing data generators do not consider multi-source EM task-related desiderata which we cover in our work.

Active Learning for EM. There is a large body of research on active learning for two-source EM [3,13,14,16], with recent work turning the focus to deep learning [14,20]. Active deep learning-based methods rely on transfer learning [14] or large randomly sampled sets [20] for model initialization and assume a pre-labeled development set for hyperparameter optimization [14,20]. Contrary to these methods, we evaluate and compare the performance of active learning methods that rely on symbolic features and traditional classification models, involve less annotation effort, and do not rely on a pre-labeled set for model initialization and optimization.

Meduri et al. [18] compare various symbolic active learning methods for the two-source EM task and show that random forest classifiers with committee- and margin-based query strategies achieve fast convergence and close to passive learning results. However, using a margin-based query strategy is shown to significantly underperform the committee-based strategy HeALER [3] in the case of multi-source EM tasks [25]. In our recent work [25], we proposed ALMSER, an active learning method for multi-source EM which exploits graph signals for boosting the query strategy and training the learner [25]. The evaluation results on five multi-source EM tasks showed that combining both graph-boosted components of ALMSER outperforms HeALER while exploiting the graph signals only as part of the query strategy does not perform better than HeALER for all tasks. In our current work, we analyze how the profiling characteristics of a multi-source EM task affect the performance of different query strategies, including the strategies used by HeALER and ALMSER.

3 Profiling Dimensions for Multi-source EM Tasks

In this section, we define three dimensions for profiling multi-source EM tasks: entity overlap, value heterogeneity, and value pattern overlap. Below, we present how each dimension is calculated and discuss its relevance to active learning.

Entity Overlap. The dimension of entity overlap (EO) refers to the ratio of real-world entities that appear in more than two sources over the entities of the multi-source task that appear in exactly two or more sources. Transforming the multi-source task into a correspondence graph with the edges denoting matches between the nodes-records, the dimension of EO is calculated as $\frac{|CC_{size>2}|}{|CC_{size\geq2}|}$, where CC are the connected components of the correspondence graph. An EO of 0 indicates that all entities are represented by records appearing in a maximum of two of the data sources while an EO of 1 indicates that all entities are represented by records in at least three data sources. In the multi-source task of Fig. 1 the EO is 1, as both entities appear in all four data sources. We expect a multi-source task with an EO level of 0 to offer low-quality graph signals. Given that the maximum size of connected components in that case is two nodes, i.e. records, no additional information can be extracted from the correspondence graph considering different graph signals such as graph transitivity [25]. In such settings, non-graph-based query strategies are expected to overperform graph-based ones.

Value Heterogeneity. The dimension of value heterogeneity (VH) captures how heterogeneous the identifying attribute values of the records that appear in different data sources and describe the same real-world entity are. As identifying attributes, we define the combination of attributes that are useful for distinguishing real-world entities of a specific domain. The heterogeneity of values may derive from different surface forms, e.g. *iphone 4s phone* vs. *4s iphone* as well as spelling errors, e.g. *apple* vs. *applle*. We compute VH as the ratio of entities that are represented by records with dissimilar values in at least one of their identifying attributes to all entities. In the example of Fig. 1, the VH is 1, as both entities are represented by records with different values either in the *name* or in the *brand* attributes. We expect that multi-source EM tasks with a low level of VH are easy to solve. Considering that for such tasks the matching and non-matching pairs are almost perfectly separable, the learner can reach a high prediction accuracy even with a small number of labeled record pairs. In contrast, given a task with high VH, we expect that a small number of labeled record pairs can lead to the overfitting of the learner. In that case, exploiting the correspondence graph for directing the query strategy to pick record pairs that are likely falsely predicted by the overfitted learner, can be helpful.

Value Pattern Overlap. The dimension of value pattern overlap (VPO) refers to the amount of groups of data sources adhering to the same attribute value patterns. The overlap of value patterns results from similar lexical patterns or types of spelling errors within the record values of the data sources. For example, within the e-commerce phone product domain, different e-shops may share one of the following lexical patterns for representing the names of smartphones: [model] [model generation] *e.g. i-phone 4s* or [model] [model generation] [product type], *e.g. i-phone 4s smartphone*. Pairs of data sources with overlapping value patterns can form groups of matching tasks sharing the same matching patterns. We illustrate this observation with the example of Fig. 1. The data sources A and B of the multi-source EM task contain the same value pattern for the name attribute [model] [model generation], while the brand value is in both sources misspelled. The name attribute values of the data sources C and D adhere to the pattern: [model] [model generation] [product type]. Consequently, we can consider that in this example task there exist two groups of data sources adhering to the same value patterns, i.e. [A, B] and [C, D]. Combining the data sources across the two groups pairwise, i.e. A-C, A-D, B-C and B-D, results in two source-tasks with the same matching pattern, while in total three matching patterns emerge for covering all two-source tasks, as shown in Fig. 1d.

We calculate the value pattern overlap as $\frac{1}{G_{VPO}}$, with G_{VPO} indicating the number of groups of data sources having the same value pattern. Following the example of Fig. 1 and considering that there exist two groups of data sources with the same value pattern, the VPO level is computed to be 0.5. A VPO of 1 indicates that all data sources contain records with the same value pattern and therefore construct pairwise matching tasks with the same underlying matching patterns. On the other hand, a value pattern overlap of 0 indicates that the records of each data source contain different value patterns and therefore the pairwise matching tasks contain distinct underlying matching patterns. We expect those query strategies that can identify groups of matching tasks that share similar matching patterns and distribute the queries so that all groups are covered, can outperform query strategies that ignore the grouping information.

It is worth noting that the calculation of the three profiling dimensions requires knowledge of the actual labels of the record pairs. While this allows us to analyze the impact of the profile of a multi-source task on the performance of different active learning methods, it does not enable the upfront selection of active learning methods, which is out of the scope of our work.

4   ALMSERgen: A Multi-source EM Task Generator

In order to enable the systematic analysis and comparison of active learning methods applied on multi-source EM tasks with different characteristics, we develop ALMSERgen, a multi-source EM task generator. ALMSERgen takes as input a set of records and generates a multi-source EM task by replicating the input record set and injecting transformations along the three dimensions explained in Sect. 3. In the following, we present each component of ALMSERgen along with Fig. 2 which provides an illustrated example of curating a multi-source EM task given a pre-defined configuration.

Step 1: Complement Initial Set. Depending on the domain and the integration task at hand, different attributes might be relevant for matching. For example, for the task of matching phone records, one might consider that the combination of phone name and phone brand identifies a distinct phone, while in more fine-grained matching tasks the phone colour might also be important. We call the set of attributes that is useful for distinguishing real-world entities of a specific domain, *identifying attributes*, and they are given as input to ALMSERgen. Considering that the input set of records may not contain enough examples for the identifying attributes to show, ALMSERgen artificially activates the identifying attributes by replicating 20% of the input records and replacing a subset of the identifying attribute values with random non-identical values of the same attribute. The non-identifying attribute values are simply copied from the original record to the replicated records. In the example of Fig. 2 the input set contains three records. Given that the identifying attributes are configured to be *name* and *brand*, ALMSERgen generates the additional records *2.iphone 4s - htc* and *5.galaxy s21 - apple* which represent phone entities different from the ones that the records 1 and 4 represent.

Step 2: Distribute Records over Sources. Next, the entity overlap level (EO) of the multi-source task is fixed. Given a pre-defined EO level value $\in [0, 1]$,

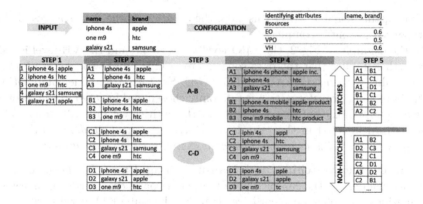

Fig. 2. Example of multi-source EM task curation with ALMSERgen.

we iterate over all initial entities (IE) produced in Step 1 and add a subset of them, the amount of which equals to $EO \times |IE|$ to more than two data sources. In order to decide in how many more than two sources the selected entities should be added, we follow a power-law distribution, i.e. most entities are contained in a few sources while a few entities are contained in all sources. Therefore, given that an entity is selected to be added in more than 2 sources, the probability that it is added in x data sources, is $1/x$, with $x > 2$. In the illustrated example, the EO level is set to 0.6, i.e. 60% of the five entities produced in Step 1, are added to more than two sources: the entity with id 1 which is added in 4 sources and the entities with ids 2 and 3 which are added in 3 sources.

Steps 3–4: Inject Groups of Patterns. In the next step, the levels of value pattern overlap (VPO) and value heterogeneity (VH) are fixed. These two dimensions are interwoven, considering that VH defines how many records across all data sources contain heterogeneous representations for the same real-world entity and VPO controls the similarity of the value patterns of the records across all data sources. Given the pre-defined VPO level, ALMSERgen creates groups of data sources to which the same value pattern will be injected. The same value pattern is injected in the records of the groups representing a subset of entities, the amount of which is $VH \times |IE|$, with IE being the initial entities generated in Step 1 and VH being the value heterogeneity level in the range of $[0, 1]$.

A value pattern comprises of distinct combinations of attributes and value transformations. ALMSERgen offers the following value transformations, similar to existing data generators for entity matching [12,28]: 1. Addition of random characters, 2. Deletion of random characters, 3. Modification of random characters, 4. Shuffling and modification on word level, 5. Shuffling of words, 6. Addition of random words, 7. Subtraction of $(5/10/20)\%$ of the value, and 8. Addition of $(5/10/20)\%$ of the value. Transformations 1–6 are performed on string attributes, while transformations 7–8 are applied only on numerical attributes. Finally, for the transformations 1–4, a level of severity in the range of $[0.1, 0.5]$ is randomly picked, i.e. maximum of 50% of the characters can be modified or deleted, in order to ensure that the identity of each entity is not completely altered and remains distinguishable. After this step, the curation of the data sources of the multi-source setting is completed.

In the example of Fig. 2, the VPO level is set to 0.5 and the VH level is 0.6. This further implies that the data sources are grouped into two groups of over-lapping value patterns, G1: A-B and G2: C-D. For each group one combination of attribute-value transformation is randomly chosen and injected in the records describing 60% of the entities of Step 1, i.e. the entities with ids 1,3, and 5. The value pattern injected in the records of G1 is *addition of random words*. The value pattern for G2 is *deletion of random characters with severity 0.2*.

Step 5: Derive Matching and Non-matching Pairs. In the final step, ALMSERgen derives the complete set of matching pairs considering all pairwise combinations of replicated records referring to the same real-world entity, e.g. A1-B1. For deriving hard non-matching pairs, we extract all combinations of records and their corresponding negative examples injected in Step 1, e.g. A1-C2. Additionally, we randomly pick easy non-matching record pairs, e.g. A1-C3, until the ratio of matching to non-matching pairs is 1/3.

5 Experimental Setup and Analysis

In this section, we present the details of our experimental setup, including the ALMSERgen configuration as well as the active learning setup and query strate-gies used in our analysis. Next, we present the active learning results on the generated tasks and discuss our main findings on the performance of different active learning methods with respect to the profiling characteristics of the tasks. Finally, we verify our findings using five benchmark tasks from the related work. The code and tasks used for all experiments are publicly available.[1]

5.1 Experimental Setup

ALMSERgen Configuration. We provide a set of 1000 deduplicated song records as input to ALMSERgen. The input data set is a subset of the last.fm song data set.[2] Each song record is described with the following four attributes: title, release, artist, and country. We configure all of the four attributes as iden-tifying ones and set the number of curated data sources for each generated multi-source EM task to 6. We iterate in steps of 0.2 in the range [0.0, 1.0] for the dimensions of entity overlap (EO) and value pattern overlap (VPO) and in steps of 0.1 in the range [0.2, 0.8] for the value heterogeneity (VH) dimension. The defined ranges and steps result in the curation of 252 multi-source matching tasks. Generating a single task with ALMSERgen takes approx-imately 50 seconds. Generating the continuum of 252 multi-source tasks requires 3.5 hours on a Linux server with Intel Xeon 2.2 GHz processor.

---

[1] https://github.com/wbsg-uni-mannheim/ALMSER-GEN.
[2] http://millionsongdataset.com/lastfm/.

Active Learning Setup. We consider a pool-based active learning setting and similarity-based features for representing the record pairs, similar to many related works [3,13,18,25]. In such a setting, a pool of unlabeled record pairs is available to the active learning query strategy which assesses the informativeness considering a set of criteria. The most informative record pair is selected, annotated as matching or non-matching and added to the labeled set. The labeled set is used for training the *learner*, i.e. a classification model.

We initialize the pool with 70% of the matching and non-matching record pairs resulting from the final step of ALMSERgen and remove all labels. The remaining 30% of the record pairs are used as a test set. We allow 200 iterations for each active learning experiment. The average size of the pool across the 252 multi-source tasks is 11,290 pairs. In each iteration, one record pair of the pool is selected for annotation, i.e. 200 record pairs (<2% of the complete pool on average) have been labeled in total by the end of each experimental run. If the query strategy assigns the maximum informativeness score to more than one record pair of the pool, one of them is randomly selected for annotation. We use a random forest classifier as learner and measure the F1 score of its predictions on the test set after each iteration. We conduct three runs for each multi-source task and each active learning method. Finally, we report the area under the mean F1 curve for iterations 50 to 200 on the test set, which we abbreviate with F1-AUC. F1-AUC is calculated as the definite integral between two points, e.g. iteration 50 to 200, and is typically used for measuring the overall performance of an active learning method across multiple iterations [19,31]. A larger area under the mean F1 curve signifies overall better results in terms of F1 score.

Active Learning Query Strategies. In our experiments, we compare the performance of three active learning methods which only differ with respect to the query strategy. In the following, we present the three active learning query strategies which we compare in our analysis.

HeALER is a committee-based active learning method developed by Chen et al. [3]. The query strategy of HeALER uses a committee of five heterogeneous classification models to evaluate the informativeness of all pool record pairs. In every active learning iteration, each classification model in the committee is trained on the current labeled set. Next, it is applied on the record pairs of the pool and votes its predictions, i.e. every record pair in the pool receives five votes. The record pairs with the maximum disagreement calculated with vote entropy, are considered to be the most informative.

ALMSER is a graph-based active learning method introduced in our previous work [25] that is tailored to the multi-source EM task. The query strategy of ALMSER exploits the correspondence graph of the multi-source matching setting in order to select record pairs that are likely falsely predicted by the learner. In each active learning iteration, the learner is trained on the current labeled set and predicts *matching* or *non-matching* pseudo-labels for all record pairs in the pool. The pseudo-labels together with the labeled set are used to construct a correspondence graph, with the edges of the graph denoting matching relations between the nodes-records. A sequence of cleansing steps is applied in order to remove likely false matching edges. Finally, considering graph transitivity the pool record pairs are assigned graph-inferred labels. The query strategy of ALMSER assigns binary informativeness scores to the pool record pairs: 1 if there is a conflict between the learner and the graph-inferred prediction, otherwise 0.

While committee-based query strategies like HeALER, aim to select instances for which the committee of models produces non-confident predictions, the query strategy of ALMSER uses the correspondence graph to pick instances that are most likely predicted wrong by the learner. These disagreements between the graph-inferred labels and the learner pseudo-labels can hint towards matching patterns that are not covered yet by the learner.

ALMSERgroup. A multi-source matching task can contain groups of two-source matching tasks sharing the same underlying matching patterns, as explained in Sect. 3. We hypothesize that exploiting such grouping information can direct the active learning strategy to select record pairs covering all underlying matching patterns of the complete multi-source task with a smaller amount of annotations. We illustrate our hypothesis with the example of Fig. 1. The pairwise combinations of the four data sources result in six matching tasks which given the underlying matching patterns can be grouped into three groups, as shown in Fig. 1d. In such a setting, the active learning query strategy should distribute the queries for labeling over the tasks A-B, C-D and any of the {A-C, A-D, B-C, B-D}, as the latter have all the same underlying matching pattern. However, to the best of our knowledge, none of the existing active learning query strategies for entity matching exploits such grouping information.

In order to investigate whether the labeling effort can be further reduced by exploiting such grouping signals, we develop ALMSERgroup, a variation of the ALMSER query strategy. ALMSERgroup filters the pool to only include record pairs belonging to matching tasks that are representative of a cluster of similar matching tasks. We explain below how representative tasks are selected. In this way, ALMSERgroup avoids picking record pairs for annotation from similar tasks. During active learning, the ALMSER query strategy is applied using the reduced pool. In the case of no disagreements between the learner predictions and the graph-inferred labels among the record pairs of the reduced pool, HeALER is used as a fallback query strategy.

In order to identify two-source tasks with similar matching patterns in an unsupervised way, we first compute the task relatedness (TR) between all pairs of two-source tasks, a metric introduced by Thirumuruganathan et al. [32]. TR calculates how similar two tasks are by training a logistic regression classifier to predict the task from which each record pair originates. A high prediction quality signifies that the two tasks are dissimilar, while a low prediction quality signifies that the tasks are similar and are expected to have the same underlying matching patterns. We measure the prediction quality of the classifier using the Matthews correlation coefficient (MCC) and calculate the TR score as $1 - MCC$, similar to [32]. Given the TR scores of each pairwise combination of two-source tasks,

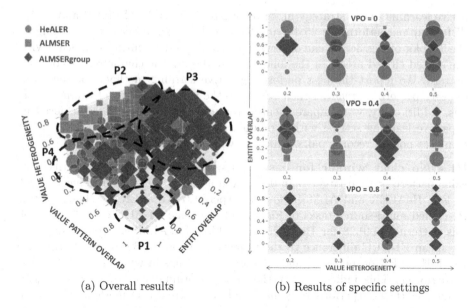

(a) Overall results   (b) Results of specific settings

Fig. 3. Outperforming AL methods per task. The size of the markers indicates the F1-AUC difference to the runner-up method.

we cluster them such that the overall mean TR score of all clusters is maximized. We determine the optimal number of clusters by penalizing the overall mean TR score with a penalty factor α multiplied by the number of clusters. In this way, we prefer smaller amounts of clusters over larger ones which results in a smaller pool of representative record pairs for the query strategy to choose from. Finally, we identify the most representative two-source tasks of each cluster, considering their TR to all other tasks of the same cluster, and select only the record pairs of the representative tasks for initializing the unlabeled pool.

5.2 Analysis of Experimental Results of Generated Tasks

We compare the results of three active learning methods using the HeALER, ALMSER, and ALMSERgroup query strategies on the 252 generated tasks and identify which signals are relevant for query selection given the profiling characteristics of the tasks. Throughout our analysis, we use the 2D and 3D scatter plots of Fig. 3 which indicate the winning active learning method for each generated task with different colours and markers. The size of the markers shows the difference of the winning method to the second-best method in terms of F1-AUC for iterations 50 to 200, i.e. large dots signify clear winners while smaller dots indicate winning methods that are only slightly better than the runners-up.

Figure 3a shows the overall comparison results of the three active learning methods on the continuum of the 252 multi-source tasks along the three dimensions described in Sect. 3: value heterogeneity (VH), value pattern overlap (VPO), and entity overlap (EO). In 41.6% of the tasks, HeALER is the winning active learning query strategy in terms of F1-AUC, while ALMSER and ALM-SERgroup outperform for 25.4% and 33% of the tasks respectively. Looking at the 3D plot of Fig. 3a, we can observe four main patterns which we indicate with the dotted circled areas. In the following, we discuss the characteristics of each pattern. We report the best performing active learning methods for the tasks of every pattern by relating their results to the runner-up active learning methods. Additionally, we compare them to the upper bound F1 scores achieved in a passive learning setting with a random forest classifier being trained on the complete pool of records pairs, which we will refer to as *passive F1*.

P1 - No clear winner for easy tasks. For all tasks with an entity and value pattern overlap larger than 0.6 as well as a low value heterogeneity of 0.3 or less, HeALER and ALMSERgroup outperform ALMSER, as shown in the P1-circled multi-source tasks of Fig. 3a. The average F1-AUC over all tasks of this pattern is 140.28 for HeALER and 140.70 for ALMSERgroup. However, the mean F1-AUC difference to the runner-ups is only 0.35 for the settings in which HeALER outperforms and 0.68 for the settings in which ALMSERgroup outperforms. This indicates that the best performing methods are not clear winners as they outperform only marginally the second best method. The mean passive F1 score for all tasks adhering to this pattern is 0.983 while the mean F1 of the best performing active learning methods at the final 200th iteration is 0.961. We consider such tasks rather easy to solve as the high overlap of mostly homogeneous entity records eases the discovery of the few distinct matching patterns, i.e. selecting one matching record pair for annotation can help the classifier to learn the underlying pattern of many other record pairs at once.

P2 - Graph signals are helpful for tasks with high value heterogeneity. In 71.6% of the tasks with a value heterogeneity level larger than 0.5, ALMSER overperforms with a mean F1-AUC difference of 2.95, given that the value pattern overlap level is 0.6 or below. The mean passive F1 score for all tasks of this pattern is 0.888 while the mean F1 of the best performing active learning methods at the final 200th iteration is 0.828. Such tasks are harder to solve as they contain heterogeneous value representations for a large number of entities, while the low value pattern overlap level signifies that there exist many different underlying matching patterns. Exploiting the signals from the correspondence graph leads to the faster discovery of all underlying matching patterns in comparison to committee-based query strategies. However, this observation only holds when there exists a minimum entity overlap, i.e. $EO > 0.0$. For multi-source tasks with $EO = 0$, i.e. all entities are represented by one record in a maximum of two data sources, the correspondence graph does not have a rich structure as the maximum component size is 2. Therefore exploiting graph signals cannot lead to the selection of informative query candidates. This causes the ALMSER query strategy to underperform in 88% of the generated tasks with $EO = 0$.

P3 - Grouping signals are helpful for tasks with low value heterogeneity and high value pattern overlap. In 55.5% of the tasks with a value heterogeneity lower than 0.5 and a value pattern overlap larger than 0.5, ALM-

SERgroup is the winning active learning strategy with a mean F1-AUC difference to the runner-up method of 1.52. However, ALMSERgroup does not deliver better results over HeALER for multi-source tasks with low value pattern overlap.

We illustrate and further analyze this observation with Fig. 3b depicting the winning strategies for tasks with a value heterogeneity level of 0.5 or lower and three different value pattern overlap levels: 0.0, 0.4, and 0.8. We can see that, for the multi-source matching tasks where the value pattern overlap is 0, i.e. different underlying matching patterns exist in each two-source task of the setting, HeALER outperforms ALMSERgroup in 66% of the settings. The mean F1-AUC difference to the runner-up method is 3.30 while for the tasks where ALMSER-group outperforms the mean F1-AUC difference to the runner-up method is 1.30. With the increase of the value pattern overlap level, we can observe that the grouping signal starts contributing to the query selection strategy. For VPO = 0.4, HeALER outperforms in 54% of the tasks with a mean F1-AUC difference to the runner-up method of 1.93, while the mean F1-AUC difference for the settings where ALMSERgroup is the best performing query strategy is 2.31. Finally, ALMSERgroup performs the best in 58.3% of the tasks when the value pattern overlap level is 0.8 with a mean F1-AUC difference to the second-best method of 2.26, while HeALER outperforms in 37% of the tasks with a marginal F1-AUC difference of 0.98.

P4 - Graph and grouping signals are not needed for tasks with low value heterogeneity and low pattern overlap. In 89.5% of the tasks with a value heterogeneity of 0.5 or lower, the HeALER and ALMSERgroup query strategies outperform ALMSER independently from the other two dimensions. This indicates that graph signals do not contribute in the case of multi-source tasks with a low value heterogeneity. The F1-AUC difference to the runner-up methods is 2.26 and 1.71 for HeALER and ALMSERgroup, respectively. In terms of F1 scores, the tasks of this pattern lie between the results of the tasks in P1 and P2. The mean passive F1 is 0.941 and the mean F1 of the best performing active learning methods at the 200th iteration, is 0.91.

As already introduced in the analysis of P3, the contribution of grouping signals is positively related to the value pattern overlap level, i.e. grouping signals contribute less for tasks with a low value pattern overlap level. More concretely, we observe that in 67% of the tasks with a value heterogeneity and a value pattern overlap of 0.5 or lower, ALMSERgroup underperforms the other two methods. In order to investigate the reasons that grouping signals do not contribute to tasks with a low value pattern overlap, we perform a two-step analysis: First, we evaluate how representative the metric of task relatedness is for finding groups of two-source matching tasks with similar patterns, and second, we evaluate to which extent ALMSERgroup selects representative two-source tasks covering all distinct matching patterns of each multi-source task.

For the first part of our analysis, we calculate the cosine similarity of the naive transfer learning (NTL) and the task relatedness (RLTD) scores for each combination of two-source matching tasks of all multi-source tasks. A high naive transfer learning score between a pair of two-source tasks, e.g. A-B and C-D, indicates they have the same underlying matching patterns, as a model trained on the record pairs of task A-B, performs well when applied on task C-D. A high similarity between the NTL scores and the RLTD scores implies that the second is a good unsupervised approximation of the first and can therefore lead to the discovery of groups of similar matching tasks. We find that higher VPO levels lead to the higher similarity of NTL and RLTD scores: for tasks with VPO = 1.0 the similarity of the NTL and RLTD scores is 0.81, while it drops to 0.75 and 0.69, for tasks with VPO = 0.6 and VPO = 0.2, respectively. Therefore, we can conclude that task relatedness can more efficiently lead to finding and grouping similar two-source tasks in the case of multi-source EM tasks with high VPO.

For the second part of our analysis, we evaluate in how many of the multi-source tasks ALMSERgroup selects a sufficient subset of two-source tasks to query from, i.e. a sufficient subset contains at least one two-source task per group of tasks with similar matching patterns. Similar to the previous finding, we observe that ALMSERgroup better identifies sufficient subsets of two-source tasks to query from for higher VPO levels: ALMSERgroup selects a sufficient subset of two-source tasks in 100% and 88% of multi-source tasks with VPO = 1.0 and VPO = 0.8, respectively. Additionally, we observe that for high VPO levels ALMSERgroup achieves a large candidate reduction: for VPO 1.0, ALMSERgroup only selects candidates from a maximum of 4 out of the 15 two-source tasks in 90% of the multi-source tasks. This further explains why ALMSERgroup generally outperforms HeALER and ALMSER for tasks with high pattern overlap. In contrast, with the decrease of the VPO level, it is harder for ALMSERgroup to identify all relevant two-source tasks to query from. For example, ALMSERgroup only identifies a sufficient subset of two-source tasks for 28% and 14% of the multi-source tasks with VPO = 0.4 and VPO = 0.2, respectively.

5.3   Analysis of Experimental Results of Benchmark Tasks

In this section, we verify our findings concerning the impact of the profile of multi-source EM tasks on the performance of the three active learning methods using the HeALER, ALMSER and ALMSERgroup strategies, on five benchmark tasks. The benchmark tasks cover the domains music, products, and restaurants and have been previously been used in the related work [25,27]. The tasks are described in detail in [25].

Table 1 contains profiling information for the benchmark tasks along the three profiling dimensions. We compute the value heterogeneity and the entity overlap as described in Sect. 3. For estimating the value pattern overlap level, we use the naive transfer learning scores of a random forest classifier for all pairs of two-source tasks of each benchmark multi-source task and extract the smallest subset of two-source tasks that best generalizes over all two-source tasks.

We present the active learning results for the five benchmark multi-source tasks in Table 1 and report the F1-AUC for iterations 50–200, the F1-AUC difference of the outperforming to the runner-up method as well as the mean F1 scores of three experimental runs for specific active learning snapshots at the

Table 1. Profile and active learning results of benchmark multi-source EM tasks.

Task	VH EO VPO	Method	F1-AUC	F1-AUC diff.	F1@85	F1@150	F1@200
		HeALER	135.62		0.893	0.912	0.918
computers	0.40 0.44 1.0	**ALMSER**	**138.96**	1.32	0.921	0.932	0.937
		ALMSERgroup	137.64		0.904	0.931	0.931
		HeALER	127.66		0.841	0.850	0.866
computers_mut	0.43 0.44 0.8	ALMSER	128.95	1.24	0.824	0.879	0.883
		ALMSERgroup	**130.19**		0.864	0.877	0.883
		HeALER	**140.07**		0.931	0.941	0.945
MusicBrainz	0.19 0.50 0.6	ALMSER	138.37	1.70	0.913	0.930	0.934
		ALMSERgroup	137.02		0.888	0.926	0.918
		HeALER	**132.43**		0.857	0.895	0.908
MusicBrainz_mut	0.14 0.50 0.4	ALMSER	131.38	1.05	0.868	0.889	0.896
		ALMSERgroup	127.19		0.820	0.879	0.888
		HeALER	**138.51**		0.921	0.927	0.937
restaurants	0.14 0.35 0.8	ALMSER	138.21	0.30	0.918	0.923	0.926
		ALMSERgroup	137.48		0.913	0.920	0.921

85th, 150th and final 200th iteration. We observe that ALMSER and ALMSER-group outperform HeALER for the computers and computers_mut tasks. The profiling dimensions of these tasks lie between patterns P2 and P3: graph signals contribute due to the rather high value heterogeneity (see column VH in Table 1) while grouping signals contribute due to the high value pattern overlap (column VPO) level.

In comparison to HeALER, we observe that graph and grouping signals contribute until the 200th iteration while the differences in F1 score of ALMSER and ALMSERgroup appear only during the earlier iterations. After the 150th iteration both ALMSER and ALMSERgroup converge to similar results. The Musicbrainz and MusicBrainz_mut tasks verify the pattern P4 of our analysis. Given the low value heterogeneity and value pattern overlap levels of the tasks, graph and grouping signals are not helpful for improving the active learning results over HeALER. Finally, pattern P1 of our analysis is confirmed by the results of the restaurants task which has a low value heterogeneity and a high value pattern overlap. Although HeALER outperforms the other two methods for this task in terms of F1-AUC, the F1-AUC difference to the runner-up method is only 0.30, indicating that there is no clear winner for the task.

6   Conclusion

This paper explored the impact of the characteristics of multi-source EM tasks on the performance of three active learning methods which utilize different types of signals for selecting record pairs for labeling. We based our analysis on a continuum of 252 generated multi-source matching tasks and additionally verified our findings using five benchmark tasks. Our findings showed that all methods perform equally well for easy multi-source EM tasks, characterized by a high entity overlap and homogeneous attribute values. With the increase of the value heterogeneity of records describing the same entity, group signals were shown to improve the active learning performance, given that there exist a few groups of two-source matching tasks sharing the same underlying matching patterns. Finally, exploiting graph signals as part of the query strategy was shown to improve the active learning performance for tasks containing large amounts of matching records with heterogeneous attribute values.

References

1. Achichi, M., Cheatham, M., et al.: Results of the ontology alignment evaluation initiative 2017. In: Proceedings of OM 2017–12th ISWC Workshop on Ontology Matching, pp. 61–113 (2017)
2. Bellare, K., Curino, C., Machanavajihala, A., et al.: WOO: a scalable and multi-tenant platform for continuous knowledge base synthesis. PVLDB 6(11), 1114–1125 (2013)
3. Chen, X., Xu, Y., Broneske, D., Durand, G.C., Zoun, R., Saake, G.: Heterogeneous committee-based active learning for entity resolution (HeALER). In: Welzer, T., Eder, J., Podgorelec, V., Kamišalić Latifić, A. (eds.) ADBIS 2019. LNCS, vol. 11695, pp. 69–85. Springer, Cham (2019). https://doi.org/10.1007/978-3-030-28730-6_5
4. Christen, P.: Data Matching: Concepts and Techniques for Record Linkage, Entity Resolution, and Duplicate Detection. Data-Centric Systems and Applications (2012)
5. Christophides, V., Efthymiou, V., et al.: An overview of end-to-end entity resolution for big data. ACM Comput. Surv. (CSUR) 53(6), 1–42 (2020)
6. Elmagarmid, A., Ipeirotis, P., et al.: Duplicate record detection: a survey. IEEE Trans. Knowl. Data Eng. 19(1), 1–16 (2007)
7. Ferrara, A., Montanelli, S., Noessner, J., Stuckenschmidt, H.: Benchmarking matching applications on the semantic web. In: Antoniou, G., et al. (eds.) ESWC 2011. LNCS, vol. 6644, pp. 108–122. Springer, Heidelberg (2011). https://doi.org/10.1007/978-3-642-21064-8_8
8. Halevy, A., Rajaraman, A., Ordille, J.: Data integration: the teenage years. In: Proceedings of VLD, pp. 9–16 (2006)
9. Heath, T., Bizer, C.: Linked Data: Evolving the Web Into a Global Data Space. Synthesis Lectures on the Semantic Web. Morgan & Claypool Publishers (2011)
10. Hildebrandt, K., Panse, F., et al.: Large-scale data pollution with Apache spark. IEEE Trans. Big Data 6(2), 396–411 (2020)
11. Huang, J., Hu, W., Li, H., Qu, Y.: Automated comparative table generation for facilitating human intervention in multi-entity resolution. In: Proceedings of SIGIR, pp. 585–594 (2018)
12. Ioannou, E., Rassadko, N., Velegrakis, Y.: On generating benchmark data for entity matching. J. Data Semant. 2(1), 37–56 (2013)
13. Isele, R., Bizer, C.: Active learning of expressive linkage rules using genetic programming. J. Web Semant. 23, 2–15 (2013)
14. Kasai, J., Qian, K., et al.: Low-resource deep entity resolution with transfer and active learning. In: Proceedings of ACL, pp. 5851–5861 (2019)
15. Konda, P., et al.: Magellan: toward building entity matching management systems over data science stacks. PVLDB 13, 1581–1584 (2016)

16. Konyushkova, K., Raphael, S., Fua, P.: Learning active learning from data. In: Proceedings of NIPS, p. 4228–4238 (2017)
17. Köpcke, H., Thor, A., Rahm, E.: Evaluation of entity resolution approaches on real-world match problems. VLDB Endow. **3**(1–2), 484–493 (2010)
18. Meduri, V., Popa, L., et al.: A comprehensive benchmark framework for active learning methods in entity matching. In: Proceedings of SIGMOD, pp. 1133–1147 (2020)
19. Mozafari, B., Sarkar, P., Franklin, M., Jordan, M., Madden, S.: Scaling up crowd-sourcing to very large datasets: a case for active learning. VLDB Endow. **8**(2), 125–136 (2014)
20. Nafa, Y., et al.: Active deep learning on entity resolution by risk sampling. Knowl.-Based Syst. **236**, 107729 (2022)
21. Nentwig, M., Hartung, M., Ngonga Ngomo, A.C., Rahm, E.: A survey of current link discovery frameworks. Semant. Web **8**(3), 419–436 (2017)
22. Ngonga Ngomo, A.-C., Lyko, K.: EAGLE: efficient active learning of link specifica-tions using genetic programming. In: Simperl, E., Cimiano, P., Polleres, A., Corcho, O., Presutti, V. (eds.) ESWC 2012. LNCS, vol. 7295, pp. 149–163. Springer, Hei-delberg (2012). https://doi.org/10.1007/978-3-642-30284-8_17
23. Papadakis, G., Ioannou, E., Thanos, E., Palpanas, T.: The four generations of entity resolution. Synthesis Lect. Data Manage. **16**(2), 1–170 (2021)
24. Primpeli, A., Bizer, C.: Profiling entity matching benchmark tasks. In: Proceedings of CIKM, pp. 3101–3108 (2020)
25. Primpeli, A., Bizer, C.: Graph-boosted active learning for multi-source entity res-olution. In: Hotho, A., et al. (eds.) ISWC 2021. LNCS, vol. 12922, pp. 182–199. Springer, Cham (2021). https://doi.org/10.1007/978-3-030-88361-4_11
26. Qian, K., Popa, L., Sen, P.: Active learning for large-scale entity resolution. In: Proceedings of CIKM, pp. 1379–1388 (2017)
27. Saeedi, A., Peukert, E., Rahm, E.: Comparative evaluation of distributed clus-tering schemes for multi-source entity resolution. In: Kirikova, M., Nørvåg, K., Papadopoulos, G.A. (eds.) ADBIS 2017. LNCS, vol. 10509, pp. 278–293. Springer, Cham (2017). https://doi.org/10.1007/978-3-319-66917-5_19
28. Saveta, T., Daskalaki, E., Flouris, G., Fundulaki, I., Herschel, M., Ngomo, A.-C.N.: LANCE: piercing to the heart of instance matching tools. In: ISWC 2015. LNCS, vol. 9366, pp. 375–391. Springer, Cham (2015). https://doi.org/10.1007/978-3-319-25007-6_22
29. Settles, B.: Active Learning: Synthesis Lectures on Artificial Intelligence and Machine Learning. Morgan & Claypool Publishers (2012)
30. Shen, W., DeRose, P., Vu, L., et al.: Source-aware entity matching: a compositional approach. In: Proceedings of ICDE, pp. 196–205 (2007)
31. Sherif, M.A., Dreßler, K., Ngomo, A.C.N.: LIGON-link discovery with noisy ora-cles. In: Proceedings of Ontology Matching Workshop (ISWC), pp. 48–59 (2020)
32. Thirumuruganathan, S., Parambath, S.A.P., et al.: Reuse and adaptation for entity resolution through transfer learning. arXiv preprint arXiv:1809.11084 (2018)
33. Ye, Y., Talburt, J.: Generating synthetic data to support entity resolution educa-tion and research. J. Comput. Sci. Coll. **34**(7), 12–19 (2019)

Optimal ABox Repair w.r.t. Static $\mathcal{EL}$ TBoxes: From Quantified ABoxes Back to ABoxes

Franz Baader[ID], Patrick Koopmann[ID], Francesco Kriegel[✉][ID],
and Adrian Nuradiansyah[ID]

Theoretical Computer Science, TU Dresden, Dresden, Germany
{franz.baader,patrick.koopmann,francesco.kriegel,
adrian.nuradiansyah}@tu-dresden.de

Abstract. Errors in Description Logic (DL) ontologies are often detected when a reasoner computes unwanted consequences. The question is then how to repair the ontology such that the unwanted consequences no longer follow, but as many of the other consequences as possible are preserved. The problem of computing such optimal repairs was addressed in our previous work in the setting where the data (expressed by an ABox) may contain errors, but the schema (expressed by an $\mathcal{EL}$ TBox) is assumed to be correct. Actually, we consider a generalization of ABoxes called quantified ABoxes (qABoxes) both as input for and as result of the repair process. Using qABoxes for repair allows us to retain more information, but the disadvantage is that standard DL systems do not accept qABoxes as input. This raises the question, investigated in the present paper, whether and how one can obtain optimal repairs if one restricts the output of the repair process to being ABoxes. In general, such optimal ABox repairs need not exist. Our main contribution is that we show how to decide the existence of optimal ABox repairs in exponential time, and how to compute all such repairs in case they exist.

1 Introduction

Description Logics (DLs) [2] are a successful family of logic-based knowledge representation languages, which are employed in various application domains, but arguably their most prominent success was the adoption of the DL-based language OWL[1] as the standard ontology language for the Semantic Web. A DL knowledge base (aka ontology) consists of a TBox and an ABox. In the former, concepts can be used to state terminological constraints as so-called general concept inclusions (GCIs). For example, the concept $\exists parent.(Famous \sqcap Rich)$ describes individuals that have a parent that is both famous and rich, and the GCI $\exists friend.Famous \sqsubseteq Famous$ states that individuals that have a famous friend are famous themselves. The expressiveness of a DL depends on which constructors for building concepts are available. The concepts in our example use the

---

[1] https://www.w3.org/OWL/.

P. Groth et al. (Eds.): ESWC 2022, LNCS 13261, pp. 130–146, 2022.
https://doi.org/10.1007/978-3-031-06981-9_8

constructors conjunction ($\sqcap$) and existential restriction ($\exists r.C$), which together with the top concept ($\top$) are the ones available in the DL $\mathcal{EL}$, to which we restrict our attention here. While being quite inexpressive, $\mathcal{EL}$ is nevertheless frequently used for building ontologies,[2] and it has the advantage over more expressive DLs that reasoning is polynomial w.r.t. $\mathcal{EL}$ ontologies. In the ABox, one can relate named individuals with concepts and with each other. For example, the concept assertion ($\exists parent.Rich$)(BEN) states that Ben has a rich parent, and the role assertion $friend(BEN, JOHN)$ says that Ben has John as friend. If concept assertions are restricted to employing only concept names, like $Famous(JOHN)$, rather than complex concepts, then the ABox is called simple. DL systems provide their users with inference services that automatically derive implicit consequences such as instance relationships. For example, given the ABox assertions and the GCI introduced above, we can derive that Ben is famous, i.e., that the assertions $Famous(BEN)$ follows from this ontology.

Although DL reasoners are usually sound (i.e., only derive instance relationship that indeed follow from the ontology), a computed consequence may still be incorrect in the application domain, due to the fact that the modelling of the domain in the ontology is erroneous. The question is then how to repair the ontology such that one gets rid of the unwanted consequences, but retains as many consequences as possible. Classical repair approaches that are based on removing axioms from the ontology [8, 11, 15, 16, 18, 19] are not optimal since, by removing large axioms, one may also lose information that does not contribute to the unwanted consequence. For example, if the concept assertion for John is ($Famous \sqcap Rich$)($JOHN$) rather than just $Famous(JOHN)$, then to get rid of the consequence $Famous(BEN)$ we need to remove the whole assertion, and thus unnecessarily also lose the information that John is rich.

Extending on our previous work in [5, 7], we investigated in [3] how to compute optimal repairs in a setting where the ABox may contain errors, but the TBox is assumed to be a correct $\mathcal{EL}$ TBox, and thus remains unchanged. More precisely, we consider a generalization of ABoxes called quantified ABoxes (qABoxes) both as input for and as result of the repair process since this allows us to retain more consequences. Such a qABox is a simple ABox where, however, some of the individuals are anonymized, which is formally expressed by existentially quantifying over them. In [3], we introduce two different notions of repair, depending on which entailment relation between qABoxes is considered: classical logical entailment or IQ-entailment, where the latter retains as many instance relationships as possible (but not necessarily answers to conjunctive queries). For the IQ case, we show that optimal IQ-repairs always exist and can be computed in exponential time. In the worst case, such repairs may be exponentially large and there may be exponentially many of them. Reusing an example from the introduction of [3], let us assume that the input ABox contains the information that Ben has a parent, Jerry, that is both rich and famous, that the TBox contains the GCI $Famous \sqsubseteq Rich$, and that we want to remove the consequence ($\exists parent.(Rich \sqcap Famous)$)($BEN$). Using the optimized repair

---

[2] For example, the large medical ontology SNOMED CT is an $\mathcal{EL}$ ontology.

approach of [3], we obtain the following qABox as one of the optimal IQ-repairs:
$\exists\{y\}.\{parent(BEN, y), Rich(y), Famous(JERRY), Rich(JERRY)\}$.

The advantage of using qABoxes rather than ABoxes for repair is that more information can be retained (e.g. the fact that Ben has a rich parent). The disadvantage is that, though anonymized individuals are part of the OWL standard, DL systems usually do not accept them as input. Thus, the question arises whether one can also obtain optimal repairs if one restricts the output of the repair process to being ABoxes. In the above example, the qABox obtained as an optimal IQ-repair can actually be expressed by an ABox with complex concept assertions: $\{(\exists parent.Rich)(BEN), Famous(JERRY), Rich(JERRY)\}$.

However, this is not always the case. As an example, consider the ABox $\mathcal{A} :=$ $\{parent(BEN, JERRY), Rich(JERRY)\}$ and the TBox $\mathcal{T} := \{\exists parent.Rich \sqsubseteq$ $Famous, Famous \sqsubseteq \exists friend.Famous, \exists friend.Famous \sqsubseteq Famous\}$, which together imply that Ben is famous. Assume that Ben wants to get rid of this consequence. The repair approach of [3] yields the following qABox as an optimal IQ-repair: $\exists\{x, y\}.\{parent(BEN, x), Rich(JERRY), friend(BEN, y), friend(y, y)\}$. This qABox retains the information that Ben has a parent (but not that Jerry is this parent) and that Ben is the starting point of an infinite $friend$-chain, i.e., Ben belongs to the concepts $C_n := (\exists friend.)^n \top$ for all $n \geq 1$. The latter is the reason why this qABox cannot be expressed by an IQ-equivalent ABox, which in turn is the reason why there is no optimal ABox repair. The culprit is obviously the cycle $friend(y, y)$. However, such cycles need not always cause problems. In fact, if we remove the third GCI $\exists friend.Famous \sqsubseteq Famous$ from the TBox, then the following qABox is an optimal IQ-repair:

$$\exists\{x, y\}.\{parent(BEN, x), Rich(JERRY),$$
$$friend(BEN, y), friend(y, y), Famous(y)\}.$$

This qABox can be expressed by an ABox that is IQ-equivalent to it w.r.t. the given TBox: $\{(\exists parent.\top)(BEN), Rich(JERRY), (\exists friend.Famous)(BEN)\}$. The reason is that, due to the existence of a famous friend of Ben, the GCI $Famous \sqsubseteq \exists friend.Famous$ now yields the infinite $friend$-chain.

These examples demonstrate that optimal ABox repairs may not always exist, and that it is not obvious to see when they do. The main contribution of the present paper is that we show how to decide the existence of optimal ABox repairs in exponential time, and how to compute all such repairs in case they exist. There may exist exponentially many such repairs, and each one may in the worst case be of double-exponential size. Our approach for showing these results roughly proceeds as follows. First, we observe that classical entailment between a qABox and an ABox coincides with so-called IRQ-entailment, which is slightly stronger than IQ-entailment by additionally taking role assertions between named individuals into account. Then, we show that both the canonical and the optimized IQ-repairs of [3] cannot only be used to obtain all optimal IQ-repairs, but also to compute all optimal IRQ-repairs. Subsequently, we introduce the notion of an optimal ABox approximation of a given qABox, and prove that the set of optimal ABox approximations of all optimal IRQ-repairs yields all optimal ABox repairs. A given qABox may not have an optimal ABox approximation, but if it does, then this approximation is unique up to equivalence and of at most exponential size. Then we investigate the problem of deciding the existence of optimal ABox approximations. The first step is to transfer the qABox into a specific form, called pre-approximation, which is saturated w.r.t. the TBox and consists of the original role assertions between named individuals and for each named individual a a sub-qABox $\mathcal{B}_a$. We prove that the original qABox has an optimal ABox approximation iff all the named individuals a have a most specific concept C_a in $\mathcal{B}_a$ w.r.t. the TBox. The optimal ABox approximation is then obtained by replacing each $\mathcal{B}_a$ with $C_a(a)$ in the pre-approximation. We can then use the results stated in [20] to test the existence of the msc in polynomial time[3] and to generate the at most exponentially large msc. Given that the optimal IRQ-repairs may be of exponential size, this yields the complexity upper bounds for testing the existence and computing optimal ABox repairs mentioned above. Due to space constraints, we cannot give complete proofs of all our results. They can be found in [4].

2 Preliminaries

We start with introducing the DL $\mathcal{EL}$ as well as TBoxes and (quantified) ABoxes. Then we consider the entailment relations relevant for this paper.

The name space available for defining $\mathcal{EL}$ concepts and ABox assertions is given by a *signature* Σ, which is the disjoint union of sets Σ_O, Σ_C, and Σ_R of *object names*, *concept names*, and *role names*. Starting with concept names and the top concept $\top$, $\mathcal{EL}$ *concepts* are defined inductively: if C, D are $\mathcal{EL}$ concepts and r is a role name, then $C \sqcap D$ (conjunction) and $\exists r.C$ (existential restriction) are also $\mathcal{EL}$ concepts. An $\mathcal{EL}$ *general concept inclusion (GCI)* is of the form $C \sqsubseteq D$, an $\mathcal{EL}$ *concept assertion* is of the form $C(u)$, and a *role assertion* is of the form $r(u,v)$, where C, D are $\mathcal{EL}$ concepts, $r \in \Sigma_R$, and $u, v \in \Sigma_O$. An $\mathcal{EL}$ *TBox* is a finite set of $\mathcal{EL}$ GCIs and an $\mathcal{EL}$ *ABox* is a finite set of $\mathcal{EL}$ concept assertions and role assertions. Such an ABox is called *simple* if all its concept assertions are of the form $A(u)$ with $A \in \Sigma_C$. A *quantified ABox (qABox)* is of the form $\exists X.\mathcal{A}$ where X is a finite subset of Σ_O and $\mathcal{A}$ is a simple ABox, which we call the *matrix* of $\exists X.\mathcal{A}$. We call the elements of X *variables* and the other object names occurring in $\mathcal{A}$ *individuals*.[4] The set of individual names occurring in $\exists X.\mathcal{A}$ is denoted with $\Sigma_I(\exists X.\mathcal{A})$, and the set of all object names (including the variables) with $\Sigma_O(\exists X.\mathcal{A})$.

The semantics of the syntactic entities introduced above can either be defined directly using interpretations, or by a translation into first-order logic (FO). For the sake of brevity, we choose the latter approach (see [3] for the former). In the translation, the elements of Σ_O, Σ_C, and Σ_R are respectively viewed as constant

---

[3] The proof for this polynomiality result in [20] is actually incorrect, but we show how to correct it.

[4] The variables correspond to what we have called anonymized individuals in the introduction, and the individuals to what we have called named individuals.

symbols, unary predicate symbols, and binary predicate symbols. $\mathcal{EL}$ concepts C are inductively translated into FO formulas $\phi_C(x)$ with one free variable x:

- concept A for $A \in \Sigma_C$ is translated into $A(x)$ and $\top$ into $A(x) \vee \neg A(x)$ for an arbitrary $A \in \Sigma_C$;
- if C, D are translated into $\phi_C(x)$ and $\phi_D(x)$, then $C \sqcap D$ is translated into $\phi_C(x) \wedge \phi_D(x)$ and $\exists r. C$ into $\exists y. (r(x, y) \wedge \phi_D(y))$, where $\phi_D(y)$ is obtained from $\phi_D(x)$ by replacing the free variable x by a different variable y.

GCIs $C \sqsubseteq D$ are translated into sentences $\phi_{C \sqsubseteq D} := \forall x. (\phi_C(x) \to \phi_D(x))$ and TBoxes $\mathcal{T}$ into $\phi_{\mathcal{T}} := \bigwedge_{C \sqsubseteq D \in \mathcal{T}} \phi_{C \sqsubseteq D}$. Concept assertions $C(u)$ are translated into $\phi_C(u)$, role assertions $r(u, v)$ stay the same, and ABoxes $\mathcal{A}$ are translated into the conjunction $\phi_{\mathcal{A}}$ of the translations of their assertions. For a quantified ABox $\exists X. \mathcal{A}$, the elements of X are viewed as first-order variables rather than constants, and its translation is $\exists \vec{x}. \phi_{\mathcal{A}}$, where $\vec{x}$ is the tuple of the variables in X in arbitrary order.

Let α, β be (q)ABoxes, concept inclusions, or concept assertions (possibly not both of the same kind), and $\mathcal{T}$ an $\mathcal{EL}$ TBox. Then we say that α *entails* β *w.r.t.* $\mathcal{T}$ (written $\alpha \models^{\mathcal{T}} \beta$) if the implication $(\phi_\alpha \wedge \phi_{\mathcal{T}}) \to \phi_\beta$ is valid according to the semantics of FO. Furthermore, α and β are *equivalent w.r.t.* $\mathcal{T}$ (written $\alpha \equiv^{\mathcal{T}} \beta$), if $\alpha \models^{\mathcal{T}} \beta$ and $\beta \models^{\mathcal{T}} \alpha$. In case $\mathcal{T} = \emptyset$, we will sometimes write $\models$ instead of $\models^{\emptyset}$. If $\emptyset \models^{\mathcal{T}} C \sqsubseteq D$, then we also write $C \sqsubseteq^{\mathcal{T}} D$ and say that C *is subsumed by* D *w.r.t.* $\mathcal{T}$; in case $\mathcal{T} = \emptyset$ we simply say that C is subsumed by D. If $\exists X. \mathcal{A} \models^{\mathcal{T}} C(a)$, then a is called an *instance of* C w.r.t. $\exists X. \mathcal{A}$ and $\mathcal{T}$. For ABoxes, the instance relation is defined analogously. Entailment between qABoxes w.r.t. an $\mathcal{EL}$ TBox is NP-complete, but the subsumption and the instance problem are polynomial [7].

Note that ABoxes are a special case of qABoxes. For simple ABoxes, this is the case where $X = \emptyset$. For general ABoxes, one can express complex concept assertions by introducing existentially quantified variables (e.g., $\{(A \sqcap \exists r.B)(a)\}$ is equivalent to $\exists \{x\}. \{A(a), r(a, x), B(x)\}$). For this reason, the entailment relations defined below for qABoxes are also well-defined for ABoxes.

IQ-Entailment. If one is mainly interested in asking instance queries, i.e., in what kind of instance relations a qABox entails, then the following weaker form of entailment can be used [3,7]. We say that the qABox $\exists X. \mathcal{A}$ IQ-*entails* the qABox $\exists Y. \mathcal{B}$ w.r.t. the $\mathcal{EL}$ TBox $\mathcal{T}$ (written $\exists X. \mathcal{A} \models^{\mathcal{T}}_{IQ} \exists Y. \mathcal{B}$) if every concept assertion $C(a)$ entailed w.r.t. $\mathcal{T}$ by the latter is also entailed w.r.t. $\mathcal{T}$ by the former. Whenever we compare two qABoxes $\exists X. \mathcal{A}$ and $\exists Y. \mathcal{B}$, we follow [7] and assume without loss of generality that they are *renamed apart*, which means that X is disjoint with $\Sigma_O(\exists Y. \mathcal{B})$ and Y is disjoint with $\Sigma_O(\exists X. \mathcal{A})$, and we further assume that the two qABoxes speak about the same set of individual names $\Sigma_I := \Sigma_I(\exists X. \mathcal{A}) \cup \Sigma_I(\exists Y. \mathcal{B})$.

For the case of an empty TBox, it was shown in [7] that $\exists X. \mathcal{A} \models^{\emptyset}_{IQ} \exists Y. \mathcal{B}$ iff there is a simulation from $\exists Y. \mathcal{B}$ to $\exists X. \mathcal{A}$. A *simulation* from $\exists Y. \mathcal{B}$ to $\exists X. \mathcal{A}$ is a relation $\mathfrak{S} \subseteq \Sigma_O(\exists Y. \mathcal{B}) \times \Sigma_O(\exists X. \mathcal{A})$ such that $(a, a) \in \mathfrak{S}$ for each $a \in \Sigma_I$ and, for each $(u, v) \in \mathfrak{S}$, $A(u) \in \mathcal{B}$ implies $A(v) \in \mathcal{A}$ and $r(u, u') \in \mathcal{B}$

implies that there exists an object $v' \in \Sigma_0(\exists X.\mathcal{A})$ such that $(u', v') \in \mathfrak{S}$ and $r(v, v') \in \mathcal{A}$. Since checking the existence of a simulation can be done in polynomial time [10], the simulation characterization of IQ-entailment shows that IQ-entailment between qABoxes can be decided in polynomial time if $\mathcal{T} = \emptyset$ [7].

$\sqcap$-**rule.** If $(C_1 \sqcap \ldots \sqcap C_n)(t) \in \mathcal{A}$, then remove this assertion from $\mathcal{A}$ and add the assertions $C_1(t), \ldots, C_n(t)$ to $\mathcal{A}$.

$\exists$-**rule.** If $(\exists r.C)(t) \in \mathcal{A}$, then remove this assertion from $\mathcal{A}$, add the two assertions $r(t, x_C)$ and $C(x_C)$ to $\mathcal{A}$, and add x_C to X if it is not already there.

$\sqsubseteq$-**rule.** If $t \in \Sigma_0(\exists X.\mathcal{A})$, $C \sqsubseteq D \in \mathcal{T}$, $\mathcal{A} \models C(t)$, and $\mathcal{A} \not\models D(t)$, then add the assertion $D(t)$ to $\mathcal{A}$.

The $\sqcap$-rule has higher precedence than the $\exists$-rule, and the latter has higher precedence than the $\sqsubseteq$-rule.

Fig. 1. The IQ-saturation rules from [3].

To extend these results to the case of a non-empty TBox, the notion of an IQ-saturation is introduced in [3]. The saturation rules given in Fig. 1 add new variables and assertions to the qABox if the existence of a corresponding element and the validity of the assertion is implied by the TBox. To be more precise, for each existential restriction $\exists r.C$ occurring in $\mathcal{T}$, a fresh variable x_C not contained in the initial qABox is introduced. When applying the $\exists$-rule to an assertion of the form $(\exists r.C)(t)$, this variable is always used for the successor object. As pointed out in [3], IQ-saturation (i.e., the exhaustive application of the IQ-saturation rules) terminates in polynomial time and generates a qABox $\mathsf{sat}_{\mathsf{IQ}}^{\mathcal{T}}(\exists X.\mathcal{A})$, which can be seen as a qABox representation of what is called the *canonical model* in [13, Sect. 5.2]. IQ-entailment for qABoxes w.r.t. an $\mathcal{EL}$ TBox is now characterized in [3] as follows.

Theorem 1 ([3]). *Let $\mathcal{T}$ be an $\mathcal{EL}$ TBox and $\exists X.\mathcal{A}$ and $\exists Y.\mathcal{B}$ qABoxes. Then the following statements are equivalent:*

- $\exists X.\mathcal{A} \models_{\mathsf{IQ}}^{\mathcal{T}} \exists Y.\mathcal{B}$,
- $\mathsf{sat}_{\mathsf{IQ}}^{\mathcal{T}}(\exists X.\mathcal{A}) \models_{\mathsf{IQ}}^{\emptyset} \exists Y.\mathcal{B}$,
- *there is a simulation from $\exists Y.\mathcal{B}$ to $\mathsf{sat}_{\mathsf{IQ}}^{\mathcal{T}}(\exists X.\mathcal{A})$.*

Since the IQ-saturation can be computed in polynomial time, this clearly shows that IQ-entailment for qABoxes w.r.t. an $\mathcal{EL}$ TBox can also be decided in polynomial time.

IRQ-Entailment. If we are not only interested in implied concept assertions, but also in implied role assertions, then IQ-entailment is not sufficient. Instead, we must use IRQ-entailment. We say that the qABox $\exists X.\mathcal{A}$ IRQ-*entails* the qABox $\exists Y.\mathcal{B}$ w.r.t. the $\mathcal{EL}$ TBox $\mathcal{T}$ (written $\exists X.\mathcal{A} \models_{\mathsf{IRQ}}^{\mathcal{T}} \exists Y.\mathcal{B}$) if every concept or role assertion entailed w.r.t. $\mathcal{T}$ by the latter is also entailed w.r.t. $\mathcal{T}$ by the former.

It is easy to see that a qABox cannot entail a role assertion involving a variable, and it can only entail a role assertion between individuals if its matrix contains this assertion. This yields the following characterization of IRQ-entailment, which shows that IRQ-entailment can be decided in polynomial time.

Proposition 2. *Let* $\mathcal{T}$ *be an* $\mathcal{EL}$ *TBox and* $\exists X.\mathcal{A}$ *and* $\exists Y.\mathcal{B}$ *qABoxes. Then the following statements are equivalent:*

- $\exists X.\mathcal{A} \models_{\mathsf{IRQ}}^{\mathcal{T}} \exists Y.\mathcal{B}$,
- $\exists X.\mathcal{A} \models_{\mathsf{IQ}}^{\mathcal{T}} \exists Y.\mathcal{B}$ *and* $r(a,b) \in \mathcal{B}$ *implies* $r(a,b) \in \mathcal{A}$ *for all* $r \in \Sigma_\mathsf{R}$ *and* $a,b \in \Sigma_\mathsf{I}$.

Since ABoxes consist of concept and role assertions, we obtain the following characterization of entailment between a qABox and an ABox, which implies that this entailment can be decided in polynomial time.

Proposition 3. *Let* $\mathcal{T}$ *be an* $\mathcal{EL}$ *TBox,* $\exists X.\mathcal{A}$ *a qABox, and* $\mathcal{B}$ *an ABox. Then* $\exists X.\mathcal{A} \models^{\mathcal{T}} \mathcal{B}$ *iff* $\exists X.\mathcal{A} \models_{\mathsf{IRQ}}^{\mathcal{T}} \mathcal{B}$.

3   Optimal ABox Repairs and Approximations

We first introduce the notion of an optimal repair w.r.t. an entailment relation, and show that the approaches for computing optimal IQ-repairs described in [3] can also be used to compute optimal IRQ-repairs. Then, we define optimal ABox approximations and show some useful properties for them. Finally, we introduce optimal ABox repairs, and describe how optimal ABox approximations can be used to obtain them from optimal IRQ-repairs.

3.1   Optimal IQ- and IRQ-Repairs

We start by recalling the definition of optimal repairs given in [3], but consider IRQ as an additional entailment relation.

Definition 4. *Let* $\mathcal{T}$ *be an* $\mathcal{EL}$ *TBox and* $\mathsf{QL} \in \{\mathsf{IRQ}, \mathsf{IQ}\}$.

- *An* $\mathcal{EL}$ *repair request is a finite set of* $\mathcal{EL}$ *concept assertions.*
- *Given a qABox* $\exists X.\mathcal{A}$ *and an* $\mathcal{EL}$ *repair request* $\mathcal{R}$, *a* QL-*repair of* $\exists X.\mathcal{A}$ *for* $\mathcal{R}$ *w.r.t.* $\mathcal{T}$ *is a qABox* $\exists Y.\mathcal{B}$ *such that* $\exists X.\mathcal{A} \models_{\mathsf{QL}}^{\mathcal{T}} \exists Y.\mathcal{B}$ *and* $\exists Y.\mathcal{B} \not\models^{\mathcal{T}} C(a)$ *for all* $C(a) \in \mathcal{R}$.
- *Such a repair* $\exists Y.\mathcal{B}$ *is optimal if there is no* QL-*repair* $\exists Z.\mathcal{C}$ *of* $\exists X.\mathcal{A}$ *for* $\mathcal{R}$ *w.r.t.* $\mathcal{T}$ *such that* $\exists Z.\mathcal{C} \models_{\mathsf{QL}}^{\mathcal{T}} \exists Y.\mathcal{B}$ *and* $\exists Y.\mathcal{B} \not\models_{\mathsf{QL}}^{\mathcal{T}} \exists Z.\mathcal{C}$.

Two qABoxes are QL-*equivalent* if they QL-entail each other, and $\exists X.\mathcal{A}$ *strictly* QL-*entails* $\exists Y.\mathcal{B}$ if $\exists X.\mathcal{A} \models_{\mathsf{QL}}^{\mathcal{T}} \exists Y.\mathcal{B}$ and $\exists Y.\mathcal{B} \not\models_{\mathsf{QL}}^{\mathcal{T}} \exists X.\mathcal{A}$. We say that a set $\mathfrak{R}$ of QL-repairs of $\exists X.\mathcal{A}$ for $\mathcal{R}$ w.r.t. $\mathcal{T}$ QL-*covers* all QL-repairs of $\exists X.\mathcal{A}$ for $\mathcal{R}$ w.r.t. $\mathcal{T}$ if for every QL-repair $\exists Y.\mathcal{B}$ of $\exists X.\mathcal{A}$ for $\mathcal{R}$ w.r.t. $\mathcal{T}$ there exists an element $\exists Z.\mathcal{C}$ of $\mathfrak{R}$ such that $\exists Z.\mathcal{C} \models_{\mathsf{QL}}^{\mathcal{T}} \exists Y.\mathcal{B}$. It is easy to see that such a covering set $\mathfrak{R}$ must contain, up to QL-equivalence, all optimal QL-repairs of $\exists X.\mathcal{A}$ for $\mathcal{R}$ w.r.t. $\mathcal{T}$, and thus one can obtain from it, up to QL-equivalence, the set of all optimal QL-repairs of $\exists X.\mathcal{A}$ for $\mathcal{R}$ w.r.t. $\mathcal{T}$ by removing elements that are strictly QL-entailed by another element. Clearly, this set still QL-covers all QL-repairs of $\exists X.\mathcal{A}$ for $\mathcal{R}$ w.r.t. $\mathcal{T}$.

In [3], two ways of computing such a covering set for IQ-repairs are described, the canonical IQ-repairs and the optimized IQ-repairs (see Proposition 8 and Theorem 14). Since these covering sets are of at most exponential cardinality, their elements are of at most exponential size, and IQ-entailment can be decided in polynomial time, this shows that, up to IQ-equivalence, the set of all optimal IQ-repairs can be computed in exponential time.

The canonical (optimized) IQ-repairs also yield covering sets for the IRQ case. The reason is basically that the approaches for constructing them introduced in [3] do not generate new role assertions between individuals and preserve as many of them as possible, although this is not required for IQ-entailment.

Proposition 5. *Let $\mathcal{T}$ be an $\mathcal{EL}$ TBox, $\exists X.\mathcal{A}$ a qABox, and $\mathcal{R}$ an $\mathcal{EL}$ repair request. If $\mathfrak{R}$ is the set of all canonical or all optimized IQ-repairs obtained from this input according to the definitions in [3], then $\mathfrak{R}$ is a set of IRQ-repairs of $\exists X.\mathcal{A}$ for $\mathcal{R}$ w.r.t. $\mathcal{T}$ that IRQ-covers all IRQ-repairs of $\exists X.\mathcal{A}$ for $\mathcal{R}$ w.r.t. $\mathcal{T}$. In particular, up to IRQ-equivalence, the set of optimal IRQ-repairs can be computed in exponential time, and it IRQ-covers all IRQ-repairs of $\exists X.\mathcal{A}$ for $\mathcal{R}$ w.r.t. $\mathcal{T}$.*

Note that, though we have the same covering set $\mathfrak{R}$ in the IQ and in the IRQ case, the sets of optimal repairs obtained from it by removing strictly entailed elements need not coincide since different entailment relations are used during this removal. Since the requirements for IQ entailment are weaker than for IRQ entailment, it could be that a qABox may be removed from $\mathfrak{R}$ in the IQ case, but must be retained in IRQ case. Also notice that the proposition need not hold for arbitrary IQ-covering sets. Its proof uses properties of the canonical and the optimized IQ-repairs that need not hold for arbitrary covering sets.

Example 6. Consider the qABox $\exists\{x\}.\mathcal{A}$ for $\mathcal{A} = \{A(a), r(a,x), r(x,x)\}$, assume that the TBox is empty, and that the repair request is $\{A(a)\}$. An optimal IQ-repair $\exists\{x\}.\mathcal{A}'$ can be obtained from this qABox by removing the assertion $A(a)$ from $\mathcal{A}$, and this is also an optimal IRQ-repair. However, the ABox $\{r(a,a)\}$ is also an optimal IQ-repair since it is IQ-equivalent to $\exists\{x\}.\mathcal{A}'$, but it is not even an IRQ-repair since it is not IRQ-entailed by $\exists\{x\}.\mathcal{A}$.

3.2 Optimal ABox Approximations

Given a qABox we are now interested in finding an ABox that approximates it as closely as possible in the sense that a minimal amount of information is lost. In the definition below, we use classical entailment. But note that, according to Proposition 3, this coincides with IRQ-entailment.

Definition 7. *Given a qABox* $\exists X.\mathcal{A}$ *and an* $\mathcal{EL}$ *TBox* $\mathcal{T}$, *we call an* $\mathcal{EL}$ *ABox* $\mathcal{B}$ *an* ABox approximation *of* $\exists X.\mathcal{A}$ *w.r.t.* $\mathcal{T}$ *if* $\exists X.\mathcal{A} \models^{\mathcal{T}} \mathcal{B}$. *The ABox approximation* $\mathcal{B}$ *of* $\exists X.\mathcal{A}$ *w.r.t.* $\mathcal{T}$ *is* optimal *if there is no ABox approximation* $\mathcal{C}$ *of* $\exists X.\mathcal{A}$ *w.r.t.* $\mathcal{T}$ *such that* $\mathcal{C} \models^{\mathcal{T}} \mathcal{B}$, *but* $\mathcal{B} \not\models^{\mathcal{T}} \mathcal{C}$.

Such an optimal ABox approximation need not exist. The qABox $\exists\{x\}.\mathcal{A}'$ with $\mathcal{A}' = \{r(a,x), r(x,x)\}$ is an example for this case. In fact, this qABox entails $((\exists r.)^n\top)(a)$ for all $n \geq 1$, which is not possible for an ABox entailed by $\exists\{x\}.\mathcal{A}'$ since such an ABox cannot contain role assertions and can contain only finitely many concept assertions. However, if an optimal ABox approximation exists, then it is unique up to equivalence. This is an easy consequence of the fact that the union of two ABox approximations is again an ABox approximation.

Proposition 8. *If* $\mathcal{B}_1$ *and* $\mathcal{B}_2$ *are optimal ABox approximations of the qABox* $\exists X.\mathcal{A}$ *w.r.t. the* $\mathcal{EL}$ *TBox* $\mathcal{T}$, *then* $\mathcal{B}_1$ *and* $\mathcal{B}_2$ *are equivalent w.r.t.* $\mathcal{T}$.

Optimal ABox approximations can now be characterized as follows.

Theorem 9. *The ABox* $\mathcal{B}$ *is an optimal ABox approximation of* $\exists X.\mathcal{A}$ *w.r.t.* $\mathcal{T}$ *iff* $\exists X.\mathcal{A}$ *and* $\mathcal{B}$ *are* IRQ-*equivalent.*

Proof. First, assume that $\exists X.\mathcal{A}$ and $\mathcal{B}$ are IRQ-equivalent w.r.t. $\mathcal{T}$. Then $\exists X.\mathcal{A} \models^{\mathcal{T}} \mathcal{B}$ by Proposition 3, and thus $\mathcal{B}$ is an ABox approximation of $\exists X.\mathcal{A}$ w.r.t. $\mathcal{T}$. If $\mathcal{C}$ is another ABox approximation of $\exists X.\mathcal{A}$ w.r.t. $\mathcal{T}$, then $\exists X.\mathcal{A} \models^{\mathcal{T}} \mathcal{C}$ by definition, and thus $\mathcal{B} \models^{\mathcal{T}} \mathcal{C}$ due to the assumed IRQ-equivalence. This shows optimality of $\mathcal{B}$.

Second, assume that $\mathcal{B}$ is an optimal ABox approximation of $\exists X.\mathcal{A}$ w.r.t. $\mathcal{T}$ that is not IRQ-equivalent with $\exists X.\mathcal{A}$. Then there is either a role assertion that belongs to $\mathcal{A}$, but not to $\mathcal{B}$, or a concept assertion that is entailed w.r.t. $\mathcal{T}$ by $\exists X.\mathcal{A}$, but not by $\mathcal{B}$. Adding this assertion to $\mathcal{B}$ yields an ABox $\mathcal{B}'$ that is an ABox approximation of $\exists X.\mathcal{A}$ w.r.t. $\mathcal{T}$. In addition, it satisfies $\mathcal{B}' \models^{\mathcal{T}} \mathcal{B}$, but not $\mathcal{B} \models^{\mathcal{T}} \mathcal{B}'$, which contradicts the assumed optimality of $\mathcal{B}$. $\square$

An approach for deciding whether a given qABox has an optimal ABox approximation, and for computing it in case it exists, will be described in Sect. 4. But first, we show how optimal ABox approximations can be used to compute optimal ABox repairs.

3.3 Optimal ABox Repairs

The repair approaches developed in [3] in general yield quantified ABoxes as output, even if the input is an ABox. We are now interested in producing repairs that are ABoxes. The approach developed below does not require the input to be an ABox. It actually assumes that the input is a qABox, which means that input ABoxes first need to be transformed into equivalent qABoxes.

Definition 10. *Let $\mathcal{T}$ be an $\mathcal{EL}$ TBox, $\exists X.\mathcal{A}$ a qABox, and $\mathcal{R}$ an $\mathcal{EL}$ repair request. We call an $\mathcal{EL}$ ABox $\mathcal{B}$ an ABox repair of $\exists X.\mathcal{A}$ for $\mathcal{R}$ w.r.t. $\mathcal{T}$ if $\exists X.\mathcal{A} \models^{\mathcal{T}} \mathcal{B}$ and $\mathcal{B} \not\models^{\mathcal{T}} C(a)$ for all $C(a) \in \mathcal{R}$. The ABox repair $\mathcal{B}$ of $\exists X.\mathcal{A}$ for $\mathcal{R}$ w.r.t. $\mathcal{T}$ is optimal if there is no ABox repair $\mathcal{C}$ of $\exists X.\mathcal{A}$ for $\mathcal{R}$ w.r.t. $\mathcal{T}$ such that $\mathcal{C} \models^{\mathcal{T}} \mathcal{B}$, but $\mathcal{B} \not\models^{\mathcal{T}} \mathcal{C}$.*

Our approach for computing optimal ABox repairs proceeds as follows: first, we compute the set of all optimal IRQ-repairs of $\exists X.\mathcal{A}$, and then ABox-approximate the elements of this set. In the following, if we say that $\mathfrak{R}$ is the set of optimal IRQ-repairs of a qABox, we mean that, for every optimal IRQ-repair, $\mathfrak{R}$ contains one element of its IRQ-equivalence class. Also, for a given qABox $\exists Y.\mathcal{B}$, we define

$$\mathsf{Oapp}^{\mathcal{T}}(\exists Y.\mathcal{B}) := \begin{cases} \{\mathcal{C}\} & \text{for an optimal ABox approx. } \mathcal{C} \text{ of } \exists Y.\mathcal{B} \text{ w.r.t. } \mathcal{T}, \\ \emptyset & \text{if no optimal ABox approx. of } \exists Y.\mathcal{B} \text{ w.r.t. } \mathcal{T} \text{ exists.} \end{cases}$$

Theorem 11. *Let $\exists X.\mathcal{A}$ be a qABox, $\mathcal{T}$ an $\mathcal{EL}$-TBox, $\mathcal{R}$ an $\mathcal{EL}$ repair request, and $\mathfrak{R}$ the set of optimal IRQ-repairs of $\exists X.\mathcal{A}$ for $\mathcal{R}$ w.r.t. $\mathcal{T}$. Then the set*

$$\bigcup_{\exists Y.\mathcal{B} \in \mathfrak{R}} \mathsf{Oapp}^{\mathcal{T}}(\exists Y.\mathcal{B})$$

consists of all optimal ABox repairs of $\exists X.\mathcal{A}$ for $\mathcal{R}$ w.r.t. $\mathcal{T}$ up to equivalence.

Proof. First, assume that the ABox $\mathcal{C}$ belongs to the union defined in the statement of the theorem. Then $\exists X.\mathcal{A} \models^{\mathcal{T}}_{\mathsf{IRQ}} \exists Y.\mathcal{B} \models^{\mathcal{T}} \mathcal{C}$ for some qABox $\exists Y.\mathcal{B} \in \mathfrak{R}$ that has $\mathcal{C}$ as an optimal ABox approximation. This implies that $\mathcal{C}$ does not entail any of the concept assertions in $\mathcal{R}$ (since $\exists Y.\mathcal{B}$ does not) and that $\exists X.\mathcal{A} \models^{\mathcal{T}} \mathcal{C}$. Thus, $\mathcal{C}$ is an ABox repair of $\exists X.\mathcal{A}$ for $\mathcal{R}$ w.r.t. $\mathcal{T}$. It remains to show that it is optimal. Assume to the contrary that $\mathcal{C}'$ is an ABox repair of $\exists X.\mathcal{A}$ for $\mathcal{R}$ w.r.t. $\mathcal{T}$ such that $\mathcal{C}' \models^{\mathcal{T}} \mathcal{C}$, but $\mathcal{C} \not\models^{\mathcal{T}} \mathcal{C}'$. Since $\mathcal{C}$ and $\exists Y.\mathcal{B}$ are IRQ-equivalent by Theorem 9, this is a contradiction to the fact that $\exists Y.\mathcal{B}$ is an optimal IRQ-repair of $\exists X.\mathcal{A}$ for $\mathcal{R}$ w.r.t. $\mathcal{T}$ since $\mathcal{C}'$ would then be a better IRQ-repair.

Second, assume that the ABox $\mathcal{C}$ is an optimal ABox repair of $\exists X.\mathcal{A}$ for $\mathcal{R}$ w.r.t. $\mathcal{T}$. Then $\mathcal{C}$ is also an IRQ-repair of $\exists X.\mathcal{A}$ for $\mathcal{R}$ w.r.t. $\mathcal{T}$, and thus Proposition 5 yields that there is an optimal IRQ-repair $\exists Y.\mathcal{B} \in \mathfrak{R}$ such that $\exists X.\mathcal{A} \models^{\mathcal{T}}_{\mathsf{IRQ}} \exists Y.\mathcal{B} \models^{\mathcal{T}}_{\mathsf{IRQ}} \mathcal{C}$. We know by Proposition 3 that the second IRQ-entailment is in fact an entailment, and thus $\mathcal{C}$ is an ABox approximation of $\exists Y.\mathcal{B}$. It remains to show that it is optimal. Assume to the contrary that $\mathcal{C}'$ is an ABox approximation of $\exists Y.\mathcal{B}$ such that $\exists Y.\mathcal{B} \models^{\mathcal{T}} \mathcal{C}' \models^{\mathcal{T}} \mathcal{C}$, but $\mathcal{C} \not\models^{\mathcal{T}} \mathcal{C}'$. But then $\mathcal{C}'$ is an ABox repair of $\exists X.\mathcal{A}$ for $\mathcal{R}$ w.r.t. $\mathcal{T}$ (since $\exists Y.\mathcal{B}$ is a repair) that is better than $\mathcal{C}$, which contradicts our assumption that $\mathcal{C}$ is optimal. $\square$

Once we have developed a method for computing the sets $\mathsf{Oapp}^{\mathcal{T}}(\exists Y.\mathcal{B})$, this theorem shows how to compute the set of all optimal ABox repairs of a given qABox. Such a method will be introduced in the next section. Before doing this, we want to point out that, in contrast to the set of optimal IRQ-repairs, which covers all IRQ-repairs, the set of optimal ABox repairs in general does not cover all ABox repairs.

Example 12. Consider the ABox $\mathcal{A} = \{A(a), r(a,b), B(b)\}$, the TBox $\mathcal{T} = \{B \sqsubseteq \exists r.B, \exists r.B \sqsubseteq B\}$ and the repair request $\mathcal{R} = \{(A \sqcap \exists r.B)(a)\}$. There are basically three options for IRQ-repairing $\mathcal{A}$: remove $A(a)$, remove $B(b)$, or remove $r(a,b)$. Since things implied by the TBox must also be taken into account, these three options yield the following optimal IRQ-repairs of $\mathcal{A}$ for $\mathcal{R}$ w.r.t. $\mathcal{T}$:[5] $\mathcal{B}_1 = \{r(a,b), B(b)\}$ as well as $\exists\{x\}.\mathcal{B}_i$ for $i = 2, 3$, where $\mathcal{B}_2 = \{A(a), r(a,b), r(b,x), r(x,x)\}$ and $\mathcal{B}_3 = \{A(a), B(b), r(a,x), r(x,x)\}$. Of these three, $\mathcal{B}_1$ is already an ABox, and thus its own optimal ABox approximation, whereas the other two have no optimal ABox approximation. However, they have non-optimal ABox approximations, which are not necessarily covered by $\mathcal{B}_1$. For example, $\{A(a), r(a,b), (\exists r.\exists r.\top)(b)\}$ is an ABox approximation of $\exists\{x\}.\mathcal{B}_2$ and an ABox repair of $\mathcal{A}$ for $\mathcal{R}$ w.r.t. $\mathcal{T}$, but since it contains $A(a)$, it is not entailed by $\mathcal{B}_1$.

4 Computing Optimal ABox Approximations

In this section, we assume that $\exists X.\mathcal{A}$ is a qABox and $\mathcal{T}$ an $\mathcal{EL}$ TBox. We will develop an approach for deciding whether $\exists X.\mathcal{A}$ has an optimal ABox approximation w.r.t. $\mathcal{T}$, which in the affirmative case also yields such an optimal approximation.

The first step is to saturate $\exists X.\mathcal{A}$ using the IQ-saturation rules of Fig. 1. In the following, let $\underline{\mathsf{sat}}^{\mathcal{T}}_{\mathsf{IQ}}(\exists X.\mathcal{A})$ denote a (fixed) qABox obtained by applying the IQ-saturation rules exhaustively to $\exists X.\mathcal{A}$. Note that the size of $\underline{\mathsf{sat}}^{\mathcal{T}}_{\mathsf{IQ}}(\exists X.\mathcal{A})$ is polynomial in the size of the input $\exists X.\mathcal{A}$ and $\mathcal{T}$, and that $\exists X.\mathcal{A}$ and $\underline{\mathsf{sat}}^{\mathcal{T}}_{\mathsf{IQ}}(\exists X.\mathcal{A})$ are IQ-equivalent w.r.t. $\mathcal{T}$ by Theorem 1. In addition, it is easy to see that these two qABoxes contain the same individuals and the same role assertions between individuals. Thus, they are even IRQ-equivalent w.r.t. $\mathcal{T}$. As before, we use Σ_{I} to denote set of individuals of $\exists X.\mathcal{A}$.

In the next step, we transform $\underline{\mathsf{sat}}^{\mathcal{T}}_{\mathsf{IQ}}(\exists X.\mathcal{A})$ into a new qABox, called pre-approximation, whose matrix basically consists of the union of ABoxes $\mathcal{B}_a$ for each $a \in \Sigma_{\mathsf{I}}$, extended with the role assertions between individuals in $\mathcal{A}$. Each ABox $\mathcal{B}_a$ contains a as the only individual name, and further contains a fully anonymized copy of the saturation $\underline{\mathsf{sat}}^{\mathcal{T}}_{\mathsf{IQ}}(\exists X.\mathcal{A})$, which is connected with a by indispensable role assertions.

Definition 13. *We call a role assertion $r(a,u)$ in $\underline{\mathsf{sat}}^{\mathcal{T}}_{\mathsf{IQ}}(\exists X.\mathcal{A})$ for $a \in \Sigma_{\mathsf{I}}$ indispensable if there is no role assertion $r(a,b)$ for $b \in \Sigma_{\mathsf{I}}$ such that there is a simulation from $\underline{\mathsf{sat}}^{\mathcal{T}}_{\mathsf{IQ}}(\exists X.\mathcal{A})$ to itself that contains (u,b).*

Since an individual always simulates itself, only role assertion $r(a,u)$ where u is a variable can be indispensable. We are now ready to define the pre-approximation.

---

[5] The IQ-repairs computed by the approaches in [3] would contain more assertions, which are however redundant for IRQ-entailment w.r.t. $\mathcal{T}$.

Definition 14. *The* pre-approximation $\mathsf{pre\text{-}approx}^{\mathcal{T}}_{\mathsf{IRQ}}(\exists X.\mathcal{A})$ *of* $\exists X.\mathcal{A}$ *w.r.t.* $\mathcal{T}$ *is defined as the quantified ABox* $\exists Y.\mathcal{B}$, *where*

$$Y := \{\, u' \mid u \text{ is an object name occurring in } \underline{\mathsf{sat}}^{\mathcal{T}}_{\mathsf{IQ}}(\exists X.\mathcal{A}) \,\},$$

$$\mathcal{B} := \bigcup\{\, \mathcal{B}_a \mid a \text{ is an individual name in } \Sigma_{\mathsf{I}} \,\}$$
$$\cup \{\, r(a,b) \mid r(a,b) \text{ occurs in } \underline{\mathsf{sat}}^{\mathcal{T}}_{\mathsf{IQ}}(\exists X.\mathcal{A}) \text{ where } a,b \in \Sigma_{\mathsf{I}} \,\},$$

$$\mathcal{B}_a := \{\, A(a) \mid A(a) \text{ occurs in } \underline{\mathsf{sat}}^{\mathcal{T}}_{\mathsf{IQ}}(\exists X.\mathcal{A}) \,\}$$
$$\cup \{\, r(a,u') \mid r(a,u) \text{ occurs in } \underline{\mathsf{sat}}^{\mathcal{T}}_{\mathsf{IQ}}(\exists X.\mathcal{A}) \text{ and is indispensable} \,\}$$
$$\cup \{\, A(u') \mid A(u) \text{ occurs in } \underline{\mathsf{sat}}^{\mathcal{T}}_{\mathsf{IQ}}(\exists X.\mathcal{A}) \,\}$$
$$\cup \{\, r(u',v') \mid r(u,v) \text{ occurs in } \underline{\mathsf{sat}}^{\mathcal{T}}_{\mathsf{IQ}}(\exists X.\mathcal{A}) \,\}.$$

Obviously, the pre-approximation can be computed in polynomial time. In addition, it is IRQ-equivalent to $\underline{\mathsf{sat}}^{\mathcal{T}}_{\mathsf{IQ}}(\exists X.\mathcal{A})$ [4].

Lemma 15. *The* q*ABoxes* $\underline{\mathsf{sat}}^{\mathcal{T}}_{\mathsf{IQ}}(\exists X.\mathcal{A})$ *and* $\mathsf{pre\text{-}approx}^{\mathcal{T}}_{\mathsf{IRQ}}(\exists X.\mathcal{A})$ *are* IRQ-*equivalent w.r.t. the empty TBox* $\emptyset$, *and thus also w.r.t.* $\mathcal{T}$.

Since we already know that $\exists X.\mathcal{A}$ and $\underline{\mathsf{sat}}^{\mathcal{T}}_{\mathsf{IQ}}(\exists X.\mathcal{A})$ are IRQ-equivalent w.r.t. $\mathcal{T}$, this shows that $\exists X.\mathcal{A}$ is IRQ-equivalent to its pre-approximation w.r.t. $\mathcal{T}$. Consequently, an ABox $\mathcal{C}$ is an optimal ABox approximation of $\exists X.\mathcal{A}$ w.r.t. $\mathcal{T}$ iff it is one of the pre-approximation w.r.t. $\mathcal{T}$.

To test whether $\mathsf{pre\text{-}approx}^{\mathcal{T}}_{\mathsf{IRQ}}(\exists X.\mathcal{A})$ has an optimal ABox approximation w.r.t. $\mathcal{T}$, it is sufficient to check whether, for all $a \in \Sigma_{\mathsf{I}}$, the individual a has a most specific concept in $\mathcal{B}_a$ w.r.t. $\mathcal{T}$.

Definition 16. *Let* $\mathcal{C}$ *be an* $\mathcal{EL}$ *ABox,* $\mathcal{T}$ *an* $\mathcal{EL}$ *TBox, and* a *an individual name. The* $\mathcal{EL}$ *concept* C *is a* most specific concept *(msc) of* a *in* $\mathcal{C}$ *w.r.t.* $\mathcal{T}$ *if* $\mathcal{C} \models^{\mathcal{T}} C(a)$ *and* $\mathcal{C} \models^{\mathcal{T}} D(a)$ *implies* $C \sqsubseteq^{\mathcal{T}} D$ *for all* $\mathcal{EL}$ *concepts* D.

The most specific concept need not exist, but if it does, then it is unique up to equivalence w.r.t. $\mathcal{T}$. The ABox $\mathcal{C} := \{r(a,a)\}$ is a simple example where the msc of a does not exist w.r.t. the empty TBox. In fact, $\mathcal{C} \models (\exists r.)^n \top(a)$ for all $n \geq 1$, and it is easy to see that no $\mathcal{EL}$ concept can be subsumed by these infinitely many concepts. Note, however, that $\mathcal{C}$ has an optimal ABox approximation since it is itself an ABox. In this case, the pre-approximation is $\{r(a,a)\} \cup \mathcal{B}_a$ where $\mathcal{B}_a = \{r(a',a')\}$. There is no role assertion $r(a,a')$ since $r(a,a)$ is not indispensable. While a' does not have an msc in $\mathcal{B}_a$, this is not what we are interested in. We want to know whether a has one, and the answer is "yes" since $\top$ is an msc of a in $\mathcal{B}_a$. The problem of testing for the existence of and computing the msc in $\mathcal{EL}$ was investigated in [20], where the following result is stated.

Proposition 17 ([20]). *Let* $\mathcal{C}$ *be an* $\mathcal{EL}$ *ABox,* $\mathcal{T}$ *an* $\mathcal{EL}$ *TBox, and* a *an individual name. It can be decided in polynomial time whether* a *has a most specific concept in* $\mathcal{C}$ *w.r.t.* $\mathcal{T}$, *and if the msc exists, then it can be computed in exponential time.*

The main idea underlying the proof of this proposition (rephrased into the setting of the present paper) is to *unravel* the IQ-saturation of $\mathcal{C}$ w.r.t. $\mathcal{T}$ into a concept C_k an increasing number k of steps, starting from a. After each step, one tests whether the ABox $\{C_k(a)\}$ IQ-entails $\exists X.\mathcal{C}$ w.r.t. $\mathcal{T}$, where X consists of the object names in $\mathcal{C}$ different from a. In case this test succeeds, the concept C_k is the msc of a in $\mathcal{C}$ w.r.t. $\mathcal{T}$. This yields an effective test for the existence of the msc since the following can be shown: there is a polynomial p such that the entailment test succeeds after at most $p(|\mathcal{C}|, |\mathcal{T}|)$ steps iff the msc exists.

For example, for the ABox $\mathcal{C}^{(1)} = \{r(a, a)\}$ and the TBox $\mathcal{T}^{(1)} = \emptyset$, the 0-step unraveling is $C_0^{(1)} = \top$, the 1-step unraveling is $C_1^{(1)} = \exists r.\top$, the two-step unraveling is $C_2^{(1)} = \exists r.\exists r.\top$, etc. It is easy to that there is no k such that the entailment test succeeds. Thus, it does not succeed for $k(\mathcal{C}^{(1)}, \mathcal{T}^{(1)})$, which shows that a does not have an msc. If instead we consider the ABox $\mathcal{C}^{(2)} = \{A(a), r(a, b), s(a, b), r(b, c), s(b, c), B(c)\}$ w.r.t. $\mathcal{T}^{(2)} = \emptyset$, then the 0-step unraveling is $C_0^{(2)} = A$, the 1-step unraveling is $C_1^{(2)} = A \sqcap \exists r.\top \sqcap \exists s.\top$, the 2-step unraveling is $C_2^{(2)} = A \sqcap \exists r.(\exists r.B \sqcap \exists s.B) \sqcap \exists s.(\exists r.B \sqcap \exists s.B)$, and the 3-step unraveling is identical to $C_2^{(2)}$. The entailment test succeeds for $k = 2$. It is easy to see that, whenever the unraveling becomes stable (which happens if no cycle in the ABox is reachable from a), then the entailment test succeeds. However, a reachable cycle in the ABox need not prevent the existence of the msc. For example, the individual a has the msc $\exists r.B$ in $\mathcal{C}^{(3)} = \{r(a, b), r(b, b), B(b)\}$ w.r.t. $\mathcal{T}^{(3)} = \{B \sqsubseteq \exists r.B\}$.

As sketched until now, this method for deciding the existence of the msc does not yield a polynomial-time decision procedure. The reason is that, though the bound $k(\mathcal{C}, \mathcal{T})$ on the number of steps is polynomial, the unraveled concepts C_k may become exponential even for $k \le k(\mathcal{C}, \mathcal{T})$, as can be seen using an obvious generalization of our example ABox $\mathcal{C}^{(2)}$. This problem can be avoided by employing structure-sharing, which can be realized by representing the ABoxes $\{C_k(a)\}$ by IQ-equivalent qABoxes. In our second example, the ABox $\{C_2^{(2)}(a)\}$ can be represented by the more compact IQ-equivalent qABox $\exists \{x, y\}.\{A(a), r(a, x), s(a, x), r(x, y), s(x, y), B(y)\}$ (see the definition of the k-unraveling in [1] for how such an unraveling with structure sharing can be defined in general). It is easy to see that the qABoxes representing the ABoxes $\{C_k(a)\}$ are of polynomial size. Since IQ-entailment between qABoxes is polynomial, this yields the polynomiality result stated in the proposition. Note, however, that the msc obtained this way is still an unraveled concept C_k without structure sharing, and thus may be of exponential size.

The following theorem shows that existence of the optimal ABox approximation can be reduced to existence of the msc (see [4] for the proof).

Theorem 18. *Let $\exists X.\mathcal{A}$ be a qABox with set of individuals Σ_1, let $\mathcal{T}$ be an $\mathcal{EL}$ TBox, and let $\mathcal{B}_a$ for all $a \in \Sigma_1$ be the ABoxes introduced in Definition 14. Then $\exists X.\mathcal{A}$ has an optimal ABox approximation w.r.t. $\mathcal{T}$ iff, for all individuals*

$a \in \Sigma_I$, the msc of a in $\mathcal{B}_a$ w.r.t. $\mathcal{T}$ exists. If the latter condition is satisfied and C_a are these most specific concepts, then the following ABox is an optimal ABox approximation of $\exists X. \mathcal{A}$ w.r.t. $\mathcal{T}$:

$$\{ C_a(a) \mid a \in \Sigma_I \} \cup \{ r(a,b) \mid r(a,b) \text{ occurs in } \underline{\mathsf{sat}}_{\mathsf{IQ}}^{\mathcal{T}}(\exists X. \mathcal{A}) \text{ where } a,b \in \Sigma_I \}.$$

In particular, the existence of an optimal ABox approximation can be tested in polynomial time and such an optimal approximation can be computed in exponential time if it exists.

5   Computing Optimal ABox Repairs

We can now reap the benefits from the results shown in the previous two sections. Given a qABox $\exists X. \mathcal{A}$, an $\mathcal{EL}$ TBox $\mathcal{T}$, and an $\mathcal{EL}$ repair request $\mathcal{R}$, we can compute the set $\mathfrak{R}$ of optimal IRQ-repairs of $\exists X. \mathcal{A}$ for $\mathcal{R}$ w.r.t. $\mathcal{T}$ in exponential time. More precisely, by Proposition 5 this set contains at most exponentially many repairs, each of which has at most exponential size. Theorem 11 then says that the set of all optimal ABox repairs of $\exists X. \mathcal{A}$ for $\mathcal{R}$ w.r.t. $\mathcal{T}$ (up to equivalence) consists of the optimal ABox approximations w.r.t. $\mathcal{T}$ of those elements of $\mathfrak{R}$ for which such an optimal approximation exists. Finally, Theorem 18 shows how to decide existence of such optimal approximations and how to compute them if they exist. Since the elements of $\mathfrak{R}$ are already of exponential size, existence can be tested in exponential time and the size of the computed approximations is at most double-exponential.

Theorem 19. *Let $\exists X. \mathcal{A}$ be a qABox, $\mathcal{T}$ an $\mathcal{EL}$-TBox, and $\mathcal{R}$ an $\mathcal{EL}$ repair request. Then the existence of an optimal ABox repair of $\exists X. \mathcal{A}$ for $\mathcal{R}$ w.r.t. $\mathcal{T}$ can be decided in exponential time, and the set of all such repairs can be computed in double-exponential time. This set contains at most exponentially many elements, each of which has at most double-exponential size.*

If the given qABox does not have an optimal repair or if we are looking for a repair not covered by an optimal one, our approach can also be used to compute non-optimal ABox repairs. In fact, consider an optimal IRQ-repair that does not have an optimal ABox approximation. Then there are individuals a whose msc in $\mathcal{B}_a$ does not exist. Following [17], we can then use the role-depth bounded msc instead, which is basically obtained by unraveling up to a fixed bound k on the role-depth (i.e., the maximal nesting of existential restrictions). This way, we can produce a set of (possibly) non-optimal ABox repairs, which covers all ABox repairs whose concept assertions satisfy this bound on the role depth.

There are also cases where the existence of the optimal ABox approximation of the optimal IRQ-repairs is guaranteed. In fact, if the qABox is acyclic and the TBox is cycle-restricted (i.e., there is no concept C such that $C \sqsubseteq^{\mathcal{T}} \exists r_1. \cdots \exists r_k.C$, as defined in [3]), then the optimal IRQ-repairs are acyclic, which implies that the ABoxes $\mathcal{B}_a$ in the pre-approximations are also acyclic. Consequently, all optimal IRQ-repairs have an optimal ABox approximation. The following corollary is an easy consequence of this observation.

Corollary 20. *Let $\exists X.\mathcal{A}$ be an acyclic qABox, $\mathcal{T}$ a cycle-restricted $\mathcal{EL}$-TBox, and $\mathcal{R}$ an $\mathcal{EL}$ repair request. Then the set of optimal ABox repairs of $\exists X.\mathcal{A}$ for $\mathcal{R}$ w.r.t. $\mathcal{T}$ is non-empty, and it* IRQ-*covers all ABox repairs of $\exists X.\mathcal{A}$ for $\mathcal{R}$ w.r.t. $\mathcal{T}$.*

6   Conclusion

Traditional repair approaches for DL-based ontologies, which compute maximal subsets of the ontology that do not have the unwanted consequences, are syntax-dependent and thus may remove too many consequences. Recently developed syntax-independent approaches for repairing DL ABoxes [3,5,7] compute optimal repairs that do not lose consequences unnecessarily, but they have the disadvantage that they produce quantified ABoxes rather than traditional ABoxes. In this paper we show how to overcome this problem by developing methods for computing optimal repairs that are traditional ABoxes. These methods are based on the computation of optimal IRQ-repairs, by adapting the approaches in [3] for computing optimal IQ-repairs, and then optimally approximating these qABoxes with ABoxes.

A perceived disadvantage of our approach could be that optimal ABox repairs need not exist, and even if they do, they need not cover all ABox repairs. However, by Corollary 20 this problem does not occur if the ABox is acyclic and the TBox is cycle-restricted. To see how often this corollary applies in practice, we checked the 80 large ontologies used in the experiments in [3]: 62 have cycle-restricted TBoxes, and of those only 7 have cyclic ABoxes. Thus, our Corollary 20 applies to 55 of the 80 ontologies considered in [3].

Another disadvantage could be the potentially double-exponential size of optimal ABox repairs. However, the first exponential comes from the computation of the optimal IQ-repairs, and the experiments in [3] indicate that this exponential blow-up does not occur in practice if the optimized approach for computing IQ-repairs is used. We do not yet have experimental results regarding the possible exponential blow-up due to the computation of ABox approximations, but would be surprised if this happened often in practice.

What is called "repair" in the DL community is closely related to what is called "contraction" in the Belief Change community. For classical repairs and also for the gentle repairs of [6], this connection was investigated in [14]. It would be interesting to see whether this investigation can be extended to our optimal ABox repairs. The original intention underlying our repair approach is that the ontology engineer chooses one of the computed optimal repairs as the new, repaired ABox. Alternatively, one could try to adapt the different repair semantics employed in inconsistency-tolerant query answering [9,12] from classical repairs to our optimal repairs.

Acknowledgements. This work was partially supported by the AI competence center ScaDS.AI Dresden/Leipzig and the Deutsche Forschungsgemeinschaft (DFG), Grant 430150274, and Grant 389792660 within TRR 248.

References

1. Baader, F.: A graph-theoretic generalization of the least common subsumer and the most specific concept in the description logic $\mathcal{EL}$. In: Hromkovič, J., Nagl, M., Westfechtel, B. (eds.) WG 2004. LNCS, vol. 3353, pp. 177–188. Springer, Heidelberg (2004). https://doi.org/10.1007/978-3-540-30559-0_15
2. Baader, F., Horrocks, I., Lutz, C., Sattler, U.: An Introduction to Description Logic. Cambridge University Press, Cambridge (2017). https://doi.org/10.1017/9781139025355
3. Baader, F., Koopmann, P., Kriegel, F., Nuradiansyah, A.: Computing optimal repairs of quantified ABoxes w.r.t. static $\mathcal{EL}$ TBoxes. In: Platzer, A., Sutcliffe, G. (eds.) CADE 2021. LNCS (LNAI), vol. 12699, pp. 309–326. Springer, Cham (2021). https://doi.org/10.1007/978-3-030-79876-5_18
4. Baader, F., Koopmann, P., Kriegel, F., Nuradiansyah, A.: Optimal ABox repair w.r.t. static $\mathcal{EL}$ TBoxes: from quantified ABoxes back to ABoxes (extended version). LTCS-Report 22–01, Chair of Automata Theory, Institute of Theoretical Computer Science, Technische Universität Dresden, Dresden, Germany (2022). https://doi.org/10.25368/2022.65
5. Baader, F., Kriegel, F., Nuradiansyah, A.: Privacy-preserving ontology publishing for $\mathcal{EL}$ instance stores. In: Calimeri, F., Leone, N., Manna, M. (eds.) JELIA 2019. LNCS (LNAI), vol. 11468, pp. 323–338. Springer, Cham (2019). https://doi.org/10.1007/978-3-030-19570-0_21
6. Baader, F., Kriegel, F., Nuradiansyah, A., Peñaloza, R.: Making repairs in description logics more gentle. In: Principles of Knowledge Representation and Reasoning: Proceedings of the Sixteenth International Conference, KR 2018, Tempe, Arizona, 30 October–2 November 2018, pp. 319–328. AAAI Press (2018). https://aaai.org/ocs/index.php/KR/KR18/paper/view/18056
7. Baader, F., Kriegel, F., Nuradiansyah, A., Peñaloza, R.: Computing compliant anonymisations of quantified ABoxes w.r.t. $\mathcal{EL}$ policies. In: Pan, J.Z., et al. (eds.) ISWC 2020. LNCS, vol. 12506, pp. 3–20. Springer, Cham (2020). https://doi.org/10.1007/978-3-030-62419-4_1
8. Baader, F., Suntisrivaraporn, B.: Debugging SNOMED CT using axiom pinpointing in the description logic $\mathcal{EL}^+$. In: Proceedings of the Third International Conference on Knowledge Representation in Medicine, Phoenix, Arizona, USA, 31 May–2 June 2008. CEUR Workshop Proceedings, vol. 410. CEUR-WS.org (2008). http://ceur-ws.org/Vol-410/Paper01.pdf
9. Bienvenu, M., Bourgaux, C.: Inconsistency-tolerant querying of description logic knowledge bases. In: Pan, J.Z., et al. (eds.) Reasoning Web 2016. LNCS, vol. 9885, pp. 156–202. Springer, Cham (2017). https://doi.org/10.1007/978-3-319-49493-7_5
10. Henzinger, M.R., Henzinger, T.A., Kopke, P.W.: Computing simulations on finite and infinite graphs. In: 36th Annual Symposium on Foundations of Computer Science, Milwaukee, Wisconsin, USA, 23–25 October 1995, pp. 453–462. IEEE Computer Society (1995). https://doi.org/10.1109/SFCS.1995.492576
11. Kalyanpur, A., Parsia, B., Horridge, M., Sirin, E.: Finding all justifications of OWL DL entailments. In: Aberer, K., et al. (eds.) ASWC/ISWC -2007. LNCS, vol. 4825, pp. 267–280. Springer, Heidelberg (2007). https://doi.org/10.1007/978-3-540-76298-0_20
12. Lembo, D., Lenzerini, M., Rosati, R., Ruzzi, M., Savo, D.F.: Inconsistency-tolerant query answering in ontology-based data access. J. Web Semant. **33**, 3–29 (2015). https://doi.org/10.1016/j.websem.2015.04.002

13. Lutz, C., Wolter, F.: Deciding inseparability and conservative extensions in the description logic $\mathcal{EL}$. J. Symb. Comput. **45**(2), 194–228 (2010). https://doi.org/10.1016/j.jsc.2008.10.007

14. Matos, V.B., Guimarães, R., Santos, Y.D., Wassermann, R.: Pseudo-contractions as gentle repairs. In: Lutz, C., Sattler, U., Tinelli, C., Turhan, A.-Y., Wolter, F. (eds.) Description Logic, Theory Combination, and All That. LNCS, vol. 11560, pp. 385–403. Springer, Cham (2019). https://doi.org/10.1007/978-3-030-22102-7_18

15. Meyer, T.A., Lee, K., Booth, R., Pan, J.Z.: Finding maximally satisfiable terminologies for the description logic $\mathcal{ALC}$. In: Proceedings, The Twenty-First National Conference on Artificial Intelligence and the Eighteenth Innovative Applications of Artificial Intelligence Conference, 16–20 July 2006, Boston, Massachusetts, USA, pp. 269–274. AAAI Press (2006). http://www.aaai.org/Library/AAAI/2006/aaai06-043.php

16. Parsia, B., Sirin, E., Kalyanpur, A.: Debugging OWL ontologies. In: Proceedings of the 14th International Conference on World Wide Web, WWW 2005, Chiba, Japan, 10–14 May 2005. pp. 633–640. ACM (2005). https://doi.org/10.1145/1060745.1060837

17. Peñaloza, R., Turhan, A.-Y.: A practical approach for computing generalization inferences in $\mathcal{EL}$. In: Antoniou, G., et al. (eds.) ESWC 2011, Part I. LNCS, vol. 6643, pp. 410–423. Springer, Heidelberg (2011). https://doi.org/10.1007/978-3-642-21034-1_28

18. Schlobach, S., Cornet, R.: Non-standard reasoning services for the debugging of description logic terminologies. In: IJCAI-03, Proceedings of the Eighteenth International Joint Conference on Artificial Intelligence, Acapulco, Mexico, 9–15 August 2003, pp. 355–362. Morgan Kaufmann (2003). http://ijcai.org/Proceedings/03/Papers/053.pdf

19. Schlobach, S., Huang, Z., Cornet, R., van Harmelen, F.: Debugging incoherent terminologies. J. Autom. Reason. **39**(3), 317–349 (2007). https://doi.org/10.1007/s10817-007-9076-z

20. Zarrieß, B., Turhan, A.: Most specific generalizations w.r.t. general $\mathcal{EL}$-TBoxes. In: IJCAI 2013, Proceedings of the 23rd International Joint Conference on Artificial Intelligence, Beijing, China, 3–9 August 2013, pp. 1191–1197. IJCAI/AAAI (2013). http://www.aaai.org/ocs/index.php/IJCAI/IJCAI13/paper/view/6709

Ensemble-Based Fact Classification
with Knowledge Graph Embeddings

Unmesh Joshi[(⊠)] and Jacopo Urbani

Department of Computer Science, Vrije Universiteit Amsterdam, Amsterdam, The Netherlands
u.n.joshi@vu.nl, jacopo@cs.vu.nl

Abstract. Numerous prior works have shown how we can use Knowledge Graph Embeddings (KGEs) for ranking unseen facts that are likely to be true. Much less attention has been given on how to use KGEs for fact classification, i.e., mark unseen facts either as true or false. In this paper, we tackle this problem with a new technique that exploits ensemble learning and weak supervision, following the principle that multiple weak classifiers can make a strong one. Our method is implemented in a new system called DuEL. DuEL post-processes the ranked lists produced by the embedding models with multiple classifiers, which include supervised models like LSTMs, MLPs, and CNNs and unsupervised ones that consider subgraphs and reachability in the graph. The output of these classifiers is aggregated using a weakly supervised method that does not need ground truths, which would be expensive to obtain. Our experiments show that DuEL produces a more accurate classification than other existing methods, with improvements up to 72% in terms of F_1 score. This suggests that weakly supervised ensemble learning is a promising technique to perform fact classification with KGEs.

1 Introduction

Knowledge Graphs (KGs) [26] have emerged as the de-facto standard to share large amounts of factual knowledge on the Web. A fundamental problem that concerns KGs is *link prediction*, i.e., the problem of predicting potential missing links in a KG.

Recently, numerous works [25,40] have shown that Knowledge Graph Embeddings (KGEs) models can be used to identify the top k completions for link patterns (e.g., ⟨London, capitalOf, ?⟩). This operation is useful to identify a smaller set of promising links, but more work is needed for selecting the correct ones. Consider, for instance, the case of a human KG curator who is searching for missing links. An embedding model can help they to identify the k most promising links, but in practice only a small fraction of such a subset is indeed correct. To recognize those, an additional evaluation is needed, which might be time consuming if it were conducted manually. This problem would be solved, or at least reduced, if we have a procedure that directly classifies potential links with a binary true/false label. Such a procedure could be used to implement a fully automated KG completion pipeline, or at least would lift the burden of interpreting ranked lists of potential completions off the user.

Surprisingly, performing fact classification with KGEs is a problem that is not yet well studied. So far, the research on KGEs has primarily focused on the model

P. Groth et al. (Eds.): ESWC 2022, LNCS 13261, pp. 147–164, 2022.
https://doi.org/10.1007/978-3-031-06981-9_9

construction, using ranking as the main evaluation metric and leaving the task of making fact classification as future work [35,42]. This makes existing KGEs models (e.g., TransE [5], ComplEx [38], RotatE [36], RDF2Vec [29]) not suitable in their current form. So far, the only proposal for performing fact classification with KGEs is to simply label all the top k completions as correct and all the others as incorrect [35]. In practice, however, this approach does not work well because not all correct completions appear in the top ranked positions; thus a small k would affect recall while a large one would affect precision. Another approach would be to include additional background knowledge such as ontologies to filter out incorrect links. For instance, if we know that a property is functional, then at most one of the k completions should be marked as correct. Unfortunately, such additional knowledge is not always available; thus we consider the setting where we do not have it.

In general, we can identify two main key challenges for performing fact classification using KGEs. *First*, KGs can be very incomplete and this affects negatively the accuracy of predictions. *Second*, it is hard to produce training samples, negative in particular, because KGs are built under Open World Assumption (OWA), thus potential links can be either missing or incorrect. One could address this problem by manually annotating the top k completions, but this is a time consuming operation which may require human experts.

The two challenges above increase significantly the difficulty of designing a single procedure, e.g., a supervised classifier, that relies solely on the embeddings to produce the classification. Fortunately, there is a well-known alternative approach in Machine Learning called *ensemble learning* that is designed to address precisely the cases when we do not have a classifier that is accurate enough. The idea behind ensemble learning is conceptually simple: instead of focusing on a single classifier, we can use multiple ones, following the principle that multiple weak classifiers can make a strong one.

Ensemble learning is a technique that has been successfully applied in multiple domains (e.g., see overview at [47]), but it has never been applied for fact classification with KGEs. In this paper, we cover this gap with a new ensemble learning method called DuEL (Dual Embedding-based Link prediction), that is specifically designed for fact classification with KGEs. With ensemble learning, the challenge is to identify a suitable set of classifiers and aggregation technique to exploit their predictive power as much as possible. Next to this, in our context we also need to face the problem that we lack ground truths to train any supervised classifier and aggregation model.

We address the aforementioned problems as follows. For the selection of a suitable set of classifiers, we considered state-of-the-art neural architectures, which can be seen as a natural choice for this type of problems. In particular, we selected three different models: an LSTM network [17], a convolutional neural network (CNN) [13], and a multi-layer perceptron (MLP) [4]. We selected these models because they interpret the input in different ways (e.g., with an LSTM it is a one-by-one sequence while with a CNN multiple facts are fed at the same time), hence each of them can capture signals that the others might miss. To function properly, however, all three models require ground truths for training, which we do not have. To fix this problem, we created (possibly wrong) training data assuming Closed World Assumption (CWA), which states that everything that is not in the KG is false by definition. A consequence of this assumption is that the training data might contain many false negatives. Hence, the classifier is trained with a bias towards rejecting potential good completions, which favors precision but harms recall. To mitigate this problem, we include two additional unsupervised classifiers which leverage subgraph embeddings [18] and shared paths in the KG, respectively. These two classifiers tend to have a higher recall. Therefore, they are a good complement to the first three classifiers.

For aggregating the classifiers' outputs, using a supervised classifier is problematic because we do not have ground truths. An alternative could be to rely on unsupervised techniques like majority voting. However, such approaches would not consider possible differences of the classifiers' accuracies, or latent correlations between them. To exploit those, we can leverage recent weakly supervised techniques that combine the outputs of classifiers without ground truths [12,28]. In the literature, these models have been shown to be very effective for making predictions with noisy data (e.g., see the Snorkel project [27]). We show here that they are also valuable for performing link prediction, which is a problem for which they have not been applied yet.

While our solution is conceptually simple, our experiments using multiple embedding models confirmed that DuEL was able to perform a fact classification that is much more accurate than currently possible, with the major benefit that our solution can be trained only with the content of the (incomplete) KG and without high-quality manual annotations. For instance, DuEL outperformed existing methods producing predictions with an F_1 of 0.60 and 0.51 on FB15k237 and DBpedia50 respectively, which are two well-known benchmarks, with improvements that range between 72% and 24% against the second best approach.

2 Link Prediction with KGEs

A KG can be seen as a directed labeled (multi)graph $\mathcal{K} = (\mathcal{V}, \mathcal{E}, \mathcal{R})$ where vertices in $\mathcal{V}$ represent entities and every edge in $\mathcal{E}$ denotes a semantic relation labeled with type $r \in \mathcal{R}$. Given $h, t \in \mathcal{V}$ and $r \in \mathcal{R}$, we write $\langle h, r, t \rangle$ to indicate the edge from h to t which is labeled with r, e.g., $\langle$London, capitalOf, UK$\rangle$. Throughout, we often refer to edges as *links*. We also introduce the expression *link pattern*, denoted $\langle h, r, ? \rangle$ ($\langle ?, r, t \rangle$), to refer to the set of all links from h (to t) with label r. Finally, we say that e is a *valid completion* for $\langle h, r, ? \rangle$ ($\langle ?, r, t \rangle$) in $\mathcal{K}$ if $\langle h, r, e \rangle \in \mathcal{K}$ ($\langle e, r, t \rangle \in \mathcal{K}$).

We assume that $\mathcal{K}$ is incomplete in the sense that some links are missing. This assumption implies the existence of another KG $K' = (\mathcal{V}, \mathcal{E}', \mathcal{R})$ where $\mathcal{E}' \supset \mathcal{E}$ is the set of all true links with a label in $\mathcal{R}$ between the entities in $\mathcal{V}$. Our goal is to predict all and only the links in $\mathcal{E}' \setminus \mathcal{E}$.

An *embedding* is a vector in $\mathbb{R}^d$ where $d > 0$ and an *embedding model* (or *model* for short) is a set of embeddings. Several techniques have been proposed to construct embedding models that are suitable for link prediction (e.g., [5,7,24,32,36,38,41,44]). In this paper, we consider three techniques: ComplEx [38], RotatE [36], and TransE [5], which we selected as examples of factorization models (ComplEx) and translational models (RotatE, TransE). ComplEx and RotatE are among the techniques that returned the best performance according to [30] while TransE is included as it is one of the oldest and most frequently used techniques.

All three techniques assign a vector of d numbers to every entity in $\mathcal{V}$ and every relation in $\mathcal{R}$, effectively creating models with $(|\mathcal{V}| + |\mathcal{R}|) \times d$ parameters. In the case of TransE, the numbers are real, while with RotatE and ComplEx the numbers are complex. These techniques first define a suitable scoring function for a candidate link $\langle h, r, t \rangle$. Then, the models are trained with different loss functions that combine the scoring functions of true and false links. With TransE, the scoring function is:

$$f_{tr}(\langle h, r, t \rangle) = ||\mathbf{h} + \mathbf{r} - \mathbf{t}|| \tag{1}$$

where $||\cdot||$ is the L1 norm, $\mathbf{h}$, $\mathbf{r}$, and $\mathbf{t}$ are the vectors associated to h, r, and t respectively (we follow convention of denoting the embeddings in boldface). With ComplEx, it is:

$$f_{co}(\langle h, r, t \rangle) = Re(\langle \mathbf{r}, \mathbf{h}, \bar{\mathbf{t}} \rangle) \tag{2}$$

where $\langle \cdot \rangle$ applied to vectors is the generalized dot product, $Re(\cdot)$ is real component, and $\bar{\ }$ is the conjugate for complex vectors. With RotatE, it is:

$$f_{ro} = ||\mathbf{h} \circ \mathbf{r} - \mathbf{t}|| \tag{3}$$

where $\circ$ denotes the element-wise product.

In the literature, empirical evaluations have shown that embedding models return higher scores for true links than for false ones [25,30,40]. This observation suggests a straightforward way to do link prediction, that is, to rank every entity t_i according to $f(\langle h, r, t_i \rangle)$, and consider the k entities with the highest ranks as potential valid completions. However, accepting indiscriminately all k is likely to yield a low precision because in practice many of the top k entities are not valid completions. One solution would be to reduce the k to retain only the most likely completions, but this would lower the recall since many correct completions will be missed. To improve the both precision and recall, we are called to critically look at the ranked list of entities and translate the numerical scores into binary decisions.

3   Our Proposal

Let $\mathcal{K}$ be the input KG and $\mathcal{K}'$ be the (unknown) KG with all the true links. DuEL is designed to predict all the links in $\mathcal{K}'$ with a given label r that either start from or end to a given entity e. This equals to finding all valid completions for a pattern that is either of the form $\langle ?, r, e \rangle$ or $\langle e, r, ? \rangle$ in $\mathcal{K}'$. Thus, from now on we will assume that the input is a link pattern p and an embedding model M of $\mathcal{K}$, while the output is the set of valid completions for p in $\mathcal{K}'$.

As an example, Fig. 1 gives a graphical overview of the functioning of DuEL with the pattern $p = \langle ?, \texttt{locatedIn}, \texttt{UK} \rangle$. The first step consists of computing the top k ranked entities for p with M, which are not valid completions in $\mathcal{K}$ (if they are, then the links are already known). Let us call E the set of such entities. DuEL considers only the entities in E, which are the most likely completions, and ignore all others.

Next, DuEL makes a binary decision for every entity $e \in E$, thus establishing the truth value of links (e.g., if the decision for $e = \texttt{London}$ is positive, then the link $\langle \texttt{London}, \texttt{locatedIn}, \texttt{UK} \rangle$ is correct). Each binary decision is a two-step process. First, multiple classifiers independently label every candidate entity. Then, the labels are aggregated to formulate a final correct/incorrect prediction.

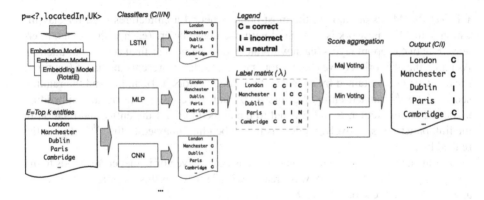

Fig. 1. Schematic overview of DuEL

3.1 Classifiers

In principle, DuEL can be configured to use an arbitrary number of classifiers. In general, we would like to have some classifiers that complement each other. For instance, some can learn to make the predictions based on the latent representations of the entities, and others that more generally consider the structure of $\mathcal{K}$. The classifiers do not need to be conceptually simple or fast to execute. For our purposes, they can be arbitrarily sophisticated as long as they do not require human input. With this desiderata in mind, we introduce five types of classifiers, **C1**,...,**C5**.

The classifiers **C1**,...,**C3** are supervised models. Thus, they require training data. Unfortunately, obtaining ground truth annotations involves a human intervention, which can be expensive. To give an idea, deciding whether a link was true often took more than a minute during the creation of our gold standard. To avoid this problem, we decided to use the content of $\mathcal{K}$ to label the training samples, effectively operating under CWA.

Since this data contains an approximation of the true labels, we train multiple classifiers hoping that mistakes will be corrected during a collective evaluation. Each classifier approaches the problem from a different perspective: **C1** is a MLP, one of the most conventional choices for classification; **C2** is an LSTM, thus it views the classification as a sequence labeling problem; **C3** is a CNN, thus it relies on the convolutional operator to perform a collective prediction of the top-k at once.

The classifiers **C4** and **C5** are added because they do not rely on supervised models: **C4** relies on ranked lists of subgraph embeddings while **C5** considers shared paths between the entities. Since they do not base their decisions using training data created under CWA, we expect them to have a higher recall than the first three which should have instead a higher precision.

For a given set E as input, every classifier returns the set $F := \{\langle e, l \rangle \mid e \in E\}$ where $l \in \{\text{CORRECT}, \text{INCORRECT}\}$. We create two classifiers of each type, one for patterns of the form $\langle ?, r, t \rangle$ and another for $\langle h, r, ? \rangle$. Below, we describe each classifier in more detail.

C1 (MLP). MLPs are among the most popular neural architectures used for classification. Our MLP network is structured with two dense layers (each with n units) interleaved by two dropout layers (each with rate r) and sigmoid as final activation function. As input, the network receives the vectors $t_1, \ldots, t_k$ that represents the list of k entities. Each vector t_i is obtained by concatenating three vectors a, b, and c, and the ranking score of the ith ranked entity. Vectors a, b, and c are created using the embedding model M. In particular, a and b equal to the embeddings of the entity and relation in the link pattern, respectively, while c is the embedding mapped to the ith entity in the ranked list.

For instance, suppose that we want to construct the vector t_i that corresponds to the entity e_i, $p = \langle ?, r, t \rangle$, and M was created with TransE. In this case, $a := t$, $b := r$, and $c := e_i$ and the score is $f_{tr}(\langle e_i, r, t \rangle)$.

C2 (LSTM). Using an LSTM for classifying the top-k answers entities is not a usual approach. This is because LSTM is a sequence model and in theory the truth value of each answer does not depend on the ones that are before or after it. However, we observed that there are some regularities in the number and positions of true links in the ranked list. For instance, if $p = \langle \text{Ferrary}, \text{isA}, ? \rangle$, then it is likely that the valid completions are few and concentrated in the top positions (since typically the number of classes is limited). However, if $p = \langle ?, \text{isA}, \text{Student} \rangle$, then the valid completions are (probably) many more. These observations hint that the sequence of completions is a useful asset for making binary predictions.

As input, the LSTM with n hidden units receives a sequence of vectors $t_1, \ldots, t_k$ that represents the list of k entities. Each t_i is constructed in the same manner as done for **C1**. Since our task is classification, we add, on top of the LSTM, one extra layer with n units, followed by a dropout layer with rate r and a final dense layer with a sigmoid activation function to produce a binary classification.

C3 (CNN). CNN networks are a very popular type of deep neural networks used primarily for image processing and other types of problems, like sentence classification [21], sentiment analys [9], or text ranking [33].

We construct a network with a single 2D convolutional layer, parametrized by a kernel of size $s_1 \times s_1$. We chose a 2D layer instead of a 1D layer because with the former we can model the interactions between the concatenated embeddings. As input, we provide a 2D matrix obtained by concatenating the $t_1, \ldots, t_k$ vectors. The convolutional layer returns an output with k channels, which is passed to a max pooling layer, parametrized by another kernel of size $s_2 \times s_2$ for further down sampling. This layer returns a 1D vector with k elements, which is post-processed by a sigmoid activation function such that it returns k binary predictions.

C4 (Subgraph Embeddings). This classifier uses subgraph embeddings, which were recently introduced by [18]. Subgraphs embeddings are created by aggregating the embeddings of the entities contained in them, and are meant to quickly provide an approximate ranking of the top k entities. In this context, subgraphs are defined as set of entities that share the same neighbour with edges with the same label. Subgraphs can be of two types, depending on the direction of the edges to the common neighbour. Let us recall that $\mathcal{K} = (\mathcal{V}, \mathcal{E})$. We denote with $S_{\langle ?,r,t \rangle} := \{h \mid \langle h,r,t \rangle \in \mathcal{E}\}$ the subgraph with all entities with outgoing edges to t which are labeled with r. Analogously, $S_{\langle h,r,? \rangle} := \{t \mid \langle h,r,t \rangle \in \mathcal{E}\}$ denotes the subgraph with incoming edges from h. Subgraph embeddings are constructed by averaging the embeddings of their entities. With TransE, this equals to

$$\mathbf{S}_l := \frac{\sum_{e \in S_l} \mathbf{e}}{|S_l|} \tag{4}$$

With ComplEx and RotatE, we take the average of the real ($Re(\cdot)$) and imaginary ($Im(\cdot)$) parts, respectively.

$$\mathbf{S}_l := \langle \frac{\sum_{e \in S_l} Re(\mathbf{e})}{|S_l|}, \frac{\sum_{e \in S_l} Im(\mathbf{e})}{|S_l|} \rangle \tag{5}$$

The average embeddings computed by Eqs. 4 and 5 allow us to apply the scoring function using the subgraph embeddings rather than the embeddings of potential completions. Once the classifier receives as input p and E, it first ranks all the subgraphs with the scoring function of M and p. This operation produces a list $\mathcal{S} = \langle S_1, S_2, \ldots \rangle$ of subgraphs. Then, it retains in $\mathcal{S}$ only the top j subgraphs, where j is a threshold value that is dynamically computed using statistics from $\mathcal{K}$ [18].

Finally, the classifier labels every entity $e \in E$ as follows. If e appears as a member of any subgraph in $\mathcal{S}$, then it is labeled as CORRECT. Otherwise, e is labeled as INCORRECT. Notice that this classifier, unlike the previous three, does not require a training phase.

C5 (Shared Paths). The previous four classifiers rely on KG embeddings to make their predictions. In contrast, this classifier does not use embeddings, but considers instead shared paths between potential valid completions in $\mathcal{K}'$ (i.e., the entities in E) and any valid completion in $\mathcal{K}$.

First, let us assume that the input link pattern p is of the form $\langle ?,r,t \rangle$ (the other case is analogous). Then, let $\mathcal{P}_{a,b}$ be the set of all paths between a and b in $\mathcal{K}$ of maximum length 2 (note that the direction of the edges is not taken into account). Furthermore, let $\mathcal{P}_p = \{q \mid q \in \mathcal{P}_{a,t}, \langle a,r,t \rangle \in \mathcal{K}\}$ the set of all paths between t and any valid completion of p in $\mathcal{K}$. This classifier will label every entity $e \in E$ as CORRECT if there is a path $p \in \mathcal{P}_{e,t}$ and another path $q \in \mathcal{P}_p$ which differ only on the first entity in the paths. In other words, entity e is marked as CORRECT if it is connected to t with the same path as one of the valid completions in $\mathcal{K}$. Otherwise, e is labeled as INCORRECT.

3.2 Aggregation

The output of classifiers can be aggregated in several ways. Two techniques which are often used are *Min Voting* and *Majority Voting*. With the first technique, we label an entity as CORRECT if at least one classifier has labeled it as CORRECT. With the second one, we pick the label chosen by the majority of the classifiers (ties are broken arbitrarily). A disadvantage of these techniques is that they do not consider latent correlations between the classifiers. To include those, we can use several approaches that were originally introduced for building machine learning models without using ground truth annotations [12,28].

The problem of aggregating without ground truths can be modeled as follows. The input consists of a set data points $\mathcal{X}$. The goal consists of labelling each data point $X \in \mathcal{X}$ with a vector $\mathbf{Y} = [Y_1, \ldots, Y_t]^T$ of t categorical task labels. We assume that we do not have any ground truth that we can use for training. To recover from the lack of such data, we consider a set of *sources* that provide approximate labels for (a subset of) the t tasks. If they cannot provide a label, then the sources can *abstain*.

The sources might be potentially correlated and have an unknown accuracy. We can estimate those considering the observed agreement and disagreement rates of the emitted labels. To this end, we can construct a matrix λ of noisy labels produced by the sources, and then compute a *label model* $P_\mu(\mathbf{Y}|\lambda)$ where μ is the vector of parameters that encodes the correlations and accuracies. Then, we can use the label model to output a single probabilistic label vector $\tilde{\mathbf{Y}}$ from the noisy labels of an unseen data point X. The problem translates into computing the parameters μ. One way consists of estimating μ from the inverse covariance matrix among the sources [28]. Another approach consists of breaking down the original problem into a set of smaller problems that considers subsets of three sources. The advantage is that the subproblems have closed-form solutions, thus the parameters can be computed without iterative solutions [12].

In our context, we have a single task, the categorical task label represents the binary prediction CORRECT and INCORRECT, the data points are the potential completions for p (i.e., E), and there are five sources, **C1**, ..., **C5**. Finally, since one of the strengths of such methods is to consider that a source might abstain, we slightly change the labeling of our classifiers as follows. Instead of predicting either CORRECT or INCORRECT, we introduce one extra label NEUTRAL. Then, we modify the output of **C1**, **C2**, and **C3** introducing two threshold values τ_1 and τ_2. If the output of the sigmoid is lower (higher) than τ_1 (τ_2) then the label is INCORRECT (CORRECT). Otherwise, if it is between τ_1 and τ_2, then it is NEUTRAL.

A problem that arise with our threshold-based approach is that we must find good values for τ_1 and τ_2. In practice, we observed that using grid search using a small held-out validation dataset yields satisfactory performance. For **C4** and **C5**, we replace INCORRECT with NEUTRAL, thus simulating a prediction under OWA where missing links are not automatically considered as incorrect.

To aggregate the classifiers' output, we first compute the label model by applying the classifiers on every potential completion and use their output to create the label matrix λ. After computing μ, either exploiting the covariance [28] or the decomposition in triplets [12], we compute λ for an unlabeled completion e, pass it to the label model and use the value of the returned $\tilde{\mathbf{Y}}$ as the final (binary) label of e.

It is important to note that none of the components in the pipeline of DuEL needs large volumes of manually annotated data, with a consequent benefit in terms of scalability. The classifiers **C4**, **C5** do not require training, **C1**, **C2**, and **C3** are trained using the true links of $\mathcal{K}$, and the training of the label model only considering the provided λ.

	FB15k237	DBpedia50
# Entities in $\mathcal{K}$	14k	50k
# Edges in $\mathcal{K}$	298k	32k
Parameters embedding models		
ComplEx: d (dimension) 256, lr (learning rate) 0.1, regularizer 0.95		
RotatE: d 400, lr 0.5, γ (margin) 6, sampling temperature 2.0		
TransE: d 200, lr 0.1, γ 1, batch size 1000		
Statistics on the training datasets for C1, C2, and C3		
# Training data points for patterns $\langle ?, r, t \rangle$	56k	11k
# Training data points for patterns $\langle h, r, ? \rangle$	93k	29k

(a) KGs, model parameters, and training data

Parameter name	Min	Max
n (# hidden units)	10	500
d (drop out)	0.0	0.8
s_1 and s_2 (kernel sizes)	2	6
τ_1	0.01	0.8
τ_2	0.01	0.8

(b) Ranges for grid search to determine value hyperparameters

Fig. 2. Statistics, model parameters, and ranges used for grid search

4 Evaluation

KGs. As inputs, we considered datasets commonly used in the related literature. We selected FB15k237, a subset of Freebase used as benchmark in many works (e.g., [18,36,38]), and DBpedia50, a subset of DBpedia used in [30]. We used the parameters for the embedding models reported as optimal by [30]. Details are in Fig. 2a. The embedding models were trained with Adagrad [10] for 1000 epochs.

Below, we set $k = 10$ ($|E|$) as default since this is a common threshold used for evaluating ranked lists of entities (*hit@10*). The experiments were performed on a machine with 64GB RAM and two 8-core CPU 2.4 GHz. The code and other experimental data is available online[1].

Training C1, C2, and C3. These classifiers are trained under CWA using training data that was automatically generated. Every pattern and list of k completions is a data point used to train the networks. Since there is a variable number of link patterns of different types, the number of data points depends on the KG and type of pattern. Figure 2a reports the size of the training data sets. Training occurred by minimizing the binary cross entropy with Adam [19] for 10 epochs.

We performed grid search to find the optimal values for n (hidden units), r (dropout rate), s_1 and s_2 (kernel sizes), τ_1, and τ_2 optimizing for the best F_1 on the validation dataset. The ranges considered for the search are reported in Fig. 2b. We observed that the values $n = 100$, $r = 0.2$, $s_1 = 3$, $s_2 = 2$, $\tau_1 = 0.2$ and $\tau_2 = 0.6$ work well with the models and KGs.

Gold Standard. To test the performance of DuEL, we cannot rely on the content of the input KG since it is, by definition, incomplete. In particular, we cannot use a held-out dataset as it is typically done for evaluating the ranking capabilities of embedding models because such a dataset would contain only (some) links which we know are true (and not the ones which we know are false). To perform a more complete evaluation, we created a gold standard with data annotated by humans. We randomly selected

---

[1] https://github.com/karmaresearch/duel.

Statistics on the gold standard	FB15k237	DBpedia50
# Annotated links	~3900	~3600
Ratio CORRECT links for $\langle ?, r, t \rangle$ in top k entities	14%	12%
Ratio CORRECT links for $\langle h, r, ? \rangle$ in top k entities	11%	3%

(a) Ratio of correct links in the gold standards with FB15k237 and DBpedia50

FB15k237	DBpedia50
tv actor	producer of film
influenced by	birth place
graduate student	director of film
award category	soccer team
award nomination	starring in film
netflix genre	creative writer
film performance	country
film language	is part of
person profession	film language
music artist	music genre

(b) Ten popular relations (ranked from the most popular to the least one) on FB15k237 and DBpedia50

Fig. 3. Details about gold standard

```
Processed queries: 260 Head: 139 Tail: 111
Annnotated answers: 3908

Query #2921 Type HEAD
? /medicine/symptom/symptom_of [peritonitis http://www.wikidata.org/entity/Q223102 (/m/011zdm)]
Search on Google: https://www.google.com/search?hl=en&q=peritonitis
Answers:
    0. [peritonitis http://www.wikidata.org/entity/Q223102 (/m/011zdm)] (8838) methods=['transe']
    https://www.google.com/search?hl=en&q=peritonitis https://en.wikipedia.org/w/index.php?search=peritonitis
    1. [jaundice http://www.wikidata.org/entity/Q133244 (/m/0hgxh)] (9229) methods=['transe', 'complex', 'rotate']
    https://www.google.com/search?hl=en&q=jaundice https://en.wikipedia.org/w/index.php?search=jaundice
    2. [fatigue http://www.wikidata.org/entity/Q9690 (/m/01j6t0)] (4633) methods=['transe', 'complex', 'rotate']
    https://www.google.com/search?hl=en&q=fatigue https://en.wikipedia.org/w/index.php?search=fatigue
```

Fig. 4. Screenshot annotation interface

250 and 150 previously unseen link patterns of both types (50/50) for FB15k237 and DBpedia50 respectively, retaining 50 patterns of each type to construct a small validation dataset. Then, for every link pattern and embedding model, we manually annotated the top $k = 10$ entities that correspond to links that are not in $\mathcal{K}$, consulting external sources to verify the correctness of the links. The annotations were performed by two human annotators who independently annotated the links using a special web interface. Figure 4 reports a screenshot of (part of) the interface. The interface shows, for a given query, which are the top ranked answers provided by the three embedding models. Additional links to Google and Wikipedia are provided to help the human annotator to decide whether a particular answer is correct.

In total, the annotators labeled about 3900 links for FB15k237 and 3600 links for DBpedia50. Figure 3a reports the rate of CORRECT links in both datasets while Fig. 3b reports the 10 most popular relations annotated in each dataset. Since the task of the annotators is to verify whether the fact is true, the degree of subjectivity is low. This is confirmed by a high Cohen's score: With FB15k237 is it 0.8137 while with DBpedia50 it is 0.869, which indicate a nearly perfect agreement between the annotators. Notice that the size of the gold standard is much smaller than the size of training data used to train the classifiers since the former requires a manual annotation while the latter can be

	Method	ComplEx			RotatE			TransE		
		P	R	F_1	P	R	F_1	P	R	F_1
FB15k237	*RankClass*	**0.727**	0.229	0.348	**0.747**	0.232	0.353	**0.682**	0.221	0.334
	DeepPath	0.193	0.210	0.198	0.203	0.207	0.205	0.247	0.210	0.202
	AnyBurl	0.217	0.257	0.208	0.199	0.253	0.223	0.211	0.258	0.221
	C1	0.619	0.436	0.505	0.592	0.435	0.501	0.529	0.296	0.371
	C2	0.517	0.589	0.539	0.531	0.624	0.560	0.544	0.388	0.451
	C3	0.704	0.267	0.384	0.674	0.284	0.392	0.564	0.271	0.347
	C4	0.356	**0.883**	0.508	0.357	0.563	0.416	0.337	**0.876**	0.487
	C5	0.394	0.651	0.490	0.406	0.655	0.499	0.382	0.651	0.480
	MinV	0.317	*1.000*	0.481	0.324	*1.000*	0.490	0.308	*1.000*	0.471
	MajV	0.560	0.588	0.567	0.559	0.513	0.523	0.548	0.492	0.514
	Super.	0.565	0.537	0.550	0.595	0.486	0.533	0.522	0.507	0.513
	DuEL (M)	0.527	0.710	**0.599**	0.467	**0.824**	**0.569**	0.491	0.641	**0.537**
	DuEL (S)	0.528	0.728	0.594	0.472	0.740	0.540	0.504	0.586	0.530
DBpedia50	*RankClass*	0.516	0.353	0.403	0.740	0.273	0.381	0.183	0.096	0.120
	C1	0.620	0.105	0.175	**0.920**	0.070	0.129	0.375	0.027	0.051
	C2	**0.650**	0.287	0.350	0.500	0.005	0.010	0.000	0.000	0.000
	C3	0.354	0.415	0.363	0.551	0.326	0.314	0.240	0.205	0.215
	C4	0.162	0.112	0.132	0.253	0.094	0.137	0.135	0.109	0.121
	C5	0.187	0.288	0.222	0.259	0.305	0.280	0.151	0.209	0.175
	MinV	0.164	*1.000*	0.277	0.323	*1.000*	0.469	0.165	*1.000*	0.275
	MajV	0.520	0.149	0.210	0.917	0.052	0.095	0.344	0.050	0.087
	Super.	0.583	0.086	0.147	0.917	0.068	0.125	**0.400**	0.018	0.035
	DuEL (M)	0.555	0.457	**0.501**	0.557	**0.499**	**0.505**	0.191	0.751	0.274
	DuEL (S)	0.399	**0.625**	0.433	0.665	0.353	0.461	0.186	**0.961**	**0.301**

Fig. 5. Performance on gold standard (P, R and F_1 denote Precision, Recall, F_1 scores, respectively). The best results are marked in boldface

automatically computed from the input KG. Also, notice that the ratio of correct links is fairly low, especially in DBpedia50. With such a low ratio, a supervised classifier trained with such data can achieve a high accuracy by simply returning always the label INCORRECT. Our method does not suffer from this problem since the aggregation does not need ground truths.

4.1 Performance of Link Prediction

Baselines. We consider three external baselines and several alternative approaches to the default pipeline, yielding a comparison against 10 other methods. The first external baseline, which we name *RankClassify*, is presented by [35] and it is, as far as we know, the only method that uses embeddings to perform binary predictions. The technique consists of marking as correct all the answers in the top k positions, where k is fine tuned upfront on a validation dataset. The second baseline is the state-of-the-art method proposed by [43]. This method, which we label *DeepPath*, does not use embeddings. Instead, it uses reinforcement learning to learn reasoning paths on the KG. We configured it to do *fact prediction* on FB15k237 and mark the top k ranked facts as correct, similarly as before. The third baseline is the reinforcement learning version of AnyBURL [22,23], which learns rules bottom-up for link prediction. AnyBURL is executed using the default parameters mentioned in the online documentation and trained for 1000 s. On DBpedia50, we report only the performance with RankClassify as it was the external baseline with the highest F_1.

To compare against more methods, we also apply the five classifiers in isolation. Finally, we compare our weak supervision approach (with the covariance [28]) denoted DuEL (M) and with the triplets [12], denoted DuEL (S)) against two unsupervised and one supervised alternatives. The unsupervised ones are the *minority* and *majority voting*, which are popular choices. The supervised one is a random forest which uses the scores of the classifiers as input features and the validation dataset as training labels.

Results. Figure 5 reports the precision, recall, and F_1 of the CORRECT predicted links with our gold standard. We make *six* main observations.

Observation 1. We observe that DuEL returns the highest F_1 in all cases. The improvement is 0.09 (DBpedia50, ComplEx) and 0.01 (FB15k237, RotatE) points better than the second-best non-DuEL result in the best and worst cases, respectively. If we compare against existing approaches in the literature, then the gain increases to 0.25 (FB15k237, ComplEx) and 0.1 (DBpedia50, ComplEx). We find remarkable that both DuEL (M) and denoted DuEL (S) achieve superior or comparable performance than the fully supervised model despite they do not make use of ground truth annotations.

Observation 2. The best performance is obtained with ComplEx, but the differences with the other models are not large, except for DBpedia50 where TransE does not perform as well as the others. A lower performance of TransE should be expected since it is a model that is often outperformed by the other two (e.g., see [30]). In general, we conclude that our approach generalizes well. Therefore, it can be used with different embedding models.

Observation 3. If we compare C1,...,C5, then we notice that C1, C2, and C3 mostly return a higher precision than the other two, as we expected. In contrast, C4 and C5 tend to return a higher recall. In some cases, some classifiers used in isolation returned remarkable performance. For instance, C2 with RotatE and FB15k237 is close to return the best result. In other cases, the classifiers perform very poorly. For instance, C2 with TransE and DBpedia50 never returned any positive answers.

Observation 4. There is not a classifier that is clearly outperforming the others, and they all contribute to improve the performance. To obtain further evidence, we performed an ablation study where we executed DuEL (M) excluding, each time, the labels of one classifier. Figure 6 reports the obtained F_1 during such a study with some representative KGs, models, and types of patterns. In this case, if the performance loss that we get when we remove one classifier is large, then it means that it provided a significant contribution. We observe that with ComplEx, FB15k237, and $\langle ?, r, t \rangle$ patterns, C3, C4, and C5 gave the most significant contribution. In contrast, with $\langle h, r, ? \rangle$

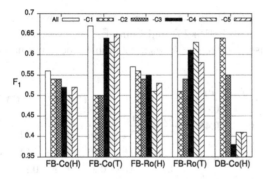

Fig. 6. Ablation study where the F_1 is computed with all classifiers but one. H refers to performance with patterns of the form $\langle ?, r, t \rangle$ while T with $\langle h, r, ? \rangle$. Other abbreviations: FB = FB15k237, DB = DBpedia50, Co = ComplEx, Ro = RotatE

patterns, **C1** and **C2** are more important. These differences highlight the benefit of weak supervision that takes all classifiers into account.

Observation 5. It is interesting to compare the results of DuEL (M) against the ones of DuEL (S). Both methods return similar scores, but DuEL (M) has slightly better performance. Thus, we select it as default choice. However, DuEL (S) has the advantage that it is much faster. Thus, it is a good alternative for a large-scale deployment or for context where a timely prediction is needed.

Observation 6. Fact classification appears to be a hard problem. The absolute performance of external baselines is low, with F_1 scores that range from 0.1 to 0.4. This is partly due to the fact that these methods were mainly designed for ranking and not for classifying. The best F_1 values that we obtained with DuEL are between 0.5 and 0.6. This is due to the fact that many links are hard to label if we have only the KG as input. We observed higher F_1 values for $\langle h, r, ? \rangle$ patterns than for $\langle ?, r, t \rangle$, which is expected since there are typically fewer tails than heads. Despite the absolute values of the F_1 are not very high, the relative improvement brought by DuEL is significant. With DBpedia50 and TransE, DuEL returns an F_1 score that is 2.5 times better than existing techniques but this likely to be due to the low quality of the embedding model. With better models like ComplEx and RotatE, the relative improvement is still significant as it ranges between 72% and 24%. We believe that the performance can be further improved by including more sources, but this is a topic that deserves a dedicated study.

Hyperparameter Tuning. We show the effect of some parameters on the overall performance. We report the results only with ComplEx and FB15k237 since they are representative of the other cases. Figure 7a shows the effect of the number of units in the networks of **C1** and **C2** (the tuning that we performed with grid search included more values than the shown ones). We observe that the impact of this parameter is limited since the performance does not change significantly. Figure 7b reports the F_1 if we set different τ_1 and τ_2. In this case, we observe that the performance changes significantly, and this makes τ_1 and τ_2 two important parameters. Figure 7c shows the impact of changing the kernel sizes in **C3**. We notice that also here the impact is noticeable.

Finally, Fig. 7d shows how the performance varies if we change the top k considered completions with FB15k237 and ComplEx. As expected, we observe that the performance decreases with higher k since the problem becomes harder.

	n	P	R	F_1
C1	100	0.53	0.39	0.45
C1	200	**0.54**	**0.41**	**0.46**
C1	500	0.51	0.40	0.46
C2	10	0.50	0.48	0.53
C2	100	0.54	**0.58**	**0.56**
C2	200	**0.55**	0.52	0.53

(a)

C1		C2		C3		F_1
τ_1	τ_2	τ_1	τ_2	τ_1	τ_2	
0.01	0.1	0.01	0.1	0.01	0.1	0.49
0.01	0.6	0.01	0.6	0.01	0.6	0.51
0.2	0.6	0.2	0.6	0.2	0.6	**0.57**
0.2	0.8	0.2	0.8	0.2	0.8	0.55

(b)

s_1	s_2	P	R	F_1
2	1	**0.60**	0.26	0.36
3	2	0.49	**0.37**	**0.41**
6	3	0.57	0.27	0.37

(c)

Top k	P	R	F_1	Top k	P	R	F_1
$k = 1$	**0.62**	**0.96**	**0.76**	$k = 5$	0.55	0.79	0.65
$k = 3$	0.61	0.87	0.73	$k = 10$	0.53	0.71	0.60

(d)

Fig. 7. Figures (a–c): Performance while changing multiple hyperparameters, with ComplEx, and FB15k237. The best results are marked in boldface. Figure (d): Performance with different k

5   Related Work

Ranking vs. Classifying. The problem that the prevailing evaluation paradigm of KGEs is based on ranking rather than classifying has been empirically studied in [42] and in [35]. Both works focus on the analysis of current methods rather than proposing new ones like ours. Previous work addressed the problem of a ranking-based evaluation by creating negative samples and measuring the accuracy [34], or with new metrics [42]. In contrast, we use a manually annotated dataset.

Link Prediction on KGs. Links can be predicted also using rules, which can be either mined from KGs [14,23] or learned with differentiable models [39]. These approaches propose themselves as alternatives to KGEs for ranking promising set of links. Therefore, they can be used as additional classifiers within our pipeline.

Another technique for finding new links is logic-based reasoning. In particular, rule-based reasoning based on Datalog [1] can compute new facts (links) in KGs with billions of edges [6]. Also, a recent work uses BERT [8] for link prediction [46], leveraging the labels of the entities, thus the language model. Our work differs from them because ours does not depend on external knowledge, like rules or language models.

(Knowledge) Graph Embeddings. In our work, we considered three representative embedding models but there are many more that could be considered. A class of embedding models that has yield good results is the one that employs Graph Neural Networks

(GNNs) [32]. In principle, our technique could also be used considering the embeddings produced by a GNN, but it is interesting, as future work, to study whether it is possible to exploit the graph-like structure of the GNN to produce more sophisticated classifiers. Another way to exploiting KGEs for fact classification could rely on a probabilistic interpretation of their scoring function. For instance, [11] has shown that it is possible to give to DistMult a probabilistic interpretation, which in turn can be used for producing a binary classification. In this case, we can employ techniques used for calibrating the probabilities, like the ones presented in [37] and [31] to improve the accuracy of classification.

If we broad our horizon and consider also different types of graphs, then it is useful to mention a recent overview of (unlabeled) graph embeddings is given at [15]. An important task for these techniques consists of classifying the nodes (instead of links), e.g., [16,20]. To achieve the best results, some techniques use semi supervision [20,45]. We conclude by mentioning that there are several fully supervised techniques for binary link prediction designed for social networks [2].

6 Conclusion

We addressed the problem of performing (binary) fact classification with KGEs. Existing KGEs methods were designed and evaluated for link prediction via ranking and not via classification. There is an emerging consensus that this is an important limitation and that future methods should be evaluated also on classification next to ranking [3,35,42]. To make the problem worse, using KGEs for classification is not trivial also because embeddings may be too noisy if they are not sufficiently trained and we lack large volumes of ground truths to train effectively supervised classifiers on top of them.

Our proposal is the first of its kind. Instead of proposing yet another (embedding) model, the main novelty of our contribution is to show how we can leverage and combine the power of existing methods following the well-established paradigm of ensemble learning. By aggregating the output of multiple classifiers together, we are able to correct mistakes that may be due to noisy embeddings. Moreover, the aggregation takes place in a weakly-supervised manner without using ground truths.

Our experiments confirm the value of our approach. Although the absolute F_1 values show that we have not yet reached human-like levels, the improvement brought by our method is significant. For instance, with our approach the F_1 improved of 72% against the second best with FB15k237 and ComplEx (0.599 vs 0.348) and of 33% against the second best with DBpedia50 and RotatE (0.505 vs 0.381). These improvements indicate that ensemble learning methods are promising techniques to implement fact classification with KGEs.

In practice, we believe that DuEL can be used in several pipelines for knowledge extractions. For instance, it can be used to further assist human curators to further populate KGs or to cover the last mile to implement a fully automated end-to-end embedding-based system for KG completion. In this latter case, more work is needed in order to further improve the accuracy. One possible extension is to include additional classifiers, e.g., based on ontological constraints. Alternatively, it is worthwhile to study whether we can further reduce a potential bias introduced by training some of our classifiers under CWA. Another topic for future work could aim at combining the rankings produced by different embedding models. Moreover, it is interesting to study how we can include the knowledge that can be extracted from textual corpora, or to investigate whether we can build classifiers for some specific relations.

References

1. Abiteboul, S., Hull, R., Vianu, V.: Foundations of Databases, vol. 8. Addison-Wesley, Reading (1995)
2. Al Hasan, M., Zaki, M.J.: A survey of link prediction in social networks. In: Aggarwal, C. (ed.) Social Network Data Analytics, pp. 243–275. Springer, Boston (2011). https://doi.org/10.1007/978-1-4419-8462-3_9
3. van Bakel, R., Aleksiev, T., Daza, D., Alivanistos, D., Cochez, M.: Approximate knowledge graph query answering: from ranking to binary classification. In: GKR, pp. 107–124 (2021)
4. Bishop, C.: Pattern Recognition and Machine Learning. Springer, Heidelberg (2006)
5. Bordes, A., Usunier, N., Garcia-Duran, A., Weston, J., Yakhnenko, O.: Translating embeddings for modeling multi-relational data. In: NIPS, pp. 2787–2795 (2013)
6. Carral, D., Dragoste, I., González, L., Jacobs, C., Krötzsch, M., Urbani, J.: VLog: a rule engine for knowledge graphs. In: Ghidini, C., et al. (eds.) ISWC 2019. LNCS, vol. 11779, pp. 19–35. Springer, Cham (2019). https://doi.org/10.1007/978-3-030-30796-7_2
7. Dettmers, T., Minervini, P., Stenetorp, P., Riedel, S.: Convolutional 2D knowledge graph embeddings. In: AAAI, pp. 1811–1818 (2018)
8. Devlin, J., Chang, M., Lee, K., Toutanova, K.: BERT: pre-training of deep bidirectional transformers for language understanding. In: NAACL, pp. 4171–4186 (2019)
9. Dos Santos, C., Gatti, M.: Deep convolutional neural networks for sentiment analysis of short texts. In: COLING, pp. 69–78 (2014)
10. Duchi, J., Hazan, E., Singer, Y.: Adaptive subgradient methods for online learning and stochastic optimization. J. Mach. Learn. Res. **12**, 2121–2159 (2011)
11. Friedman, T., Van den Broeck, G.: Symbolic querying of vector spaces: probabilistic databases meets relational embeddings. In: UAI, pp. 1268–1277 (2020)
12. Fu, D.Y., Chen, M.F., Sala, F., Hooper, S.M., Fatahalian, K., Ré, C.: Fast and three-rious: speeding up weak supervision with triplet methods. In: ICML, pp. 3280–3291 (2020)
13. Fukushima, K., Miyake, S.: Neocognitron: a self-organizing neural network model for a mechanism of visual pattern recognition. In: Amari, S., Arbib, M.A. (eds.) Competition and Cooperation in Neural Nets. Lecture Notes in Biomathematics, vol. 45, pp. 267–285. Springer, Heidelberg (1982). https://doi.org/10.1007/978-3-642-46466-9_18
14. Galárraga, L., Teflioudi, C., Hose, K., Suchanek, F.M.: Fast rule mining in ontological knowledge bases with AMIE+. VLDB J. **24**(6), 707–730 (2015)
15. Goyal, P., Ferrara, E.: Graph embedding techniques, applications, and performance: a survey. Knowl.-Based Syst. **151**, 78–94 (2018)
16. Grover, A., Leskovec, J.: node2vec: scalable feature learning for networks. In: KDD, pp. 855–864 (2016)
17. Hochreiter, S., Schmidhuber, J.: Long short-term memory. Neural Comput. **9**(8), 1735–1780 (1997)
18. Joshi, U., Urbani, J.: Searching for embeddings in a haystack: link prediction on knowledge graphs with subgraph pruning. In: WWW, pp. 2817–2823 (2020)
19. Kingma, D.P., Ba, J.: Adam: a method for stochastic optimization. arXiv:1412.6980 (2017)

20. Kipf, T.N., Welling, M.: Semi-supervised classification with graph convolutional networks. arXiv:1609.02907 (2017)
21. Lai, S., Xu, L., Liu, K., Zhao, J.: Recurrent convolutional neural networks for text classification. In: AAAI, pp. 2267–2273 (2015)
22. Meilicke, C., Chekol, M.W., Fink, M., Stuckenschmidt, H.: Reinforced anytime bottom up rule learning for knowledge graph completion. arXiv:2004.04412 (2020)
23. Meilicke, C., Chekol, M.W., Ruffinelli, D., Stuckenschmidt, H.: Anytime bottom-up rule learning for knowledge graph completion. In: IJCAI, pp. 3137–3143 (2019)
24. Nguyen, D.Q., Nguyen, T.D., Nguyen, D.Q., Phung, D.: A novel embedding model for knowledge base completion based on convolutional neural network. In: NAACL, pp. 327–333 (2018)
25. Nickel, M., Murphy, K., Tresp, V., Gabrilovich, E.: A review of relational machine learning for knowledge graphs. Proc. IEEE **104**(1), 11–33 (2016)
26. Noy, N., Gao, Y., Jain, A., Narayanan, A., Patterson, A., Taylor, J.: Industry-scale knowledge graphs: lessons and challenges. Commun. ACM **62**(8), 36–43 (2019)
27. Ratner, A., Bach, S.H., Ehrenberg, H., Fries, J., Wu, S., Ré, C.: Snorkel: rapid training data creation with weak supervision. VLDB J. **29**(2), 709–730 (2020)
28. Ratner, A., Hancock, B., Dunnmon, J., Sala, F., Pandey, S., Ré, C.: Training complex models with multi-task weak supervision. In: AAAI, pp. 4763–4771 (2019)
29. Ristoski, P., Paulheim, H.: RDF2Vec: RDF graph embeddings for data mining. In: Groth, P. (ed.) ISWC 2016. LNCS, vol. 9981, pp. 498–514. Springer, Cham (2016). https://doi.org/10.1007/978-3-319-46523-4_30
30. Ruffinelli, D., Broscheit, S., Gemulla, R.: You CAN teach an old dog new tricks! On training knowledge graph embeddings. In: ICLR (2020)
31. Safavi, T., Koutra, D., Meij, E.: Evaluating the calibration of knowledge graph embeddings for trustworthy link prediction. In: EMNLP, pp. 8308–8321 (2020)
32. Schlichtkrull, M., Kipf, T.N., Bloem, P., van den Berg, R., Titov, I., Welling, M.: Modeling relational data with graph convolutional networks. In: Gangemi, A., et al. (eds.) ESWC 2018. LNCS, vol. 10843, pp. 593–607. Springer, Cham (2018). https://doi.org/10.1007/978-3-319-93417-4_38
33. Severyn, A., Moschitti, A.: Learning to rank short text pairs with convolutional deep neural networks. In: SIGIR, pp. 373–382 (2015)
34. Socher, R., Chen, D., Manning, C.D., Ng, A.: Reasoning with neural tensor networks for knowledge base completion. In: NIPS, pp. 926–934 (2013)
35. Speranskaya, M., Schmitt, M., Roth, B.: Ranking vs. classifying: measuring knowledge base completion quality. In: AKBC (2020)
36. Sun, Z., Deng, Z.H., Nie, J.Y., Tang, J.: RotatE: knowledge graph embedding by relational rotation in complex space. In: ICLR (2019)
37. Tabacof, P., Costabello, L.: Probability calibration for knowledge graph embedding models. In: ICLR (2019)
38. Trouillon, T., Welbl, J., Riedel, S., Gaussier, E., Bouchard, G.: Complex embeddings for simple link prediction. In: ICML, pp. 2071–2080 (2016)
39. Wang, P.W., Stepanova, D., Domokos, C., Kolter, J.Z.: Differentiable learning of numerical rules in knowledge graphs. In: ICLR (2019)
40. Wang, Q., Mao, Z., Wang, B., Guo, L.: Knowledge graph embedding: a survey of approaches and applications. IEEE Trans. Knowl. Data Eng. **29**(12), 2724–2743 (2017)
41. Wang, Q., Wang, B., Guo, L.: Knowledge base completion using embeddings and rules. In: IJCAI, pp. 1859–1865 (2015)
42. Wang, Y., Ruffinelli, D., Gemulla, R., Broscheit, S., Meilicke, C.: On evaluating embedding models for knowledge base completion. In: The 4th Workshop on Representation Learning for NLP, pp. 104–112 (2019)

43. Xiong, W., Hoang, T., Wang, W.Y.: DeepPath: a reinforcement learning method for knowledge graph reasoning. In: EMNLP, pp. 564–573 (2017)
44. Yang, B., Yih, W., He, X., Gao, J., Deng, L.: Embedding entities and relations for learning and inference in knowledge bases. In: ICLR (2015)
45. Yang, Z., Cohen, W.W., Salakhutdinov, R.: Revisiting semi-supervised learning with graph embeddings. In: ICML, pp. 40–48 (2016)
46. Yao, L., Mao, C., Luo, Y.: KG-BERT: BERT for knowledge graph completion. arXiv:1909.03193 (2019)
47. Zhou, Z.H.: Ensemble learning. In: Zhou, Z.H. (ed.) Machine Learning, pp. 181–210. Springer, Singapore (2021). https://doi.org/10.1007/978-981-15-1967-3_8

The Problem with XSD Binary Floating Point Datatypes in RDF

Jan Martin Keil$^{(\boxtimes)}$ (iD) and Merle Gänßinger (iD)

Heinz Nixdorf Chair for Distributed Information Systems,
Institute for Computer Science, Friedrich Schiller University Jena,
Jena, Germany
{jan-martin.keil,merle.gaenssinger}@uni-jena.de

Abstract. The XSD binary floating point datatypes are regularly used for precise numeric values in RDF. However, the use of these datatypes for knowledge representation can systematically impair the quality of data and, compared to the XSD decimal datatype, increases the probability of data processing producing false results. We argue why in most cases the XSD decimal datatype is better suited to represent numeric values in RDF. A survey of the actual usage of datatypes on the relevant subset of the December 2020 Web Data Commons dataset, containing 19 453 060 341 literals from real web data, substantiates the practical relevancy of the described problem: 29%–68% of binary floating point values are distorted due to the datatype.

Keywords: Data Quality · Datatypes · Floating Point Numbers · Knowledge Graphs · Numerical Stability · RDF · XSD

1 Introduction

The Resource Description Framework (RDF) is the fundamental building block of knowledge graphs and the Semantic Web. In RDF, values are represented as literals. A literal consists of a lexical form, a datatype, and possibly a language tag. The RDF standard [1] recommends to use XML Schema Definition Language (XSD) built-in datatypes [2]. For numeric values, this includes the primitive types *decimal, double* and *float* as well as all variations of integer[1] which are derived from decimal.

The datatype decimal allows the representation of numbers with arbitrary precision, whereas the datatypes float and double allow the representation of binary floating point values of limited range and precision [2]. However, in practice, the binary floating point datatypes are regularly used for precise numeric values, although the datatype cannot accurately represent these values. For example, out of nine unit ontologies selected in a comparison study [3], five ontologies (OM 1, OM 2, QU, QUDT, SWEET) used XSD binary floating point

---

[1] *integer, long, int, short, byte, nonNegativeInteger, positiveInteger, unsignedLong, unsignedInt, unsignedShort, unsignedByte, nonPositiveInteger,* and *negativeInteger.*

© The Author(s), under exclusive license to Springer Nature Switzerland AG 2022
P. Groth et al. (Eds.): ESWC 2022, LNCS 13261, pp. 165–182, 2022.
https://doi.org/10.1007/978-3-031-06981-9_10

values and only two ontologies or knowledge graphs (OBOE, Wikidata) used `xsd:decimal` values for unit conversion factors. Even a popular ontology guideline [4] and a World Wide Web Consortium (W3C) working group note [5] use binary floating point datatypes for precise numeric values in examples.

In general, binary floating point numbers are meant to approximate decimal values in a fixed length binary representation to limit memory consumption and increase computation speed. In RDF, however, binary floating point numbers are defined to represent the exact value of the binary representation: Binary floating point values do not approximate typed decimals, as in programming languages, but typed decimals are abbreviations for exact binary floating point values. This causes ambiguity about the intended meaning of numeric values. We show that 29%–68% of the floating point values in real web data are distorted due to the datatype. With regard to the growing use of RDF for the representation of data, including research data, this ambiguity is concerning.

Further, the use of binary floating point datatypes for precise numeric values regularly causes rounding errors in the values actually represented, compared to typed values provided as decimals. Subsequently, error accumulation may significantly falsify the result of processing these values. Disasters, such as the Patriot Missile Failure [6], which resulted in 28 deaths, illustrate the potential impact of accumulated errors in real world applications. The increasing relevance of knowledge graphs for real-world applications calls for general awareness of these issues in the Semantic Web community.

In this paper, we discuss advantages and disadvantages of different numeric datatypes. We demonstrate the practical relevance of the outlined problem with a survey of the actual usage of datatypes on the relevant subset of the December 2020 Web Data Commons dataset, containing 19 453 060 341 literals from real web data. We aim to raise awareness of the implications of datatype selection in RDF and to enable a more informed choice in the future. This work is structured as follows: In Sect. 2, we give an overview of relevant standards and related work, followed by a comparison of the properties of the binary floating point and decimal datatypes in Sect. 3. In Sect. 4, we discuss the implications of the datatype properties in different use cases. An approach for automatic problem detection is outlined in Sect. 5. In Sect. 6, we present a survey on the use of datatypes in the World Wide Web that demonstrates the practical relevance of the outlined problem. Finally, we indicate approaches for the general mitigation of the problem in Sect. 7.

2 Background

Each datatype in RDF consists of a lexical space, a value space, and a lexical-to-value mapping. This is compatible with datatypes in XSD [1].

Value space: the set of values for a datatype [1,2].

Lexical space: the prescribed set of strings, which the lexical mapping for a datatype maps to values of that datatype. The members of the lexical space are **lexical representations (lexical forms)** of the values to which they are mapped [1,2].

Lexical mapping (lexical-to-value mapping): a prescribed relation which maps from the lexical space of a datatype into its value space [1,2].

RDF reuses the XSD datatypes with only a few exceptions and additions of non-numeric datatypes [1]. For non-integer numbers, XSD provides the datatypes decimal, float and double. The XSD datatype **decimal** (xsd:decimal) represents a subset of the real numbers [2].

Value space of xsd:decimal: the set of numbers that can be obtained by dividing an integer by a non-negative power of ten: $\frac{i}{10^n}$ with $i \in \mathbb{Z}, n \in \mathbb{N}_0$, precision is not reflected [2].

Lexical space of xsd:decimal: the set of all decimal numbers with or without a decimal point [2].

Lexical mapping of xsd:decimal: set i according to the decimal digits of the lexical representation and the leading sign, and set n according to the position of the period or 0, if the period is omitted. If the sign is omitted, "+" is assumed [2].

The XSD datatype **float** (xsd:float) is aligned with the IEEE 32-bit binary floating point datatype [7][2], the XSD datatype **double** (xsd:double) is aligned to the IEEE 64-bit binary floating point datatype [7]. Both represent subsets of the rational numbers. They only differ in their three defining constants [2].

Value space of xsd:float (xsd:double): the set of the special values *positiveZero, negativeZero, positiveInfinity, negativeInfinity,* and *notANumber* and the numbers that can be obtained by multiplying an integer m whose absolute value is less than 2^{24} (double: 2^{53}) with a power of two whose exponent e is an integer between -149 (double: -1074) and 104 (double: 971): $m \cdot 2^e$ [2].

Lexical space of xsd:float (xsd:double): the set of all decimal numbers with or without a decimal point, numbers in exponential notation, and the literals INF, +INF, -INF, and NaN [2].

Lexical mapping of xsd:float (xsd:double): set either the according numeric value (including rounding, if necessary), or the according special value. An implementation might choose between different rounding variants that satisfy the requirements of the IEEE specification.

Numbers with a fractional part of infinite length, like the rational number $\frac{1}{3} = 0.\bar{3}$ or the irrational number $\sqrt{2} = 1.4142\ldots$, are not in the value space of xsd:float or xsd:double, as a number of finite length multiplied or divided by two is always a number of finite length again. Consequently, a finite decimal with sufficient precision can exactly represent every possible numeric value or lexical representation of an xsd:float or xsd:double, except of the special values *positiveInfinity, negativeInfinity,* and *notANumber*. In contrast, a finite binary floating point value can not exactly represent every possible decimal value.

---

[2] As the XSD recommendation refers to IEEE 754-2008 version of the standard, we do not refer to the subsequent IEEE 754-2019 version.

Some serialization or query languages for RDF provide a shorthand syntax for numeric literals without explicit datatype specification. In Turtle, TriG and SPARQL a number without fraction is an `xsd:integer`, a number with fraction is an `xsd:decimal`, and a number in exponential notation is an `xsd:double` [8–10]. In JSON-LD a number without fractions is an `xsd:integer` and a number with fraction is an `xsd:double`, to align with the common interpretation of numbers in JSON [11]. However, this is not necessary to comply with the JSON specifications [12,13]. The serialization languages RDF/XML, N-Triples, N-Quads, and RDFa do not provide a shorthand syntax for numeric literals [14–17]. Other languages for machine-readable annotation of HTML, which are regularly mapped to RDF, i.e. Microformats[3], and Microdata[4], do not incorporate explicit datatypes.

In addition to the core XSD datatypes, a W3C working group note introduces the `precisionDecimal` datatype [18]. It is aligned to the IEEE decimal floating-point datatypes [7] and represents a subset of real numbers. It retains precision and permits the special values *positiveZero, negativeZero, positiveInfinity, negativeInfinity*, and *notANumber*. Further, it supports exponential notation. The precision and exponent values of the `precisionDecimal` datatype are unbounded, but can be restricted in derived datatypes to comply with an actual IEEE decimal floating-point datatype. However, even though the RDF standard permits the use of `precisionDecimal`, it does not demand its support in compliant implementations [1]. Therefore, RDF frameworks can not be expected to support `precisionDecimal`.

Another W3C working group note addresses the selection of proper numeric datatypes [5]. It identified three relevant use cases of numeric values: count, measurement, and constant. According to the note, the appropriate datatypes are (derived datatypes of) `xsd:integer` for counts, `xsd:float` or `xsd:double` for measurements, and `xsd:decimal` for constants.

The common vocabulary schema.org[5] defines the alternative numeric datatypes `schema:Integer` and `schema:Float` and their super datatype `schema:Number`. A usage note restricts the lexical space of `schema:Number` to the digits 0 to 9 and at most one full stop. No further restrictions of the lexical or value space are made. `schema:Number` is directly in the range of 91 properties and `schema:Integer` is directly in the range of 47 properties. `schema:Float` is not directly in the range of any property.

The digital representation or computation of numerical values can cause numerical problems: An *overflow error* occurs, if a represented value exceeds the maximum positive or negative value in the value space of a datatype [19]. An *underflow error* occurs, if a represented value is smaller than the minimum positive or negative value different from zero in the value space of a datatype [19]. A *rounding error* occurs, if a represented value is not in the value space of a datatype. It is then represented by a nearby value in the value space that

---

[3] https://microformats.org.

[4] https://html.spec.whatwg.org/multipage/microdata.html.

[5] http://schema.org, current version 13.0.

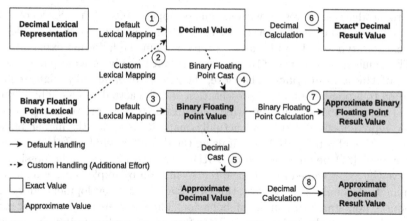

Fig. 1. Possible processing paths of numeric literals depending on their datatype.

is determined by a rounding scheme [19]. A *cancellation* is caused by the subtraction of nearly equal values and eliminates leading accurate digits. This will expose errors in the input values of the subtraction [19]. *Error accumulation* is the insidious growth of errors due to the use of a numerically instable sequence of operations [19].

3   Properties of Binary Floating Point and Decimal Datatypes in RDF

Binary floating point and decimal datatypes in the context of RDF have individual properties, which make them more or less suitable for specific use cases:

xsd:float and xsd:double permit the use of **positive and negative infinite values**. xsd:decimal supports neither positive nor negative infinite values.

xsd:float and xsd:double permit the **exponential notation**. Especially in the case of numbers with many leading or trailing zeros, this is more convenient and less error-prone to read or write for humans. xsd:decimal does not permit the exponential notation. There is no actual reason for this limitation. For example, Wikibase[6] also accepts exponential notation for xsd:decimal.[7] The *XML Schema Working Group* decided against allowing exponential notation for xsd:decimal, as the requirement to have a decimal datatype permitting exponential notation was already met by precisionDecimal [20], which, however, has been dropped in the later process of the XSD standardization [21]. To our knowledge, this has not been considered during the RDF standardization.

Figure 1 presents the possible processing paths of numeric literals depending on their datatype. It shows that only decimal lexical representations can be

---

[6] https://wikiba.se/, SPARQL endpoint example: https://query.wikidata.org.

[7] Example: **SELECT** ("1e-9"^^<http://www.w3.org/2001/XMLSchema#decimal> AS ?d) WHERE {}

used to produce exact result without custom operations. ① to ⑧ denote the mapping operations, cast operations, and calculation operations on values of different datatypes and will be used as references in the following explanations.

The value spaces of xsd:float and xsd:double only provide partial **coverage of the lexical space**. Therefore, the lexical mapping (③) might require rounding to a possible binary representation and the actual value might slightly differ from the lexical representation. For example, xsd:float has no exact binary representation of 0.1 and actually maps it to a slightly higher binary representation of 0.100 000 001 4..., if using the default *roundTiesToEven* rounding scheme [7]. Depending on the used RDF framework, it might be possible to preserve the exact value of the lexical representation by implementing a custom mapping to decimal (②). However, this causes additional development effort and introduces non standard compliant behavior. The value space of xsd:decimal covers all values in the lexical space. Therefore, the lexical mapping (①) always provides the exact numeric value described in the lexical representation without any rounding. All three datatypes, xsd:float, xsd:double, and xsd:decimal, do not cover the precision reflected by the lexical representation. For example, literals with the lexical representations 0.5 and 0.50 are considered equal although their lexical representations reflect different precision. The only discussed datatype that preserves the reflected precision is precisionDecimal.

The **accuracy of calculations** based on xsd:float or xsd:double literals (⑦) is limited, as a properly implemented RDF framework will use binary floating point arithmetic by default. For example, this happens during the execution of SPARQL queries that include arithmetic functions or aggregations. Therefore, the calculations might be affected by various numeric problems, i.e. underflow errors, overflow errors, rounding errors, cancellation, and error accumulation. Calculations based on xsd:decimal literals (⑥) will by default use a decimal arithmetic with arbitrary precision. Thus, they might only be affected by rounding errors in case of (intermediate) results with a fractional part of infinite length, as well as accumulations of these rounding errors. This different behavior is demonstrated in Fig. 2. Depending on the used RDF framework, it might be possible to cast between the datatypes (④ and ⑤). However, a value cast from binary floating point to decimal (⑤) is still affected by the rounding error of the floating point value caused by the lexical mapping. Subsequent calculations (⑧) will still result in approximate results only. In contrast, the results of calculations based on a value cast from decimal to floating point (④) and based on an initial floating point value (③) do not differ, if the same rounding method is used. The SPARQL query in Fig. 2 and the according result provided by Wikibase demonstrate differing numerical problems of the datatypes. Other SPARQL endpoints, i.e. Virtuoso 8.3[8] and Apache Fuseki 5.16.0[9], provide similar results.

---

[8] https://virtuoso.openlinksw.com/.
[9] https://jena.apache.org/.

```
PREFIX xsd: <http://www.w3.org/2001/XMLSchema#>
SELECT ?datatype
  (xsd:decimal(STRDT("0.1", ?datatype)) AS ?rounded)
  (xsd:decimal(STRDT("1", ?datatype) / STRDT("3", ?datatype)) AS
    ?roundedInfinit)
  (xsd:decimal(STRDT("1.0000001", ?datatype) - STRDT("1.0000000",
    ?datatype)) AS ?cancellation)
  (STRDT("100000000000000000000000000000000000000", ?datatype) * STRDT("1",
    ?datatype) AS ?overflow)
  (STRDT("0.000000000000000000000000000000000000000000000001", ?datatype) *
    STRDT("1", ?datatype) AS ?underflow)
WHERE {VALUES ?datatype {xsd:float xsd:decimal}}
```

datatype	xsd:float	xsd:decimal
rounded	0.10000000149011612	0.1
roundedInfinit	0.3333333432674408	0.33333333333333333333
cancellation	0.00000011920928955078125	0.0000001
overflow	Infinity	100000000000000000000000000000000000000
underflow	0.0	0.000000000000000000000000000000000000000000000001

Fig. 2. Top: A SPARQL query that demonstrates differing numerical problems of the datatypes xsd:float and xsd:decimal. Bottom: The corresponding query output (transformed), as on http://query.wikidata.org.

4 Implications for the Selection of Numeric Datatypes

The traditional use case of RDF is knowledge representation. The XSD floating point datatypes provide two advantages for knowledge representation compared to xsd:decimal: Firstly, the permitted representation of positive and negative infinite might be needed in some cases. Secondly, the exponential notation eases the representation of very large and very small values and reduces the risk of typing errors due to missing or additional zeros. This would not be an issue in case of proper user interface support. But popular tools, like WebProtégé and Protégé Desktop[10], do not help the user here. Further, projects that manipulate their RDF documents under version control using SPARQL UPDATE queries, custom generation scripts and manual edits do not have such a user interface at all.

However, in most cases, knowledge concerned with numbers deals with exact decimal numbers or intervals of decimal values. Intervals are typically described with two exact decimal numbers, either with a minimum value and a maximum value (e.g. [0.05, 0.15]) or a value and a measurement uncertainty (e.g. 0.1±0.05) [22]. The binary floating point datatypes do not allow the accurate representation of exactly known or defined numbers in many cases. In addition, they entail the risk to fool data curators into believing that they stated the exact number, as the lexical representation on first sight appears to be exact. This becomes even more critical, if xsd:double was used unintentionally due to a shorthand syntax in Turtle, TriG, SPARQL, or JSON-LD. This way, the use of binary floating point datatypes produces ambiguity in the data: The intended meaning could be either the actually represented number in the value space or the verbatim interpretation

---

[10] https://protege.stanford.edu/.

of the lexical representation. This ambiguity counteracts the basic ideas behind the Semantic Web and Linked Open Data to ease understanding and reuse of data. Therefore, binary floating point datatypes are not suitable to fulfill the requirements for knowledge representation.

In consequence, the knowledge cannot be used for exact calculations without programming overhead. The possible small rounding errors of binary floating point input values might accumulate to significant errors in calculation results. Disasters, as the Patriot Missile Failure [6], illustrate the potential impact of accumulated errors in real world applications.

This contradicts a W3C working group note [5], stating that binary floating point datatypes are appropriate for measurements. It provided the following example representation of a measurement in the interval of 73.0 to 73.2:

```
_:w eg:value     "73.1"^^xsd:float .
_:w eg:errorRange "0.1"^^xsd:float .
```

However, if using the default *roundTiesToEven* rounding scheme [7], this example actually represents a measurement in the interval 72.999 998 472 6... to 73.199 998 475 6..., as 73.1 and 0.1 are not in the value space of xsd:float.[11] In consequence, the actual represented error interval does not cover the points between 73.199 998 475 6... and 73.2. A common solution for this problem is the use of different rounding schemes for the calculation of the upper and lower bound of the interval (outward rounding) [23]. Unfortunately, this is not provided in current RDF frameworks and causes additional programming effort. The example shows that also in case of measurements binary floating point datatypes have clear disadvantages compared to xsd:decimal.

Further, the use of binary floating point values in RDF restricts the selection of the used arithmetic for calculations, as it causes an implementation overhead for the application of decimal arithmetic with arbitrary precision. It must be mentioned that calculations using decimal arithmetic with arbitrary precision probably are significantly slower, compared to calculations using binary floating point arithmetic with limited precision. Hence, floating point calculations are better suited for many use cases. However, in certain cases they are not. Therefore, the selection of an arithmetic must be up to the application, not to the input data, as applications might widely vary regarding the required accuracy and the numerical conditioning of the underlying problem.

The same problem arises in use cases that involve the comparison of values, like instance-based ontology matching or ontology based data validation, because comparison values become blurred due to rounding. For example, if using the default *roundTiesToEven* rounding scheme, an upper bound of "0.1"^^xsd:float in a constraint still permits a value of 0.100000001. Thus, the use of binary floating point datatypes for knowledge representation can systematically impair the quality of data and increases the probability of false results of data processing.

---

[11] Lexical mappings (*roundTiesToEven* rounding scheme): 73.1 → 73.099 998 474 1... and 0.1 → 0.100 000 001 4..., Interval calculations: 73.099 998 474 1... ± 0.100 000 001 4....

In other use cases, RDF might be used for the exchange of initially binary floating point values, as computational results or the output of analog-to-digital converters. If the data to exchange are binary floating point values, the original value can only contain values with an exact binary representation and corruption of data with rounding is impossible. Thus, the use of floating point datatypes for the exchange of computational results is reasonable.

5 Automatic Distortion Detection

The automatic detection of quality issues is key to an effective quality assurance. Therefore, RDF editors, like Protégé, or evaluation tools, like the OntOlogy Pitfall Scanner! [24], would ideally warn data curators, if the use of binary floating point datatypes would distort numeric values.

A simple test can be implemented by comparing the results of the default mapping to a binary floating point value (③ in Fig. 1) followed by a cast to decimal (⑤ in Fig. 1) and a custom mapping to a decimal value (② in Fig. 1). The SPARQL query in Fig. 3 demonstrates the approach.

```
PREFIX xsd: <http://www.w3.org/2001/XMLSchema#>
SELECT
  (xsd:decimal(?xsdFloatValue) AS ?xsdFloatSemantic)
  (?xsdDecimalValue AS ?xsdDecimalSemantic)
  (xsd:decimal(?xsdFloatValue) != ?xsdDecimalValue AS ?distorted)
WHERE {
  VALUES ?lexical {"1" "0.1" "0.5"}
  BIND(STRDT(?lexical, xsd:float) AS ?xsdFloatValue)
  BIND(STRDT(?lexical, xsd:decimal) AS ?xsdDecimalValue)
}
```

xsdFloatSemantic	xsdDecimalSemantic	distorted
0.5	0.5	false
0.10000000149011612	0.1	true
1	1	false

Fig. 3. Top: A SPARQL query that demonstrates an approach to detect number distortion. Bottom: The corresponding query output, as on http://query.wikidata.org.

6 Datatype Usage Survey

To determine the practical relevancy of the described problem, we conducted a survey of the actual usage of datatypes. The survey is based on the December 2020 edition[12] of the Web Data Commons dataset [25]. The Web Data Commons dataset provides in several N-Quads files the embedded RDF data of 1.7e9 HTML documents extracted from all 3.4e9 HTML documents contained

---

[12] http://webdatacommons.org/structureddata/#results-2020-1.

in the September 2020 Common Crawl archive,[13] a freely available web crawl archive. We selected it because of its large size, an expected large proportion of literals, the uniform access of the whole corpus, its heterogeneous original sources (15e6 domains), and a good reflection of RDF usage by a wide range of people. The December 2020 Web Data Commons dataset is divided into data extracted from embedded JSON-LD, RDFa, Microdata, and several Microformats. We only considered data from embedded JSON-LD (7.7e8 URLs, 3.2e10 triples) and RDFa (4.1e8 URLs, 5.9e9 triples), as Microdata and Microformats do not incorporate explicit datatypes.

We created a Java program based on Apache Jena to stream and analyze the relevant parts of the Web Data Commons dataset. The dataset replicates malformed IRIs or literals as they appeared in the original source. To avoid parsing failures of whole files due to single malformed statements, each line was parsed independently and failures were logged separately. Overall, about 4.5e7 failures occurred. The main reasons for failures were malformed IRIs and illegal character encodings. Transaction mechanisms were used to ensure the consistency of the resulting dataset in case of temporary failures of involved systems. Per source type, dataset file, property, and datatype we measured:

1. **UnpreciseRepresentableInDouble:** the number of lexicals that are in the lexical space but not in the value space of `xsd:double`.
2. **UnpreciseRepresentableInFloat:** the number of lexicals that are in the lexical space but not in the value space of `xsd:float`.
3. **UsedAsDatatype:** the total number of literals with the datatype.
4. **UsedAsPropertyRange:** the number of statements that specify the datatype as range of the property.
5. **ValidDecimalNotation:** the number of lexicals that represent a number with decimal notation and whose lexical representation is thereby in the lexical space of `xsd:decimal`, `xsd:float`, and `xsd:double`.
6. **ValidExponentialNotation:** the number of lexicals that represent a number with exponential notation and whose lexical representation is thereby in the lexical space of `xsd:float`, and `xsd:double`.
7. **ValidInfOrNaNNotation:** the number of lexicals that equals either `INF`, `+INF`, `-INF` or `NaN` and whose lexical representation is thereby in the lexical space of `xsd:float`, and `xsd:double`.
8. **ValidIntegerNotation:** the number of lexicals that represent an integer number and whose lexical representation is thereby in the lexical space of `xsd:integer`, `xsd:decimal`, `xsd:float`, and `xsd:double`.

Unfortunately, the lexical representation of `xsd:double` literals from embedded JSON-LD was normalized during the creation of the Web Data Commons dataset to always use exponential notation with one integer digit and up to 16 fractional digits.[14] This is a legal transformation according to the definition

---

[13] https://commoncrawl.org/2020/10/september-2020-crawl-archive-now-available/.

[14] https://github.com/jsonld-java/jsonld-java/blob/v0.13.1/core/src/main/java/com/github/jsonldjava/core/RDFDataset.java#L673.

Table 1. The number of datatype occurrences in the Web Data Commons December 2020 dataset from RDFa and embedded JSON-LD sources (Measure 3) in absolute numbers and relative to the total number of literals in the source type (Measure 3). Only the top ten, as well as selected further datatypes are shown.

RDFa		Embedded JSON-LD	
Datatype	Occurrences (rel)	Datatype	Occurrences (rel)
rdf:langString	3 179 161 585 (.68)	xsd:string	11 277 500 571 (.76)
xsd:string	1 305 371 136 (.28)	xsd:integer	2 021 243 795 (.14)
xsd:dateTime	102 987 223 (.02)	schema:Date	1 313 408 439 (.09)
rdf:XMLLiteral	62 337 177 (.01)	xsd:double	101 959 406 (.01)
xsd:integer	21 547 053 (.00)	xsd:boolean	26 144 338 (.00)
xsd:float	1 025 753 (.00)	schema:DateTime	25 002 464 (.00)
use:sku	729 858 (.00)	rdf:langString	12 934 431 (.00)
xsd:date	507 454 (.00)	xsd:float	90 895 (.00)
xsd:boolean	348 334 (.00)	xsd:dateTime	12 260 (.00)
schema:Date	246 995 (.00)	rdf:HTML	5785 (.00)
xsd:decimal	8288 (.00)	xsd:decimal	1 (.00)
xsd:double	234 (.00)	schema:Number	0 (.00)
schema:Number	0 (.00)	schema:Integer	0 (.00)
schema:Integer	0 (.00)	schema:Float	0 (.00)
schema:Float	0 (.00)		

of xsd:double, as the represented value is preserved. However, this limits the use of the according *Valid...* and *Unprecise...* measures. At the same time, this demonstrates that the use of xsd:float or xsd:double might easily cause the loss of information due to legal transformation, if information is only reflected in the lexical representation.

The resulting dataset consists of a CSV file containing the measurement results (5.4e7 lines, 0.6 GiB compressed, 11.0 GiB uncompressed). The analysis was conducted with Python scripts. The tool [26], the resulting dataset [27], and the analysis scripts [28] are freely available for review and further use under permissive licenses.

For the analysis, we first applied some data cleaning: Some properties and datatypes were regularly denoted by IRIs in the http scheme as well as in the https scheme. To enable proper aggregation, the scheme of all IRIs in the dataset were unified to http. Further, the omission of namespace definitions in the source websites causes the occurrence of prefixed names instead of full IRIs. All prefixes in datatypes that occurred at least for one datatype more than 1000 times and all prefixes in properties that occurred at least for one property more than 1000 times have been replaced with the actual namespace, if we found a resource with a matching local name and matching default vocabulary prefix during a web search or in other used properties or datatypes. Rarer prefixes have not

Table 2. The number of property occurrences with XSD or schema.org numerical datatypes in the Web Data Commons December 2020 dataset from RDFa and embedded JSON-LD sources (Measure 3) in absolute numbers and relative to the total number of numeric literals in the source type (Measure 3). Only the top ten are shown.

RDFa		Embedded JSON-LD	
Property	Occurrences (rel)	Property	Occurrences (rel)
sioc:num_replies	21 391 187 (.95)	schema:position	893 910 601 (.42)
gr:hasCurrencyValue	525 491 (.02)	schema:width	448 036 253 (.21)
gr:hasMinValue	137 018 (.01)	schema:height	446 308 779 (.21)
gr:amountOfThisGood	94 978 (.00)	schema:price	71 045 655 (.03)
gr:hasMaxValue	52 772 (.00)	schema:commentCount	65 723 049 (.03)
vcard:latitude	49 428 (.00)	schema:ratingValue	26 261 677 (.01)
vcard:longitude	49 428 (.00)	schema:longitude	17 096 852 (.01)
gr:hasValue	25 800 (.00)	schema:latitude	17 093 196 (.01)
dv:count	24 672 (.00)	schema:bestRating	16 333 042 (.01)
dv:price	23 936 (.00)	schema:userInteractionCount	13 347 182 (.01)

been replaced because of the high effort, the susceptibility to errors caused by ambiguity, and the lack of significance for the results. Further, we did not clean other kinds of typos like missing or duplicated # or / after the namespace, as these errors could also not easily be fixed by applications with, e.g., a maintained list of widely used prefixes.

Overall, we processed 14 778 325 375 literals from embedded JSON-LD and 4 674 734 966 literals from RDFa. Table 1 shows the number of occurrences of the most frequent datatypes.[15] Table 2 shows the most frequently used properties that occurred with numerical datatypes from XSD or schema.org. Although the use of the schema.org numeric datatypes instead of XSD numeric datatypes is expected by the definition of many schema.org properties, including widely used properties, like schema:position or schema:price, we found zero occurrences of schema.org numeric datatypes. The most probable reason is the existence of shorthand syntaxes for XSD numeric datatypes. In contrast, the usage of schema.org temporal datatypes schema:Date and schema:DateTime in JSON-LD exceeds the usage of XSD temporal datatypes by orders of magnitude. This emphasizes the importance of shorthand syntaxes for the choice of datatypes.

As shown in Table 1, the occurrences of xsd:float in RDFa and xsd:double in embedded JSON-LD surpass the occurrences of xsd:decimal by orders of

---

[15] Prefixes used for results presentation: dcterms: http://purl.org/dc/terms/, dv: http://rdf.data-vocabulary.org/#, gr: http://purl.org/goodrelations/v1#, rdf: http://www.w3.org/1999/02/22-rdf-syntax-ns#, rev: http://purl.org/stuff/rev#, schema: http://schema.org/, use: http://search.yahoo.com/searchmonkey-datatype-use/, vcard: http://www.w3.org/2006/vcard/ns#, xsd: http://www.w3.org/2001/XMLSchema#.

Table 3. The number of property occurrences with `xsd:float` in RDFa and with `xsd:double` in embedded JSON-LD in the Web Data Commons December 2020 dataset (Measure 3) in absolute numbers and relative to the total number of literals with the same datatype in the same source type (Measure 3). Only the top ten are shown.

`xsd:float` in RDFa		`xsd:double` in Embedded JSON-LD	
Property	Occurrences (rel)	Property	Occurrences (rel)
`gr:hasCurrencyValue`	516 256 (.50)	`schema:price`	49 740 982 (.49)
`gr:hasMinValue`	134 954 (.13)	`schema:longitude`	17 055 600 (.17)
`gr:amountOfThisGood`	94 978 (.09)	`schema:latitude`	17 053 362 (.17)
`gr:hasMaxValue`	52 772 (.05)	`schema:ratingValue`	9 928 412 (.10)
`vcard:latitude`	49 428 (.05)	`schema:lowPrice`	2 240 110 (.02)
`vcard:longitude`	49 428 (.05)	`schema:highPrice`	1 840 080 (.02)
`gr:hasValue`	25 800 (.03)	`schema:value`	1 776 255 (.02)
`dv:price`	23 086 (.02)	`schema:worstRating`	311 374 (.00)
`dv:average`	21 038 (.02)	`schema:position`	240 577 (.00)
`rev:rating`	20 970 (.02)	`schema:minPrice`	197 850 (.00)

magnitude. Remarkably, we did find only one single occurrence[16] of `xsd:decimal` among 14 778 325 375 literals from valid triples in embedded JSON-LD sources in the whole Web Data Commons December 2020 dataset. Table 3 shows properties that most frequently occurred with `xsd:float` in RDFa and with `xsd:double` in embedded JSON-LD. We manually classified the top ten properties using their definitions, if found, and the local names. Based on these figures, at least 62% for `xsd:float` in RDFa and 54% for `xsd:double` in embedded JSON-LD represent (monetary) amounts, position numbers or single rating values (later refereed to as T10NIFP literals), which are not initially binary floating point values. At least 33% for `xsd:float` in RDFa and 35% for `xsd:double` in embedded JSON-LD represent geolocation values, arbitrary quantity values or aggregated values, which might but do not need to origin from initially binary floating point values. `rev:rating` and `schema:ratingValue` cannot be assigned unambiguously to these categories. This shows that binary floating point numbers are regularly used for not initially binary floating point values.

As expected, because embedded RDF is not the proper place for vocabulary definitions, we found only few cases of property range definitions (Measure 4). They are limited to 54 unique property-datatype-pairs with two to 153 occurrences and for properties from only five different namespaces. This does not allow to draw further conclusions.

Table 4 shows the number of occurrences of different notations. Except for `xsd:double` in embedded JSON-LD, which is affected by normalization, exponential notation is only little used in the binary floating point datatypes. Special

---

[16] https://web.archive.org/web/20200919100939/https://open.nrw/dataset/telefonver zeichnis-alphabetisch-oktober-2019-odp.

Table 4. The number of numeric notations occurrences in the lexical representation of literals per numeric datatype in the Web Data Commons December 2020 dataset (Measures 5, 6, 7, 8) in absolute numbers and relative to the total number of literals with the same datatype (Measure 3). The notation of `xsd:double` in embedded JSON-LD was normalized during the dataset generation.

Embedded JSON-LD

Datatype	Notation			
	Integer	Decimal	Exponential	Inf / NaN
`xsd:decimal`	0 (.00)	1 (1)	0 (.00)	0 (.0)
`xsd:double`	0 (.00)	0 (.00)	101 959 382 (1)	24 (.0)
`xsd:float`	35 951 (.40)	24 837 (.27)	4252 (.05)	0 (.0)
`xsd:integer`	2 021 243 613 (1)	0 (.00)	0 (.00)	0 (.0)
`xsd:long`	36 (1)	0 (.00)	0 (.00)	0 (.0)

RDFa

Datatype	Notation			
	Integer	Decimal	Exponential	Inf / NaN
`xsd:decimal`	89 (.01)	7349 (.89)	0 (.00)	0 (.0)
`xsd:double`	26 (.11)	208 (.89)	0 (.00)	0 (.0)
`xsd:float`	353 851 (.34)	643 206 (.63)	0 (.00)	4 (.0)
`xsd:int`	16 751 (.86)	0 (.00)	0 (.00)	0 (.0)
`xsd:integer`	21 507 446 (1)	38 (.00)	0 (.00)	0 (.0)
`xsd:nonNegativeInteger`	585 (1)	0 (.00)	0 (.00)	0 (.0)
`xsd:positiveInteger`	6 (1)	0 (.00)	0 (.00)	0 (.0)

values occurred only in even more rare cases. From that, we conclude that the notation or needed special values are not the crucial consideration behind using binary floating point datatypes.

The number of lexical representations that are not precisely representable in binary floating point datatypes is presented in Table 5. 33% of the represented `xsd:float` values in RDFa and 24% in embedded JSON-LD differ from lexical representations. In embedded JSON-LD the initial lexical representation of 69% of the `xsd:double` values must either have contained more then 17 significant digits or already been differing from the represented value. Referring to the most common properties used with `xsd:double` in embedded JSON-LD, shown in Table 3, the frequent occurrence of values with more then 17 significant digits is implausible. All together, this shows that 29%–68% of the values with binary floating point datatype in real web data are distorted due to the datatype.[17]

---

[17] $\frac{\sum_{\text{T1ONIFP literals}} \text{Measures 1\&2}}{\sum_{\text{xsd:double,xsd:float literals}} \text{Measure 3}} \approx 0.29$, $\frac{\sum_{\text{xsd:double,xsd:float literals}} \text{Measures 1\&2}}{\sum_{\text{xsd:double,xsd:float literals}} \text{Measure 3}} \approx 0.68$.

Table 5. The number of lexical representation occurrences without exact representation in the value space of per numeric datatype in the Web Data Commons December 2020 dataset xsd:float and xsd:double (Measures 1, 2) in absolute numbers and relative to the total number of literals with the same datatype (Measure 3). The notation of xsd:double in embedded JSON-LD was normalized during the dataset generation.

	Embedded JSON-LD		RDFa	
	Unprecise In			
Datatype	xsd:float	xsd:double	xsd:float	xsd:double
xsd:decimal	0 (.00)	0 (.00)	3087 (.37)	3087 (.37)
xsd:double	69 648 087 (.68)	69 646 819 (.68)	58 (.25)	58 (.25)
xsd:float	21 750 (.24)	21 750 (.24)	339 583 (.33)	338 676 (.33)
xsd:int	-	-	0 (.00)	0 (.00)
xsd:integer	7 564 635 (.00)	996 (.00)	1492 (.00)	38 (.00)
xsd:long	2 (.06)	0 (.00)	-	-
xsd:nonNegativeInteger	-	-	136 (.23)	0 (.00)
xsd:positiveInteger	-	-	0 (.00)	0 (.00)

7   Conclusion

Binary floating point numbers are meant to approximate decimal values to reduce memory consumption and increase computation speed. However, in RDF, decimal representations are used to approximate binary floating point numbers. This way, the use of XSD binary floating point datatypes in RDF can systematically impair the quality of data and produces ambiguity in represented knowledge. Our survey reveals that a considerable proportion of real web data is distorted due to the datatype. Further, its use restricts the choice of the arithmetic in standards compliant implementations and can falsify the results of data processing. This can cause serious impacts in real world applications.

As a second outcome, our survey indicates that shorthand syntaxes for literals are a major cause for the choice of inappropriate datatypes. We conclude that the datatypes and shorthand syntaxes in current RDF related standards encourage the distortion of numeric values. We recommend an overhaul of relevant parts of the standards to make RDF well suited for numeric data.

A radical solution that requires no update of existing data would be the deprecation and replacement of xsd:float and xsd:double with an extended mandatory xsd:decimal datatype in RDF. The extended xsd:decimal datatype should additionally permit exponential notation and the special values *positiveInfinity*, *negativeInfinity*, and *notANumber* to cover the whole lexical space and value space of xsd:float and xsd:double. We recommend to declare it as default datatype in the different serialization and query languages for numbers in decimal and exponential notation. It should also be used for interpretation instead of the deprecated datatypes, if these are used in existing data. One or several additional new datatypes with hexadecimal lexical representations should be used for the actual representation of binary floating point values. However, this radical solution would make a decision for existing data in favor of the verbatim interpretation of the lexical representation. Thus, in (presumable not occurring) cases of an intended representation of e.g. 0.100 000 001 4... with `"0.1"^^xsd:float`, existing data would get distorted.

A more cautious mitigation of the problem should tackle the disadvantages of `xsd:decimal`: It would be desirable to introduce in RDF mandatory support for (a) an exponential notation for the decimal datatype, and (b) a decimal datatype that supports infinite values, like `precisionDecimal`, to eliminate these disadvantages. Further, binary floating point datatypes should only be used for numeric values if (a) a representation of infinity is required, or (b) the original source provides binary floating point values. In general, `xsd:decimal` must become the first choice for the representation of numbers. Semantic Web teaching materials should clearly name the disadvantages of the binary floating point datatypes, shorthand syntaxes should in future prioritize the decimal datatype, and Semantic Web tools should hint to use `xsd:decimal`.

Acknowledgments. Many thanks to Alsayed Algergawy, Felicitas Löffler, Samira Babalou, Sheeba Samuel, Sirko Schindler, Eberhard Zehendner, and the first author's supervisor Birgitta König-Ries, as well as 10 anonymous reviewers for very helpful comments on earlier drafts of this manuscript.

Author contributions. Study conception and design, analysis and interpretation of results, and draft manuscript preparation were performed by Jan Martin Keil. Data collection was performed by Merle Gänßinger and Jan Martin Keil. All authors read and approved the final manuscript.

References

1. W3C RDF Working Group: RDF 1.1 Concepts and Abstract Syntax. W3C Recommendation. W3C, 25 February 2014. http://www.w3.org/TR/2014/REC-rdf11-concepts-20140225/
2. W3C XML Schema Working Group: W3C XML Schema Definition Language (XSD) 1.1 Part 2: Datatypes. W3C Recommendation. W3C, 5 April 2012. http://www.w3.org/TR/2012/REC-xmlschema11-2-20120405/
3. Keil, J.M., Schindler, S.: Comparison and evaluation of ontologies for units of measurement. Semant. Web **10**(1), 33–51 (2019). https://doi.org/10.3233/SW-180310
4. Noy, N.F., McGuinness, D.L.: Ontology development 101: a guide to creating your first ontology. Technical report KSL-01-05/SMI-2001-0880. Stanford Knowledge Systems Laboratory and Stanford Medical Informatics, March 2001. http://www.ksl.stanford.edu/people/dlm/papers/ontology-tutorial-noy-mcguinness-abstract.html
5. W3C Semantic Web Best Practices and Deployment Working Group: XML Schema Datatypes in RDF and OWL. W3C Working Group Note. W3C, 14 March 2006. https://www.w3.org/TR/2006/NOTE-swbp-xsch-datatypes-20060314/
6. Patriot Missile Defense: Software Problem Led to System Failure at Dhahran, Saudi Arabia. Technical report GAO/IMTEC-92-26. General Accounting Office, Information Management and Technology Division, 20 p., 4 February 1992. https://www.gao.gov/products/IMTEC-92-26

7. IEEE: IEEE 754–2008 Standard for Floating-Point Arithmetic. Standard 754, 70 p., 29 August 2008. https://doi.org/10.1109/IEEESTD.2008.4610935

8. Beckett, D., Berners-Lee, T., Prud'hommeaux, E., Carothers, G.: RDF 1.1 turtle: terse RDF triple language. W3C Recommendation. W3C, 25 February 2014. https://www.w3.org/TR/2014/REC-turtle-20140225/

9. Bizer, C., Cyganiak, R.: RDF 1.1 TriG: RDF dataset language. W3C Recommendation. W3C, 25 February 2014. https://www.w3.org/TR/2014/REC-trig-20140225/

10. W3C SPARQL Working Group: SPARQL 1.1 Query Language. W3C Recommendation. W3C, 21 March 2013. https://www.w3.org/TR/2013/REC-sparql11-query-20130321/

11. Sporny, M., Longley, D., Kellogg, G., et al.: JSON-LD 1.1: a JSON-based Serialization for Linked Data. W3C Recommendation. W3C, 16 July 2020. https://www.w3.org/TR/2020/REC-json-ld11-20200716/

12. Bray, T.: The JavaScript Object Notation (JSON) data interchange format. Standard 8259, 16 p., December 2017. https://doi.org/10.17487/RFC8259

13. ECMA International: ECMA-404, The JSON Data Interchange Format. Standard (2017). https://ecma-international.org/publications/standards/Ecma-404.htm

14. W3C RDF Working Group: RDF 1.1 XML Syntax. In: Gandon, F., Schreiber, G. (eds.) W3C Recommendation, 25 February 2014. https://www.w3.org/TR/2014/REC-rdf-syntax-grammar-20140225/

15. Beckett, D.: RDF 1.1 N-triples: a line-based syntax for an RDF graph. W3C Recommendation. W3C, 25 February 2014. https://www.w3.org/TR/2014/REC-n-triples-20140225/

16. W3C RDF Working Group: RDF 1.1 N-quads: a line-based syntax for RDF datasets. W3C Recommendation. W3C, 25 February 2014. https://www.w3.org/TR/2014/REC-n-quads-20140225/

17. W3C RDFa Working Group: RDFa Core 1.1 - Third Edition: Syntax and processing rules for embedding RDF through attributes. W3C Recommendation. W3C, 17 March 2015. https://www.w3.org/TR/2015/REC-rdfa-core-20150317/

18. W3C XML Schema Working Group: An XSD datatype for IEEE floating-point decimal. W3C Working Group Note. W3C, 9 June 2011. https://www.w3.org/TR/2011/NOTE-xsd-precisionDecimal-20110609/

19. Higham, N.J.: Accuracy and Stability of Numerical Algorithms, 2nd edn., p. xxvii + 663. SIAM (2002). https://doi.org/10.1137/1.9780898718027

20. W3C XML Schema Working Group: RQ-28 Allow scientific notation for decimals (scientific-notn), 11 February 2006. https://www.w3.org/Bugs/Public/show_bug.cgi?id=2853

21. W3C XML Schema Working Group: W3C XML Schema Definition Language (XSD) 1.1 Part 2: Datatypes. W3C Candidate Recommendation. W3C, 21 July 2011. https://www.w3.org/TR/2011/CR-xmlschema11-2-20110721/

22. International Vocabulary of Metrology. Basic and general concepts and associated terms. JCGM 200:2012 (JCGM 200:2008 with minor corrections). Joint Committee for Guides in Metrology (2012)

23. Neumaier, A.: Introduction to Numerical Analysis, p. 366. Cambridge University Press, Cambridge (2012)

24. Poveda-Villalón, M., Gómez-Pérez, A., Suárez-Figueroa, M.C.: OOPS! (ontology pitfall scanner!): an on-line tool for ontology evaluation. Int. J. Semant. Web Inf. Syst. **10**(2), 7–34 (2014). https://doi.org/10.4018/ijswis.2014040102

25. Meusel, R., Petrovski, P., Bizer, C.: The WebDataCommons microdata, RDFa and microformat dataset series. In: Mika, P., et al. (eds.) ISWC 2014. LNCS,

vol. 8796, pp. 277–292. Springer, Cham (2014). https://doi.org/10.1007/978-3-319-11964-9_18

26. Gänßinger, M., Keil, J.M.: RDF property and datatype usage scanner v1.0.0 (2021). https://doi.org/10.5281/zenodo.6258887

27. Keil, J.M., Gänßinger, M.: Web data commons (December 2020) property and datatype usage dataset (2022). https://doi.org/10.5281/zenodo.6205111

28. Keil, J.M.: Web data commons (December 2020) property and datatype usage analysis scripts (2022). https://doi.org/10.5281/zenodo.6264286

DCWEB-SOBA: Deep Contextual Word Embeddings-Based Semi-automatic Ontology Building for Aspect-Based Sentiment Classification

Roos van Lookeren Campagne, David van Ommen, Mark Rademaker, Tom Teurlings, and Flavius Frasincar$^{(\boxtimes)}$ (iD)

Erasmus University Rotterdam, PO Box 1738, 3000 DR Rotterdam, The Netherlands
{471085rl,483808do,467978mr,482163tt}@student.eur.nl,
frasincar@ese.eur.nl

Abstract. In this paper, we propose the use of deep contextualised word embeddings to semi-automatically build a domain sentiment ontology. Compared to previous research, we use deep contextualised word embeddings to better cope with various meanings of words. A state-of-the-art hybrid method is used for aspect-based sentiment analysis, called HAABSA++, to evaluate our obtained ontology on the SemEval-2016 restaurant dataset. We achieve a prediction accuracy of 81.85% for the hybrid model with our ontology, which outperforms the hybrid model with other considered ontologies. Furthermore, we find that the ontology obtained from our proposed domain sentiment ontology builder, called DCWEB-SOBA, on itself improves the accuracy for the conclusive cases from 83.04% to 84.52% compared to the ontology builder based on non-contextual word embeddings, WEB-SOBA.

Keywords: Ontology learning · Contextual word embeddings · Aspect-based sentiment analysis · Hybrid method

1 Introduction

Aspect-based sentiment analysis (ABSA) determines the sentiment relating to the aspects (features) of products or services in (Web) text [15]. ABSA provides businesses more insight into which specific aspects of a product or service need to be improved upon [17].

ABSA involves two steps. First, the aspects are identified and categorised (aspect detection), thereafter the sentiments for the identified aspects are gauged (sentiment analysis) [17]. In this paper we perform these steps with state-of-the-art hybrid models for sentiment classification, using benchmark data in which the aspects are given. These models use a domain sentiment ontology to predict sentiment. Whenever the ontology reasoner is unable to predict the sentiment, a neural attention model serves as a backup solution [16,19,20].

P. Groth et al. (Eds.): ESWC 2022, LNCS 13261, pp. 183–199, 2022.
https://doi.org/10.1007/978-3-031-06981-9_11

There are various methods to obtain a domain sentiment ontology. The most general method is to manually build one [16,17]. Although this method has good performance, lexicalisations are limited and building the ontology is time-consuming. Additionally, the ontology is manually constructed for a specific domain and therefore hard to transfer to another domain. Automatically constructed ontologies, suggested by [2], shorten building time, but unsupervised building results in less accuracy. [22] proposes to semi-automatically build a domain sentiment ontology from a domain-specific corpus, which could produce more extensive ontologies in a time-efficient manner. For these ontologies, user input is required to control the builder on mistakes. [9] extends the work proposed in [22] by using non-contextual word embeddings instead of word co-occurrence for ontology building and achieves a higher accuracy in a hybrid model.

The objective of this research is to improve the performance of the sentiment classification by semi-automatically building a domain sentiment ontology from a domain-specific sentiment corpus based on deep contextual word embeddings. These deep contextual word embeddings allow the model to cope with polysemy, which is not considered by non-contextual word embeddings. Two forms of deep contextual word embeddings are BERT [5] and ELMo [13]. Compared to ELMo, BERT is easily applicable for a wide range of Natural Language Processing (NLP) tasks and can consider both left and right contexts simultaneously, while ELMo can only take the concatenation of the left and right context representations. Unlike other research on contextual word embeddings, our approach can serve as a refined way of aspect-based sentiment analysis by accounting for word meanings in context.

We adopt the ontology structure posed in [9] so that our work contributes to previous research in four ways. First, we use deep contextualised word embeddings obtained from BERT to deal with polysemy. Second, our ontology builder considers adverbs to carry a sentiment. For example, 'carefully' conveys a positive sentiment in the context of 'carefully prepared'. Additionally, we use an extra set of BERT word embeddings that are sentiment-aware instead of using external data sources to make word embeddings sentiment-aware. Last, we select relevant words for the domain sentiment ontology using a novel threshold function. These four extensions will allow the ontology to be built time-efficiently without the cost of losing accuracy. We use the ontology obtained from our ontology builder in a state-of-the-art hybrid approach for aspect-based sentiment analysis (HAABSA++) [19], using LCR-Rot-hop++ as a backup model to obtain a sentiment when our ontology is inconclusive.

This paper is structured as follows. In Sect. 2, we briefly discuss some related work, followed by an overview of the used datasets in Sect. 3. In Sect. 4 first, the structure of the ontology is explained, followed by an explanation of our proposed approach for building a domain sentiment ontology. The performance of our proposed solution is evaluated in Sect. 5. Last, in Sect. 6, we draw our conclusions and provide suggestions for future research.The source code is available from: https://github.com/RoosVanLookeren/DCWEB-SOBA. .

2   Related Work

In this section, we discuss relevant literature to our work. First, we look into hybrid approaches for ABSA in Sect. 2.1. Second, different domain sentiment ontology builders are outlined in Sect. 2.2. Last, in Sect. 2.3, we briefly describe various deep contextual word embeddings.

2.1   Hybrid Models

The research of [16] is among the first showing that a hybrid approach to aspect-based sentiment analysis algorithms outperforms other state-of-the-art approaches. As input for sentiment prediction, [16] proposes to manually construct a domain sentiment ontology. In case the ontology does not specify a sentiment value for an aspect a Bag-of-Words model trained with a multi-class Support Vector Machine (SVM) is used to complement the ontology. This work started more research for optimising the machine learning-based backup models.

[20] introduces a hybrid approach for aspect-based sentiment analysis, called HAABSA, using the domain sentiment ontology of [16] and a Left-Center-Right neural network with an iterative rotatory attention mechanism (LCR-Rot-hop) as a backup method. By iterating the rotary attention mechanism multiple times, the model focuses better on the relevant sentiment words related to a given aspect. This results in a better accuracy as the interaction between aspect and relevant context is better captured.

Following the HAABSA model, [19] introduces HAABSA++, which extends HAABSA in two directions. First, non-contextual GloVe word embeddings are replaced by deep contextual word embeddings, BERT [5] and ELMo [13], to deal with word semantics in the text. Second, the authors introduce hierarchical attention by adding an extra attention layer to the attention mechanism of [20], enabling the model to distinguish the importance of the high-level input sentence representations. The adjustment of the rotatory attention mechanism to a hierarchical architecture is called LCR-Rot-hop++. This paper shows that exploiting BERT word embeddings in LCR-Rot-hop++ results in better performance compared to ELMo word embeddings and outperforms LCR-Rot-hop.

In our paper we introduce the next evolution of hybrid models by using deep contextual word-embeddings to build the domain sentiment ontology. We implement our obtained domain sentiment ontology in HAABSA++ using LCR-Rot-hop++ as a backup model, due to the proven success of this method [19].

2.2   Ontology Building Approaches

Besides improvements of the backup model in hybrid approaches, research has been done in refining the first part of hybrid models, more precisely the ontology. Ideally an ontology is built in a time-efficient manner and is as accurate as possible for the considered task. However, accuracy and time-efficiency do not always go hand in hand. Therefore a good ontology needs to be built time-efficiently while remaining able to predict the sentiments accurately.

One solution for building ontologies is to integrate existing knowledge resources as proposed by OntoSenticNet 2 [6]. While such an approach is able to model commonsense sentiment, it provides the Domain concept that can be used to extend the ontology with domain sentiment representations. For building domain sentiment representations, often a data-driven approach is used due to the availability of domain text, which is also the approach pursued here.

To decrease the building time of a domain sentiment ontology, [4, 22] suggest to build a domain sentiment ontology semi-automatically, which requires manual input of the user to control for possible mistakes made by the ontology builder. [22] proposes SOBA, in which word co-occurrence is used to deal with word semantics. Besides word co-occurrence, [4] also employs synsets in an ontology builder, called SASOBUS, to enable a fair and reliable comparison of words, while simultaneously capturing their meaning. These two methods give comparable results to the manually constructed ontology, yet a significant reduction in constructing time is achieved.

Better performance is attained when using non-contextual word embeddings for the automated part of the domain sentiment ontology building. In this method, named WEB-SOBA [9], word2vec [12] word embeddings are exploited. The CBOW model is used, which can detect syntactic and semantic word similarities. When the ontology obtained from WEB-SOBA is used in HAABSA, it performs better than the manually constructed domain sentiment ontology. WEB-SOBA reduces both computing and user time required to construct the ontology compared to SOBA and SASOBUS, given that the non-contextual word embeddings for a specific domain are already made. For this research, we use the manually constructed ontology, SOBA, and WEB-SOBA as benchmarks. We do not compare our ontology with SASOBUS as [9] already showed that WEB-SOBA outperforms SASOBUS on accuracy and time-efficiency.

2.3 Deep Contextual Word Embeddings

Word embeddings are vector word representations. In this paper we make a clear distinction between non-contextual word embeddings and deep contextual word embeddings. The difference is that deep contextual word embeddings will include the context of the word in the representation, while non-contextual word embeddings will not. This means that while non-contextual word embeddings (e.g., word2vec or GloVe) have difficulty coping with polysemy, deep contextual word embeddings (e.g., ELMo or BERT) can capture different meanings of a single word, as they assign a unique vector per instance of a word in its context.

Recently, two new game-changing types of machine-learning models that create deep contextual word embeddings were introduced. These are deep contextual word embeddings, in the sense that they are a function of many internal layers of the neural model. The first model is ELMo, the model captures information about the entire input sentence using multiple bidirectional Long Short-Term Memory (LSTM) layers. A big advantage of the second model, BERT [5], is that it can pre-train deep bidirectional vector representations from an unlabelled text by considering both left and right context simultaneously, while

ELMo can only take the concatenation of the separate contexts-based representations. BERT is an open-sourced model and a version of the pre-trained model trained on a massive dataset is publicly available. Next to that, BERT can efficiently be fine-tuned for a wide range of tasks with only one additional output layer, this makes BERT applicable in many situations. For our research, we therefore propose to use BERT to construct deep contextual word embeddings.

3 Data

To build a domain sentiment ontology we need a domain-specific corpus to determine the domain-specific terms. Earlier built ontologies use a restaurant domain dataset as domain-specific corpus. We opt to use such a dataset as well since it simplifies the comparison of our ontology with other ontologies. The dataset we use is the Yelp Open Dataset, which consists of 8,635,403 reviews of different businesses [21]. After filtering out the restaurant reviews, around 4,700,000 reviews of more than 500,000 restaurants remain in the dataset. We use the BERT base model (uncased), which is trained on the BooksCorpus (800M words) and Wikipedia (2500M words) [5] to create pre-trained word embeddings for 2000 reviews containing 200.000 unique words due to computational limitations.

When a sentiment word is negated in its context (e.g., 'not'), the polarity of this word embedding is reversed. To ensure that the word embedding is only determined by the correct polarity of the word, we remove the sentences in which a word from the negation set appears. The set of negation words used is $\mathcal{NW}$ = {'not', 'never', 'nothing', 'don't', 'doesn't', 'didn't', 'can't', 'wouldn't'}. In 10.4% of the sentences of the dataset one of these negation words is present. Consequently, we use 89.6% of the review sentences to create our ontology.

To evaluate the ontology we use the standard dataset from SemEval-2016 Task 5 [14], containing restaurant reviews. The data is divided into a training set of 1879 sentences and a test set of 650 sentences. Every sentence contains opinions regarding specific aspects. Each aspect is labelled with a polarity from the set $\mathcal{P}$ = {negative, neutral, positive}. Furthermore, an entity (category) E (e.g., restaurant, food, service) and attribute A (e.g., prices, quality, general) pair is attached to each aspect. All implicit aspects are not used in the analysis because the machine learning method assumes aspects to be explicitly present.

4 Methodology

In this section, we explain the methods used in our ontology builder, to which we refer as Deep Contextual Word Embedding-Based Semi-Automatic Ontology Builder for Aspect-Based Sentiment Analysis (DCWEB-SOBA). First, we explain which BERT word embeddings we use to detect word semantics in Sect. 4.1. Thereafter, we discuss the methods used in the four stages of our ontology builder. The first stage comprises the construction of the skeletal structure and the initialisation of the ontology, which is described in Sect. 4.2. Then we explain the methods used to select the relevant terms in Sect. 4.3 and to classify sentiment terms in Sect. 4.4. The last stage involves hierarchical clustering of aspect terms to assign them to ontology classes, explained in Sect. 4.5.

4.1 Word Embeddings

The BERT model [5] has been shown to properly model language and handle different tasks. The model can, after pre-training, be post-trained and fine-tuned. In this section, we evaluate to what extent the model has to be post-trained and fine-tuned. We use the dimension reduction method *t-SNE* for a two-dimensional visualisation of the word embeddings for the evaluation. Throughout this research, we run all BERT models on a 2.3 GHz Intel Core i5 CPU with 8 GB RAM in combination with a GPU Tesla V100-SXM2. We use the AdamW optimiser [8] to update the weights during post-training and fine-tuning, as it has been proven to be one of the fastest optimisers in training neural networks [11].

Polysemy-Aware Word Embeddings. The pre-trained BERT model is able to distinguish semantics and polysemous words in our corpus. Figure 1 shows the word embeddings of the word 'turkey' for two different synsets. In half of the sentences 'turkey' is used in the meaning of an animal, denoted by 'Turkey#A', and in the other half of the sentences 'turkey' is used in the meaning of country, denoted by 'Turkey#B'. It can be seen in the plot that the pre-trained model can position these words near similar words. For the comparison, we use word embeddings of 'pizza' and 'Italy'. The pre-trained BERT model can be post-trained to capture the language characteristics in a corpus. However, we find that post-training, using 50.000 reviews from the Yelp dataset, results in a worse separation of the synsets of a word, possibly due to the small size of our domain corpus, therefore, we choose to use the pre-trained word embeddings for term selection and aspect term hierarchical clustering.

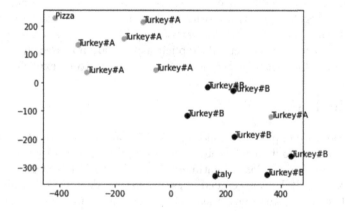

Fig. 1. Pre-trained BERT word embeddings for aspects 'turkey', 'pizza', and 'Italy': polysemous words are well-separated.

Sentiment-Aware Word Embeddings. The pre-trained BERT models position sentiment mention word embeddings based on more characteristics than only their polarity, causing words of a different polarity to be in the same vicinity. This can cause a problem for the sentiment-based part of our method. Figure 2 shows the representation of the pre-trained word embeddings of negative (black) and positive (grey) sentiment words. It can be seen that word embeddings do not separate the sentiment words very well based on polarity.

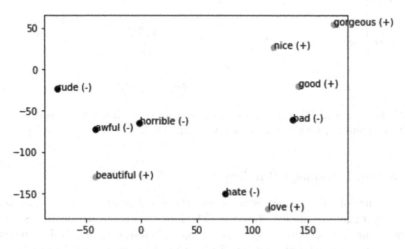

Fig. 2. Pre-trained BERT word embeddings for sentiment words: positive (+) and negative (−) words are not well-separated.

The problem can be solved by fine-tuning the pre-trained BERT model on sentiment classification. The polarity of a sentiment word will determine the position of the word embedding from the created model. This is done on the task of classifying reviews as either positive or negative. Each review in the Yelp dataset is labeled with a star rating between 1 and 5. We treat 4-star and 5-star reviews as positive reviews ($y = 1$) and 1-star and 2-star reviews as negative reviews ($y = 0$). The binary variable y denotes the label for the classification task. During fine-tuning, the model learns the sentiment value of words, and readjusts its word embeddings accordingly. The fine-tuned BERT model is trained on 100.000 reviews from the Yelp dataset.

Figure 3 shows that the fine-tuned model can better separate words on their polarity than the pre-trained model, shown in Fig. 2. Therefore we conclude that these word embeddings are sentiment-aware. In this paper, we refer to fine-tuned word embeddings when sentiment-aware word embeddings are used and refer to pre-trained word embeddings when we mention word embeddings.

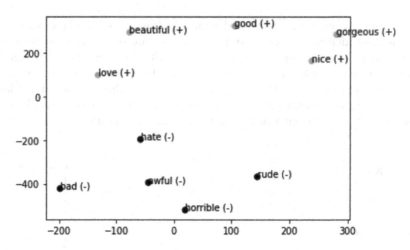

Fig. 3. Fine-tuned BERT word embeddings for sentiment words without post-training: positive (+) and negative (−) words are well-separated.

4.2 Skeletal Ontology Building

The structure of the ontology is based on [16], which contains two main classes, namely SentimentValue and Mention. The first class contains subclasses denoting sentiment. The second class is divided into subclasses distinguishing the part-of-speech of a term and the aspect. Every relevant term to our domain being a noun, verb, adjective, or adverb gets placed in, respectively, the Entity, Action, Property, or Modifier classes. We refer to these subclasses as <Pos>.

Each aspect has the format of CATEGORY#ATTRIBUTE pair. The Mention class has each category, denoted as <Cat>, and attribute, denoted as <Att>, as subclasses, resulting in the aspect subclasses: Restaurant, Location, Food, Drinks, Prices, Experience, Service, Ambiance, Quality, Style, and Options. The aspect subclasses are linked through one or multiple aspect properties to corresponding CATEGORY#ATTRIBUTE pairs. Table 1 shows all possible pairs.

Table 1. Combinations of categories and attributes in the ontology.

Categories	Attributes				
	General	Prices	Quality	Style&Options	Miscellaneous
Ambience	x				
Drinks		x	x	x	
Experience					x
Food		x	x	x	
Location	x				
Service	x				
Restaurant	x	x		x	

Note, the attribute *Style&Options* is split through lexical properties into two separate aspect subclasses, namely `Style` and `Options`. We use these representations instead of `Style&Options` as there does not exist a word embedding for *Style&Options*, which is required in the next steps of the ontology builder. Furthermore, we have attributes *Miscellaneous* and *General*, as these attributes are too generic, they do not appear as subclasses. Figure 4 presents an overview of the `Mention` class structure (subclasses have suffix 'Mention' removed to avoid cluttering). Each <Cat/Att><Pos>Mention is a subclass of <Cat/Att>Mention. For instance, `ServiceEntityMention` is a subclass of `ServiceMention` (<Cat>) and `Entity` (<Pos>), which in their turn are subclasses of `Mention`. Furthermore, each <Cat/Att><Pos>Mention has two subclasses <Cat/Att>Positive<Pos> and <Cat/Att>Negative<Pos>. These subclasses have `Positive` or `Negative` as their parent class.

The `SentimentValue` class consists of two subclasses, namely `Positive` and `Negative`, which refer to their corresponding sentiment shown in Fig. 5. Neutral sentiments are disregarded in this research. These sentiments are hard to interpret since they carry inherent ambiguity. We consider three types of sentiment words. Type-1 sentiment words (generic sentiment) carry only one polarity irrespective of the context and aspect (e.g., the term 'bad' is always negative). Type-1 sentiment words are assigned to the `GenericPositive` and `GenericNegative`, which are subclasses of `Positive` and `Negative`, respectively, corresponding to their polarity. Furthermore, Type-1 words are also subclasses of their corresponding part-of-speech classes, called `GenericPositive`<Pos> and `GenericNegative`<Pos>. Next, polarity consistent words that can not be used for every aspect are Type-2 sentiment words. For example, the word 'delicious' is a Type-2 sentiment word only applicable for the aspects 'food' and 'drinks' and always denotes a positive sentiment. Therefore, 'delicious' will appear as lexicalisation in the classes `FoodMention` and `DrinksMention`. Last, the polarity of Type-3 sentiment words depends on the aspect and context. For instance, in the context of price the word 'cheap' carries positive sentiment, while 'cheap' regarding style carries negative sentiment.

User Intervention. We initialise the skeletal structure of the ontology with some Type-1 sentiment words of different part-of-speeches and various sentiment polarities (e.g., 'good', 'bad', 'poorly', and 'hate'). This is necessary because these generic sentiment words are less likely to be extracted in the term selection as these are not specific to a `Mention` class. For each generic sentiment word, we suggest the 15 most similar words to the user, based on the *cosine similarity* of the word embeddings. These can be accepted or rejected. In preliminary experiments, we find that adding the 15 closest words results in the best trade-off between the quality of the accepted words and the class lexical coverage.

Our word embedding model creates a unique vector for each word instance. For comparisons, however, it is more efficient to limit the comparisons to one representative vector for a synset of a word. Therefore, we average the vectors for each instance of an initialised generic sentiment word. Most of the generic

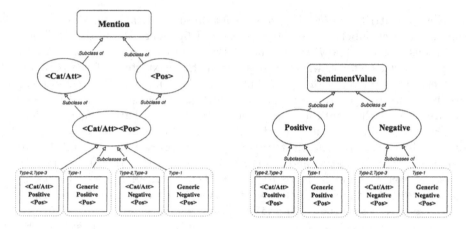

Fig. 4. Mention class structure.   **Fig. 5.** SentimentValue class structure.

sentiment words only have one meaning (i.e., 'hate', 'poorly', 'expensive', 'excellent'), so for these words we do not lose relevant information.

4.3 Term Selection

The goal of the term selection is to determine the relevant terms for our domain sentiment ontology. The part-of-speech tagger of Stanford NLP Processing Group [18] is used in our research to extract all verbs, nouns, adjectives, and adverbs from the corpus. In contrast to [9], we consider adverbs as well. After tagging, we use a pre-defined stopwords list [3] to filter out all stopwords as those do not contribute to our analysis.

First, we average all word embeddings referring to the same meaning of a word to improve computation time. This is done by averaging the vectors of the lexical representations of each aspect in the Mention classes, $\mathcal{A} = \{$Restaurant, Location, Food, Drinks, Price, Experience, Service, Ambiance, Quality, Style, Options$\}$. In our restaurant domain corpus, these aspects do not have ambiguous semantics, therefore we can average all corresponding vectors without losing semantic information.

Second, we calculate for each instance of a word the *Mention Class Similarity* (*MCS*) value using the *cosine similarity* as follows:

$$MCS_i = \max_{a \in \mathcal{A}} \left(\frac{v_i \cdot v_a}{\|v_i\| \cdot \|v_a\|} \right), \qquad (4.1)$$

where v_i is the vector of word i in the domain-specific word embedding model and v_a is the averaged vector of all lexical representations of the aspect a in the set $\mathcal{A}$. *MCS* indicates the Mention class to which the representation of the word embedding is the closest.

Next, all the word embeddings with an *MCS* value below a threshold are eliminated. Words that are allocated to the same Mention class according to their MCS have their embeddings averaged and named as *word<mention>*, where *word* is the currently considered word and *mention* is an element of $\mathcal{A}$. This averaging procedure is done for MCS values higher than the previous threshold. A good threshold value is pivotal, as otherwise the average vector will not accurately represent a cluster for one synset of a word. We find in preliminary research that a threshold of 0.68 gives a good balance between accuracy and coverage. We define terms as being the refined words after averaging vectors and assigning new word names to them.

After the word embeddings are reduced to relevant terms, we can efficiently determine which terms are suggested to the user. We compute again the MCS value, as this value can differ after averaging. It is used to determine the order in which the terms are suggested. In order to suggest the right amount of words to the user, the MCS values are compared against the threshold value specific to each of the four part-of-speeches. The threshold function is defined as follows:

$$ TH_{pos}^{*} = \max_{TH_{pos}} \left(\frac{accepted}{n+1} \right), \tag{4.2} $$

where TH_{pos} is the threshold score for a <Pos> class, n is the number of suggested terms, and *accepted* denotes the number of accepted terms. The value of TH_{pos}^{*} is increasing in the number of accepted terms. We divide by $n+1$ to avoid the case of a threshold value of 1. This would happen if we divided by n and the first and only suggested term is accepted. The effect of the penalty diminishes for larger n. The threshold will be set to the optimal ratio of accepted terms for a <Pos> class, denoted by TH_{pos}^{*}.

User Intervention. For all accepted terms that are a noun or verb, the user is asked whether the term is a *Sentiment Mention* or *Aspect Mention*. We only consider adjectives and adverbs to indicate a sentiment, as these words can describe the *Aspect Mention*. When a term is a *Sentiment Mention*, the user is asked whether it is a Type-1, 2, or 3 *Sentiment Mention*. For each accepted word, we add all similar words that have a *cosine similarity* larger than a threshold to the ontology as well. Preliminary research results in a threshold of 0.7 which gives a good balance between accuracy and coverage.

4.4   Sentiment Term Clustering

Now that we have determined the relevant *Sentiment Mention* terms, we identify their polarity and class. As the sentiment-aware word embeddings separate the terms well based on polarity, we can calculate the negative and positive scores of our *Sentiment Mention* terms in the following way:

$$ PS_i = \max_{p \in \mathcal{P}} \left(\frac{v_i \cdot v_p}{\|v_i\| \cdot \|v_p\|} \right) \qquad NS_i = \max_{n \in \mathcal{N}} \left(\frac{v_i \cdot v_n}{\|v_i\| \cdot \|v_n\|} \right), \tag{4.3} $$

where PS and NS are the positive and negative scores, respectively, for $term_i$. $\mathcal{P}$ and $\mathcal{N}$ are a set of, respectively, positive and negative words of different kinds of sentiment intensity. The set of positive and negative words are, respectively, $\mathcal{P} = \{$'good', 'decent', 'great', 'tasty', 'fantastic', 'solid', 'yummy', 'terrific'$\}$ and $\mathcal{N} = \{$'bad', 'awful', 'horrible', 'terrible', 'poor', 'lousy', 'shitty', 'horrid'$\}$. For each word we use the averaged vectors that are made from the sentiment-aware word embeddings. Finally, v_i denotes the vector of term i, and v_p and v_n are the vectors of a term in the positive and negative set, respectively. The largest score determines the predicted polarity of the word.

User Intervention. For each *Sentiment Mention* term the user decides whether the suggested polarity is correct. Type-2 and Type-3 sentiment words are aspect-specific and can therefore be used in multiple <Cat/Att>Mention classes. To assign these words to multiple classes, the user is asked to check whether the *Sentiment Mention* term belongs to the <Cat/Att>Mention class to which the term has the highest cosine similarity. If the user accepts this <Cat/Att>Mention, the second most similar <Cat/Att>Mention is suggested. This procedure continues until the user rejects adding a *Sentiment Mention* term to a <Cat/Att>Mention class.

4.5   Aspect Term Hierarchical Clustering

In this section, we explain the hierarchical clustering procedure for *Aspect Mention* terms in two steps. First, the *Aspect Mention* terms are allocated to their corresponding cluster of the Mention classes, $\mathcal{A} = \{$Restaurant, Location, Food, Drinks, Price, Experience, Service, Ambiance, Quality, Style, Options$\}$. This is done by representative clustering based on the *cosine similarity* between the vector of a term and a vector of the 'base'. The 'base' is made up of the averaged vectors of the lexical representation of the aspects in the Mention class, in the same way as previously described in Sect. 4.3.

Next, we create a hierarchy for each cluster in the 'base', using agglomerative hierarchical clustering [1]. Terms start in single clusters and throughout the clustering process they are merged together. In each iteration, clusters with the lowest *Average Linkage Clustering* (*ALC*) value get merged together. *ALC* is determined as follows:

$$ALC(A, B) = \frac{1}{|A| \cdot |B|} \sum_{a \in A} \sum_{b \in B} d(a, b), \qquad (4.4)$$

where $d(a, b)$ is the Euclidean distance between vectors a and b, and a is in cluster A and b is in cluster B. We find by using the 'elbow' method that the optimal depth in our dendogram to create the preferred hierarchy equals 3 subclasses.

User Intervention. After a term is clustered to a 'base' cluster, the user decides whether the term is correctly placed in the Mention class belonging to 'base' cluster by either accepting or rejecting. Whenever a term is rejected, the user has to specify the correct Mention class. In this way, all terms start in the right 'base' cluster before performing the hierarchical clustering.

5   Evaluation

In this section, we evaluate our ontology on the SemEval-2016 dataset using the DRANZIERA evaluation protocol "Open" setting [7] against three benchmarks: the manual constructed ontology [16], SOBA [22], and WEB-SOBA [9]. First, in Sect. 5.1 we evaluate our ontology against the benchmarks in terms of descriptive characteristics, time-efficiency, accuracy, and conclusiveness. Thereafter in Sect. 5.2, our ontology is used in a hybrid approach to evaluate the performance for aspect-sentiment classification. The machine used for the evaluation methods is a 2.1 GHz Intel Core i3-10110U CPU with 16 GB RAM.

5.1   Ontology Building Results

We consider an ontology as good if it is accurate and conclusive. A model is conclusive in the case that it is able to make a prediction of the sentiment for an aspect. The higher the coverage of the domain ontology, the more conclusive an ontology is. The accuracy indicates the percentage at which this prediction is correct. Additionally, an ontology should be built time-efficiently.

Table 2 shows that DCWEB-SOBA requires more user time to construct an ontology than WEB-SOBA, as the user has to consider more suggested words due to the addition of adverbs and the inclusion of different synsets of a word. Consequently, more classes and lexicalisations are added to the ontology resulting in higher coverage of the domain. The overall building time (user and computing time) of an ontology with DCWEB-SOBA is shorter since the construction of the fine-tuned BERT model only takes 120 min where the construction of word embeddings in WEB-SOBA takes 300 min. To summarise, DCWEB-SOBA does well at balancing the amount of user time and attained coverage compared to other ontology builders.

Table 2. Comparison statistics for the different ontologies.

	Manual	SOBA	WEB-SOBA	DCWEB-SOBA
Classes	365	470	376	539
Lexicalisations	374	1087	348	485
User time (minutes)*	420	90	40	60
Computing time (minutes)**	0	90	30 (+300)	30 (+120)

*User time is the time spent on the user interventions.

**Computing time is the time required to construct the ontology excluding user time.

Additionally, we evaluate the performance of the ontologies obtained from the ontology builders on the SemEval-2016 test dataset. Table 3 shows that our obtained ontology has a fairly high accuracy for its conclusive cases compared to the other semi-automatically built ontologies, only SOBA performs better. Furthermore, we can see that our ontology is conclusive in 49.69% of the cases which is a significant increase from the 35.39% of WEB-SOBA. This indicates that using deep contextual word embeddings results in a more accurate and more conclusive ontology than an ontology built with non-contextual word embeddings like WEB-SOBA.

Table 3. Comparison statistics for the percentage of conclusive cases.

	Manual	SOBA	WEB-SOBA	DCWEB-SOBA
Conclusive (%)	61.85	**64.62**	35.39	49.69
Accuracy for conclusive cases (%)	**86.82**	85.48	83.04	84.52

5.2 Hybrid Setting Results

Table 4 shows that SOBA performs best on conclusiveness and the accuracy on these predictions, however, [9] shows that SOBA is outperformed by WEB-SOBA in the hybrid setting of HAABSA. Therefore, we aim to show that DCWEB-SOBA can outperform other ontology builders in the state-of-the-art HAABSA++ setting, first proposed in [19]. For LCR-Rot-hop++, the backup model in HAABSA++, the following hyperparameters can be used to reproduce our analysis: the learning rate, the keep probabilities, and the momentum, are set to 0.001, 0.7, and 0.085, respectively. The model is trained on 150 iterations on the train SemEval-2016 dataset.

To analyse the overall performance of DCWEB-SOBA in this setting, we use the weighted average of the accuracy for conclusive and inconclusive cases. We observe in Table 4 that the ontology builders using word embeddings, WEB-SOBA and DCWEB-SOBA, reach a lower accuracy for their ontology but a higher accuracy for its backup model similar to the results in the HAABSA setting in [9].

Furthermore, the DCWEB-SOBA ontology gives the highest accuracy when used in HAABSA++ compared to the other ontologies with an accuracy of 81.85%. This indicates that the use of word embeddings is essential for the HAABSA++ model. DCWEB-SOBA scores better on the accuracies of both the ontology and backup model than WEB-SOBA, this further emphasises the importance of deep contextual word embeddings in sentiment classification.

Table 4. Comparison statistics for each ontology, LCR-Rot-hop++ and the two combined.

| | Ontology | | LCR-Rot-hop++ | Combined |
	Conclusiveness	Accuracy	Accuracy	Accuracy
Manual	61.85%	86.82%	72.98%	81.54%
SOBA	64.62%	**85.48%**	73.91%	81.38%
WEB-SOBA	35.39%	83.04%	78.81%	80.31%
DCWEB-SOBA	49.69%	84.52%	**79.20%**	**81.85%**

6   Conclusion

In this research, we propose DCWEB-SOBA to construct a semi-automatically built ontology using deep contextual word embeddings for aspect-based sentiment analysis. We hypothesise that by using deep contextual word embeddings which can deal with semantics and polysemy, we can improve the performance of sentiment classification based on its accuracy and conclusiveness. The proposed methodology makes use of contextual word embeddings at its various steps: skeletal ontology building, term selection, sentiment term clustering, and aspect term hierarchical clustering.

DCWEB-SOBA achieves higher accuracy compared to WEB-SOBA when we measure the sentiment prediction accuracy on conclusive cases. The accuracy increases from 83.04% to 84.52%. Additionally, DCWEB-SOBA is able to predict the sentiment for more aspects as it is conclusive in 49.69% of the cases, compared to 35.39% for WEB-SOBA. When we use the ontology obtained from DCWEB-SOBA in a hybrid approach, HAABSA++, we achieve an accuracy of 81.85% which outperforms the other used ontologies. This shows that using deep contextual word embeddings increases the performance in a hybrid approach for aspect-based sentiment analysis compared to non-contextual word embeddings.

Our research highlights the importance of the usage of deep contextual word embeddings in sentiment classification. For future work, we suggest the usage of Facebook's RoBERTa, introduced in [10], for creating contextual word embeddings. This model has been trained on a dataset more than ten times as large as BERT and is solely trained on the Masked Language Modeling task. In addition, RoBERTa makes use of dynamic masking: training by masking different words for every epoch. These factors make us believe that the usage of RoBERTa in a hybrid approach to aspect-based sentiment analysis is very promising.

Furthermore, the fine-tuned model can be improved by training on separating Type-3 sentiment words. This can be done by training the model on data with aspect ratings instead of review rating. As a consequence, we expect the fine-tuned model to be trained on less noisy data and result in a model which is able to cope better with aspect-based sentiment analysis.

References

1. Behnke, L.: (2012). https://github.com/lbehnke/hierarchical-clustering-java
2. Blaschke, C., Valencia, A.: Automatic ontology construction from the literature. Genome Inform. **13**, 201–213 (2002)
3. Bleier, S.: (2000). https://gist.github.com/sebleier/554280
4. Dera, E., Frasincar, F., Schouten, K., Zhuang, L.: SASOBUS: semi-automatic sentiment domain ontology building using Synsets. In: Harth, A., et al. (eds.) ESWC 2020. LNCS, vol. 12123, pp. 105–120. Springer, Cham (2020). https://doi.org/10.1007/978-3-030-49461-2_7
5. Devlin, J., Chang, M.W., Lee, K., Toutanova, K.: BERT: pre-training of deep bidirectional transformers for language understanding. In: 2019 Conference of the North American Chapter of the Association for Computational Linguistics: Human Language Technologies (NAACL-HLT 2019), pp. 4171–4186. ACL (2019)
6. Dragoni, M., Donadello, I., Cambria, E.: OntoSenticNet 2: Enhancing reasoning within sentiment analysis. IEEE Intell. Syst. **37**(1) (2022)
7. Dragoni, M., Tettamanzi, A., da Costa Pereira, C.: DRANZIERA: an evaluation protocol for multi-domain opinion mining. In: 10th International Conference on Language Resources and Evaluation (LREC 2016), pp. 267–272. ELRA (2016)
8. Gugger, S., Howard, J.: AdamW and super-convergence is now the fastest way to train neural nets. fast.ai (2018)
9. ten Haaf, F., et al.: WEB-SOBA: word embeddings-based semi-automatic ontology building for aspect-based sentiment classification. In: Verborgh, R., et al. (eds.) ESWC 2021. LNCS, vol. 12731, pp. 340–355. Springer, Cham (2021). https://doi.org/10.1007/978-3-030-77385-4_20
10. Liu, Y., et al.: RoBERTa: a robustly optimized BERT pretraining approach. arXiv preprint arXiv:1907.11692 (2019)
11. Loshchilov, I., Hutter, F.: Decoupled weight decay regularization. In: 7th International Conference on Learning Representations (ICLR 2019). OpenReview.net (2019)
12. Mikolov, T., Sutskever, I., Chen, K., Corrado, G.S., Dean, J.: Distributed representations of words and phrases and their compositionality. In: 27th Annual Conference on Neural Information Processing Systems (NIPS 2013), pp. 3111–3119. Curran Associates (2013)
13. Peters, M.E., et al.: Deep contextualized word representations. In: 2018 Conference of the North American Chapter of the Association for Computational Linguistics: Human Language Technologies (NAACL-HLT 2018), pp. 2227–2237. ACL (2018)
14. Pontiki, M., et al.: SemEval-2016 task 5: aspect-based sentiment analysis. In: 10th International Workshop on Semantic Evaluation (SemEval 2016), pp. 19–30. ACL (2016)
15. Schouten, K., Frasincar, F.: Survey on aspect-level sentiment analysis. IEEE Trans. Knowl. Data Eng. **28**(3), 813–830 (2015)
16. Schouten, K., Frasincar, F.: Ontology-driven sentiment analysis of product and service aspects. In: Gangemi, A., et al. (eds.) ESWC 2018. LNCS, vol. 10843, pp. 608–623. Springer, Cham (2018). https://doi.org/10.1007/978-3-319-93417-4_39
17. Schouten, K., Frasincar, F., de Jong, F.: Ontology-enhanced aspect-based sentiment analysis. In: Cabot, J., De Virgilio, R., Torlone, R. (eds.) ICWE 2017. LNCS, vol. 10360, pp. 302–320. Springer, Cham (2017). https://doi.org/10.1007/978-3-319-60131-1_17

18. Toutanova, K., Manning, C.D.: Enriching the knowledge sources used in a maximum entropy part-of-speech tagger. In: 2000 Joint SIGDAT Conference on Empirical Methods in Natural Language Processing and Very Large Corpora (EMNLP 2010), pp. 63–70. ACL (2000)

19. Truşcă, M.M., Wassenberg, D., Frasincar, F., Dekker, R.: A hybrid approach for aspect-based sentiment analysis using deep contextual word embeddings and hierarchical attention. In: Bielikova, M., Mikkonen, T., Pautasso, C. (eds.) ICWE 2020. LNCS, vol. 12128, pp. 365–380. Springer, Cham (2020). https://doi.org/10.1007/978-3-030-50578-3_25

20. Wallaart, O., Frasincar, F.: A hybrid approach for aspect-based sentiment analysis using a lexicalized domain ontology and attentional neural models. In: Hitzler, P., Fernández, M., Janowicz, K., Zaveri, A., Gray, A.J.G., Lopez, V., Haller, A., Hammar, K. (eds.) ESWC 2019. LNCS, vol. 11503, pp. 363–378. Springer, Cham (2019). https://doi.org/10.1007/978-3-030-21348-0_24

21. Yelp (2019). https://www.yelp.com/dataset

22. Zhuang, L., Schouten, K., Frasincar, F.: SOBA: semi-automated ontology builder for aspect-based sentiment analysis. J. Web Semant. **60**, 100–544 (2020)

Never Mind the Semantic Gap: Modular, Lazy and Safe Loading of RDF Data

Eduard Kamburjan[✉], Vidar Norstein Klungre, and Martin Giese

Department of Informatics, University of Oslo, Oslo, Norway
{eduard,vidarkl,martingi}@ifi.uio.no

Abstract. Any attempt at a tight integration between semantic technologies and object oriented programming will invariably stumble over the gap between the two underlying object models. We illustrate how this *semantic gap* manifests from the point of view of data retrieval with SPARQL. We present a novel mechanism to load data from RDF knowledge graphs into object-oriented languages that gives static guarantees about the data access and modularly integrates the mapping between the program and the RDF view with the class definition in the program. This allows us to preserve the separation of concerns between the class system of RDF (geared towards domain modeling and data), and that of the program (geared towards typing and code reuse). Loading of RDF can be performed lazily, when required by the program, based on *query-futures* – subqueries that are only evaluated if and when the data is accessed. We formulate a Liskov principle for the mapping queries to characterize when they respect the subclass relation. Moreover, we provide tool support to detect when the user-provided mapping would cause the loading mechanisms to result in data structures that manifest the semantic gap.

1 Introduction

Motivation. Despite the important role that Semantic Web technologies can play in modern software applications, their integration with programming languages remains a challenge. The main challenge is the so-called *impedance mismatch* [1, 2] or *semantic gap* [3] between the object model of RDF, geared towards data-driven tasks, and the object model of programming languages, geared towards typeability and modularity.

This impedance mismatch for RDF manifests when mapping RDF into the class system of the program: to load data from an RDF store, one executes some SPARQL query, manually traverses the results and creates objects on the go to perform computation later on. This turns data loading into a fragile, work-intensive and highly error-prone task: (a) There is no type safety mechanism. (b) The mapping between the OO class and the RDF pattern is not modular—the retrieving queries quickly grow in size and are overwhelming for the programmer. (c) While most endpoints support lazy iterators, they do not support lazy data structures within one answer: If the application is only interested in parts of the loaded data, but only decides so after the query is formulated (e.g., based on prior answers) it is not possible to postpone loading. (d) Finally, depending on the data model and retrieving query, one node in the RDF graph may be mapped to different objects, if it occurs in several answers.

In this work, we investigate *lazy, modular* and *type-safe* loading of data from RDF graphs into an object-oriented (OO) programming language. Instead of fighting the impedance mismatch, we embrace it and keep it under control: we clearly describe the structure for which such a mapping can be defined and give static analyses to warn the programmer about unintuitive effects.

Approach. We solve these challenges by providing tool support to the programmer: Our mapping tightly couples OO classes with graph patterns in RDF using a SPARQL query per OO-class. This query describes how to *construct* an object from an RDF graph, it does not establish a perpetual link between RDF nodes and OO objects. Moving the mapping from the point where the data *is used* to the program point where the data structures *are defined* simplifies modeling and allows *reuse* of queries. Reuse, in turn, is critical for maintenance. Without reuse, maintenance is aggravated if multiple queries perform similar data retrieval tasks, but are scattered over the program or dynamically manipulated.

Using the query containment-based typing mechanism presented in [4], we directly address type safety and include support for OO inheritance: we always load the most specific class possible and give a static analysis that checks uniqueness of this class. Furthermore, we give a Liskov principle [5] to statically check whether the retrieval query of a subclass correctly refines the retrieval query of the superclass. We give another analysis to inform developers whether multiple objects constructed during data access correspond to the same RDF node.

Integrating the mapping into the OO class structure enables us to perform *lazy evaluation*. Lazy evaluation only retrieves data if the computation indeed requires it. This is implemented as follows: When loading a class, its query is automatically executed. If a class has a field of another class type, this second query is lazily evaluated: The field is initialized with a *future* [6,7]. A future is a placeholder that contains the inner SPARQL query that is only evaluated when the future is explicitly accessed during a computation.

It is crucial to our approach that we do *not* relate *concepts* of RDF and OO to each other directly, i.e., do not enforce a one-to-one correspondence between OO objects and RDF individuals. Systems that try to close this *semantic gap* [3] fail to address the fundamental differences in assumptions and modeling techniques in RDF/OWL and OO, most prominently the open-world assumption and multiple inheritance. Instead, we embrace the differences in modeling and give the programmer a systematic and safe way to close the gap specifically for their application. Thus, our system can be used to both relate the OO and RDF representation of some objects, but also as a type-safe container to load the results of queries, where such a relation is not desirable.

We give an informal example in Sect. 2 and preliminaries in Sect. 3. Modular mappings are described in Sect. 4 and inheritance in Sect. 5. Lazy loading is introduced in Sect. 6 and evaluated in Sect. 7. Related work is given in Sect. 8.

```
1 List<Nodes> it =
2 query("SELECT * WHERE { ?o :id ?id; :stamp ?stamp; :back ?w1; :front ?w2.
3         ?w1 :wheelId ?wId1; :stamp ?last1. ?w2 :wheelId ?wId2; :stamp ?last2.
4         FILTER(?wId1 != ?wId2}.");
5 Int i = it.next().get("id"); //dynamic cast to Int
6 Bike bike = new Bike(i, ...);
```

Fig. 1. Dynamic data access with SPARQL.

2 Running Example

Before we formalize our approach, we give an informal example that shows the
targeted application. We do not present advanced features, such as typing and
inheritance here. The example is given in the LMOL language introduced in Sect. 6,
where the exact syntax and runtime semantics are introduced. Consider the
following domain model about bikes and an application that loads all bikes into
its class structure. At a later point, the application will then use the data to
perform some computation on the wheels. This is illustrated in Fig. 1

$$\exists \mathtt{hasWheel}.\top \sqsubseteq \mathtt{Bike} \qquad \mathtt{back}, \mathtt{front} \sqsubseteq \mathtt{hasWheel} \qquad \top \sqsubseteq \forall \mathtt{id}.\mathtt{Int}$$

$$\exists \mathtt{id}.\top \sqcup \exists \mathtt{stamp}.\top \sqsubseteq \mathtt{Bike} \sqcup \exists \mathtt{hasWheel}^{-1}.\mathtt{Bike} \qquad \top \sqsubseteq \forall \mathtt{stamp}.\mathtt{Int}$$

We observe the following: (a) the data access in line 5 requires dynamic
typing, (b) The connection between classes and data is established by a a query
that is not modular w.r.t. nested classes, and (c) we may retrieve too much data:
if the computation on a bike stops after the first wheel, then the second wheel
should be not loaded in the first place. Note that it may not be known at the
time the data is loaded which computations will be performed on it, thus, it is
not always possible to adjust the query. Even when this is the case, it leads to
the situation that several queries are manually optimized versions of the same
data retrieval.

Our idea is twofold: We (1) annotate the class declaration with a SPARQL
query to retrieve instances of the class, and (2) use futures for nested class
structures. The following shows an annotated Wheel class and how to use it.

```
1 class Wheel anchor ?o
2   (Int wheelId, Int last) //id is the id of the wheel, not the IRI
3 end retrieve SELECT ?wheelId ?last
4           WHERE{ ?o :wheelId ?wheelId; :stamp ?last. }
5 ... List<Wheel> it = load Wheel();
```

There is *no* Wheel class declared in the RDF vocabulary – the annotated query
models *retrieval*. The **load** statement returns an iterator over all results of the
query (in some preconfigured KB). Note that already here, we introduce a separation of concerns in the language: data modeling is concentrated on the class,
while computations can be performed with **load** as an encapsulation mechanism.
Indeed, it is not visible to the programmer what kind of KB access is performed.
The following code block shows the annotation for the Bike class.

```
1 class Bike anchor ?o(
2   Int id,  Int last, //id is the id of the bike, not the IRI
3   link(?o :front ?front) QFut<Wheel> front,
4   link(?o :back ?back) QFut<Wheel> back)
5 end retrieve SELECT ?id ?last WHERE { ?o :id ?id; :stamp ?last. }
6 ...
7   List<Bike> it = load Bike();  Bike bike = it[0];
8   List<Wheel> it = load bike.front;
9   Wheel w = it[0];
```

The `Wheel` instances are *not* described by the query. Their retrieval is already described in the `Wheel` class. The **link** annotation makes their loading lazy: When the **load** `Bike()` statement in line 8 is executed, only the query of `Bike` is executed. The fields of the `Wheel` class are initialized with *futures*: containers that contain the delayed query of `Wheel`. It is only in line 9, when `bike.front` is accessed, that the query below is executed. Note that it is enriched with information from the original query. This is crucial to correctly connect the nodes.

```
1 SELECT ?wheelId ?last WHERE{ run:obj1 :front ?w1.
2                              ?w1 :id ?wheelId; :last ?last. }
```

The annotations for lazy loading are illustrated in the figure to the right. The circles denote the parts of a graph retrieved by a single query and the thick edges are the **link** queries between the queries of the single classes. Lazy KB access aims to (1) make KBs more usable by reducing the load on the programmer and (2) allow a more flexible control over data loading by delaying the exploration of certain parts of the KB. The goal is not to replace all 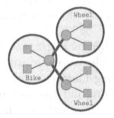 possible usages of queries and we stress that the lazy loading mechanism subtly changes the way the KB is accessed in two ways: First, by delaying the query, we take control away from the query planner and give it to the program. While this reduces the possibilities to optimize on the query level, it allows the programmer to be more flexible in its data modeling. Secondly, lazy evaluation may retrieve too many objects in the first step, as it is not known whether the next query will succeed or not. For example, the original query above will not return bikes with only one wheel, while lazy loading will do so and create a `Bike` object with one `Wheel` field evaluating to `null`. We address this by allowing to flatten the queries of nested classes to one overall query, but note that this is not possible for (mutually) recursive class structures.

3 Preliminaries

Semantic Web. We assume that the reader is familiar with the basics of established technologies of RDF KBs and SPARQL, and only repeat basic notation.

A knowledge base $\mathcal{K} = (\mathcal{T}, \mathcal{A})$ is a pair of a TBox $\mathcal{T}$ and an ABox $\mathcal{A}$. We represent the ABox as a set of triples. A query is denoted as $Q(\overline{x})$, where x ranges over variables. Given a query $Q(\overline{x})$, we say that $\overline{x}$ are its answer variables. The

$$T ::= \texttt{C} \mid \texttt{List<C>} \mid \texttt{Int} \mid \ldots \qquad Prog ::= \overline{\texttt{Class}}\ \textbf{main}\ s\ \textbf{end} \qquad \text{Types and Programs}$$

$$Class ::= \textbf{class}\ \texttt{C}\ (\overline{\texttt{T f}})\ \textbf{end} \qquad s ::= \texttt{T v:=rhs;} \mid \texttt{e.f:=rhs;} \mid s\ s \qquad \text{Classes and Statements}$$

$$rhs ::= \textbf{new}\ \texttt{C}(\overline{\texttt{f = e}}) \mid e \qquad e ::= \texttt{null} \mid \texttt{v} \mid \texttt{n} \mid \texttt{e.f} \mid \texttt{e[e]} \qquad \text{Expressions}$$

Fig. 2. Surface Syntax. The notation $\overline{\cdot\cdot}$ denotes lists, f ranges over fields, v over variables, n over literals and C over class names.

query Q may contain non-answer variables as well. If it has two answer variables, then we say that Q is binary. Given two queries Q_1, Q_2, we define their conjunction $Q_1 \wedge Q_2$ as the query that returns the intersection of answers to Q_1 and Q_2. We say that a query Q_1 is contained in another query Q_2 over a TBox $\mathcal{T}$ under an entailment regime er, written $Q_1 \sqsubseteq^{\mathcal{T}}_{er} Q_2$, if for every ABox $\mathcal{A}$ each answer to Q_1 over $(\mathcal{T}, \mathcal{A})$ under er is also an answer to Q_2 over $(\mathcal{T}, \mathcal{A})$ under er. We say that two queries are equivalent, written $Q_1 \equiv^{\mathcal{T}}_{er} Q_2$, if they contain each other. Furthermore, a query is said to be unsatisfiable under a TBox $\mathcal{T}$ if it has no possible answer under $\mathcal{T}$. Given two binary queries $Q_1(?a, ?b)$, $Q_2(?c, ?d)$ which have no variable in common, we define the concatenation $Q_1(?a, ?b) \circ Q_2(?c, ?d) = Q_3(?a, ?d) = Q_1(?a, ?x) \wedge Q_2(?x, ?d)$. We say that a binary query $Q_1(?a, ?b)$ is the inverse of another binary query $Q_2(?a, ?b)$ if $Q_1(?a, ?b) \equiv^{\mathcal{T}}_{er} Q_2(?b, ?a)$.

Programming Model. We use a minimal object-oriented programming language to illustrate our approach. While the implementation works on a full programming language with methods and additional statements, such as branching or loops, and the method itself can be adapted on top of any OO language, we give here only the minimal fragment to focus on the interaction between OO and RDF in a minimal setting. In this section, we give the basic structure of the language, while the later sections will extend it with a modular mapping (Sect. 4), inheritance (Sect. 5), and lazy loading (Sect. 6).

Definition 1 (Syntax). *The syntax of our base language is given in Fig. 2.*

A program is a set of classes, each defined by a set of fields and a name, and a main block, which is a sequence of statements. As statements we only consider assignments to fresh variables and fields, as well as object creation and sequence. For types we assume at least integers and parametric lists (to store the results of data loading). Expressions are standard, e[e'] is list access.

The runtime semantics is defined using a standard Structured Operational Semantics (SOS) [8], i.e., a set of rewrite rules on runtime configurations. A runtime configuration contains the class table, the created objects and the statement that remains to execute. A rewrite rule takes a runtime configuration and transforms it by executing the next statement.

Definition 2 (Runtime Semantics). *A configuration* conf *is defined by:*

$$\mathsf{conf} ::= (\mathcal{K}, \mathsf{CT})\ s\ (\sigma, \mathsf{obs}) \qquad \mathsf{obs} ::= \mathbf{obj}(X, C, \rho) \mid \mathsf{obs}, \mathsf{obs}$$

where $\mathcal{K}$ is a knowledge base, CT is the class table, a map from class names to the set of their fields, σ is a map from the variables in the main block to the literals or object identifiers, s is a statement or the special symbol ϵ for termination, and obs is a list of created objects. An object $\mathbf{obj}(X, C, \rho)$ has an identifier X, a class c and a store ρ that maps all fields of c to literals or object identifiers. We say that an object is well-formed if ρ respects the annotated type, i.e., maps each field to a literal of the fitting data type or to an object of the correct class.

Malformed objects can lead to undefined operations and our type system ensures *statically* that they do not occur at runtime.

An SOS-rule has the form $\mathrm{conf}_1 \to \mathrm{conf}_2$, optionally with some conditional premises that have to hold for the rule to be executable. Additionally, the function $[\![\cdot]\!]^\sigma_{\mathrm{obs}}$ evaluates an expression to a literal or object identifier, given a certain local store and a set of objects. We give only the rule for assignment of side-effect free expressions to local variables here, which we need in the following.

$$(\text{assign-local}) \quad \frac{[\![e]\!]^\sigma_{\mathrm{obs}} = l}{(\mathcal{K}, \mathsf{CT}) \; \mathsf{T} \; v := e; \; s \; (\sigma, \mathrm{obs}) \to (\mathcal{K}, \mathsf{CT}) \; s \; (\sigma[v \mapsto l], \mathrm{obs})}$$

4 Modular Loading

We extend the syntax with annotations that instantiate object instances from RDF graphs and add a statement to construct and execute the query for a class.

Definition 3 (Syntax of MOL). *The syntax of MOL is the one of Definition 1, with the rule for classes replaced, and the rule for expression extended by the following, where Q ranges over SPARQL queries and x over SPARQL variables.*

Class ::= class c anchor x ($\overline{[\mathsf{link(Q)}]? \; \mathsf{T} \; \mathsf{f}}$) end retrieve Q   rhs ::= ... | load c()

The class definition now contains a query annotated with retrieve, which maps graph patterns to object instances. Additionally, it contains an anchor variable, which must occur in the retrieve query and is used to construct queries for nested classes. Finally, the link clause of each field of class-type links the graph pattern of this class with the graph pattern of the class of the field in question.

We write anchor_C for the anchor variable of a class, query_C for its retrieve query and $\mathrm{link}_{C,f}$ for the linking query. We assume the following, easy to check, syntactical restrictions: (1) query_C has a connected graph pattern containing anchor_C and one variable $?v_f$ for each field c.f that has a data-type (2) $\mathrm{link}_{C,f}(\mathrm{anchor}_C, ?v_f)$ is a binary SPARQL query with a connected graph pattern. The set of fields f with linking queries in c is denoted $\mathrm{cf}(c)$. For our mapping, all fields of the class, as well as the fields of objects referred to are mapped to one query variable. This is not possible in general: for example, consider class Link anchor ?o (link(?o :next ?x) Link x) end retrieve ?o a :C – there is no bound on the retrieved object, which can be an arbitrarily long list, so there is no a priori bound on the number on variables for the query. We can identify classes for which we can construct a finite query as *forward-cycle-free* in their structure.

A class is forward-cycle-free, if all cycles caused by link clauses between classes can be resolved by considering one of the clauses as the inverse of the others. Forward-cycle-freedom is only needed for eager queries.

Definition 4 (Retrieval Trees and Forward-Cycle-Free Classes). *Let C be the set of classes in a program P and $\mathfrak{G}(P) = (C, L)$ be its retrieval tree with edges $L \subseteq C \times Q \times C$. An edge $(C_1, \mathbb{Q}, C_2)$ is part of L, if C_1 has a field of type C_2 or* `List<`C_2`>` *with link query $\mathbb{Q}$. Let $\mathfrak{G}(P, \mathbb{C})$ be the subgraph of $\mathfrak{G}(P)$ that is reachable from* $\mathbb{C}$.

Given a cycle $e_1, \ldots, e_n$, we say that edge e_1 is backwards *if $\mathbb{Q}_1$ is the inverse of $\mathbb{Q}_2 \circ \cdots \circ \mathbb{Q}_n$.[1] We say that $\mathbb{C}$ is* forward-cycle-free *if every cycle in $\mathfrak{G}(P, \mathbb{C})$ contains a backwards edge that does not originate in $\mathbb{C}$, and removal of all the backwards edges turns $\mathfrak{G}(P, \mathbb{C})$ into a directed acyclic graph, and does not make any node unreachable from $\mathbb{C}$. We denote this graph with $\mathfrak{R}(P, \mathbb{C})$.*

We omit the P parameter if the program is understood. We can now define the eager queries for forward-cycle-free classes.

Definition 5 (Eager Queries for Forward-Cycle-Free Classes). *The eager query* $\mathsf{eq}(\mathbb{c})$ *of a forward-cycle-free class $\mathbb{c}$ is defined as follows. We set*

$$\mathsf{eq}(\mathbb{c}) = \mathsf{query}_\mathbb{C} \wedge \bigwedge_{(\mathtt{f},\mathtt{D}) \in \mathtt{cf}(\mathbb{C})} \left(\mathsf{link}_{\mathbb{C},\mathtt{f}}[?v_\mathtt{f} \setminus v_{\mathtt{f},\mathtt{D}}] \wedge \mathsf{eq}(\mathtt{D})[\mathsf{anchor}_\mathtt{D} \setminus v_{\mathtt{f},\mathtt{D}}] \right)$$

where all variables $v_{\mathtt{f},\mathtt{D}}$ are fresh, i.e., do not occur anywhere else. Additionally, we get a set of equalities for the form `this.e` $= x$ *for each backwards edge removed in $\mathfrak{R}(\mathbb{c})$, that maps the field of the source of the backlink to the query variable that is used for the target node of the backwards link.*

Example 1. The eager query of class `Bike` in Sect. 2 results in a query that is transformed into the one of Fig. 1 by substituting the nested queries. To illustrate forward-cycle-free classes, consider the following variant of the `Wheel` class.

```
1 class Wheel anchor ?o(
2   Int wheelId, Int last,
3   link(?bike :wheel ?o) Bike bike)
4 end retrieve SELECT ?wheelId ?last WHERE{?o :wheelId ?wheelId; :stamp ?last.}
```

Here, the `Wheel` class has a link to a `Bike` instance. However, analyzing the link clauses, we can see that this is a backwards edge: `?bike` is always instantiated with the anchor of the outer query when loading `Bike`. Thus, the additional equalities are `this.front.bike` $=?o$ and `this.back.bike` $=?o$.

---

[1] We assume that we can reorder the cycle so that the potential backwards edge is always as index 1. We remind that the anchor variable is always the first answer variable of a link query, so the concatenation is indeed well-defined.

After constructing the query, it remains to show how we construct an object from a query result. This is done by constructing a store from the variables of the query that are mapped back to their fields. The construction is straightforward, but technically intricate. For the sake of readability, we give it in the technical report [9] and define the signature here.

Definition 6. *Let* c *be a class and* RS *a set of answers to its eager query. We denote with* rs2ob(RS, c) *the objects (cf. Definition 2) created when instantiating* RS.

If neither c nor any class in $\Re(c)$ has a field of list type, then each answer rs $\in$ RS corresponds to one instantiated object. To define rs2ob(RS, c), we only need to map back from query variables to the field they are associated with and apply the equations generated during query construction for all fields that have no query associated with them. Every class has implicitly a field string uri that is instantiated with the URI of the node mapped from the anchor variable. We require that no blank nodes are mapped to anchors.

To connect with the runtime semantics, we add a rule for **load** statements that connect query construction and subsequent object instantiation.

Definition 7 (Runtime Semantics). *The runtime semantics of* MOL *is the one of Definition 2, extended with the following rule:*

$$\text{(eager)} \frac{\text{obs}', X = \text{listOf}(\text{obs}'') \quad \text{obs}'' = \text{rs2ob}(\text{RS}, c) \quad \text{RS} = \text{ans}(\mathcal{K}, \text{eq}(C))}{(\mathcal{K}, \text{CT})v := \textbf{load } c(); \ s(\sigma, \text{obs}) \rightarrow (\mathcal{K}, \text{CT})v := X; \ s(\sigma, \text{obs}, \text{obs}', \text{obs}'')}$$

The rule executes the query in the first premise. It then creates objects for all results (obj'') and stores them in a list via listOf. The listOf function returns a pair of objects implementing the list (obj') and the name of the head object of the list (X). The **load** statement is then reduced to an assignment of this head object to the target variable.

We can ensure that all loaded objects are well-formed, i.e., respect the declared types of their field, by checking whether each query variable respects each declared type. To do so, we check whether the query restricted to this variable is contained in the query retrieving all elements of the declared type, respective its OWL equivalent. The TBox is used to approximate the data statically. The proof follows directly from the typing theorem for semantically lifted programs [4].

Theorem 1 (Safety). *Let* c *be a class. Let* $\varphi = \text{eq}(c)$ *be its eager query and* V *the set of variables within* φ *that correspond to data-typed fields. If for each* v *with data type* D, *the query containment* $\varphi(v) \sqsubseteq_{er}^{\mathcal{T}} D$ *holds, then each object created from a result from a KB respecting* $\mathcal{T}$ *of* eq(c) *is well-formed.*

Finally, we investigate the case where one node occurs in different results and, thus, corresponds to multiple constructed objects. Consider, e.g., the class **class** C **anchor** ?o Int i; **retrieve** ?o P ?i and the data set o1 P 1. o1 P 2.. This touches a core aspect of the relation between objects, nodes and queries: is an object a container for the results or is it in a bijective relation with some RDF-class? Instead of forcing the developer down one of these roads, we can characterize the situations and provide feedback about the annotated mapping.

Theorem 2 (Bijective Instantiation for Essentially Functional Classes). *A forward-cycle-free class* C *is* essentially functional *in a program* P, *if for all paths in* $\mathfrak{R}(P, C)$ *starting in* C, *the concatenation of the labeling queries is functional, and all data properties of reachable classes are functional.*

In the list retrieved from the retrieval query of an essentially functional class, then there is only one object per node in the answers for the outermost anchor.

The bijection is established for one execution of one query, not globally. As we are only interested in safe and modular loading, we also do not change the knowledge graph if the instances are manipulated by the program. Similarly, new objects created with **new** are not written into the KB. However, as we will see later, the combination with semantic lifting [10] allows us to write as well.

5   Inheritance

We have so far neglected a core element of class-oriented programming: inheritance. Inheritance is, besides the RDFS meta model that defines rdfs:Class recursively and uses meta-classes, one of the critical points where OO and OWL class models diverge. Most programming languages forbid, or at least restrict, multiple inheritance, especially diamond inheritance, which causes problems for *methods*. Consequently, in OO, one object cannot be an element of several classes which are not subclasses of each other. It is, thus, out of question to try to reconcile the class model of Java-like languages with the class model of OWL. In this section, we extend our programming language to handle inheritance and give two static analyses that catch modeling errors in the retrieving queries.

Definition 8 (Syntax of MOL$^+$). *The syntax of* MOL$^+$ *is based on the syntax of* MOL, *with the definition of classes in Definition 3 replaced by the following:*

$$\mathsf{Class} ::= \mathbf{class}\ \mathsf{C}\ [\mathbf{extends}\ \mathsf{C}]?\ \mathbf{anchor}\ \mathsf{X}\ (\overline{[\mathsf{link}(\mathsf{R})]?\ \mathsf{T}\ \mathsf{f}})\ \mathbf{end}\ \mathbf{retrieve}\ \mathsf{Q}$$

The only change of the syntax is the addition of the **extends** clause. Semantically, the only change is the generation of the class table CT, which for any class now also includes the fields of all its superclasses. Thus, query(c) must have the variable for the fields of all its superclasses as well. If D has a clause **extends** C, then we write $\mathsf{D} \leq \mathsf{C}$. For the transitive closure, we write $\leq^*$.

Retrieval Query. Retrieval for a class that has subclasses must respect these subclasses and construct the most specific class, not necessarily the general one written in the program. For example, consider the following program and ABox.

```
1  class C anchor ?o () end retrieve {?o a :Q}
2  class D anchor ?o extends C (Int j)
3     end retrieve {?o a :Q. ?o :j ?j.}
4  class E anchor ?o extends C (Int k)
5     end retrieve {?o a :Q. ?o :k ?k.}
```

obj1 a :Q,   obj1 :j 1,

obj2 a :Q,   obj2 :k 1,

When executing **load** C(), one would expect that obj1 is loaded as a D instance, because we can retrieve data for the j field, and, analogously obj2 as a E instance. Running the C query, however does not detect this – it is necessary to adapt our expansion of the retrieval query. Intuitively, we run the queries of the subclasses in OPTIONAL clauses and check during object construction whether a given subclass can be instantiated, by checking whether all *variables* belonging to this subclass are instantiated. The idea is to put the additional fields in optional clauses of the query and check whether they are instantiated. If they are, one can downcast the created object.

Definition 9 (Runtime Semantics of MOL⁺). *Let* C *be a class with subclasses* $(D_i)_{i \in I}$. *We define the query* inheq(c) = query(c) $\wedge \bigwedge_{i \in I}$ OPTIONAL(inheq(D_i)). *The object* rs2ob⁺(rs, c) *retrieved from a result* rs *of* inheq(c) *is constructed as follows. Let* $(E_i)_{i \in J} = \{E \leq^* C\}$ *be the set of subclasses of* C, *such that all fields of* E_i *correspond to an assigned variable in* rs.

rs2ob⁺(rs, c) = rs2ob(rs, E) *for some* E $\in (E_i)_{i \in J}$ *where* rs2ob(rs, E) *is defined.*

The runtime semantics of MOL⁺ *is the one of* LMOL, *except that every occurrence of* rs2ob *is replaced by* rs2ob⁺, *and every occurrence of* query *by* inheq.

Two unintuitive effects can occur during the retrieval. First, the object instantiation is nondeterministic, and second, it may violate behavioral subtyping.

Unique Retrieval. Nondeterminism occurs if two optionals corresponding to unrelated classes are instantiated. E.g., the ABox {o3 a :Q; :j 1; :k 1.} used with the above code. Object o3 can be retrieved as *both* D and E, but in our class model it is not possible for an object to be both. There are four solutions: (1) Introduce a new language mechanism that allows objects to have the fields of two classes without a subtyping relation between them. (2) Define a preference relation, e.g., say that instantiating D is always preferred over instantiating E. (3) Retrieve multiple objects, one for each possible instantiation. (4) Take the least specific class of all possible instantiations, i.e., here: C.

We deem (1) as unpractical as it changes the programming language according to a specific use case and therefore counteracting our aim of easy to use integration of RDF into OO. Similarly, we deem (2) as unpractical as it questionable if such a preference relation is sensible in many applications. Solution (3) leads to a one-to-many relation between loaded objects and anchored nodes, which we consider undesirable. We, thus, use (4), but give a static analysis that detects the situation where this design decision may play a role: If the conjunction of the queries for each pair of subclass is unsatisfiable, then the most specific constructed class is uniquely determined.

Theorem 3 (Unique Retrieval). *Let C be a class with subclasses $(D_i)_{i \in I}$. If for all D_j, D_k with $j, k \in I, j \neq k$ the query* query(D_j) $\wedge$ query(D_k) *is unsatisfiable, then the function* rs2ob$^+$(rs, C) *in Definition 9 is deterministic.*

If the check fails, we can precisely give feedback which two clauses overlap and give the programmer detailed feedback where the mapping between RDF and the MOL$^+$ class models fails.

Behavioral Subtyping. Given a class C with a direct subclass D, we must relate their **retrieve** clauses, such that each retrieved object is also a C object. This is essentially a case of the Liskov principle [5] of behavioral subtyping: if a property $\varphi(x)$ holds for all instances x of class C, then $\varphi(y)$ must also hold for all instances y of all subclasses of C. In our case, the properties in question are all data class invariants over the fields of the superclass. For example, consider the above example again, but with the definition of D changed to the following:

```
1 class D anchor ?o extends C (Int j) end retrieve ... {?o a :R. ?o :j ?j.}
```

Here, if :R is not a sub-class of :Q (in the sense of RDF), then running the query for D will retrieve objects that are not retrieved by the query for C *even when restricted to the fields of* C.

The Liskov principle for MOL$^+$ is reducible to query containment. Given a TBox $\mathcal{T}$, we check, if for all KBs respecting some TBox $\mathcal{T}$ and for all classes, C, D, with $D \leq C$, the query of the subclass does not add new instances when restricted to the fields of the superclass.

Theorem 4 (Behavioral Subtyping for MOL$^+$). *Let* C, D *be two classes* $D \leq C$ *and $\bar{f}_C$ is the set of fields in the superclass. If for each such pair of classes the query containment* eq(C)($\bar{f}_C$) $\sqsubseteq^{\mathcal{T}}$ eq(D)($\bar{f}_C$) *holds, then*

$$\{\text{rs2ob}(\text{rs}, C) \mid \text{rs} \in \text{ans}(\mathcal{K}, \text{eq}(C))\} \subseteq \{\text{rs2ob}(\text{rs}, C) \mid \text{rs} \in \text{ans}(\mathcal{K}, \text{eq}(D))\}$$

6 Lazy Loading

We now extend MOL$^+$ to LMOL by introducing a lazy loading mechanism. Lazy loading splits the eager query into several subqueries, of which some are delayed and only executed on demand, as has advantages for usability and performance.

First, it gives the programmer very precise control over the used data. Instead of loading all possible data that *may* be used, it enables to load data as it is indeed used. In our running example, it may depend on the data loaded for the Bike instance whether the front or back wheel must be investigated. This condition may not be (easily) encodable in the query, or indeed not be known upfront and depend on user input or data loaded from other sources. For example, the following program accesses three bikes, but only *two* wheels: the front wheel of the second result and the back wheel of the third one.

```
List<Bike> l := load Bike(); l[0].id; l[1].front.id; l[2].back.id
```

It is easy to see how in a more complex language one can decide which wheel to access based on prior data. Lazy loading can thus solve the problem of loading data that is not required in the application.

Second, lazy loading decouples modeling the data mapping from writing the query for a specific optimization: the programmer can be more generous in data modeling, as the specialization to a specific computation occurs at runtime.

Syntactically, we add futures to the types and a statement to resolve them.

Definition 10 (Syntax of LMOL). *The syntax of* LMOL *is the syntax of* MOL$^+$ *of Definition 8, with the following extensions for types and right-hand-side expressions:*

$$\text{T} ::= \ldots \mid \text{QFut<C>} \qquad \text{rhs} ::= \ldots \mid \textsf{load } e$$

Intuitively, a future is a delayed expression, in our case a query. We use explicit futures here [11] that must be explicitly resolved. For resolving, we reuse the **load** keyword: a **load** e expression takes an expression e of QFut<C> type and returns an expression of List<C> type – the results of executing the delayed query.[2] A future field cannot be a backwards edge.

Next, we augment the runtime with the required elements for futures. *Anchor maps* keep track of the variable instantiations of the previously executed queries.

Definition 11 (Runtime Configurations of LMOL and Lazy Queries). *Runtime configurations are extended as follows. Let* F *be the set of future identifiers, a subset of the object identifiers, and* A *the set of anchor map identifiers, which is disjoint to the set of object identifiers. Let* Q *range over SPARQL queries and object identifiers.*

$$\text{conf} ::= (\mathcal{K}, \text{CT}) \; s \; (\sigma, \text{obs?}, \text{futs?}, \text{acs?})$$
$$\text{futs} ::= \textbf{fut}(F, A, \text{Q}) \mid \text{futs futs} \qquad \text{acs} ::= \textbf{ac}(A, \mathcal{A}) \mid \text{acs acs}$$

where $\mathcal{A}$ *are maps from RDF literals to object identifiers. Additionally,* σ *and all stores* ρ *may map to future identifiers* F*. The lazy query of a class is defined analogous to its eager query, except that none of the queries of the fields with future type are executed. Only the linking queries are executed.*

$$\text{lq(c)} = \textsf{query}_\text{C} \wedge \bigwedge_{(\textf{f,D}) \in \textsf{cf(C)}} \left(\textsf{link}_{\textsf{C,f}}[?v_\textf{f} \setminus v_{\textf{f,D}}] \right)$$

Object instantiation is analogous and described in the technical report. The main differences are that (1) the fields of future type are instantiated with runtime futures, whose query is the query of the class enhanced with the link query where the anchor variable is replaced by its instantiation, and that (2) an anchor map is used as an additional parameter. The anchor map keeps track of *all* instantiations

---

[2] We refrain from introducing (a) expressions for resolved futures and (b) lazy loading of the result list. Both is standard and orthogonal to lazy loading *within* one result.

so far. Instantiation for lazy class loading $\text{lrs2ob}(\text{RS}, \text{c})$ thus takes an answer set RS and the class c as input and returns a set of objects, a set of futures and an anchor map. Lazy instantiation for the inner queries, $\text{llrs2ob}(\text{RS}, \text{c}, \mathcal{A})$ takes additionally an anchor map as input. It returns a set of objects and a set of futures, as well as a modified anchor map with added bindings.

Definition 12 (Runtime Semantics of LMOL). *The rule for* load $c()$ *is almost the same as* (eager), *except that we use* lq *instead of* eq. *It is given in the technical report. The rule for* load e *is as follows:*

$$
\text{(lazye)} \quad \frac{\llbracket e \rrbracket^{\sigma}_{\text{obs}} = F \quad RS = \text{ans}(\mathcal{K}, Q) \quad \text{obs}', X = \text{listOf}(\text{obs}'') \quad \text{obs}'', \text{futs}', \text{acs}' = \text{llrs2ob}(RS, c, \mathcal{A})}{\begin{array}{c} (\mathcal{K}, \text{CT})\text{v} := \text{load } e; \ s(\sigma, \text{obs}, \text{futs}, \textbf{fut}(F, A, Q), \text{acs}, \textbf{ac}(A, \mathcal{A})) \\ \rightarrow (\mathcal{K}, \text{CT})\text{v} := X; \ s(\sigma, \text{obs}, \text{obs}', \text{obs}'', \text{futs}, \text{futs}', \text{acs}, \text{acs}') \end{array}}
$$

The rules work analogously: First, the lazy query (either by constructing it for the class, or by reading it from the future) is executed. Then objects and futures are instantiated for its results and the anchor map is updated. Finally, the objects are stored in a list and all created constructs are added to the state.

To be clear, we do not save the result when resolving a future, so a future might be resolved multiple times. The mechanism to avoid this is straightforward [6] and would only obfuscate our contribution.

Theorem 5. *For forward-cycle-free classes* C, *such that every class in* $\mathfrak{R}(P, \text{C})$ *has only data-type and list-type fields,* MOL^+ *and* LMOL *load the same results, i.e., if all futures are resolved, then the objects in the list of the first* load $c()$ *contain the same elements, except with one* **fut** *reference in every list field.*

For general classes, LMOL can load more or less data. As an example for more data, consider that futures can be used to encode streams [12], and thereby load an unbounded number of objects within *one* result of the overall query. For less data, consider a knowledge base where the bike instances have no stored wheels. Executing the eager query will return no bikes, but executing the lazy query will return all the bikes, with empty lists for their wheels.

7 Evaluation

The on-going implementation is available as open-source software. LMOL is implemented as an interpreter that takes a LMOL file, and a RDF file for external data as input. The interpreter, including the experiments described below, is available under https://github.com/Edkamb/SemanticObjects/tree/lazy.

To assess the runtime overhead and memory consumption caused by delaying parts of the query through futures, which reduce the possibilities of the DBMS for query optimization, we run the following experiment. We generate n classes of the form **class** $C_i(\text{Int } f_i, C_{i+1} \text{ next})$ **end** with the obvious link and retrieval query, as well as a lazy version of the same form with fields $\texttt{QFut<}C_{i+1}\texttt{>}$ next. We consider two scenarios: scenario 1 uses a dataset with two loadable nodes a_i, b_i of each class c_i, with links from a_i to a_{i+1} and b_i to b_{i+1}, making two paths of length n. Scenario 2 has additional links from a_i to b_{i+1} and b_i to a_{i+1}, leading to 2^n possible paths. We evaluate how fast it is to load every possible node into the OO structure and compare three runs for each scenario: one with the eager query, and two with the lazy query, which uses the delay to remove duplicate objects based on their URI before running the next query: The first lazy run accesses only the first class, the second accesses the last class. Figure 3 shows the results. As expected, there is no performance gain for the simple scenario, but a considerable one for the one with more interconnected data, especially when only parts of the data is accessed. For a class chain of $n = 18$, eager evaluation requires 18 s, while lazy evaluation with fine-grained control requires 2.6s if only half of the chain is accessed (-85%). For higher n, eager evaluation runs out of heap space. Memory consumption behaves, as expected, like the time consumption.

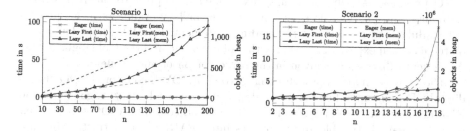

Fig. 3. Runtime comparison of eager loading and lazy loading with fine-grained control.

To confirm that our approach indeed simplifies the design of realistic queries, we remodeled the queries used to access the Slegge database of Equinor of subsurface exploration data [13]. The aim of this is to show that our system enables reuse and forward-cycle-free classes are not a strong restriction. We remodeled the 8 SPARQL queries for the main, first information need of Slegge need using LMOL in a class structure of 21 classes, each corresponding to a reusable pattern or original query. The code below is a representative excerpt. The eager queries for WBQ1 and WBQ3 correspond to one of the original Slegge queries, which all retrieve wellbores based on different criteria. WBQ1 and WBQ3 only differ in restrictions on the interval, encoded in different classes (SZ13 and DepthSZ). All other parts of the queries, e.g. the wellbore name, are shared through common superclasses.

```
1  class WBWithName extends WB anchor ?w ( String name )
2    end retrieve "?w a :Wellbore. ?w :name ?name."
3  class WBQ1 extends WBWithName anchor ?w (
4    link("?w :wellboreInterval ?int.") SZ13 int, ..) end ..
5  class WBQ3 extends WBWithName anchor ?w (
6    link("?w :wellboreInterval ?int.") DepthSZ int, ..) end ..
```

8   Related Work

LMOL is implemented on top of the semantic lifting language SMOL. Semantic lifting [10] also integrates OO and RDF: It exports a program state and allows to query it then, thus realizing data writes through state change. Consequently, an RDF graph can be changed in any way using LMOL/SMOL.

LMOL is the first language with modular queries and lazy evaluation for RDF data. In the following, we discuss other approaches that connect OO programming and RDF data. Frameworks like Apache Jena [14] or RDF4J [15] connect OO programming and RDF as well, but do not connect the object *models*.

The impedance mismatch for relational databases has been extensively studied for several decades [1] and we refer to Ireland [16] for a comprehensive discussion. It is worth noting that one of the systems to connect OO with relational databases, the LINQ [17] framework for .Net, has been extended to RDF [18]. LINQ provides its own query language, which is mapped to different storage endpoints, and provides no safety mechanisms or lazy evaluation. The query is not modular and provided at the loading statement. For RDF, impedance mismatch has been explored, starting with Goldman [19] and the Go! language [20]. An in-depth discussion is given in the survey of Baset and Stoffel [3].

The most common approach taken is to relate OWL concepts to OO classes directly. For example, Leinberger et al. [21] use a special query language to load RDF data into an OO language by relating OWL concepts to OO types. Their query language is typable and translates into SPARQL. In contrast, LMOL supports full SPARQL and does not require the user to learn an additional query language. Owl2Java [22], Agogo [23] or ActiveRDF [24] are similar approaches, suffering from the impedance mismatch. They all generate OO classes based on some RDF schema for a certain target language and establish a direct connection this way, where only certain RDF schemata are allowed. In contrast, we give a way to detect whether the defined mapping establishes a one-to-one correspondence in the results of one query (Theorem 2) to help the programmer.

While our approach embraces the semantic gap between OO and RDF object models, and the approaches so far attempt to bridge it, Eisenberg and Kanza [2] attempt a unification in a programming model that treats RDF individuals as program primitives. This essentially imports the RDF object model into the programming language, and the authors do not discuss typing. Indeed, they present their approach for Ruby, with a loose object model using dynamic duck typing.

As for type checking, Seifer et al. [25] give a system where DL concepts are types and that requires to type check the SPARQL query itself. In contrast, the entailment-based system we use is more modular: we do not have any restrictions on the used SPARQL subset. Furthermore, we do not entangle type checking in the impedance mismatch by mixing concepts and types.

Leinberger [26] gives an extended type system based on SHACL and shape containment, instead of SPARQL and query containment. That approach focuses on ensuring the existence of data that is loaded and is neither modular nor supports lazy evaluation, as the semantic gap is only considered at the interface.

9 Conclusion

We have presented a connection between RDF and OO programming that maintains and embraces the semantic gap between the underlying inheritance mechanisms and object models, and allows modular data modeling, data access and type safety checks. Our approach is the first to explore lazy evaluation of SPARQL queries within a single answer and the first to formally define a Liskov principle for a connection of OO and RDF. Our evaluation shows that lazy evaluation leads to significant performance gains for loading of complex data.

We plan to generalize our prototype to a static tool for a mainstream programming language and increase its expressive power by (1) using a `Option<C>` type that maps to optionals in the retrieval query and (2) allowing the **load** statement to take an additional parameter that defines additional constraints.

Acknowledgments. This work was supported by the RCN via PeTWIN (294600). The authors thank Dirk Walther for motivating this work and the anonymous reviewers for the constructive feedback.

References

1. Copeland, G.P., Maier, D.: Making smalltalk a database system. In: Yormark, B. (ed.) SIGMOD, pp. 316–325. ACM Press (1984)
2. Eisenberg, V., Kanza, Y.: Ruby on semantic web. In: ICDE, pp. 1324–1327. IEEE Computer Society (2011)
3. Baset, S., Stoffel, K.: Object-oriented modeling with ontologies around: a survey of existing approaches. Int. J. Softw. Eng. Knowl. Eng. **28**(11–12), 1775–1794 (2018)
4. Kamburjan, E., Kostylev, E.V.: Type checking semantically lifted programs via query containment under entailment regimes. In: Description Logics, volume 2954 of CEUR Workshop Proceedings. CEUR-WS.org (2021)
5. Liskov, B., Wing, J.M.: A behavioral notion of subtyping. ACM Trans. Program. Lang. Syst. **16**(6), 1811–1841 (1994)
6. Halstead Jr., R.H.: MULTILISP: A language for concurrent symbolic computation. ACM Trans. Program. Lang. Syst. **7**(4), 501–538 (1985)
7. Baker, H.G., Hewitt, C.: The incremental garbage collection of processes. In: Low, J. (ed.) Proceedings of the 1977 Symposium on Artificial Intelligence and Programming Languages, USA, 15–17 August 1977, pp. 55–59. ACM (1977)
8. Plotkin, G.: A structural approach to operational semantics. J. Log. Algebr. Program. 60–61 (2004)
9. Kamburjan, E., Klungre, V.N., Giese, M.: Never mind the semantic gap: modular, lazy and safe loading of RDF data (technical report). Research report 502, Department of Informatics, University of Oslo, March 2022
10. Kamburjan, E., Klungre, V.N., Schlatte, R., Johnsen, E.B., Giese, M.: Programming and debugging with semantically lifted states. In: Verborgh, R., et al. (eds.) ESWC 2021. LNCS, vol. 12731, pp. 126–142. Springer, Cham (2021). https://doi.org/10.1007/978-3-030-77385-4_8
11. de Boer, F.S., et al.: A survey of active object languages. ACM Comput. Surv. **50**(5), 76:1–76:39 (2017)

12. Azadbakht, K., de Boer, F.S., Bezirgiannis, N., de Vink, E.P.: A formal actor-based model for streaming the future. Sci. Comput. Program. **186**, 102341 (2020)
13. Hovland, D., Kontchakov, R., Skjæveland, M.G., Waaler, A., Zakharyaschev, M.: Ontology-based data access to Slegge. In: d'Amato, C., et al. (eds.) ISWC 2017. LNCS, vol. 10588, pp. 120–129. Springer, Cham (2017). https://doi.org/10.1007/978-3-319-68204-4_12
14. Apache Foundation. Apache Jena. https://jena.apache.org/
15. Eclipse Foundation. Eclipse RDF4J. https://rdf4j.org/
16. Ireland, C.J.: Object-relational impedance mismatch: a framework based approach. Ph.D. thesis, Open University, Milton Keynes, UK (2011)
17. Meijer, E., Beckman, B., Bierman, G.M.: LINQ: reconciling object, relations and XML in the .net framework. In: SIGMOD, p. 706. ACM (2006)
18. Matthew, A.: LINQtoRDF (2006). https://code.google.com/archive/p/linqtordf/
19. Goldman, N.M.: Ontology-oriented programming: static typing for the inconsistent programmer. In: Fensel, D., Sycara, K., Mylopoulos, J. (eds.) ISWC 2003. LNCS, vol. 2870, pp. 850–865. Springer, Heidelberg (2003). https://doi.org/10.1007/978-3-540-39718-2_54
20. Clark, K.L., McCabe, F.G.: Ontology oriented programming in go! Appl. Intell. **24**(3), 189–204 (2006)
21. Leinberger, M., Scheglmann, S., Lämmel, R., Staab, S., Thimm, M., Viegas, E.: Semantic web application development with LITEQ. In: Mika, P., et al. (eds.) ISWC 2014. LNCS, vol. 8797, pp. 212–227. Springer, Cham (2014). https://doi.org/10.1007/978-3-319-11915-1_14
22. Kalyanpur, A., Pastor, D.J., Battle, S., Padget, J.A.: Automatic mapping of OWL ontologies into Java. In: SEKE, pp. 98–103 (2004)
23. Parreiras, F.S., Saathoff, C., Walter, T., Franz, T., Staab, S.: APIs à gogo: automatic generation of ontology APIs. In: ICSC, pp. 342–348. IEEE Computer Society (2009)
24. Oren, E., Heitmann, B., Decker, S.: ActiveRDF: embedding semantic web data into object-oriented languages. J. Web Semant. **6**(3), 191–202 (2008)
25. Seifer, P., Leinberger, M., Lämmel, R., Staab, S.: Semantic query integration with reason. Art Sci. Eng. Program. **3**(3), 13 (2019)
26. Leinberger, M.: Type-safe programming for the semantic web. Ph.D. thesis, University of Koblenz and Landau, Germany (2021)

Improving Question Answering Quality Through Language Feature-Based SPARQL Query Candidate Validation

Aleksandr Gashkov[1], Aleksandr Perevalov[2,3], Maria Eltsova[1],
and Andreas Both[3,4(✉)]

[1] Perm National Research Polytechnic University, Perm, Russia
[2] Anhalt University of Applied Sciences, Köthen, Germany
[3] Leipzig University of Applied Sciences, Leipzig, Germany
andreas.both@htwk-leipzig.de
[4] DATEV eG, Nuremberg, Germany

Abstract. Question Answering systems are on the rise and on their way to become one of the standard user interfaces. However, in conversational user interfaces, the information quantity needs to be kept low as users expect a limited number of precise answers (often it is 1) – similar to human-human communication. The acceptable number of answers in a result list is a key differentiator from search engines where showing more answers (10–100) to the user is widely accepted. Hence, the quality of Question Answering is crucial for the wide acceptance of such systems. The adaptation of natural-language user interfaces for satisfying the information needs of humans requires high-quality and not-redundant answers. However, providing compact and correct answers to the users' questions is a challenging task. In this paper, we consider a certain class of Question Answering systems that work over Knowledge Graphs. We developed a system-agnostic approach for optimizing the ranked lists of SPARQL query candidates produced by the Knowledge Graph Question Answering system that are used to retrieve an answer to a given question. We call this a SPARQL query validation process. For the evaluation of our approach, we used two well-known Knowledge Graph Question Answering benchmarks. Our results show a significant improvement in the Question Answering quality. As the approach is system-agnostic, it can be applied to any Knowledge Graph Question Answering system that produces query candidates.

Keywords: Question Answering over Knowledge Graphs · Query Validation · Query Candidate Filtering

1 Introduction

The Web has become the major knowledge source for many people worldwide. While aiming at efficient knowledge modeling and representation, the Semantic

---

A. Gashkov and A. Perevalov—Shared first authorship–these authors contributed equally to this work.

© The Author(s), under exclusive license to Springer Nature Switzerland AG 2022
P. Groth et al. (Eds.): ESWC 2022, LNCS 13261, pp. 217–235, 2022.
https://doi.org/10.1007/978-3-031-06981-9_13

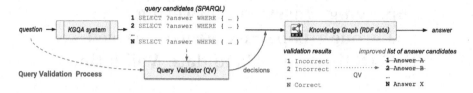

Fig. 1. General overview of the Query Validation process. The core component is the *Query Validator* intended to filter incorrect query candidates.

Web initiative was proposed and is permanently growing. The objective of this initiative is to make Web data machine-readable and machine-understandable by describing concepts, entities, and relations between them [6]. Hence, the Semantic Web may be considered as a giant Knowledge Graph (KG). In this regard, Knowledge Graph Question Answering (KGQA) systems are actively developing already for more than a decade [14,16]. These systems are bridging the gap between Linked Data and end-users by transforming natural-language (NL) questions into structured queries (e.g., represented as SPARQL[1]) to make the information accessible using NL requests.

When answering a question, KGQA systems often generate a ranked list of SPARQL queries that are considered to be capable of retrieving answers to a given question. Thereafter, a ranked Top-N of the retrieved answers is shown to the end-users (often N is 1). Thus, a *Query Candidate* is a SPARQL query generated by a KGQA system to retrieve data. An *Answer Candidate* is a result of a SPARQL Query Candidate execution which is proposed as a possible answer to a user. In this paper, we propose a *Query Validation* (QV) process that is intended to remove all queries that cannot be resolved to a correct answer from a query candidates list. This helps to reduce the number of incorrect query candidates (and therefore, answers) in the output and move the correct ones to the top of the list. In addition, unanswerable questions should be recognized, and, hence, an empty output should be presented for such questions (s.t., users are not confronted with incorrect/guessed answers).

The field of Answer Validation (AV) is well-researched for information retrieval (IR) and IR-based question answering (QA) and many approaches were proposed in the last decade [1,5,21,22,30,39,47]. However, there is just a very limited number of studies on AV and QV in the context of KGQA systems (e.g., [9,27]).

In this work, we propose a new system-agnostic QV approach that can determine whether a query candidate produced by a KGQA system is correct or not without executing the SPARQL query (see Fig. 1). The approach uses a straightforward process of converting a query candidate to NL representation and a fine-tuned classifier [11] to distinguish between correct and incorrect query candidates and, therefore, the answers. In Fig. 1, the process is visualized. To demonstrate the efficiency of the proposed QV approach, we utilize several well-known QA

---

[1] https://www.w3.org/TR/rdf-sparql-query/.

quality metrics, such as Precision@k and NDCG@k (Normalized Discounted Cumulative Gain) [37]. To tackle the dilemma of how to address the difference between "guessing" answers versus providing an empty result, we introduce a new integral metric that takes into account correct, incorrect, and also empty answer sets. In addition, we consider the quality w.r.t. the unanswerable questions and the influence of our approach on the corresponding results. Given our experimental results on one KGQA system, QAnswer [12], and two benchmarking datasets, LC-QuAD 2.0 [17] and RuBQ 2.0 [35], we demonstrate that the QV approach provides a relative improvement of Precision@1 up to 204.6% (see Table 2) as well as it is improving other metrics significantly. Moreover, the approach enabled us to obtain almost 50% of correct answers for the unanswerable questions.

To increase the reproducibility of our work, we performed evaluation of experiments with the Gerbil [43] system that provides standardized shareable links to the experiments for KGQA systems. We provide the links to the Gerbil experiments, source code, and the experimental data[2] (our experimental data is also shared as an RDF Turtle dataset) as an online appendix. This paper is structured as follows. In the next section, the related work is presented followed by Sect. 3 which introduces our approach in detail. Section 4 highlights the used QA system, QV component, datasets and data preparation process. We describe in Sect. 5 how experiments were processed in general. Section 6 estimates the quality of the query validator as well as the impact of the QV process on the QA quality. Section 7 concludes the paper and outlines future work.

2   Related Work

Techniques that tackle the task of validating the answer were applied mainly in IR-based QA, which we mentioned in Sect. 1. IR-based QA systems are often required to rank huge amounts of candidate answers [26], e.g., the incorrect answer candidates in form of textual paragraphs have to be eliminated by the AV module. In [34], e.g., the AV process is performed on the basis of Expected Answer Type, Named Entities Presence, and Acronym Checking (only if a question is about an acronym). The authors mention that sometimes AV module is "too strict", i.e., it removes also correct answers.

However, the AV and QV processes in KGQA are not well investigated in the research community. Our previous paper [19] describes the novel approach for improving the QA quality where answer candidates are filtered just by evaluating the NL input (i.e., the user's question) and output (i.e., the system's answer), accordingly, it is a system-agnostic approach. Nevertheless, it requires well-formed NL answers that are hard to compute automatically.

On the other hand, there appeared recently some approaches to semantic parsing by treating it as a problem of semantic graph generation and re-ranking [27,31,45,46]. While Yih et al. [45] introduce grounded query graph candidates using a staged heuristic search algorithm and employs a neural ranking model for

---

[2] https://doi.org/10.6084/m9.figshare.19434515.

scoring and finding the optimal semantic graph, Yu et al. [46] utilize a hierarchical representation of KG predicates in their neural query graph ranking model. A local sub-sequence alignment model with cross-attention is presented in [31]. A slot-matching model to rank query graphs for complex KGQA [27] exploits the inherent structure of query graphs and uses multiple attention scores to explicitly compare each predicate in a query graph with the NL question.

Another topic which has being attracted more and more attention of the research community in recent years is the problem of unanswerable questions [2,20,23,40,44]. However, most of them deal with Machine Reading Comprehension, not KGQA. Unfortunately, different classifications of unanswerable questions (e.g., [2,23,24,44]) consider only the situation in which a (not answered) question has an answer (an answer that is available, but could not be computed by the QA-system, for any reason) and do not investigate the case when there exists no answers to a question, e.g., "What is the capital of Mars?". These two cases differ fundamentally for the field of QA, therefore, they need different approaches to be resolved. However, to distinguish these two cases is not important for this paper. Instead, we focus on deciding whether an answer (represented as query candidate) is correct or not. For this reason, we call all these questions *unanswerable questions*.

3   Approach

Our approach is based on the general assumption that a SPARQL query is expressing a question in a formal representation which can be translated back to an NL text that should be similar to the original question.

In a KGQA system, the generation of a SPARQL query given a question can be considered as a translation from NL to the formal language (cf. Fig. 1). We consider the direct comparison of SPARQL queries and NL questions as very challenging, especially for SPARQL over Wikidata [18] because of non-intuitive URIs naming convention, therefore, we convert SPARQL query candidates to a textual representation. For this purpose, the labels stored in the KG are used to convert all Semantic Web identifiers (e.g., https://www.wikidata.org/entity/Q84) to their textual representation (e.g., "London"). We call the process of generating an NL representation *verbalization of a SPARQL query*.

In our approach we assume that *given a NL question, a KGQA system produces a list of SPARQL query candidates ranked by a relevance score*, computed internally within the system, i.e., the first *query candidate will be used to compute the answer to show to an end-user*. The goal of our QV approach is to ensure that incorrect query candidates are filtered while relying only on the user's NL question and the computed SPARQL query candidates of the considered KGQA system. Hence, our QV process is intended to distinguish query candidates resulting in correct answers from those that would result in incorrect answers. Our approach is system-agnostic and does not require executing the SPARQL queries. In the following subsections, we describe the approach in detail.

```
# What is the cause and place of John Denver's death?
PREFIX wd: <http://www.wikidata.org/entity/>
PREFIX wdt: <http://www.wikidata.org/prop/direct/>
SELECT ?cause ?place WHERE {
  wd:Q105460 wdt:P509 ?cause .
  wd:Q105460 wdt:P20 ?place .
}
```

Fig. 2. An example of a correct SPARQL query candidate (using Wikidata).

3.1 SPARQL Query Candidates Verbalization

To convert SPARQL query candidates to NL answers, we use a straight-forward process where only the WHERE clause of the SPARQL query is considered. All entities and predicates are replaced by their labels, e.g., wd:Q5 is replaced by its English label "human". All variable names are kept as they are, e.g., ?o1, ?subject, ?answer. It is worth mentioning that any other modifiers (e.g., LIMIT, ORDER BY, GROUP BY) in a query are removed in our current approach and do not influence the final NL representation[3]. Finally, all the labels are concatenated with each other in the order of appearance with the space separator.

Considering the SPARQL query presented in Fig. 2, our process computes the following *NL representation*: "John Denver cause of death ?cause John Denver place of death ?place". As many property and entity labels used in a question are often mirrored in its verbalization, our approach is based on the assumption that the QV classifier will be capable of determining such query candidate as a correct one (i.e., the user's question and the query verbalization are similar).

3.2 Query Validation Process

The intention of the proposed QV process is as follows: given a query candidates list, we are *aiming at excluding as many incorrect query candidates as possible while not removing the correct ones*. Thus, our approach increases the chances of showing a correct answer to a user of a QA system. In Fig. 3, several different QV cases are shown. All the incorrect query candidates were removed by the QV (e.g., in A') while the correct ones were left untouched (cf. A''). In an extreme case, a query candidates list contains only incorrect items (e.g., in D and E). In this case, the list should become empty after a perfect QV (cf. D' and D''). Likewise, there could be only correct query candidates in the list (not shown in Fig. 3). In this (unlikely) case, at least one query candidate should be recognized as correct by the QV.

---

[3] Measuring the impact on the verbalization regarding the QV results would be part of additional research.

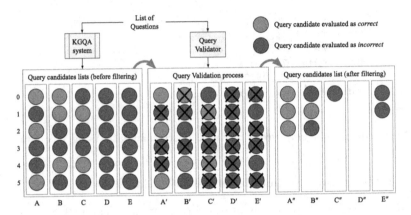

Fig. 3. An example of *query validation process* where a KGQA system proposed 6 ranked query candidates (index: 0–5) for each of the 5 questions $(A–E)$.

3.3 Measures of Query Validation Efficiency

Classification Quality of the Query Validator. To measure the quality of the query validator, we use the following well-known metrics: True Positive Rate (TPR/Recall), True Negative Rate (TNR), Balanced Accuracy (BA), Precision, and F1 score. The metrics are calculated in the machine learning setting for binary classification [33].

Question Answering Quality. To measure the efficiency of QV process, we need to take into account two types of situations that may occur – (S_A) when a query candidate list generated by a KGQA system for a given question contains records with at least one correct query candidate, (S_B) when a query candidate list contains no correct items. In addition, the questions are divided into answerable and unanswerable.

The most common way to measure QA quality is to take the answers generated by the first-ranked query candidate and compare them to the gold standard answers. In this regard, to measure the efficiency of our QV approach, we use well-known Precision, Recall, and F1 score metrics calculated in the information-retrieval setting [37]. The metrics are calculated on the answers obtained before and after QV process. Thereafter, relative improvement is computed.

As our approach influences the whole query candidate list, the other metrics that take into account ranking order have to be considered. Therefore, we utilize Precision@k and NDCG@k [37], where $k \in [1, n]$ and n is the number of query candidates provided by the considered KGQA system.

It is reasonable to use the aforementioned metrics only w.r.t. situation S_A (mentioned at the beginning of this paragraph). While considering situation S_B, we propose the new metric *Answer Trustworthiness Score* (formally defined in Eq. 1 in Sect. 6.2). This metric "gives a reward" (bonus) when a correct answer is shown to a user, and "punishes" (penalty) the score otherwise. In addition, there is a special case when the query candidate list is an empty set, although a correct answer would be available within the considered data. In this situation, the metric does not reward or punish the score, as it is considered to be an "honest" response by the QA system to provide a "don't know the answer" statement (and not "guess" an answer). However, if after the QV process the empty answer is presented to a user (instead of "guessing" an answer), the score will be higher as no punishment will be done. The intuition behind this metric is that *no answer is better than a wrong answer*[4]. Other relevant statistics should be calculated over all query candidate sets, such as average position of correct/incorrect query candidates, the average number of correct/incorrect query candidates in a list.

Finally, unanswerable questions have to be considered. For this class of questions, the expected response is defined as an empty query candidates list.

4   Material and Methods

To validate our approach, we used the state-of-the-art QA system QAnswer [12,13,15] and two well-known datasets – RuBQ 2.0 [35] and LC-QuAD 2.0 [17].

4.1   The KGQA System QAnswer

Out of many existing QA systems (e.g., DeepPavlov [10], Platypus [32], Deep-gAnswer [25] etc.), we have chosen QAnswer because of its portability, accessibility [12] and its following features: robustness, multilingualism, support for multiple KGs (including Wikidata), and it provides high precision and recall [13]. QAnswer also provides an API to ask a question and receive the corresponding ranked query candidate list (of a maximum of 60 candidates). The limitations of QAnswer as described in [13] are not essential to this paper.

4.2   Datasets Overview

QA over KGs is a substantial task that matches a user's question to a query over a KG to retrieve the correct answer [38]. After several updates of the DBpedia [3] KG, many well-known datasets (QALD [42], LC-QuAD 1.0 [41], SimpleDBpedi-aQA [4] etc.) cannot be utilized on its latest version because a QA system compiled for the inundated version has stopped returning valid requests. Moreover, not all datasets (e.g., CSQA [36]) employ SPARQL as a formal representation (which is a requirement for our work). Some datasets (e.g., VANiLLa [7]) have a structure that does not enable to retrieve an answer without ambiguity. Therefore, it was decided to use the RuBQ 2.0 [35] and LC-QuAD 2.0 [17] datasets for our purpose on this step of our research.

---

[4] Example: Assuming a user asks for the red or green wire to be cut for defusing a bomb, then a guessed answer by the QA system might have a devastating result in real life.

RuBQ 2.0 Dataset. RuBQ 2.0 is the first Russian dataset for QA over Wikidata that consists of 2,910 questions of varying complexity, their machine translations into English without any post-editing, and annotated SPARQL queries, which are essential for our approach. Here, we only use the English questions. There are 2,357 unique entities, namely 1,218 in questions and 1,250 in answers, as well as 242 unique relations in the dataset. RuBQ 2.0 is split into development (580) and test (2,330) subsets in such a way to keep a similar distribution of query types in both subsets. 510 RuBQ 2.0 questions are unanswerable, which is a new challenge for KGQA systems to make the task more realistic. The fact that RuBQ 2.0 contains unanswerable questions was a strong incentive to use them in our evaluation.

We were not able to utilize the RuBQ 2.0 data split to dev/test parts, as the number of dev samples is too small for fine-tuning our Query Validator. To obtain the new split, we joined both parts and divided the entire dataset into new train/test parts in 80/20 split (see Sect. 4.3).

LC-QuAD 2.0 Dataset. LC-QuAD 2.0 (2nd instance of the Large-Scale Complex Question Answering Dataset) with 30,000 questions, their paraphrases, and their corresponding SPARQL queries is compatible with both Wikidata and DBpedia 2018 KGs. This dataset has a good variety and reasonable complexity levels for questions (e.g., multi-fact questions, temporal questions, and questions that utilize qualifier information). This dataset consists of 21,258 unique entities and 1,310 unique relations. LC-QuAD 2.0 contains 10 different types of questions (such as boolean, dual intentions, fact with qualifiers, and others) spread over 22 unique templates.

4.3   Data Preparation Process

The process of data preparation is analogous for all datasets considered. It consists of the following steps: (1) processing the questions with the KGQA system in order to get query candidate lists for each of them, (2) executing SPARQL queries from the candidate lists on Wikidata in order to get the answer sets, (3) comparing the answer sets from the query candidates with the "gold standard" answer sets from the dataset in order to determine whether a query candidate is correct or not[5], (4) transforming query candidates to NL (according to Sect. 3.1). The sample entry of a resulting dataset in the RDF Turtle format[6] is presented in Fig. 4. The dataset is available in the online appendix.

We summarized the information on the prepared data and divided it into 3 groups (cf. Table 1).

---

[5] A query candidate is considered as correct if $F1\ score(y_{pred}, y_{true}) = 1$, where y_{pred} – is the set of answers obtained with query candidate and y_{true} is the "gold standard" answer set.

[6] https://www.w3.org/TR/turtle/.

```
fqaqc:dataset-experiment:DATASET:TRAIN:19719-0
    rdf:type fqaac:AnswerCandidate ;
    fqaqc:relatedTo <urn:benchmark:qa:DATASET:TRAIN:19719> ;
    fqaqc:hasPositionBeforeFiltering "0"^^xsd:nonNegativeInteger ;
    fqaqc:hasSPARQL "SELECT * WHERE ?s ?p '0' "^^xsd:string ;
    fqaqc:hasSPARQLResult "{\"head\":{}, \"results\":{}}"^^xsd:string ;
    fqaqc:qaF1Score "0.6"^^xsd:double ; # F1 score(goldStd, sparqlRes)
    fqaqc:hasNaturalLanguageRepresentation "sparql2text"^^xsd:string .
```

Fig. 4. The sample example from the prepared dataset in RDF Turtle format. Where fqaac is a local RDF PREFIX.

```
fqaac:DATASET:19719-0 fqaac:confidenceScore "0.95"^^xsd:double .
```

Fig. 5. Example of RDF Turtle representation of the query validator output.

Table 1. The statistics of the prepared datasets. The training subset is used only for training and validation of the QV and is not considered in this table. The testing subset is used only for the KGQA system evaluation. AQ – answerable questions in the dataset, uAQ – unanswerable questions in the dataset, Group A – questions where a correct query candidate is at the first position, Group B – questions where a correct query candidate is not at the first position, Group C – questions with zero correct query candidates. QC = ∅ – questions that were resolved with an empty query candidates list (not included in Group C).

Dataset	# AQ	# uAQ	# Group A	# Group B	# Group C	# QC = ∅
RuBQ 2.0	480	102	78	85	419	0
LC-QuAD 2.0	6001	0	958	1288	3733	22

4.4 BERT-Based Query Validator

For the QV classifier, we used the BERT model [11]. As the pre-training process of this model included *next sentence prediction task* (NSP)[7] [11], we intentionally fine-tune BERT using the same setting. While providing a question text as a first sentence and a NL representation of a query candidate as the next one, we follow the assumption that a question and a text representation form a valid pair. To create the training data for the QV classifier, we used the prepared data as follows: to each of the textual questions from the training subset, we assigned one NL representation of a randomly picked query candidate. As the output of the QV is a real value p, such as $\{p \in \mathbb{R} \mid 0 \leq p \leq 1\}$, we *empirically define a threshold for assigning a particular class label*. The target classification label $T = \{t^-, t^+\}$ equals t^+ if and only if the assigned query candidate is considered

---

[7] BERT was consciously trained to understand relationships between two consecutive sentences if the second sentence follows the first one (e.g., "[CLS] the man went to [MASK] store [SEP] he bought a gallon [MASK] milk [SEP]") because many important downstream tasks such as QA and Natural Language Inference (NLI) are based on understanding the relationship between two sentences" [11].

as correct to a given question, otherwise, the target label equals t^-. The output of the classifier is represented in the RDF Turtle data as demonstrated in Fig. 5 on page 9.

5   Experimental Setup

We conduct our experiments as follows. In *the first step*, the Query Validators QV are trained separately on the corresponding training subsets D_i, where $D = \{LC\text{-}QuAD, RuBQ\}$, $D_i \in D^8$. Hence, the *a single input of the query validator* QV_{D_i} is a pair (q, a), where $q \in D_i$ is the question text and a is the query candidate transformed to NL. To identify a specific question, we will use $q_{D_i,k}$, where k is the ID of the question in the dataset D_i (e.g., a question with id=1 of the RuBQ 2.0 dataset will be represented by $q_{RuBQ,1}$).

The output of the *target label of the query validator* is $T = \{t^-, t^+\}$, where t^- corresponds to the incorrect (q, a) pair (i.e., a is incorrect for q), and t^+ corresponds to the correct (q, a) pair. We used a balanced distribution (i.e., t^-: 50% to t^+: 50%). Table 2 presents the training data statistics of our QV models. The training data is split into two sets for training and validation (67%/33%) of the query validator.

In *the second step*, we apply the QV_{D_i} to the outputs of the QAnswer system. The outputs have been produced by feeding the questions from the test subset of D_i to the KGQA system (see Step (1) in Sect. 4.3). Thus, the output represents a ranked query candidate list $L_{D_i,q}$ produced for a given question q. L_{D_i} is the set of the query candidate lists for all $q \in D_i$ (i.e., L_{RuBQ} is referring to all candidate lists from questions of the RuBQ 2.0 dataset). $L_{RuBQ,q}$ is a specific query candidate list of the question $q \in D_{RuBQ}$. Consequently, $L_{RuBQ,q,1}$ is the query candidate at position 1 from $L_{RuBQ,q}$), where $0 \leq |L_{D_i,q}| < n$ (where n is the maximum number of available candidates in a list of query candidates) and $|L_{D_i}| = |D_i|$.

After applying the particular QVs (QV_{RuBQ} and $QV_{LC\text{-}QuAD}$) to L_{RuBQ} and $L_{LC\text{-}QuAD}$ respectively, we obtain new filtered lists of query candidates (i.e., $\hat{L}_{RuBQ}$ and $\hat{L}_{LC\text{-}QuAD}$). Hence, if the prediction of QV_{D_i} was t^-, then a query candidate is eliminated from $L_{D_i,q}$. As we track the experimental data using RDF, we are capable of obtaining such information as: position before filtering (fqaac:hasPositionBeforeFiltering), is correct (fqaac:qaF1Score = 1) for each $L_{D_i,q}$. Therefore, we calculate a set of metrics for QA quality as proposed in Sect. 3.3.

6   Evaluation and Analysis

In this section, we describe the evaluation w.r.t. the two steps described in Sect. 5. First, we evaluated the quality of the QV itself (i.e., binary classification quality). Secondly, we evaluated the impact of the QV process on the QA quality.

---

[8] The trained Query Validators are available online;
    LC-QuAD: https://huggingface.co/perevalov/query-validation-lcquad,
    RuBQ: https://huggingface.co/perevalov/query-validation-rubq.

Table 2. Quality metrics of the trained dataset-specific Query Validators.

Query Validator	$|t^-|$	$|t^+|$	TPR (Recall)	TNR	BA	Precision	F1 score
QV_{RuBQ}	9040	9040	0.9805	0.8968	0.9386	0.8874	0.9316
$QV_{\text{LC-QuAD}}$	24045	24045	0.9846	0.9854	0.9850	0.9854	0.9849

Table 3. Evaluation of QA quality improvement using the Gerbil system.

	Micro			Macro		
	Precision	Recall	F1 score	Precision	Recall	F1 score
RuBQ 2.0						
Before QV_{RuBQ}	0.0456	0.4910	0.0834	0.4531	0.4469	0.4462
After QV_{RuBQ}	0.1389	0.5000	0.2174	0.4594	0.4562	0.4505
Improvement in %	204.61	1.83	160.67	1.39	2.08	0.96
LC-QuAD 2.0						
Before $QV_{\text{LC-QuAD}}$	0.1621	0.2984	0.2100	0.5094	0.5191	0.4982
After $QV_{\text{LC-QuAD}}$	0.3561	0.3679	0.3619	0.5341	0.5495	0.5247
Improvement in %	119.68	23.29	72.33	4.85	5.86	5.32

6.1 Answer Validation Classifier Evaluation

In our task *the importance of a false negative prediction is higher than a false positive one*. If the only one correct query candidate from $L_{D_i,q}$ is eliminated by the false negative error, it will inevitably lead to 0% of QA quality, which is not the case for false positive errors. The results regarding the quality of the QV are presented in Table 2. With these results we prove that it is possible to obtain comparable classification quality w.r.t. the well-formed query candidate verbalizations[9]. The obtained Recall score shows that the classifier is capable of avoiding many false negative predictions, which matches our requirements. Thus, even by using our straight-forward query candidate verbalization method (cf. Sect. 3.1), the trained QV models can distinguish between correct and incorrect (q, a) pairs.

6.2 Question Answering Quality Improvement

In the following, the complete evaluation process is described. We applied our QVs to the outputs of the QAnswer system to remove incorrect query candidates (cf. Table 1). In the following paragraphs, by the term *before the QV*, we imply the original results from the QAnswer system (L_{D_i}). The term *after QV* implies the results after applying the QV ($\hat{L}_{D_i}$).

---

[9] In our previous study, we already compared QV's quality using different query candidate verbalization methods [19].

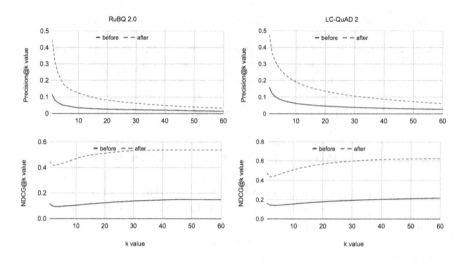

Fig. 6. Precision@k, NDCG@k before and after Query Validation process

Improvement of General Question Answering Quality. The standard metrics for QA systems are based on the computed results of the first element (i.e., in our evaluation, just the first element of QAnswer's answer candidate list is used). We used Gerbil[10] [29,43] for calculating automatically Micro/Macro Precision, Recall and F1 score from our computed data in a comparable and citable form. As the input of the evaluation, the subset of questions was used where QAnswer was computing at least one correct answer candidate for a question $q \in$ Group A $\cup$ Group B (cf. Table 1). The results in Table 3[11] show an improvement of up to 204.61% (micro precision improvement w.r.t. RuBQ 2.0) while the overall macro quality was improved for both datasets. Hence, our approach is capable of improving the answer quality of QA systems w.r.t. the given metrics.

Improvement w.r.t. Groups A and B. In this paragraph, we demonstrate the impact of our approach w.r.t. the whole list of query candidates. To do this, we also selected $L_{D_i,q}$ such that they contain at least one correct query candidate (i.e., Group A $\cup$ Group B). Thereafter, we measured Precision@k and NDCG@k, where $k \in [1, 60]$ before and after the QV process. Hence, we show the impact of our approach w.r.t. the different sizes of $L_{D_i,q}$ and not considering only the first query candidate (i.e., $L_{D_i,q,1}$) as done in the previous paragraph. The results of calculations are presented in Fig. 6. In the charts, we recognize significant improvement w.r.t. all k values (e.g., 382% w.r.t. Precision@1 (=NDCG@1) on

---

[10] http://gerbil-qa.aksw.org/gerbil/, version 0.2.3.
[11] Our results are available online. LC-QuAD 2.0:
      http://gerbil-qa.aksw.org/gerbil/experiment?id=202112080001 and
      http://gerbil-qa.aksw.org/gerbil/experiment?id=202112080002;
      RuBQ 2.0: http://gerbil-qa.aksw.org/gerbil/experiment?id=202112090005 and
      http://gerbil-qa.aksw.org/gerbil/experiment?id=202112090006.

RuBQ 2.0 and 297% on LC-QuAD 2.0 respectively). This tendency is discovered on both datasets and metrics, thus, showing that the proposed approach is not a specialized solution for a particular setting. However, this experiment does not show the impact of our method on the query candidate lists provided by QAnswer that do not contain correct candidates (i.e., Group C). In the next subsection, we discuss this aspect in detail.

Improvement w.r.t. Group C. In this paper, we follow the assumption that *no answer is better than a wrong answer* (cf. Sect. 3.3). It is driven by the observation that an incorrect answer might confuse users who would often not be able to decide if the answer is correct or incorrect. Thus, using our method, we highlight the possibility that all incorrect query candidates of a question can be removed from $L_{D_i,q}$. If instead of an incorrect answer produced by a query candidate, a system should provide an empty answer (i.e., the QA system does not "guess" but explicitly expresses the missing capabilities to answer the given question), this will lead to an improved QA quality from users' perspective. The "standard" QA metrics (e.g., Precision, NDCG) are not reflecting this behavior. For example, a QA system that would provide 50% correct answers and 50% ("guessed") incorrect answers could have the same scores as a system with 50% correct results and 50% no answer (i.e., "don't know the answer") – which is not a valid approach from our point of view. To tackle this scenario, we defined the novel metric *Answer Trustworthiness Score* (ATS) that takes into account our initial assumption. If no such metric is used, "guessing an answer" by a QA system will statistically improve the QA quality. Therefore, we define here the metric ATS that in particular takes into account the number of questions that were answered with an empty result:

$$ATS(D_i) = \frac{\sum_{q \in D_i} f(q)}{|D_i|}, \text{ where } f(q) \begin{cases} +1 & \text{if } isCorrect(L_{D_i,q,1}) = True \\ 0 & \text{else if } L_{i,q} = \varnothing \\ -1 & \text{else} \end{cases} \quad (1)$$

where $isCorrect(L_{i,q,1}) = True$ if and only if for a question d the correct answer is shown (i.e., for an empty response no answer is shown). The answer is produced by a query candidate $L_{i,q,1}$. If an unexpected empty result is shown, then the second case is triggered. The proposed metric may show a clear improvement regarding unanswerable questions. We propose to the scientific community to adopt this metric to ensure reasonable QA quality reflection. In addition, we also analyze such statistics as average correct $(\overline{j^+})$ and incorrect $(\overline{j^-})$ query candidate position, average number of correct $(\overline{L_{D_i,q}^+})$ and incorrect $(\overline{L_{D_i,q}^-})$ query candidates in a list. In Table 4 on page 14 we present the metrics for QA quality introduced in Sect. 3.3. The results demonstrate a significant difference between the values before and after QV. The values of the statistics $\overline{j^+}$ and $\overline{j^-}$ demonstrate that the positions of the correct and incorrect query candidates were shifted to the top of the list after QV. The other values of the $\overline{L_{i,d}^+}$ and $\overline{L_{i,d}^-}$ indicate that the numbers of the correct and incorrect query candidates were decreased after QV. These results are ambiguous, however, the metric proposed

Table 4. Question answering metrics before and after answer filtering.

metric	state	RuBQ 2.0	LC-QuAD 2.0
$\overline{j^+}$	Before QV	16.32	19.63
	After QV	9.16	12.93
$\overline{j^-}$	Before QV	21.98	29.74
	After QV	10.28	19.34
$\overline{L_{D_i,q}^+}$	Before QV	3.12	4.27
	After QV	2.55	3.88
$\overline{L_{D_i,q}^-}$	Before QV	43.49	57.74
	After QV	11.95	27.60
$ATS(D_i)$	Before QV	-0.31	-0.75
	After QV	-0.12	-0.71

Table 5. Evaluation of unanswerable questions from RuBQ 2.0 dataset before and after QV

	state	# questions
$L_{\text{RuBQ},q} = \emptyset$ (correct)	Before QV	0
	After QV	50
$L_{\text{RuBQ},q} \neq \emptyset$ (incorrect)	Before QV	102
	After QV	52

by us is able to disambiguate them. Given our novel metric $(ATS(D_i))$, all the values are negative. The metric results after QV were improved from -0.75 to -0.71 for LC-QuAD 2.0 and from -0.31 to -0.12 for RuBQ 2.0 dataset. Negative values signify that a QA system gives more incorrect answers rather than correct ones. Thus, the QV process decreased the number of incorrect and increased the number of correct query candidates, respectively. Hence, the trustworthiness of the considered system w.r.t. the two analyzed datasets is not good, i.e., users need to evaluate the results carefully.

Improvement w.r.t. Unanswerable Questions. The *unanswerable questions were intentionally integrated by authors of the RuBQ 2.0 dataset.* In this paragraph, we utilize this subset of questions to see if the proposed approach improves the performance of the KGQA system regarding unanswerable questions. *The correct response to an unanswerable question* $q \in D_i$ *is* $L_{D_i,q} = \emptyset$, *otherwise, the response is incorrect.* Such evaluation strategy is also supported in the original paper [24] of the RuBQ 2.0 authors. Our evaluation results of the RuBQ 2.0 dataset regarding the contained 102 unanswerable questions are shown in Table 5. As the QAnswer system's strategy is to generate a list of query candidates for any kind of question, none of the results were considered as correct before QV. After QV, all the query candidates from the respective 50 lists $L_{\text{RuBQ},q}$ were completely eliminated, and hence 50 unanswerable questions would have been answered correctly (with an empty answer).

6.3 Discussion and Limitations

We raise several questions for the discussion. Firstly, the validity of our approach is strongly depending on the labels provided by the considered KG. For the given datasets, it works surprisingly well, however, for a real-world scenario, additional methods would be required (e.g., integrating synonyms from [28]). Futhermore, this raises the question if current KGQA benchmarks already represent the variety of NL questions well enough or would require additional extensions. Secondly, in this work, we consider a query candidate as correct if the F1 score of the expected and computed results is 1. In this regard, the other options would be to consider a different threshold instead of 1 (e.g., 0.75 or 0.5). The major point of concern of the strict threshold for the real-valued measure (F1 score) for determining whether a query candidate is correct is that some gold-standard correct answers sets contain more than one item (e.g., question: "Give me all post-punk artists"). In this regard, if a query candidate could produce a query equivalent to "Give me all *German* post-punk artists", the result would be *partially correct*. Hence, the *"is correct" threshold should be adapted according to the tolerance to the partially correct answers* of the evaluation context (i.e., in our case, we have zero tolerance to incorrect answers). Thirdly, this evaluation provided in the work may be biased due to the only one KGQA system was used. Although QAnswer is well-known and used for much research, further work should cover different KGQA systems as their approaches and capabilities may vary significantly. Finally, more training techniques of the query validator should be explored, s.t., the final impact on the QA quality can be increased. At the moment, there are still additional opportunities for the QV process improvement, considering not only answerable but also unanswerable questions.

7 Conclusions

In this paper, we have proven the impact of our query candidate validation approach. It uses a NL representation of the SPARQL query that is compared by a trained model with the given question. Our approach takes into account the verbalized information of concepts, predicates, and instances that are already defined in a SPARQL query candidate. Hence, we did not create a complete nor well-formed NL representation of the SPARQL queries. However, the results of our research show significant QA quality improvements w.r.t. different aspects that are important for QA systems. In particular, our evaluation includes answerable and unanswerable questions as well as it shows that the quality of the query candidate list can be improved.

As we have shown, our method is capable of improving the quality of QA systems without knowledge about the implemented approach of a QA system. Consequently, it might be integrated in the query builder component of QA systems, or as a reusable component via the QA framework (e.g., the Qanary framework [8]) to improve the quality of answers or intermediate candidate lists. Hence, our main contribution is providing a domain-agnostic method that can be applied to any knowledge base that provides verbalization (typically available as predicate `rdfs:label`), s.t., corresponding KGQA systems increase their quality.

Looking forward, we plan to use different models, verify the presented approach on different systems and benchmarks and check the applicability of the approach to other languages.

References

1. Abdiansah, A., Azhari, A., Sari, A.K.: Survey on answer validation for Indonesian question answering system (IQAS). Int. J. Intell. Syst. Appl. **10**, 68–78 (2018). https://doi.org/10.5815/ijisa.2018.04.08
2. Asai, A., Choi, E.: Challenges in information seeking QA: unanswerable questions and paragraph retrieval. arXiv preprint arXiv:2010.11915 (2020)
3. Auer, S., Bizer, C., Kobilarov, G., Lehmann, J., Cyganiak, R., Ives, Z.: DBpedia: a nucleus for a web of open data. In: Aberer, K., et al. (eds.) ASWC/ISWC-2007. LNCS, vol. 4825, pp. 722–735. Springer, Heidelberg (2007). https://doi.org/10.1007/978-3-540-76298-0_52
4. Azmy, M., Shi, P., Lin, J., Ilyas, I.: Farewell freebase: migrating the simplequestions dataset to DBpedia. In: Proceedings of the 27th International Conference on Computational Linguistics, pp. 2093–2103 (2018)
5. Babych, S., Henn, A., Pawellek, J., Padó, S.: Dependency-based answer validation for German. In: Petras, V., Forner, P., Clough, P.D. (eds.) CLEF 2011 Labs and Workshop, Notebook Papers, 19–22 September 2011, Amsterdam, The Netherlands. CEUR Workshop Proceedings, vol. 1177. CEUR-WS.org (2011)
6. Berners-Lee, T., Hendler, J., Lassila, O.: The semantic web. Sci. Am. **284**(5), 34–43 (2001)
7. Biswas, D., Dubey, M., Rony, M.R.A.H., Lehmann, J.: VANiLLa: verbalized answers in natural language at large scale. CoRR abs/2105.11407 (2021)
8. Both, A., Diefenbach, D., Singh, K., Shekarpour, S., Cherix, D., Lange, C.: Qanary – a methodology for vocabulary-driven open question answering systems. In: Sack, H., Blomqvist, E., d'Aquin, M., Ghidini, C., Ponzetto, S.P., Lange, C. (eds.) ESWC 2016. LNCS, vol. 9678, pp. 625–641. Springer, Cham (2016). https://doi.org/10.1007/978-3-319-34129-3_38
9. Both, A., Gashkov, A., Eltsova, M.: Similarity detection of natural-language questions and answers using the VANiLLa dataset. J. Phys: Conf. Ser. **1886**(1), 012017 (2021). https://doi.org/10.1088/1742-6596/1886/1/012017
10. Burtsev, M., et al.: DeepPavlov: open-source library for dialogue systems. In: Proceedings of ACL 2018, System Demonstrations, pp. 122–127. Association for Computational Linguistics, Melbourne (2018)
11. Devlin, J., Chang, M.W., Lee, K., Toutanova, K.: BERT: pre-training of deep bidirectional transformers for language understanding. In: Proceedings of the 2019 Conference of the North American Chapter of the Association for Computational Linguistics: Human Language Technologies (Long and Short Papers), vol. 1, pp. 4171–4186. Association for Computational Linguistics, Minneapolis (2019). https://doi.org/10.18653/v1/N19-1423

12. Diefenbach, D., Both, A., Singh, K., Maret, P.: Towards a question answering system over the semantic web. Semantic Web **11**, 421–439 (2020)
13. Diefenbach, D., Giménez-García, J., Both, A., Singh, K., Maret, P.: QAnswer KG: designing a portable question answering system over RDF data. In: Harth, A., Kirrane, S., Ngonga Ngomo, A.-C., Paulheim, H., Rula, A., Gentile, A.L., Haase, P., Cochez, M. (eds.) ESWC 2020. LNCS, vol. 12123, pp. 429–445. Springer, Cham (2020). https://doi.org/10.1007/978-3-030-49461-2_25
14. Diefenbach, D., Lopez, V., Singh, K., Maret, P.: Core techniques of question answering systems over knowledge bases: a survey. Knowl. Inf. Syst. **55**(3), 529–569 (2017). https://doi.org/10.1007/s10115-017-1100-y
15. Diefenbach, D., Migliatti, P.H., Qawasmeh, O., Lully, V., Singh, K., Maret, P.: QAnswer: a question answering prototype bridging the gap between a considerable part of the LOD cloud and end-users. In: Liu, L., et al. (eds.) The World Wide Web Conference, WWW 2019, San Francisco, May 13–17, 2019, pp. 3507–3510. ACM (2019). https://doi.org/10.1145/3308558.3314124
16. Dimitrakis, E., Sgontzos, K., Tzitzikas, Y.: A survey on question answering systems over linked data and documents. J. Intell. Inf. Syst. **55**(2), 233–259 (2019). https://doi.org/10.1007/s10844-019-00584-7
17. Dubey, M., Banerjee, D., Abdelkawi, A., Lehmann, J.: LC-QuAD 2.0: a large dataset for complex question answering over Wikidata and DBpedia. In: Ghidini, C., Hartig, O., Maleshkova, M., Svátek, V., Cruz, I., Hogan, A., Song, J., Lefrançois, M., Gandon, F. (eds.) ISWC 2019. LNCS, vol. 11779, pp. 69–78. Springer, Cham (2019). https://doi.org/10.1007/978-3-030-30796-7_5
18. Erxleben, F., Günther, M., Krötzsch, M., Mendez, J., Vrandečić, D.: Introducing Wikidata to the linked data web. In: Mika, P., Tudorache, T., Bernstein, A., Welty, C., Knoblock, C., Vrandečić, D., Groth, P., Noy, N., Janowicz, K., Goble, C. (eds.) ISWC 2014. LNCS, vol. 8796, pp. 50–65. Springer, Cham (2014). https://doi.org/10.1007/978-3-319-11964-9_4
19. Gashkov, A., Perevalov, A., Eltsova, M., Both, A.: Improving the question answering quality using answer candidate filtering based on natural-language features. In: 16th International Conference on Intelligent Systems and Knowledge Engineering (ISKE 2021) (2021)
20. Godin, F., Kumar, A., Mittal, A.: Learning when not to answer: a ternary reward structure for reinforcement learning based question answering. In: Proceedings of the 2019 Conference of the North American Chapter of the Association for Computational Linguistics: Human Language Technologies (Industry Papers), Vol. 2, pp. 122–129. Association for Computational Linguistics, Minneapolis (2019). https://doi.org/10.18653/v1/N19-2016
21. Gómez-Adorno, H., Pinto, D., Vilariño, D.: A question answering system for reading comprehension tests. In: Carrasco-Ochoa, J.A., Martínez-Trinidad, J.F., Rodríguez, J.S., di Baja, G.S. (eds.) MCPR 2013. LNCS, vol. 7914, pp. 354–363. Springer, Heidelberg (2013). https://doi.org/10.1007/978-3-642-38989-4_36
22. Grappy, A., Grau, B., Falco, M., Ligozat, A., Robba, I., Vilnat, A.: Selecting answers to questions from web documents by a robust validation process. In: 2011 IEEE/WIC/ACM International Conferences on Web Intelligence and Intelligent Agent Technology, vol. 1, pp. 55–62 (2011). https://doi.org/10.1109/WI-IAT.2011.210
23. Hu, M., Wei, F., Peng, Y., Huang, Z., Yang, N., Li, D.: Read+verify: machine reading comprehension with unanswerable questions. In: Proceedings of the AAAI Conference on Artificial Intelligence, vol. 33, pp. 6529–6537 (2019)

24. Korablinov, V., Braslavski, P.: RuBQ: a Russian dataset for question answering over Wikidata. In: Pan, J.Z., Tamma, V., d'Amato, C., Janowicz, K., Fu, B., Polleres, A., Seneviratne, O., Kagal, L. (eds.) ISWC 2020. LNCS, vol. 12507, pp. 97–110. Springer, Cham (2020). https://doi.org/10.1007/978-3-030-62466-8_7

25. Lin, Y., Zhang, M., Zhang, R., Zou, L.: Deep-gAnswer: a knowledge based question answering system. In: U, L.H., Spaniol, M., Sakurai, Y., Chen, J. (eds.) APWeb-WAIM 2021. LNCS, vol. 12859, pp. 434–439. Springer, Cham (2021). https://doi.org/10.1007/978-3-030-85899-5_33

26. Magnini, B., Negri, M., Prevete, R., Tanev, H.: Is it the right answer? Exploiting web redundancy for answer validation. In: Proceedings of the 40th Annual Meeting of the Association for Computational Linguistics, pp. 425–432. Association for Computational Linguistics, Philadelphia (2002). https://doi.org/10.3115/1073083.1073154

27. Maheshwari, G., Trivedi, P., Lukovnikov, D., Chakraborty, N., Fischer, A., Lehmann, J.: Learning to rank query graphs for complex question answering over knowledge graphs. In: Ghidini, C., Hartig, O., Maleshkova, M., Svátek, V., Cruz, I., Hogan, A., Song, J., Lefrançois, M., Gandon, F. (eds.) ISWC 2019. LNCS, vol. 11778, pp. 487–504. Springer, Cham (2019). https://doi.org/10.1007/978-3-030-30793-6_28

28. Miller, G.A.: WordNet: An Electronic Lexical Database. MIT Press (1998)

29. Napolitano, G., Usbeck, R., Ngomo, A.-C.N.: The scalable question answering over linked data (SQA) challenge 2018. In: Buscaldi, D., Gangemi, A., Reforgiato Recupero, D. (eds.) SemWebEval 2018. CCIS, vol. 927, pp. 69–75. Springer, Cham (2018). https://doi.org/10.1007/978-3-030-00072-1_6

30. Pakray, P., Barman, U., Bandyopadhyay, S., Gelbukh, A.: Semantic answer validation using universal networking language. Int. J. Comput. Sci. Inf. Technol. 3(4), 4927–4932 (2012)

31. Parikh, A.P., Täckström, O., Das, D., Uszkoreit, J.: A decomposable attention model for natural language inference. In: Proceedings of the 2016 Conference on Empirical Methods in Natural Language Processing, pp. 2249–2255. Association for Computational Linguistics (2016)

32. Pellissier Tanon, T., de Assunção, M.D., Caron, E., Suchanek, F.M.: Demoing Platypus – a multilingual question answering platform for Wikidata. In: Gangemi, A., et al. (eds.) ESWC 2018. LNCS, vol. 11155, pp. 111–116. Springer, Cham (2018). https://doi.org/10.1007/978-3-319-98192-5_21

33. Powers, D.M.W.: Evaluation: from precision, recall and F-factor to ROC, informedness, markedness & correlation. J. Mach. Learn. Technol. 2(1), 37–63 (2011)

34. Rodrigo, A., Pérez-Iglesias, J., Peñas, A., Garrido, G., Araujo, L.: A question answering system based on information retrieval and validation. In: CLEF 2010 LABs and Workshops, Notebook Papers (2010)

35. Rybin, I., Korablinov, V., Efimov, P., Braslavski, P.: RuBQ 2.0: an innovated Russian question answering dataset. In: Verborgh, R., Hose, K., Paulheim, H., Champin, P.-A., Maleshkova, M., Corcho, O., Ristoski, P., Alam, M. (eds.) ESWC 2021. LNCS, vol. 12731, pp. 532–547. Springer, Cham (2021). https://doi.org/10.1007/978-3-030-77385-4_32

36. Saha, A., Pahuja, V., Khapra, M.M., Sankaranarayanan, K., Chandar, S.: Complex sequential question answering: towards learning to converse over linked question answer pairs with a knowledge graph. In: Thirty-Second AAAI Conference on Artificial Intelligence (2018)

37. Schütze, H., Manning, C.D., Raghavan, P.: Introduction to Information Retrieval. Cambridge University Press, Cambridge (2008)

38. Singh, K., et al.: Why reinvent the wheel: let's build question answering systems together. In: Proceedings of the 2018 World Wide Web Conference, pp. 1247–1256 (2018)
39. Solovyev, A.: Dependency-based algorithms for answer validation task in Russian question answering. In: Gurevych, I., Biemann, C., Zesch, T. (eds.) GSCL 2013. LNCS (LNAI), vol. 8105, pp. 199–212. Springer, Heidelberg (2013). https://doi.org/10.1007/978-3-642-40722-2_20
40. Tan, C., Wei, F., Zhou, Q., Yang, N., Lv, W., Zhou, M.: I know there is no answer: modeling answer validation for machine reading comprehension. In: Zhang, M., Ng, V., Zhao, D., Li, S., Zan, H. (eds.) NLPCC 2018. LNCS (LNAI), vol. 11108, pp. 85–97. Springer, Cham (2018). https://doi.org/10.1007/978-3-319-99495-6_8
41. Trivedi, P., Maheshwari, G., Dubey, M., Lehmann, J.: LC-QuAD: a corpus for complex question answering over knowledge graphs. In: d'Amato, C., Fernandez, M., Tamma, V., Lecue, F., Cudré-Mauroux, P., Sequeda, J., Lange, C., Heflin, J. (eds.) ISWC 2017. LNCS, vol. 10588, pp. 210–218. Springer, Cham (2017). https://doi.org/10.1007/978-3-319-68204-4_22
42. Usbeck, R., Gusmita, R.H., Ngomo, A.N., Saleem, M.: 9th challenge on question answering over linked data (QALD-9). In: Joint proceedings of the 4th Workshop on Semantic Deep Learning (SemDeep-4) and NLIWoD4: Natural Language Interfaces for the Web of Data (NLIWOD-4) and 9th Question Answering over Linked Data challenge (QALD-9) co-located with 17th International Semantic Web Conference (ISWC 2018), Monterey, 8th–9th October 2018, pp. 58–64 (2018)
43. Usbeck, R., et al.: GERBIL - general entity annotation benchmark framework. In: 24th WWW Conference (2015)
44. Yen, A.Z., Huang, H.H., Chen, H.H.: Unanswerable question correction in question answering over personal knowledge base. In: Proceedings of the AAAI Conference on Artificial Intelligence, vol. 35, pp. 14266–14275 (2021)
45. Yih, S.W., Chang, M.W., He, X., Gao, J.: Semantic parsing via staged query graph generation: question answering with knowledge base. In: Proceedings of the Joint Conference of the 53rd Annual Meeting of the ACL and the 7th International Joint Conference on Natural Language Processing of the AFNLP (2015)
46. Yu, M., Yin, W., Hasan, K.S., Santos, C.D., Xiang, B., Zhou, B.: Improved neural relation detection for knowledge base question answering. In: Proceedings of the 53rd Annual Meeting of the Association for Computational Linguistics and the 7th International Joint Conference on Natural Language Processing (Long Papers), Vol. 1, pp. 1321–1331. Association for Computational Linguistics (2017)
47. Zamanov, I., Kraeva, M., Hateva, N., Yovcheva, I., Nikolova, I., Angelova, G.: Voltron: a hybrid system for answer validation based on lexical and distance features. In: Proceedings of the 9th International Workshop on Semantic Evaluation (SemEval 2015). pp. 242–246. Association for Computational Linguistics, Denver (2015). https://doi.org/10.18653/v1/S15-2043

Learning Concept Lengths Accelerates Concept Learning in ALC

N'Dah Jean Kouagou(✉)◉, Stefan Heindorf◉, Caglar Demir◉,
and Axel-Cyrille Ngonga Ngomo◉

Paderborn University, Paderborn, Germany
nkouagou@mail.uni-paderborn.de, heindorf@uni-paderborn.de,
{caglar.demir,axel.ngonga}@upb.de

Abstract. Concept learning approaches based on refinement operators explore partially ordered solution spaces to compute concepts, which are used as binary classification models for individuals. However, the number of concepts explored by these approaches can grow to the millions for complex learning problems. This often leads to impractical runtimes. We propose to alleviate this problem by predicting the length of target concepts before the exploration of the solution space. By these means, we can prune the search space during concept learning. To achieve this goal, we compare four neural architectures and evaluate them on four benchmarks. Our evaluation results suggest that recurrent neural network architectures perform best at concept length prediction with a macro F-measure ranging from 38% to 92%. We then extend the CELOE algorithm, which learns ALC concepts, with our concept length predictor. Our extension yields the algorithm CLIP. In our experiments, CLIP is at least 7.5× faster than other state-of-the-art concept learning algorithms for ALC—including CELOE—and achieves significant improvements in the F-measure of the concepts learned on 3 out of 4 datasets. For reproducibility, we provide our implementation in the public GitHub repository at https://github.com/dice-group/LearnALCLengths.

Keywords: Concept learning · Concept length · Structured machine learning · Description logic · Learning from examples · Prediction of concept lengths

1 Introduction

Knowledge bases have recently become indispensable in a number of applications driven by machine learning [12]. For instance, the Gene Ontology (GO) [1,9], Drug-Bank [36], and the Global Network of Biomedical Relationships (GNBR) [27] are actively being used to find treatments for certain diseases [17,25]. We consider the supervised machine learning task of concept learning[1] [24] on knowledge bases in the description logic (DL) $\mathcal{ALC}$ (attributive language with complements) [32]. We focus on approaches based on refinement operators [3,13,24,28,29].

Recent works on concept learning over DLs [5,14,28] indicate that approaches based on refinement operators often fail to achieve practical runtimes on large real-world knowledge bases, which often contain millions of individuals and concepts with

---

[1] Also called class expression learning (CEL) [13]. See Sect. 2 for a formal definition.

© The Author(s), under exclusive license to Springer Nature Switzerland AG 2022
P. Groth et al. (Eds.): ESWC 2022, LNCS 13261, pp. 236–252, 2022.
https://doi.org/10.1007/978-3-031-06981-9_14

billions of assertions. As noted by Rizzo et al. [28], this is partially due to the size of the search space that needs to be explored to detect relevant concepts. In this paper, we *accelerate concept learning by predicting the length of the target concept* in advance. By these means, we can prune the search space traversed by a refinement operator and therewith reduce the overall runtime of the concept learning process. To quantify our runtime improvement, we compare our new algorithm—dubbed CLIP—against the state-of-the-art approaches CELOE [23], OCEL [24], and ELTL [7]. The price for our runtime improvement is paid in the prior training of the concept length predictor. Therefore, we also show that the prediction of concept lengths can be carried out using rather simple neural architectures.

To the best of our knowledge, no similar work has been carried out before. Hence, we hope that our findings will serve as a foundation for more investigations in this direction. In a nutshell, our contributions are as follows:

1. We design different neural network architectures for learning concept lengths.
2. We implement a length-based refinement operator to generate training data.
3. We integrate our concept length predictors into the CELOE algorithm, resulting in a new algorithm that we call CLIP. We show that CLIP achieves state-of-the-art performance in terms of F-measure while outperforming the state of the art in terms of runtime.

The remainder of the paper is organized as follows: In Sect. 2, we give a brief overview of the required background in DL, concept learning, knowledge graph embeddings, and refinement operators for DLs. We also present the notation and terminology used in the rest of the paper. Section 3 presents related work on concept learning using refinement operators. In Sects. 4 and 5, we describe our new approach for concept learning in $\mathcal{ALC}$. Our results on different knowledge bases are presented in Sect. 6. Section 7 draws conclusions from our findings and introduces new directions for future work.

2   Background

In this section, we present the background on description logics, concept learning, refinement operators, and knowledge graph embeddings. We also introduce the notation and terminology used throughout the paper.

Description Logics. Description logics [2] are a family of languages for knowledge representation. While there are more powerful variants of description logics [2,20], we focus on the description logic $\mathcal{ALC}$ ($\mathcal{A}$ttributive $\mathcal{L}$anguage with $\mathcal{C}$omplement) because it is the simplest closed description logic with respect to propositional logics. Its basic components are *concept* names (e.g., *Teacher*, *Human*), *role* names (e.g., *hasChild*, *bornIn*) and *individuals* (e.g., *Mike*, *Jack*). Table 1 introduces the syntax and semantics of $\mathcal{ALC}$ (see [24] for more details). In $\mathcal{ALC}$, concept lengths are defined recursively [24]

1. $length(A) = length(\top) = length(\bot) = 1$, for all atomic concepts A
2. $length(\neg C) = 1 + length(C)$, for all concepts C
3. $length(\exists\, r.C) = length(\forall\, r.C) = 2 + length(C)$, for all concepts C
4. $length(C \sqcup D) = length(C \sqcap D) = 1 + length(C) + length(D)$, for all concepts C and D.

Table 1. $\mathcal{ALC}$ syntax and semantics. $\mathcal{I}$ stands for an interpretation, $\Delta^{\mathcal{I}}$ for its domain.

Construct	Syntax	Semantics
Atomic concept	A	$A^{\mathcal{I}} \subseteq \Delta^{\mathcal{I}}$
Atomic role	r	$r^{\mathcal{I}} \subseteq \Delta^{\mathcal{I}} \times \Delta^{\mathcal{I}}$
Top concept	$\top$	$\Delta^{\mathcal{I}}$
Bottom concept	$\bot$	$\emptyset$
Conjunction	$C \sqcap D$	$C^{\mathcal{I}} \cap D^{\mathcal{I}}$
Disjunction	$C \sqcup D$	$C^{\mathcal{I}} \cup D^{\mathcal{I}}$
Negation	$\neg C$	$\Delta^{\mathcal{I}} \setminus C^{\mathcal{I}}$
Existential restriction	$\exists\, r.C$	$\{a^{\mathcal{I}} / \exists\, b^{\mathcal{I}} \in C^{\mathcal{I}}, (a^{\mathcal{I}}, b^{\mathcal{I}}) \in r^{\mathcal{I}}\}$
Universal restriction	$\forall\, r.C$	$\{a^{\mathcal{I}} / \forall\, b^{\mathcal{I}}, (a^{\mathcal{I}}, b^{\mathcal{I}}) \in r^{\mathcal{I}} \Rightarrow b^{\mathcal{I}} \in C^{\mathcal{I}}\}$

The pair $\mathcal{K} = (TBox, ABox)$ denotes an $\mathcal{ALC}$ knowledge base. The $TBox$ contains statements of the form $C \sqsubseteq D$ or $C \equiv D$, where C and D are concepts. The $ABox$ consists of statements of the form $C(a)$ and $R(a, b)$, where C is a concept, R is a role, and a, b are individuals. N_C and N_R are the sets of concept names and role names in $\mathcal{K}$, respectively. $\mathcal{K}_I$ stands for the set of all individuals in $\mathcal{K}$. $|.|$ denotes the cardinality function, that is, a function that takes a set as input and returns the number of elements in the set. Given a concept C, we denote the set of all instances of C by $C_{\mathcal{P}}$. $C_{\mathcal{N}}$ stands for the set of all individuals that are not instances of C.

Concept Learning. We recall the definition introduced by [24].

Definition 1. *Let $\mathcal{K}$, T, P, and N be a knowledge base, a target concept, and sets of positive and negative examples from $\mathcal{K}_I$, respectively. The learning problem is to find a concept C such that T does not occur in C, and for $\mathcal{K}' = \mathcal{K} \cup \{T \equiv C\}$, we have that $\mathcal{K}' \models P$ and $\mathcal{K}' \not\models N$.*

Since such a concept C does not always exist, we target an approximate definition in this work.

Definition 2. *Given a knowledge base $\mathcal{K}$, a set of positive examples P, and a set of negative examples N, the learning problem is to find a concept C which maximizes the F-measure F, where F is defined by $F = 2 \times \frac{Precision \times Recall}{Precision + Recall}$, with $Precision = \frac{|C_{\mathcal{P}} \cap P|}{|C_{\mathcal{P}} \cap P| + |C_{\mathcal{P}} \cap N|}$ and $Recall = \frac{|C_{\mathcal{P}} \cap P|}{|C_{\mathcal{P}} \cap P| + |C_{\mathcal{N}} \cap P|}$.*

Note that in the above definition, $C_{\mathcal{N}}$ and $C_{\mathcal{P}}$ depend on both the concept C and the learning problem. Following [24], we use the closed-world assumption (CWA) to compute $C_{\mathcal{P}}$ and $C_{\mathcal{N}}$: every individual that cannot be inferred to be an element of $C_{\mathcal{P}}$ is considered to be in $C_{\mathcal{N}}$. In this work, we are interested in finding such a concept C by using refinement operators (see the following subsection). The found concept C might not be unique for a learning problem.

Refinement Operators.

Definition 3. *A quasi-ordering $\preceq$ is a reflexive and transitive binary relation. Let $(S, \preceq)$ be a quasi-ordered space. A downward (upward) refinement operator on S is a mapping $\rho : S \rightarrow 2^S$ such that for all $C \in S$, $C' \in \rho(C)$ implies $C' \preceq C$ ($C \preceq C'$).*

Example 1. Let $\mathcal{K} = (TBox, ABox)$ be a knowledge base, with

$TBox = \{Female \sqsubseteq Human, Mother \sqsubseteq Female, Human \sqsubseteq \neg Car\};$

$ABox = \{Female(Anna), Mother(Kate), Car(Venza), hasChild(Jack, Paul)\}.$

Assume the sets of concept names N_C and role names N_R in $\mathcal{K}$ are given by:

$$N_C = \{Car, Female, Human, Mother, Parent\};$$
$$N_R = \{hasChild, manufacturedBy, marriedTo\}.$$

Let **C** be the set of all $\mathcal{ALC}$ concept expressions [30] that can be constructed from N_C and N_R (note that **C** is infinite and every concept name is a concept expression). Consider the mapping $\rho : \mathbf{C} \rightarrow 2^{\mathbf{C}}$ defined by: $\rho(C) = \{C' \in \mathbf{C} | C' \sqsubseteq C, C' \neq C\}$ for all $C \in \mathbf{C}$. ρ is clearly a downward refinement operator and we have for example,

– $\{Female, Mother, Female \sqcup Mother\} \subseteq \rho(Human);$
– $\{\exists\, marriedTo.Mother, \forall\, marriedTo.Female\} \subseteq \rho(\exists\, marriedTo.(\neg Car)).$

Refinement operators can have a number of important properties which we do not discuss in this paper (for further details, we refer the reader to [24]). In the context of concept learning, these properties can be exploited to optimize the traversal of the concept space in search of a target concept.

Knowledge Graph Embeddings. The *ABox* of a knowledge base in $\mathcal{ALC}$ can be regarded as a knowledge graph [16] (see also Sect. 4). A knowledge graph embedding function typically maps a knowledge graph to a continuous vector space to facilitate downstream tasks such as link prediction and knowledge graph completion [10,33]. We exploit knowledge graph embeddings to improve concept learning in DLs. Knowledge graph embedding approaches can be subdivided into two categories: the first category of approaches uses only facts in the knowledge graph [6,26,35], and the second category of approaches takes into account additional information about entities and relations, such as textual descriptions [34,37]. Both approaches typically initialize each entity and relation with a random vector, matrix, or tensor. Then, a scoring function is defined to learn embeddings so that facts observed in the knowledge graph receive high scores, while unobserved facts receive low scores. It can also happen that unobserved facts receive a high score, for instance, if a fact is supposed to hold but it is not observed in the knowledge graph, or if it is a logical implication of the learned patterns. For more details, we refer the reader to the surveys [10,33]. In this work, we use the Convolutional Complex Embedding Model (ConEx) [11], which has been shown to produce state-of-the-art results with fewer trainable parameters.

3   Related Work

Lehmann and Hitzler [24] investigated concept learning using refinement operators by studying combinations of possible refinement operator properties and designed their own refinement operator in $\mathcal{ALC}$. Their approach proved to be competitive in accuracy with (and in some cases, superior to) the state-of-the-art, namely inductive logic programs. Badea and Nienhuys-Cheng [3] worked on a similar topic in the DL $\mathcal{ALER}$. The evaluation of their approach on real ontologies from different domains showed promising results, but it had the disadvantage of depending on the instance data.

DL-Learner [21] is the most mature framework for concept learning. CELOE [23], OCEL [22], and ELTL [7] are algorithms implemented in DL-Learner. CELOE is an extension of OCEL that uses the same refinement operator but with a different heuristic function. It is considered the best algorithm in DL-Learner to date. The algorithm uses a soft syntactic bias in its heuristic function that balances between predictive performance and short, readable concepts. In this work, we opt for a hard syntactic bias that constrains our algorithm to generate expressions shorter than a given threshold. ELTL is designed for the simple description logic $\mathcal{EL}$. Despite its usefulness, DL-Learner suffers from performance issues in certain scenarios [cf. 14,31].

DL-FOIL [13,29] is another concept learning algorithm that uses refinement operators and progressively constructs the solution as a disjunction of partial descriptions. Each partial description covers a part of the positive examples and rules out as many negative/uncertain-membership examples as possible. DL-FOCL1–3 [28] are variants of DL-FOIL that employ meta-heuristics to help reduce the search space. The first release of DL-FOCL, also known as DL-FOCL1, is essentially based on omission rates: to check if further iterations are required, DL-FOCL1 compares the score of the current concept definition with that of the best concept obtained at that stage. DL-FOCL2 employs a look-ahead strategy by assessing the quality of the next possible refinements of the current partial description. Finally, DL-FOCL3 attempts to solve the myopia problem in DL-FOIL by introducing a local memory, used to avoid reconsidering suboptimal choices previously made. EvoLearner [14] is a concept learner based on evolutionary algorithms. In contrast, we propose to learn *concept lengths* from positive and negative examples to boost the performance of *concept learners*.

4   Concept Length Prediction

In this section, we address the following learning problem: Given a knowledge base $\mathcal{K}$, a set of positive examples P and negative examples N, predict the length of the shortest concept C that is a solution to the learning problem defined by $\mathcal{K}$, P, and N according to Definition 2. To achieve this goal, we devise a generator that creates training data for our prediction algorithm based solely on $\mathcal{K}$ and a user-given number of learning problems to use at training time.

4.1   Training Data for Length Prediction

Data Generation. Given a knowledge base, the construction of training data (concepts with their positive and negative examples) is carried out as follows:

1. Generate concepts of various lengths using the length-based refinement operator described in Algorithm 1 and 2. In this process, short concepts are preferred over long concepts, i.e., when two concepts have the same set of instances, the longest concept is left out.
2. Compute the sets $C_\mathcal{N}$ and $C_\mathcal{P}$ for each generated concept C.
3. Define a hyper-parameter $\mathbf{N} \in [1, |\mathcal{K}_I|]$ that represents the total number of positive and negative examples we want to use per learning problem.
4. Sample positive and negative examples as follows:
   - If $|C_\mathcal{P}| \geq \frac{\mathbf{N}}{2}$ and $|C_\mathcal{N}| \geq \frac{\mathbf{N}}{2}$, then we randomly sample $\frac{\mathbf{N}}{2}$ individuals from each of the two sets $C_\mathcal{P}$ and $C_\mathcal{N}$.
   - Otherwise, we take all individuals in the minority set and sample the remaining number of individuals from the other set.

Training Data Features. A knowledge graph is commonly defined as $\mathcal{G} \subseteq \mathcal{E} \times \mathcal{R} \times \mathcal{E}$, where $\mathcal{E}$ is a set of entities and $\mathcal{R}$ is a set of relations. We convert a given knowledge base $\mathcal{K}$ into a knowledge graph by converting $ABox$ statements of the form $R(a, b)$ into (a, R, b). Statements of the form $C(a)$ are converted into $(a, \mathtt{rdf\!:\!type}, C)$. In our experimental data, the $TBoxes$ contained only subsumptions $C \sqsubseteq D$ between atomic concepts C and D, which were converted into triples $(C, \mathtt{rdfs\!:\!subClassOf}, D)$. Hence, in our experiments, $\mathcal{E} \subseteq N_C \cup \mathcal{K}_I$ and $\mathcal{R} = N_R \cup \{\mathtt{rdfs\!:\!subClassOf}\}$.

The resulting knowledge graph is then embedded into a continuous vector space to serve for the prediction of concept lengths. On the vector representation of entities, we create an extra dimension at the end of the entries, where we insert $+1$ for positive examples and -1 for negative examples. Formally, we define an injective function f_C for each target concept C

$$f_C : \mathbb{R}^\mathbf{d} \longrightarrow \mathbb{R}^{\mathbf{d}+1}$$

$$\mathbf{x} = (x_1, \ldots, x_\mathbf{d}) \longmapsto \begin{cases} (x_1, \ldots, x_\mathbf{d}, 1) & \text{if } ent(\mathbf{x}) \in C_\mathcal{P}, \\ (x_1, \ldots, x_\mathbf{d}, -1) & \text{otherwise}, \end{cases} \tag{1}$$

where $\mathbf{d}$ is the dimension of the embedding space, and $ent(\mathbf{x})$ is the entity whose embedding is $\mathbf{x}$. Thus, a data point in the training, validation, and test datasets is a tuple $(M_C, length(C))$, where M_C is a matrix of shape $\mathbf{N} \times (\mathbf{d} + 1)$ constructed by concatenating the embeddings of positive examples followed by those of negative examples. Formally, assume n_1 and n_2 are the numbers of positive and negative examples for C, respectively. Further, assume that the embedding vectors of positive examples are $x^{(i)}$, $i = 1 \ldots, n_1$ and those of negative examples are $x^{(i)}$, $i = n_1 + 1, \ldots, n_1 + n_2 = \mathbf{N}$. Then, the $i - th$ row of M_C is given by

$$M_C[i, :] = \begin{cases} (x_1^{(i)}, \ldots, x_\mathbf{d}^{(i)}, 1) & \text{if } 1 \leq i \leq n_1, \\ (x_1^{(i)}, \ldots, x_\mathbf{d}^{(i)}, -1) & \text{if } n_1 + 1 \leq i \leq n_1 + n_2 = \mathbf{N}. \end{cases} \tag{2}$$

We view the prediction of concept lengths as a classification problem with classes $0, 1, \ldots, L$, where L is the length of the longest concept in the training dataset. As shown in Table 2, the concept length distribution can be imbalanced. To prevent concept

Table 2. Number of concepts per length in the training, validation, and test datasets for the four knowledge bases considered

Length	Carcinogenesis			Mutagenesis			Semantic Bible			Vicodi		
	Train	Val.	Test	Train	Val.	Test	Train	Val.	Test	Train	Val.	Test
3	3,647	405	1,013	1,038	115	288	487	54	135	3,952	439	1,098
5	782	87	217	1,156	129	321	546	61	152	2,498	278	694
6	0	0	0	0	0	0	162	18	45	335	37	93
7	1,143	127	318	1,310	146	364	104	12	29	3,597	400	999
8	0	0	0	0	0	0	0	0	0	747	83	207
9	0	0	0	0	0	0	73	8	21	0	0	0
11	0	0	0	0	0	0	41	5	11	0	0	0

length predictors from overfitting on the majority classes, we used the weighted cross-entropy loss

$$\mathcal{L}_w(\bar{y}, y) = -\frac{1}{bs} \sum_{i=1}^{bs} \sum_{k=1}^{L} w_k \mathbb{1}(k, y^i) \log(\bar{y}_k^i), \tag{3}$$

where bs is the batch size, $\bar{y}$ is the batch matrix of predicted probabilities (or scores), y is the batch vector of targets, w is a weight vector, and $\mathbb{1}$ is the indicator function. The weight vector is defined by: $w_k = 1/\sqrt{[k]}$, where $[k]$ is the number of concepts of length k in the training dataset. Table 2 provides details on the training, validation, and test datasets for each of the four knowledge bases. Though the maximal length for the generation of concepts was fixed to 15, many long concepts were equivalent to shorter concepts. As a result, they were removed from the training dataset and the longest remaining are of length 11 (see Table 2).

4.2   Concept Length Predictors

We consider four neural network architectures: Long Short-Term Memory (LSTM) [15], Gated Recurrent Unit (GRU) [8], Multi-Layer Perceptron (MLP), and Convolutional Neural Network (CNN). Recurrent neural networks (LSTM, GRU) take as input a sequence of embeddings of the positive and negative examples (all positive examples followed by all negative examples). The CNN model takes the same input as recurrent networks and views it as an image with a single channel. In contrast, the MLP model inputs the average embeddings of a set of positive and negative examples. The implementation details and the hyper-parameter setting for each of the networks are given in Sect. 6.1.

5   Concept Learner with Integrated Length Prediction (CLIP)

The intuition behind CLIP is that *if we have a reliable concept length predictor, then our concept learner only needs to test concepts of length up to the predicted length.* Figure 1 illustrates CLIP's exploration strategy.

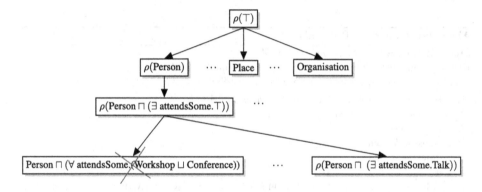

Fig. 1. CLIP search tree when the predicted length is 5. After each refinement, CLIP discards all concepts whose length is larger than the set threshold.

Refinements that exceed the predicted length are ignored during the search. In the figure, the concept Person ⊓ (∀ attendsSome. (Workshop ⊔ Conference)) is of length 7 and is therefore neither tested nor added to the search tree.

Remark 1. For concept length prediction during concept learning, we sample n_1 positive examples and n_2 negative examples from the considered learning problem such that $n_1 + n_2 = \mathbf{N}$, as described in Sect. 4.2 (4).

We implemented the intuition behind CLIP by extending CELOE's refinement operator. Our refinement operator differs from CELOE's in how it refines atomic concepts (see Algorithms 1 and 2). For example, it considers all refinements $A' \sqsubset A$ of an atomic concept A whereas CELOE's refinement operator only considers $A' \sqsubset A$ such that there is no A'' with $A' \sqsubset A'' \sqsubset A$. Omitting this expensive check allows more concepts to be tested in the same amount of time. In the following, we describe our method for refining atomic concepts and refer the reader to [23, 24] for details on CELOE. In Algorithms 1 and 2, the hyper-parameters max_length, k, and $construct\_frac$ control the refinement operator: max_length specifies how long the refinements can become (Algorithm 2, lines 8, 11, 13); k controls the number of fillers sampled without replacement for universal and existential restrictions (Algorithm 1, line 7); $construct\_frac \in (0, 1]$ specifies the fraction of constructs to be sampled (Algorithm 2, lines 1–3).

Given a knowledge base $\mathcal{K}$, the refinement of an atomic concept A is carried out as follows: (1) obtain the subconcepts *subs* of A in $\mathcal{K}$; (2) compute the negations *neg_subs* of all subconcepts of A; (3) construct existential and universal role *restrictions* where the fillers are in the set made of $\top, \bot, A$, and sample k elements from each of the sets *subs* and *neg_subs*; (4) obtain the union *constructs* of *subs*, *neg_subs*, and *restrictions*, and finally (5) Algorithm 2 returns the refinements as intersections or unions of the *subs* and *constructs* computed before, with the generated refinements having length at most max_length. The refinement operator is designed to yield numerous meaningful downward refinements from a single atomic concept.

Algorithm 1. Function REFINEHELPER

Input: Knowledge base $\mathcal{K}$, atomic concept A
Hyper-parameters: Number of subconcepts to be sampled: k, default 5
Output: Subconcepts, negated subconcepts and restrictions of A

1: subs $\leftarrow$ SUBCONCEPTS$_{\mathcal{K}}(A)$ # *Subconcepts of A in $\mathcal{K}$*
2: neg_subs $\leftarrow \{\neg C | C \in$ subs$\}$
3: restrictions $\leftarrow \{\}$
4: **if** |subs| $< k$ **then**
5:      fillers $\leftarrow \{\top, \bot, A\}$
6: **else**
7:      fillers $\leftarrow \{\top, \bot, A\} \cup$ RANDSAMPLE(subs, $n = k$) $\cup$ RANDSAMPLE(neg_subs, $n = k$)
8: **end if**
9: **for** C in fillers **do**
10:     **for** R in role names of $\mathcal{K}$ **do**
11:         restrictions $\leftarrow$ restrictions $\cup \{\exists R.C\}$
12:         restrictions $\leftarrow$ restrictions $\cup \{\forall R.C\}$
13:     **end for**
14: **end for**
15: constructs $\leftarrow$ subs $\cup$ neg_subs $\cup$ restrictions
16: **return** constructs

Algorithm 2. Function REFINEATOMICCONCEPT

Input: Knowledge base $\mathcal{K}$, atomic concept A
Hyper-parameters: Fraction of constructs to be sampled: $construct\_frac \in (0, 1]$, default 0.8;
        max concept length to be generated: max_length, default 15
Output: Set of concepts $\{C_1, \ldots, C_n\}$ which are refinements of A

1: constructs = REFINEHELPER($\mathcal{K}, A$)
2: $m \leftarrow \lfloor construct\_frac \times$ SIZEOF(constructs)$\rfloor$ # *Integer part function*
3: constructs $\leftarrow$ RANDSAMPLE(constructs, $n = m$)   # *Sample without replacement*
4: subs $\leftarrow$ SUBCONCEPTS$_{\mathcal{K}}(A)$ # *Subconcepts of A in $\mathcal{K}$*
5: result $\leftarrow$ subs   # *All subconcepts are of length 1*
6: **for** S_1 in subs **do**
7:      **for** S_2 in constructs **do**
8:          **if** $S_1 \neq S_2$ and $length(S_1 \sqcap S_2) \leq max\_length$ **then**
9:              result $\leftarrow$ result $\cup \{S_1 \sqcap S_2\}$
10:         **end if**
11:         **if** $S_1 \neq S_2$ and $S_2 \in$ subs and $length(S_1 \sqcup S_2) \leq max\_length$ **then**
12:             result $\leftarrow$ result $\cup \{S_1 \sqcup S_2\}$
13:         **else if** $S_1 \neq S_2$ and $length((S_1 \sqcup S_2) \sqcap A) \leq max\_length$ **then**
14:             result $\leftarrow$ result $\cup \{(S_1 \sqcup S_2) \sqcap A\}$ # *All refinements are downward refinements*
15:         **end if**
16:     **end for**
17: **end for**
18: **return** result

Table 3. Overview of benchmark datasets

Dataset	|Individuals|	|At. Concepts|	|Obj. Prop.|	|Data Prop.|	$\|TBox\|$	$\|ABox\|$
Carcinogenesis	22,372	142	4	15	138	96,757
Mutagenesis	14,145	86	5	6	82	61,965
Semantic Bible	724	48	29	9	51	3,211
Vicodi	33,238	194	10	2	193	149,634

6 Evaluation

Datasets. We used four datasets for our experiments: Carcinogenesis, Mutagenesis, Semantic Bible, and Vicodi. All datasets are available in our GitHub repository and described in Table 3. We conducted two sets of experiments. First, we wanted to know which neural architecture performs best at predicting concept lengths. Second, we assessed CLIP's performance w.r.t. its runtime and F-measure when compared with the state-of-the-art refinement approaches dubbed CELOE, OCEL, ELTL, and DL-Foil.

Hardware. The training of our concept length learners was carried out on a single 11 GB memory NVIDIA K80 GPU with 4 Intel Xeon E5-2670 CPUs at 2.60 GHz, and 24 GB RAM. During concept learning with CELOE, OCEL, ELTL, and CLIP, we used an 8-core Intel Xeon E5-2695 at 2.30 GHz, and 16 GB RAM to ensure a fair comparison.

6.1 Concept Length Prediction

Hyper-parameter Optimization. In our preliminary experiments on all four knowledge bases, we used a random search [4] to select fitting hyper-parameters (as summarized in Table 4). Our experiments suggest that choosing two layers for the recurrent neural networks (LSTM, GRU) is the best choice in terms of computation cost and classification accuracy. In addition, two linear layers, batch normalization, and dropout layers are used to increase the performance. The CNN model consists of two convolution layers, two linear layers, two dropout layers, and a batch normalization layer. Finally, we chose 4 layers for the MLP model with batch normalization and dropout layers. The Rectified Linear Unit (ReLU) is used in the intermediate layers of all models, whereas the sigmoid function is used in the output layers.

We ran the experiments in a 10-fold cross-validation setting with ten repetitions. Table 4 gives an overview of the hyper-parameter settings on each of the four knowledge bases considered. The number of epochs was set based on the training speed and the performance of the validation dataset. For example, on the Carcinogenesis knowledge base, most length predictors are able to reach 90% accuracy with just 50 epochs, which suggests that more epochs would probably lead to overfitting. Adam optimizer [18] is used to train the length predictors. We varied the number of examples **N** between 200 and 1000, and the embedding dimension **d** from 10 to 100, but we finally chose $\mathbf{N} = \min(1000, \frac{|\mathcal{K}_\mathcal{I}|}{2})$ and $\mathbf{d} = 40$ as best values for both classification accuracy and computation cost on the four datasets considered.

Table 4. Hyper-parameter setting

Dataset	|Epochs|	lr	d	Batch Size	N
Carcinogenesis	50	0.003	40	512	1,000
Mutagenesis	100	0.003	40	512	1,000
Semantic Bible	200	0.003	40	256	362
Vicodi	50	0.003	40	512	1,000

Table 5. Model size and training time

Model	Carcinogenesis		Mutagenesis	
	|Parameters|	Train. Time (s)	|Parameters|	Train. Time (s)
LSTM	160,208	188.42	160,208	228.13
GRU	125,708	191.16	125,708	228.68
CNN	838,968	16.77	838,248	44.74
MLP	61,681	10.04	61,681	14.29
Model	**Semantic Bible**		**Vicodi**	
	|Parameters|	Train. Time (s)	|Parameters|	Train. Time (s)
LSTM	161,012	196.20	160,409	362.28
GRU	125,512	197.86	125,909	367.55
CNN	96,684	18.43	839,377	71.95
MLP	61,933	9.56	61,744	24.61

Results. Table 5 shows the number of parameters and training time of LSTM, GRU, CNN, and MLP architectures on each of the datasets. From the table, we can observe that our concept length predictors can be trained in less than an hour and be used for efficient concept learning on corresponding knowledge bases.

In Fig. 2, we show the training curves for each model on all datasets. We can observe a decreasing loss on all knowledge bases (see Fig. 2b), which suggests that the models were able to learn. Moreover, the Gated Recurrent Unit (GRU) model outperforms the other models on all datasets, see Fig. 2c and Table 6. The input to the MLP model is the average of the embeddings of the positive and negative examples for a concept. This may have caused loss of information in the inputs. As shown in Fig. 2a, MLP curves tend to saturate in the early stages of training. We also assessed the element-wise multiplication of the embeddings and obtained similar results. However, as reflected in Table 6, all our proposed architectures outperform a random model that knows the distribution of the lengths of concepts in the training dataset. A modified version of MLP where the embedding of each example is processed independently before averaging the final output (no interaction) yielded even poorer results. This suggests that the full interaction between examples is the main factor in the increased performance of recurrent neural networks.

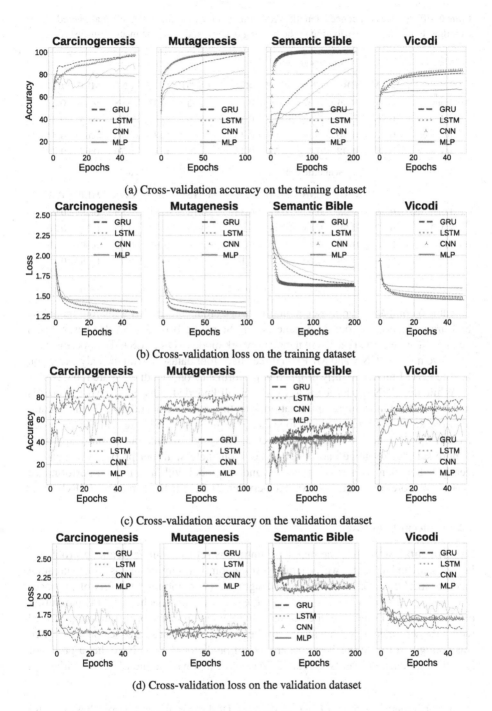

(a) Cross-validation accuracy on the training dataset

(b) Cross-validation loss on the training dataset

(c) Cross-validation accuracy on the validation dataset

(d) Cross-validation loss on the validation dataset

Fig. 2. Training and validation curves

Table 6. Effectiveness of concept length prediction. RM is a random model that makes predictions according to the length distribution in the training dataset, and F1 is the macro F-measure.

Metric	Carcinogenesis					Mutagenesis				
	LSTM	GRU	CNN	MLP	RM	LSTM	GRU	CNN	MLP	RM
Train. Acc	0.89	0.96	0.97	0.80	0.48	0.83	0.97	0.98	0.68	0.33
Val. Acc.	0.76	0.93	0.82	0.77	0.48	0.70	0.82	0.71	0.65	0.35
Test Acc.	0.92	**0.95**	0.84	0.80	0.49	0.78	**0.85**	0.70	0.68	0.33
Test F1	0.88	**0.92**	0.71	0.59	0.33	0.76	**0.85**	0.70	0.67	0.32

Metric	Semantic Bible					Vicodi				
	LSTM	GRU	CNN	MLP	RM	LSTM	GRU	CNN	MLP	RM
Train. Acc	0.85	0.93	0.99	0.68	0.33	0.73	0.81	0.83	0.66	0.28
Val. Acc.	0.49	0.58	0.44	0.46	0.26	0.55	0.77	0.70	0.64	0.30
Test Acc.	0.52	**0.53**	0.37	0.40	0.25	0.66	**0.80**	0.69	0.66	0.29
Test F1	0.27	**0.38**	0.20	0.22	0.16	0.45	**0.50**	0.45	0.38	0.20

Table 6 compares our chosen neural network architectures and a random model on the Carcinogenesis, Vicodi, Mutagenesis, and Semantic Bible knowledge bases. From the table, it appears that recurrent neural network models (GRU, LSTM) outperform the other two models (CNN and MLP) on three out of four datasets, with the only exception that the LSTM model slightly dropped in performance on Vicodi compared to CNN.

While the CNN model tends to overfit on all knowledge bases, the MLP model is unable to extract meaningful information from the average embeddings. On the Semantic Bible knowledge base, which appears to be the smallest dataset, all our proposed networks performed less well than expected. This suggests that our learning approach is more suitable for large knowledge bases. Nonetheless, all our proposed models are clearly better than a distribution-aware random model with a minimum performance (macro F1 score) difference on average between 21.25% (MLP) and 41% (GRU).

6.2 Concept Learning

Experimental Settings. The maximal runtime is set to 2 min per learning problem.[2] For all knowledge bases, we generate 100 random learning problems by (1) creating random $\mathcal{ALC}$ concepts C of maximal length 15, (2) computing the sets of instances $C_{\mathcal{P}}$ and $C_{\mathcal{N}}$, (3) providing $C_{\mathcal{P}}$ and $C_{\mathcal{N}}$ to each of the approaches, and (4) measuring the accuracy, the F-measure, the runtime, and the length of the best solution generated within the set timeout. We ran all approaches on the same hardware (see Sect. 6). CLIP was configured to use our best concept length predictor (GRU). Note that a predictor is trained for each dataset (see Table 2). Also note that we add the ELTL algorithm—a

---

[2] The implementations of OCEL and ELTL in the DL-Learner framework, which we used for our experiments, fail to consider the set threshold accurately. Hence, Table 7 contains values larger than 2 min for these two algorithms.

Table 7. Performance of CLIP compared with CELOE, OCEL, and ELTL on 100 learning problems per knowledge base. The presence of an asterisk indicates that the performance difference is significant between CLIP and the best between CELOE and OCEL. The upward arrow (↑) indicates that the higher is better, whereas the downward arrow (↓) indicates the opposite. All results are average results per knowledge base. The average time is in minutes. ELTL is shown in gray since it learns concepts in $\mathcal{EL}$ instead of $\mathcal{ALC}$ as the others do.

Metric	Carcinogenesis			
	CELOE	OCEL	ELTL	CLIP
Acc. ↑	0.78 ± 0.27	0.89 ± 0.31	0.58 ± 0.46	**0.99** $\pm$ **0.00**
F1 ↑	0.62 ± 0.46	–	0.51 ± 0.47	**0.96**∗ $\pm$ **0.10**
Runtime (min) ↓	0.93 ± 0.94	3.01 ± 0.72	0.75 ± 0.07	**0.10**∗ $\pm$ **0.09**
Length ↓	**1.69** $\pm$ **0.89**	7.81 ± 6.88	1.04 ± 0.39	2.00 ± 1.28

Metric	Mutagenesis			
Metric	CELOE	OCEL	ELTL	CLIP
Acc. ↑	0.99 ± 0.00	0.71 ± 0.45	0.37 ± 0.43	**0.99** $\pm$ **0.00**
F1 ↑	0.81 ± 0.35	–	0.29 ± 0.40	**0.93**∗ $\pm$ **0.18**
Runtime (min) ↓	0.70 ± 0.77	2.39 ± 0.18	0.29 ± 0.16	**0.07**∗ $\pm$ **0.05**
Length ↓	2.79 ± 1.17	12.63 ± 7.03	1.10 ± 0.81	**2.20** $\pm$ **1.16**

Metric	Semantic Bible			
	CELOE	OCEL	ELTL	CLIP
Acc. ↑	0.99 ± 0.02	0.66 ± 0.47	0.59 ± 0.37	**0.99** $\pm$ **0.00**
F1 ↑	0.97 ± 0.10	–	0.57 ± 0.38	**0.98** $\pm$ **0.05**
Runtime (min) ↓	0.47 ± 0.80	22.15 ± 96.55	0.09 ± 0.07	**0.06**∗ $\pm$ **0.05**
Length ↓	3.85 ± 2.44	9.54 ± 5.73	1.38 ± 1.76	**2.52**∗ $\pm$ **1.26**

Metric	Vicodi			
	CELOE	OCEL	ELTL	CLIP
Acc. ↑	0.29 ± 0.44	0.25 ± 0.43	0.28 ± 0.44	**0.99**∗ $\pm$ **0.00**
F1 ↑	0.25 ± 0.44	–	0.25 ± 0.44	**0.97**∗ $\pm$ **0.09**
Runtime (min) ↓	1.30 ± 0.71	4.78 ± 1.12	1.81 ± 0.46	**0.16**∗ $\pm$ **0.12**
Length ↓	10.79 ± 6.30	11.54 ± 6.00	11.14 ± 6.11	**1.68**∗ $\pm$ **0.98**

concept learner for the DL $\mathcal{EL}$—to investigate whether our randomly generated concepts are equivalent to concepts in a simpler description logic.

Results. Table 7 presents a comparison of the results achieved by CLIP, CELOE, OCEL, and ELTL; results are formatted *mean $\pm$ standard deviation*. Note that the table does not contain DL-FOIL because it could not solve the learning problems that we considered. For instance, the first learning problem on the Semantic Bible knowledge base targets SonOfGod ⊔ (∃ locationOf.StateOrProvince). Here, DL-FOIL was stuck on the refinement of GeographicLocation with over 5×10^3 unsuccessful trials. Similar scenarios were observed on other datasets. We also tried running DL-FOCL, but it was not possible using the documentation provided.

Our results suggest that CLIP outperforms the other three algorithms in F1 and in runtime on most datasets. The ELTL algorithm appears to be faster than CELOE and OCEL but slower than CLIP. However, its runtime performance stems from the fact that it detects concepts in the DL $\mathcal{EL}$. Since some of our learning problems can only be solved in $\mathcal{ALC}$ or a more expressive DL, ELTL performs poorly in F1 score on all datasets. This result suggests that we do not generate trivial problems.

We used a Wilcoxon Rank Sum test to check whether the difference in performance between CLIP, CELOE, and OCEL was significant. Significant differences are marked with an asterisk. The null hypothesis for our test was as follows: "the two distributions that we compare are the same". The significance level was $\alpha = 0.05$. The performance differences in F1 between CLIP and the other algorithms are significant on 3 out of 4 datasets.[3] With respect to runtimes, we significantly outperform all other algorithms on all datasets. Large time differences correspond to scenarios where CLIP detects short solution concepts while other algorithms explore longer concepts. Low time differences correspond to either simple learning problems, where all algorithms find a solution in a short period of time, or complex learning problems where CLIP explores long concepts as other algorithms.

The average runtimes of CELOE, OCEL, and CLIP across all datasets are 0.85, 8.08, and 0.1 min, respectively. Given that the average training time of the length predictor GRU is 4.10 min, we can conjecture the following: (1) the expected number of learning problems from which CLIP should be preferred over CELOE is 5, and (2) CLIP should be preferred over OCEL for any number of learning problems.

7 Conclusion and Future Work

We investigated the prediction of concept lengths in the description logic $\mathcal{ALC}$, to speed up the concept learning process using refinement operators. To this end, four neural network architectures were evaluated on four benchmark knowledge bases. The evaluation results suggest that all of our proposed models are superior to a random model, with recurrent neural networks performing best at this task. We showed that integrating our concept length predictors into a concept learner can reduce the search space and improve the runtime and the quality (F-measure) of solution concepts.

Even though our proposed learning approach was very efficient when dealing with concepts of length up to 11 (in $\mathcal{ALC}$), its behavior is not guaranteed when longer concepts are considered. Moreover, the use of generic embedding techniques might lead to suboptimal results. In future work, we plan to jointly learn the embeddings of a given knowledge graph and the lengths of its complex (long) concepts. We will also explore further network architectures such as multi-set convolutional networks [38] and neural class expression synthesis [19].

Acknowledgements. This work is part of a project that has received funding from the European Union's Horizon 2020 research and innovation programme under the Marie Skłodowska-Curie grant agreement No 860801. This work has been supported by the German Federal Ministry of Education and Research (BMBF) within the project DAIKIRI under the grant no 01IS19085B and by the German Federal Ministry for Economic Affairs and Climate Action (BMWK) within the project RAKI under the grant no 01MD19012B. The authors gratefully acknowledge the funding of this project by computing time provided by the Paderborn Center for Parallel Computing (PC²).

---

[3] Note that we ran OCEL with its default settings and F1 scores are not available.

References

1. Ashburner, M., et al.: Gene ontology: tool for the unification of biology. Nat. Genet. **25**(1), 25–29 (2000)
2. Baader, F., Calvanese, D., McGuinness, D.L., Nardi, D., Patel-Schneider, P.F. (eds.): The Description Logic Handbook: Theory, Implementation, and Applications. Cambridge University Press, Cambridge (2003)
3. Badea, L., Nienhuys-Cheng, S.-H.: A refinement operator for description logics. In: Cussens, J., Frisch, A. (eds.) ILP 2000. LNCS (LNAI), vol. 1866, pp. 40–59. Springer, Heidelberg (2000). https://doi.org/10.1007/3-540-44960-4_3
4. Bergstra, J., Bengio, Y.: Random search for hyper-parameter optimization. J. Mach. Learn. Res. **13**(2) (2012)
5. Bin, S., Bühmann, L., Lehmann, J., Ngonga Ngomo, A.C.: Towards SPARQL-based induction for large-scale RDF data sets. In: ECAI 2016, pp. 1551–1552, IOS Press (2016)
6. Bordes, A., Glorot, X., Weston, J., Bengio, Y.: A semantic matching energy function for learning with multi-relational data. Mach. Learn. **94**(2), 233–259 (2013). https://doi.org/10.1007/s10994-013-5363-6
7. Bühmann, L., Lehmann, J., Westphal, P.: DL-Learner—a framework for inductive learning on the Semantic Web. J. Web Semant. **39**, 15–24 (2016)
8. Cho, K., et al.: Learning phrase representations using RNN encoder-decoder for statistical machine translation. arXiv preprint arXiv:1406.1078 (2014)
9. Gene Ontology Consortium: The Gene Ontology (GO) database and informatics resource. Nucleic Acids Res. **32**(suppl1), D258–D261 (2004)
10. Dai, Y., Wang, S., Xiong, N.N., Guo, W.: A survey on knowledge graph embedding: approaches, applications and benchmarks. Electronics **9**(5), 750 (2020)
11. Demir, C., Ngomo, A.C.N.: Convolutional complex knowledge graph embeddings. arXiv preprint arXiv:2008.03130 (2020)
12. Deshpande, O., et al.: Building, maintaining, and using knowledge bases: a report from the trenches. In: Proceedings of the 2013 ACM SIGMOD International Conference on Management of Data, pp. 1209–1220 (2013)
13. Fanizzi, N., d'Amato, C., Esposito, F.: DL-FOIL concept learning in description logics. In: Železný, F., Lavrač, N. (eds.) ILP 2008. LNCS (LNAI), vol. 5194, pp. 107–121. Springer, Heidelberg (2008). https://doi.org/10.1007/978-3-540-85928-4_12
14. Heindorf, S., et al.: EvoLearner: Learning description logics with evolutionary algorithms. In: Proceedings of the ACM Web Conference (2022)
15. Hochreiter, S., Schmidhuber, J.: Long short-term memory. Neural Comput. **9**(8), 1735–1780 (1997)
16. Hogan, A., et al.: Knowledge graphs. Synth. Lect. Data Semant. Knowl. **12**(2), 1–257 (2021)
17. Ioannidis, V.N., et al.: DRKG-drug repurposing knowledge graph for COVID-19 (2020)
18. Kingma, D.P., Ba, J.: Adam: a method for stochastic optimization. arXiv preprint arXiv:1412.6980 (2014)
19. Kouagou, N.J., Heindorf, S., Demir, C., Ngomo, A.N.: Neural class expression synthesis. CoRR abs/2111.08486 (2021)
20. Krötzsch, M., Simancik, F., Horrocks, I.: A description logic primer. CoRR abs/1201.4089 (2012)
21. Lehmann, J.: DL-Learner: learning concepts in description logics. J. Mach. Learn. Res. **10**, 2639–2642 (2009)
22. Lehmann, J.: Learning OWL Class Expressions, vol. 22. IOS Press (2010)
23. Lehmann, J., Auer, S., Bühmann, L., Tramp, S.: Class expression learning for ontology engineering. J. Web Semant. **9**(1), 71–81 (2011)

24. Lehmann, J., Hitzler, P.: Concept learning in description logics using refinement operators. Mach. Learn. **78**(1–2), 203 (2010)
25. MacLean, F.: Knowledge graphs and their applications in drug discovery. Expert Opin. Drug Discov. **16**(9), 1057–1069 (2021)
26. Nickel, M., Tresp, V., Kriegel, H.P.: Factorizing YAGO: scalable machine learning for linked data. In: Proceedings of the 21st international conference on World Wide Web, pp. 271–280 (2012)
27. Percha, B., Altman, R.B.: A global network of biomedical relationships derived from text. Bioinformatics **34**(15), 2614–2624 (2018)
28. Rizzo, G., Fanizzi, N., d'Amato, C.: Class expression induction as concept space exploration: from DL-Foil to DL-Focl. Future Gener. Comput. Syst. **108**, 256–272 (2020)
29. Rizzo, G., Fanizzi, N., d'Amato, C., Esposito, F.: A framework for tackling myopia in concept learning on the web of data. In: Faron Zucker, C., Ghidini, C., Napoli, A., Toussaint, Y. (eds.) EKAW 2018. LNCS (LNAI), vol. 11313, pp. 338–354. Springer, Cham (2018). https://doi.org/10.1007/978-3-030-03667-6_22
30. Rudolph, S.: Foundations of description logics. In: Polleres, A., d'Amato, C., Arenas, M., Handschuh, S., Kroner, P., Ossowski, S., Patel-Schneider, P. (eds.) Reasoning Web 2011. LNCS, vol. 6848, pp. 76–136. Springer, Heidelberg (2011). https://doi.org/10.1007/978-3-642-23032-5_2
31. Sarker, M.K., Hitzler, P.: Efficient concept induction for description logics. In: AAAI, pp. 3036–3043 (2019)
32. Schmidt-Schauß, M., Smolka, G.: Attributive concept descriptions with complements. Artif. Intell. **48**(1), 1–26 (1991)
33. Wang, Q., Mao, Z., Wang, B., Guo, L.: Knowledge graph embedding: a survey of approaches and applications. IEEE Trans. Knowl. Data Eng. **29**(12), 2724–2743 (2017)
34. Wang, Z., Li, J., Liu, Z., Tang, J.: Text-enhanced representation learning for knowledge graph. In: Proceedings of International Joint Conference on Artificial Intelligent (IJCAI), pp. 4–17 (2016)
35. Weston, J., Bordes, A., Yakhnenko, O., Usunier, N.: Connecting language and knowledge bases with embedding models for relation extraction. arXiv preprint arXiv:1307.7973 (2013)
36. Wishart, D.S., et al.: DrugBank 5.0: a major update to the DrugBank database for 2018. Nucleic Acids Res. **46**(D1), D1074–D1082 (2018)
37. Xie, R., Liu, Z., Jia, J., Luan, H., Sun, M.: Representation learning of knowledge graphs with entity descriptions. In: Thirtieth AAAI Conference on Artificial Intelligence (2016)
38. Zaheer, M., Kottur, S., Ravanbakhsh, S., Póczos, B., Salakhutdinov, R., Smola, A.J.: Deep sets. In: NIPS, pp. 3391–3401 (2017)

Dihedron Algebraic Embeddings for Spatio-Temporal Knowledge Graph Completion

Mojtaba Nayyeri[1,5], Sahar Vahdati[2], Md Tansen Khan[1(✉)], Mirza Mohtashim Alam[2], Lisa Wenige[2], Andreas Behrend[3], and Jens Lehmann[1,4]

[1] University of Bonn, Bonn, Germany
nayyeri@cs.uni-bonn.de, s6mmkhan@uni-bonn.de
[2] Institute for Applied Informatics (InfAI), Dresden, Germany
{vahdati,alam,wenige}@infai.org
[3] Institute for Telecommunications (INT), TH Köln, Cologne, Germany
andreas.behrend@th-koeln.de
[4] Fraunhofer IAIS, Dresden, Germany
jens.lehmann@iais.fraunhofer.de
[5] University of Stuttgart, Stuttgart, Germany

Abstract. Many knowledge graphs (KG) contain spatial and temporal information. Most KG embedding models follow triple-based representation and often neglect the simultaneous consideration of the spatial and temporal aspects. Encoding such higher dimensional knowledge necessitates the consideration of true algebraic and geometric aspects. Hypercomplex algebra provides the foundation of a well defined mathematical system among which the Dihedron algebra with its rich framework is suitable to handle multidimensional knowledge. In this paper, we propose an embedding model that uses Dihedron algebra for learning such spatial and temporal aspects. The evaluation results show that our model performs significantly better than other adapted models.

Keywords: Knowledge graph · Embedding · Spatio-temporal

1 Introduction

Large cross-domain Knowledge Graphs (KGs), such as DBpedia [15] and Wikidata [27], leverage a triple representation of facts in the form of (h, r, t) where h, t and r refer to head and tail entities and the relation respectively. Despite the availability of huge amounts of such data, one of the major challenges is the impossibility of capturing all (true) facts of the target domain. Thus, sparsity and incompleteness are major problems of KGs. The objective of KG completion is to generate true triples that are not explicitly given in the KG. For example, the query $(PrinceWilliam, met, ?)$, with an unknown tail as "?", means *"With whom did Prince William meet?"* for which $h = VolodymyrZelensky$ is a possible answer. This leads to predict the $(PrinceWilliam, met, VolodymyrZelensky)$ triple.

P. Groth et al. (Eds.): ESWC 2022, LNCS 13261, pp. 253–269, 2022.
https://doi.org/10.1007/978-3-031-06981-9_15

Knowledge Graph Embedding (KGE) models have shown high performance for KG completion. A KGE model usually maps the entities and relations into a d dimensional vector space (e.g. $\mathbb{R}^d$). The plausibility of a triple (h, r, t) is measured via a score function $f(h, r, t)$. In this way, a KGE model performs the KG completion task by replacing the potential entities or relations in incomplete triple patterns. After measuring the plausibility, the triples with high scores are regarded to be likely true and can be added to the KG for completing it or undergoing further human inspection. This is a widely used approach for major KGE models that are only designed for triple-based KGs. However, important semantic aspects of some facts are neglected when not considering Spatio-temporal aspects. For example, for the triple $(PrinceWilliam, met, VolodymyrZelensky)$ it would be highly relevant to know the location and time of the meeting. We found that at least 13% of the resources in DBpedia fall into this category where triples are connected to additional information via time, location, or both.

However, the existing KGE models often only consume triples and are not capable of exploiting the additional spatial and temporal dimension of facts. Recent attempts in temporal knowledge graph embeddings advance consideration of temporal aspects [31], but do not consider spatial information. In those models, facts are then represented as quadruples (h, r, t, τ) where τ is the temporal information. Therefore, the respective models are capable to complete quadruples of the form $(?, r, t, \tau)$, $(h, ?, t, \tau)$ or $(h, r, ?, \tau), (h, r, t, ?)$. However, none of the current models can directly consider spatial information. Spatio-temporal facts can be represented as a quintuple (h, r, t, l, τ) where l reflects the location information (spatial). Such

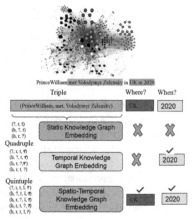

Fig. 1. Spatio-temporal treated by KGEs.

quintuples, e.g. $(PrinceWilliam, met, VolodymyrZelensky, 2020, UK)$, can simultaneously associate time and space information to a given fact. As shown in Fig. 1 time and location are excluded from most current KGE models and thus potentially valuable Spatio-temporal information remains unused for completion tasks. This might in turn reduce the performance as well as the meaningfulness and interpretability of machine learning results. Therefore, we propose a family of KGE models to fully exploit the spatial and temporal information for KG completion. We specifically take advantage of the 4D algebra of hypercomplex vectors to complete a quintuple representation in the form $(?, r, t, l, \tau), (h, ?, t, l, \tau), (h, r, ?, l, \tau), (h, r, t, ?, \tau)$ or $(h, r, t, l, ?)$. This is especially suitable because for each incomplete quintuple, four of the five elements are always present. All of these four elements (entity (h or t), relation r, location l, time τ) are assumed to be mutually independent which will be represented by 4 orthogonal bases. In this work, we employ the Dihedron algebra, which is a rich 4D algebra of hypercomplex spaces and provides a well-suited theoretical foundation for the embedding of quintuples considering time and location aspects. Our contributions are as follows: We present a) a family of novel KGE models using Dihedron algebra to capture Spatio-temporal information, b) a technique that allows existing KGE models to be adapted to include Spatio-temporal information, c) a geometric interpretation of the used algebra for the problem of encoding Spatio-temporal information, d) several Spatio-temporal KGs (ST-KGs) derived from YAGO3K, DBpedia34K, and WikiData53K.

2 Related Work

Static Knowledge Graph Embedding Most existing KGE models learn over KGs containing triples of the form (h, r, t). One of the primary KGE models is TransE [2]. For a given positive triple, TransE represents a relation r as translation from head to tail i.e. $h + r = t$ where h, r, t are embedding vectors of head, relation and tail. In order to address the limitations of TransE in encoding various relational patterns such as symmetry, transitivity etc., several variants of TransE have been proposed such as TransH [28], TransR [18], TransD [8] etc. In RotatE, each relation is represented as a rotation in the complex space. Due to algebraic characteristics of rotation, e.g. a) every rotation matrix has a unique inverse, b) composition of two rotation matrices is a rotation matrix, this model can encode various relational patterns such as inversion and composition [24]. The ComplEx model embeds KGs into a complex vector space which together with N3 regularization as in ComplEx-N3, has become one of the state of the art KGE models [13,26]. ComplEx and ComplEx-N3 efficiently model symmetric and anti-symmetric patterns. QuatE extends ComplEx to Quaternion vector representation and obtains state-of-the-art results on link prediction over static KGs when it is combined with N3 regularization i.e. QuatE-N3 [33]. The matrix representation from Dihedral groups [30] has been used for modeling each relation of a KG to represent various patterns such as skew-symmetric, inversion, and composition in static KGs. There are several other static KGE models that can be found in [3,9,17,21].

Temporal Knowledge Graph Embedding Temporal Knowledge Graph Embedding (TKGE) models focus on dynamic KGs with an additional temporal part. In this way, triple-based representation is formed as quadruple. Most of the early TKGE models have been built on top of the already existing KGEs. The HyTE [4] model is one of the early TKGEs that first projects the head, relation, and tail embeddings to the space of the timestamp. Furthermore, for the final scoring of the newly predicted facts, it employs TransE on the projected embeddings. There are several other TKGEs which have been proposed as extensions of TransE such as TTransE [14] and TA-TransE [6]. HyTE and other extensions do not consider any hypercomplex algebraic aspects that could let the model cover the spatial information beside the temporal ones. The other state of the art model among TKGEs is the ConT model that is an extension of the Tucker [1] KGE. There are also several extensions of DisMult [32] that were proposed for encoding of temporal KGs such as TDistMult [19] and TA-DistMult [6]. These models are based on recurrent neural network (RNNs) that captures the entity embeddings for head and tail parts. RE-NET is another RNN-based TKGE that captures pair-wise knowledge in the form of (head, relation) or (tail, relation) by using specific patterns from the historical information between entities [10]. However, none of these extensions exploits the algebraic aspects of the embedding models or attempts to consider the spatial aspects.

Another issue of these models is that all of them inherit the problems of the underlying base models on top of which they are extended. For example, the TKGEs that were built on top of TransE suffer from encoding of certain relational patterns. The TeRo model [31] has been recently proposed to overcome such problems of the already existing TKGEs on inference of relational patterns. Although TeRo solves the limitation of other models to some extend, it does not leverage the characteristics of hypercomplex spaces, also does not target ST-KGs. The TComplEx model [12] is temporal version of ComplEx-N3, that obtained state-of-the art performance on link prediction over temporal KGs.

Other Works related to Spatio-Temporal In our work, a spatio-temporal context is denoted as a combination of a location as cities or countries and a time slot with the granularity of year, e.g., Spain, 1982. It shall be noted that in other domains, the term "spatial" also refers to the detailed geographical aspects of locations either with satellite information or geo-coordinates information. There are different tracks of interdisciplinary research related to the later meaning of spatial data such as geospatial artificial intelligence (GeoAI) combining geography, earth science and artificial intelligence which are not directly in the scope of our work [7,16,22]. In [34], a general framework for analysing multi-source spatio-temporal data is given. Although, this approach is based on KGEs, the meaning of spatial data lies in the urban data, land maps and satellite data.

3   Dihedron Algebra

Dihedron is a hypercomplex number system, extending complex number to 4D space. The space is denoted by $\mathbb{D}$. A Dihedron number $d = x_r 1 + x_i i + x_j j + x_k k \in \mathbb{D}$ includes one real part x_r and three imaginary parts $x_i i$, $x_j j, x_k k$, where the bases are $1 = \begin{bmatrix} 1 & 0 \\ 0 & 1 \end{bmatrix}, i = \begin{bmatrix} 0 & 1 \\ \bar{1} & 0 \end{bmatrix}, j = \begin{bmatrix} 0 & 1 \\ 1 & 0 \end{bmatrix}, k = \begin{bmatrix} 1 & 0 \\ 0 & \bar{1} \end{bmatrix}$, $\bar{a} = -a$, and the following equations hold $i^2 = \bar{1}, j^2 = 1, k^2 = 1, ij = k$, $ij = k, jk = \bar{i}, ki = j, ji = \bar{k}, kj = i$, and $ik = \bar{j}$. Some other mathematical representations and operations are as follows:

Dihedron Matrix Representation A Dihedron number $d \in \mathbb{D}$ can be represented as the following matrix form

$$q = x_r 1 + x_i i + x_j j + x_k k = \begin{bmatrix} x_r + x_k & x_i + x_j \\ \bar{x}_i + x_j & x_r + \bar{x}_k \end{bmatrix} \equiv (x_r, x_i, x_j, x_k) \equiv$$
$$x_r + v_q, \ v_q = (x_i, x_j, x_k) \in \mathbb{R}^{3d} \ q \in \mathbb{D}.$$

Dihedron Product The Dihedron product (denoted by $\otimes$) between two Dihedron numbers $q_x, q_y \in \mathbb{D}$ is defined as follows

$$q_z = z_r 1 + v_{q_z} = q_x \otimes q_y = (x_r 1 + v_{q_x}) \otimes (y_r 1 + v_{q_y})$$
$$= (x_r y_r - \langle v_{q_x}, v_{q_y} \rangle)1 + (x_r v_{q_y} + y_r v_{q_x}) + v_{q_x} \times v_{q_y}, \tag{1}$$

$$\text{where } \langle v_{q_x}, v_{q_y} \rangle = x_i y_i - x_j y_j - x_k y_k, \quad v_{q_x} \times v_{q_y} = \begin{bmatrix} -x_j y_k + x_k y_j \\ x_k y_i - x_i y_k \\ x_i y_j - x_y y_i \end{bmatrix}.$$

The product incorporates all element factors into computation. We will explain the advantage of this product in modeling spatio-temporal data in the next section.

4 Proposed Approach

Here, we introduce the proposed approach dubbed ST-NewDE based on Dihedron Algebra. Let us assume the following fact *"Prince William met Volodymyr Zelensky in UK in 2020."* that contains five parts namely subject (*Prince William*), relation (*meet*), object (*Volodymyr Zelensky*), adverb of place (*UK*), and adverb of time (*2020*) . Incompleteness occurs when one of the parts is missing at a time. In a ST-KG, those could be seen as a quintuple in the form $(?, r, t, l, \tau), (h, ?, t, l, \tau), (h, r, ?, l, \tau), (h, r, t, ?, \tau)$ or $(h, r, t, l, ?)$. One efficient way towards completing such a KG is to represent such spatio-temporal queries in the vector space where four elements are embedded separately in a real vector space. Then, each of those four parts are combined to build up the query representation as 4D hypercomplex vectors. To complete the missing part ?, we employ a rich 4D algebra of hypercomplex space named Dihedron that is used to measure the similarity of the query and the plausible answers (i.e. possible entities for object), while capturing the mutual correlation between each pair elements.

Spatio-Temporal KG Let us have a spatio-temporal KG $\mathcal{K} = \{(h, r, t, l, \tau) | h, t \in \mathcal{E}, r \in \mathcal{R}, l \in \mathcal{L}, \tau \in \mathcal{T}\}$, where $\mathcal{E}, \mathcal{R}, \mathcal{L}, \mathcal{T}$ are entity, relation, location and time dictionaries respectively.

Embedding Space Each entity $(e = h, t)$, relation (r), location (l) and time (τ) are embedded into d dimensional real vector space, shown in bold i.e. $\boldsymbol{e}, \boldsymbol{r}, \boldsymbol{l}, \boldsymbol{\tau} \in \mathbb{R}^d$. Note that an entity $(e \in \mathcal{E})$ plays both roles of subject (head h) or object (tail t) in a quintuple. If an entity is in the subject role, the embedding is shown as $\boldsymbol{h}$ ($\boldsymbol{t}$ for subject role). **Incomplete Quintuples and Answers** Given a quintuple $\{(h, r, t, l, \tau)$, we split it into two parts: incomplete quintuples (IQ) and answer (A) as shown in Table 1. In this way,

Table 1. Incomplete quintuples and their answers in spatio-temporal knowledge graphs.

IQ	$(?, r, t, l, \tau)$	$(h, ?, t, l, \tau)$	$(h, r, ?, l, \tau)$	$(h, r, t, ?, \tau)$	$(h, r, t, l, ?)$
A	h	r	t	l	τ

each incomplete quintuple contains four fixed parts. Because 4D spaces contain four mutually orthogonal bases, they are the most suitable algebraic representation for such quintuples with four fixed parts. Among 4D spaces, Dihedron and Quaternion are two of the main and well-defined hypercomplex algebras [23,29]. Dihedron representation covers a wider range of geometric shapes than the Quaternion in the 4D vector space. Consequently, it is more expressive and thus better suited for encoding the complex spatio-temporal incomplete quintuples in the 4D space. Figure 2 illustrates the flexibility of Dihedron over Quaternion in a 4D vector space. From the geometric view point, Dihedron algebra represents various hyperboloid models (two-sheet, one-sheet) as well as conic surfaces and spheres. In this regard, we consider Dihedron algebra as a *true (well-suited) algebra* for modeling complex spatio-temporal incomplete quintuples.

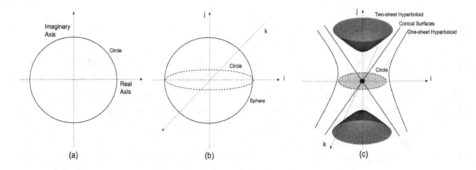

Fig. 2. Illustration of (a) Complex plane; (b) Quaternion space; (c) Dihedrons space.

Dihedron Representation of Incomplete Quintuples: We represent a 4D spatio-temporal query (incomplete quintuples) of the form $(h, r, ?, l, \tau)$ as a $d \times 4D$ vector in a Dihedron space $\mathbb{D}^d$ i.e. $q = h1 + ri + lj + \tau k$, where q is the generalization of complex numbers with three imaginary units i, j, k, and also we have $h, r, l, \tau \in \mathbb{R}^d$. We specifically focus on Dihedron algebra where the detailed description was introduced in previous section. Therefore, a Dihedron query is represented as follows

$$q = h1 + ri + lj + \tau k = \begin{bmatrix} h + \tau & r + l \\ \bar{r} + l & h + \bar{\tau} \end{bmatrix} \equiv (h, r, l, \tau) \equiv$$
$$h + v_q, \quad v_q = (r, l, \tau) \in \mathbb{R}^{3d}, \quad q \in \mathbb{D}^d.$$

Object Dihedron Completion of Quintuples: Here we show how the answers to the above incomplete quintuples are computed by a KGE using Dihedron algebra. Let t e.g. *Volodymyr Zelensky* be the answer of an incomplete quintuple $(h, r, ?, l.\tau)$ for a query e.g. *Whom Prince William (h) did meet (r) in 2020 (τ) in UK (l)?*. In order to complete the quintuple in the vector space, the similarity of the query (q) and its given answer (t) is maximized. Here we use the Dihedron product $(\otimes)$ between the query q and its answer t representations. Note that while the query q lies on the Dihedron space, its answer $t = e$ lies on the real space. To match these two representations on the same space, we add three real auxiliary (t_x, t_y, t_z) vectors to tail embedding i.e. $t = e1 + v_t$, $v_t = t_x i + t_y j + t_z k$. Using two representations for an entity e according to its role (head $h = e$ or tail $t = e1 + v_t$) enables our model to differentiate between each roles. Such representations facilitate efficient capturing of the natural role of entities per each triple for the underlying KG. After matching the spaces, now, we can apply the Dihedron product to measure the plausibility of tail t to be the answer of the query q. This leads us to define the score function as

$$S(q, t) = -\|s1 + v_s\|, \tag{2}$$

where

$$s1 + v_s = q \otimes t = (h1 + v_q) \otimes (e1 + v_t), \tag{3}$$

$$s = h.e - v_q.v_t, \quad v_s = hv_t + ev_q + v_q \times v_t, \quad v_q.v_t = rt_x - lt_y - \tau t_z,$$

$$v_q \times v_t = \begin{bmatrix} -lt_z + \tau t_y \\ \tau t_x - rt_z \\ rt_y - lt_x \end{bmatrix}. \tag{4}$$

The advantage of Eq. 2 is its efficiency in memory and time due to representing query and answer separately in Dihedron space while to compute the query, no mathematical operation (e.g., addition/subtraction/multiplication) are used. All operations are done in the query answering phase which reduces the complexity significantly.

Subject Dihedron Completion of Quintuples: The previous formulation for incomplete quintuple representation and their corresponding answers was used when tail (object) was queried i.e. $(h, r, ?, l, t)$. Here we present our formulation for an incomplete quintuple and its answer where the head (subject) is queried i.e. $(?, r, t, l, \tau)$. Let us assume that, the example question is *Who met with Volodymyr Zelensky in UK in 2020?* and the answer is *Barack Obama*. We first use the common approach from [13] of augmentation on the KG by adding reverse relations (here reverse quintuples) (t, r^{-1}, h, l, τ) for each quintuple (h, r, t, l, τ) present in the KG (train set). During testing, we use the reverse quintuple as following $q = t1 + r^{-1}i + lj + \tau k$, and the answer for this Dihedron representation is the following h where entity is attached with three auxiliary vectors i.e., $h = e1 + h_x i + h_y j + h_z k$. In this way, we preserve the basis of entity (1), relation (verb) (i), adverb of location (j) and adverb of time (k) regardless of head (subject) or tail (object) Incomplete Quintuples.

Mathematical Interpretation We highlight the advantages of our formulation from various mathematical view points namely *Spatio-temporal coordinate representation, Capturing Spatio-temporal and Relational Correlations* and *Geometric Interpretation*. *Spatio-temporal Coordinate*. We represent a query with $q = h1 + ri + lj + \tau k = h1 + v_q$, $v_q = (r, l, \tau) \in \mathbb{R}^{3d}$, $q \in \mathbb{D}^d$. In this representation, we assign distinct coordinate bases to each of the distinct quintuple elements i.e. h, r, l, τ (i, j, k in Fig. 2 part (c)). This is consistent with the nature of KGs where entity, relation, location and time are four distinct components. It is noteworthy that the orthogonality of bases in Dihedron space is related to the position of elements of a quintuple pattern which affects the order of the elements in 4D Dihedron space corresponding to **1, i, j, k**, and it is not related to the embedding vectors h, r, τ, l. In addition, this representation enables the head entity (h) to move towards the tail entity (t), by using the vector v_q. In order to determine the direction of movement, the v_q vector is not only dependent on the relation, but also on the location and time adverbs.

Capturing Spatio-Temporal and Relational Correlations While entity, relation, location and time are considered as orthogonal bases in the Dihedron space, the correlation between i) head entity-(relation, location, time) (hv_t), ii) tail entity-(relation, location, time) (ev_q), iii) head-tail entity (he), iv) relation-location rt_y, v) relation-time (rt_z), vi) location-time (lt_z), vii) location-relation (lt_x), viii) time-relation (τt_x), etc. are involved in final score calculation. Such comprehensive correlation capturing is enabled by properly formulating the query and the answer and efficiently using the Dihedron product in score calculation (see Eq. 2, 3, 4).

Geometric Interpretation (Two/Three Dimensional Subalgebras of $\mathbb{D}$) Without loss of generality, let $d = 1$ be the length of a Dihedron $h1 + ri + lj + \tau k$ is $h^2 + r^2 - l^2 - \tau^2 = c$ where c is a constant value. This representation is rich in terms of geometry and covers the following geometric objects (see part (c) of Fig. 2):

i **Circle**: Let us fix $(-l^2 - \tau^2)$ in $h^2 + r^2 - l^2 - \tau^2 = c$ to a constant v alue as $-c_1^2$. We then have a circle of the form $h^2 + r^2 = c_2$, $c_2 = c + c_1^2$.

ii **Two Sheet Hyperboloid**: Let us then fix h^2 to a constant value c_2^2. Therefore, $r^2 - l^2 - \tau^2 = c - c_2^2$. If $c - c_2^2 < 0$, then the object is one sheet Hyperboloid.

iii **One Sheet Hyperboloid**: In the previous representation, if $c - c_2^2 > 0$, the object is two sheet Hyperboloid.

iv **Conical Surface**: Following the above definition for $r^2 - l^2 - \tau^2 = c$, if $c - c_2^2 = 0$, then the object is then conical surfaces.

Regarding the mentioned advantages, our proposed model is capable of efficiently embedding the spatio-temporal KGs into a well-suited geometric space. For the other incomplete quintuples $(h, r, t, ?, \tau)$, $(h, r, t, l, ?)$ that were not discussed here, the same approach is followed. We add the potential location or time of the query part in the Eq. 4 and then compute the score using Eq. 2. If the resulted score is high, the selected location or time leads to a plausible quintuple.

5   Experiments

Here, we provide evaluation of our model against the adapted models for ST KGs. **Baseline Models.** We compare our model with several baselines and state-of-the-art KGE models. Overall, there are three categories of KGE models selected for comparison. *Static KGE Models.* Many KGE models learn over triples. The TransE [2] model has been widely used as baseline for KGEs. Although this model is simple, it obtained high performance in link prediction task on several benchmark KGs. RotatE [24] is another model that uses self-adversarial negative sampling that led to obtain state-of-the art performance on the distance based class of embeddings (i.e. the models that use distance function in their score functions). In order to have a fair comparison, we train this model without using self-adversarial negative sampling (in same setting as ours). The ComplEx model trained with N3 regularization [13] got state-of-the-art performance in most of the used static KGs. To have a fair comparison, we trained all the models (including our models and their competitors) using N3 regularization. QuatE [33] took the advantage of the Quaternion space and obtained state-of the-art performance with N3 regularization. *Temporal KGE Models.* There is less research on the topic of temporal KGEs in comparison to the work done on static KGEs. However, there are several strong baseline methods for temporal KGEs. Similar to ComplEx-N3, temporal version of this model obtained state-of-the art performance on link prediction over temporal KGs. Therefore, we chose TComplEx [12] that was trained with N3 regularization, as a competitor. For other models, we extended the three state of the art KGE models (TransE, RotatE, and QuatE) in terms of capability of encoding temporal aspects. Therefore, T-TransE,

T-RotatE and T-QuatE are our extension over existing static KGE models for temporal embedding. The formulation of the models are specified in the Table 2.

Spatio-Temporal KGE Models We are not aware of any knowledge graph embedding model over spatio-temporal KGs (ST KGs). Therefore, we extended the formulation of the above-mentioned static and temporal KGE models to learn over ST KGs. ST-TransE, ST-RotatE, ST-ComplEx, ST-QuatE are the resulting models. Their characteristics and formulations are included in Table 2.

Table 2. Specification of baseline and state of the art KGE models."Ours" denote our proposed models which are mostly extensions of already existing models for temporal and ST-KG embeddings. DyHE, T-DyHE and ST-DyHE are models we propose for the ablation study on the effect of translation and rotation in Dihedron space for static, temporal and spatio-temporal KGEs. Note that r_1, r_2 are Dihedron rotation and translation vectors. The symbols $\circ, \otimes_H$ and $\otimes_D$ denote the complex, Hamilton and Dihedron products, respectively. $Re(.)$ refers to the real part of complex numbers. (T-L) refer to time and location. $r_{1,2} = r_1 + r_2$

Model	(T-L)	Score function	Pattern	Embeddings
TransE [2]	(✗-✗)	$-\|q-t\|$	$q = h + r$	$q, h, r, t \in \mathbb{R}^d$
T-TransE (ours)	(✓-✗)	$-\|q-t\|$	$q = h + r + \tau$	$q, h, r, t, \tau \in \mathbb{R}^d$
ST-TransE (ours)	(✓-✓)	$-\|q-t\|$	$q = h + r + l + \tau$	$q, h, r, t, \tau, l \in \mathbb{R}^d$
RotatE [24]	(✗-✗)	$\|q-t\|$	$q = h \circ r$	$q, h, r, t \in \mathbb{C}^d$
T-RotatE (ours)	(✓-✗)	$-\|q-t\|$	$q = h \circ r \circ \tau$	$q, h, r, t, \tau \in \mathbb{C}^d$
ST-RotatE (ours)	(✓-✓)	$-\|q-t\|$	$q = h \circ r \circ l \circ \tau$	$q, h, r, t, \tau, l \in \mathbb{C}^d$
ComplEx [26]	(✗-✗)	$Re(q \times \bar{t})$	$q = h \times r$	$q, h, r, t \in \mathbb{C}^d$
T-ComplEx [12]	(✓-✗)	$Re(q \times \bar{t})$	$q = h \times r \times \tau$	$q, h, r, t, \tau \in \mathbb{C}^d$
ST-ComplEx	(✓-✓)	$Re(q \times \bar{t})$	$q = h \times r \times l \times \tau$	$q, h, r, t, l, \tau \in \mathbb{C}^d$
QuatE [33]	(✗-✗)	$Re(q \otimes_H \bar{t})$	$q = h \otimes_H r$	$q, h, r, t \in \mathbb{Q}^d$
T-QuatE (ours)	(✓-✗)	$Re(q \otimes_H \bar{t})$	$q = h \otimes_H r \otimes_H \tau$	$q, h, r, t, \tau \in \mathbb{Q}^d$
ST-QuatE (ours)	(✓-✓)	$Re(q \otimes_H \bar{t})$	$q = h \otimes_H r \otimes_H l \otimes_H \tau$	$q, h, r, t, l, \tau \in \mathbb{Q}^d$
DyHE (ours)	(✗-✗)	$-\|q-t\|$	$q = h \otimes_D r_{1,2}$	$q, h, r_{1,2}, t \in \mathbb{D}^d$
T-DyHE (ours)	(✓-✗)	$-\|q-t\|$	$q = h \otimes_D r_{1,2} + \tau$	$q, h, r_{1,2}, t, \tau \in \mathbb{D}^d$
ST-DyHE (ours)	(✓-✓)	$-\|q-t\|$	$q = h \otimes_D r_{1,2} + \tau + l$	$q, h, r_{1,2}, t, l, \tau \in \mathbb{D}^d$

Spatio-Temporal Knowledge Graphs We constructed three spatio-temporal KGs. The distribution shown in Fig. 3 depicts that temporal aspects mostly appear after year 1900. Former to this time, there is a steady number of temporal information.

We formed quintuples from YAGO [25] dataset namely YAGO3K by extracting triples with time and location information. Our initial analysis suggested that spatio-temporal information is connected to specific relations (i.e., wroteMusicFor, created). We have worked with YAGO3-10 [20] used in ConvE [5]. The training set of YAGO3-10 has 107,9040 triples. After extraction of time and location, we ended up with a total number of 9734 quintuples with 3619 entities, 8 relations, 422 locations and 195 time. We also constructed quintuples from DBpedia and Wikidata using dedicated SPARQL queries on the respective public SPARQL endpoints. The datasets are named DBpedia34K & WikiData53K. We utilize the DBpedia release of January 2021 which contains more than 900 million triples[1] and the Wikidata release of April 2021 which contains more than 1.26 billion statements.[2]. In the case of DBpedia, it can be assumed that at least 13% of the resources have temporal and/or spatial references. For Wikidata a comparable figure could not be determined because the number of resources in the knowledge base is much higher. Despite the timeout errors, it is safe to say that spatio-temporal information is abundant in both of the resources since we were able to extract a high number of representative triples From DBpedia, we queried more than 82,000 quintuples comprising information on multimedia items (books, music and movies), space missions, battles or buildings. Similar to the DBpedia dataset, the Wikidata dataset also contains a high number of quintuples (approx. 103,000). Overall, we gained 53849 number of entities, and 8 relations from WikiData where 296 different locations is present in the data and 627 different time information. For the obtained dataset from DBpedia, we have 34604 entities and 7 relations with 5687 different locations and 896 time information.

Evaluation Metrics We use common metrics namely Mean Reciprocal Rank (MRR) and Hits@k(k=1,3,10) for the evaluations. Here we explain each of the metrics. Given a set of test quintuple of the form (h, r, t, l, τ), we first remove the head entity and generate a query in the form of $(?, r, t, l, \tau)$. We then replace ? with all entities e' in the KG to generate n_e number of quintuples (e', r, t, l, τ), where n_e is the number of entities in the KG. We compute the score of each triple, and sort them to return the rank of the original triple (h, r, t, l, τ). We denote the left rank by r_l. With a similar procedure, the right rank r_r is computed by completion of the right query $(h, r, ?, l, \tau)$. The average rank for the ith quintuple is computed by $r_{ai} = \frac{r_l + r_r}{2}$. **MRR** is computed by taking the average of the reciprocals from the ranks in the testing triples i.e. $\sum_{i=1}^{n_t} \frac{1}{r_{ai}}$ where n_t is the number of testing triples. **Hits@k** is the percentage of testing triples that are ranked lower than k. For static and temporal KGE models, we used the following queries $(?, r, t), (h, r, ?)$ and $(?, r, t, \tau), (h, r, ?, \tau)$ and present their ranking in the Table 3. We additionally report the evaluation results on time and location completion in the Table 4.

Evaluation Results We implemented[3] all models using the Pytorch library. We used full-cross entropy loss [26] with N3 regularization [13] for training each of the models.

Hyperparameter settings: A wide array of testing has been done based on searching hyperparameters. The hyperparameters for which we achieved the best results are provided in Table 5. Throughout the experiments the *Adam* [11] optimizer has performed well and is thus used majorly. The hyperparameters mentioned in Table 5 are provided as d-dimension, B-Batch size, LR-Learning rate and R-regularization parameters.

In Table 6, we provide a selected list of incomplete quintuples that were used for an ablation study between the models. As can be seen, the results on predicted answers for each incomplete quintuple are fully correct by our proposed model where ComplEx and ST-ComplEx were not capable of correct predictions. When the query is about incompleteness of head or tail, we compared it to the ComplEx model as otherwise it

---

Table 3. Head/Tail completion results. Models without prefix consume triples. Models with "T" as prefix consume quadruples (triple plus time). Models with "ST" as prefix consume quintuple (triple plus location and time). The best results are written bold.

	YAGO3K			DBpedia34K			WikiData53K		
	MRR	Hits@1	Hits@10	MRR	Hits@1	Hits@10	MRR	Hits@1	Hits@10
TransE	0.561	0.496	0.689	0.454	0.415	0.519	0.245	0.152	0.451
T-TransE	<u>0.709</u>	0.676	<u>0.779</u>	0.501	0.475	0.544	0.396	0.337	0.538
ST-TransE	0.705	0.670	0.775	0.500	0.451	<u>0.577</u>	<u>0.565</u>	<u>0.546</u>	<u>0.599</u>
RotatE	0.564	0.503	0.688	0.461	0.425	0.521	0.246	0.153	0.458
T-RotatE	0.700	<u>0.679</u>	0.736	<u>0.505</u>	<u>0.487</u>	0.534	0.356	0.283	0.525
ST-RotatE	0.682	0.668	0.702	0.428	0.412	0.455	0.523	0.486	0.588
ComplEx	0.562	0.501	0.686	0.462	0.427	0.523	0.250	0.154	0.464
T-ComplEx	0.702	0.674	0.753	0.500	0.482	0.529	0.376	0.306	0.539
ST-ComplEx	0.689	0.668	0.727	0.450	0.424	0.497	0.532	0.495	0.594
QuatE	0.564	0.503	0.685	0.460	0.425	0.519	0.249	0.156	0.459
T-QuatE	0.694	0.675	0.722	0.500	0.481	0.531	0.358	0.285	0.525
ST-QuatE	0.690	0.675	0.714	0.502	0.485	0.528	0.515	0.478	0.576
DyHE	0.563	0.503	0.684	0.460	0.426	0.518	0.243	0.152	0.448
T-DyHE	**0.715**	**0.684**	0.775	**0.516**	**0.487**	0.564	0.377	0.318	0.517
ST-DyHE	0.704	0.665	**0.779**	0.485	0.427	**0.583**	0.568	0.550	**0.599**
ST-NewDE	0.708	0.682	0.758	**0.536**	**0.500**	0.598	0.572	0.556	**0.603**

Table 4. Location/Time completion results. The best results are written bold.

Location Completion $(h, r, t, ?, \tau)$									
	YAGO3K			DBpedia34K			WikiData53K		
	MRR	Hits@1	Hits@10	MRR	Hits@1	Hits@10	MRR	Hits@1	Hits@10
ST-TransE	0.349	0.083	0.903	0.245	0.045	0.633	0.321	0.008	0.838
ST-NewDE	**0.352**	**0.114**	**0.920**	**0.354**	**0.094**	**0.797**	**0.582**	**0.231**	**0.970**

Time Completion $(h, r, t, l, ?)$									
	YAGO3K			DBpedia34K			WikiData53K		
	MRR	Hits@1	Hits@10	MRR	Hits@1	Hits@10	MRR	Hits@1	Hits@10
ST-TransE	0.249	0.024	0.620	0.271	0.082	0.571	0.075	0.006	0.198
ST-NewDE	**0.580**	**0.241**	**0.958**	**0.432**	**0.151**	**0.792**	**0.162**	**0.020**	**0.514**

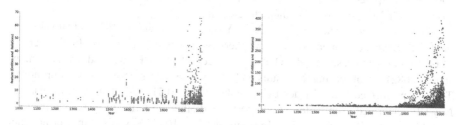

Fig. 3. Distribution of temporal information in YAGO3K and WikiData53K.

Table 5. Hyperparameter settings. Models without prefix consume triples. In this Table d = emb dimension, B = Batch size, LR = Learning rate, R = regularization parameter.

	YAGO3K				DBpedia34K				WikiData53K			
	d	B	LR	R	d	B	LR	R	d	B	LR	R
TransE	100	100	0.001	10e-11	100	100	0.001	0.001	100	100	0.001	10e-11
T-TransE	100	100	0.001	0.01	100	100	0.001	0.01	100	100	0.001	0
ST-TransE	100	100	0.001	0	100	100	0.001	0.001	100	100	0.001	0.001
RotatE	100	100	0.001	0.1	100	100	0.001	0.1	100	100	0.001	0.001
T-RotatE	100	100	0.001	0.01	100	100	0.001	0.1	100	100	0.001	0.001
ST-RotatE	100	100	0.001	0.1	100	100	0.001	0.1	100	100	0.001	0
ComplEx	100	100	0.001	0.01	100	100	0.001	0.01	100	100	0.001	10e-11
T-ComplEx	100	100	0.001	0.01	100	100	0.001	0.1	100	100	0.001	0.001
ST-ComplEx	100	100	0.001	0.001	100	100	0.001	0	100	100	0.001	0.001
QuatE	100	100	0.001	0.1	100	100	0.001	0.01	100	100	0.001	0.01
T-QuatE	100	100	0.001	0.001	100	100	0.001	0.1	100	100	0.001	0.001
ST-QuatE	100	100	0.001	0.01	100	100	0.001	0	100	100	0.001	0.001
DyHE	100	100	0.001	0.01	100	100	0.001	0.1	100	100	0.001	0.1
T-DyHE	100	100	0.001	0.01	100	100	0.001	0.1	100	100	0.001	0.1
ST-DyHE	100	100	0.001	0	100	100	0.001	0.1	100	100	0.001	0.01
ST-NewDE	100	100	0.001	0.001	100	100	0.001	0.001	100	100	0.001	0

Table 6. Example of ablation study results on predicted answers for incomplete quintuples over three datasets of YAGO3K, DBpedia34K, and WikiData53K. The correct objects are written **bold**.

Dataset	Query on head or tail parts	ST-NewDE	ComplEx
WikiData53K	(Philip_Guston, creatorOf, ?, United_States, 1972)	**Late_Afternoon**	Jules_Olitski
DBpedia34K	(Santiago_Calatrava, architectOf, ?, Maroussi, 1982)	**Olympic_Stadium_(Athens)**	SoFi_Stadium
YAGO3K	(?, created, Ulysses_(movie), Italy, 1955)	**Ennio_de_Concini**	Lasse_Hallstroem
Dataset	Query on location or time parts	ST-NewDE	ST-ComplEx
WikiData53K	(Edward_Witten, awardReceived, Alan_T._Waterman_Award, ?, 1982)	**United States**	Spain
DBpedia34K	(Nikolai_Nekrasov, authorOf, Korobeiniki_(poem), ?, 1861)	**Russia**	Serbia
YAGO3K	(Richard_Harvey, wroteMusicFor, Animal_Farm_(movie), United_States, ?)	**1999**	2010

does not consume spatial or temporal parts. For the queries about the spatial or temporal parts, we compared our model ST-NewDE against ST-ComplEx. In all of the studied cases, the ST-NewDE model predicts the correct matches.

Result Analysis: Regarding Table 3, for YAGO3K in terms of MRR and Hits@1 our model T-DyHE performed better than others by achieving MRR and Hits@1 scores of 0.715 and 0.684 respectively. Our other model ST-NewDE achieved a very similar score (0.682). By observing the results on DBpedia34K, it can be stated that, our model T-DyHE performed better by achieving higher MRR (0.516) and Hits@1 (0.487). ST-NewDE performed even better and outperformed others in terms of these two metrics by achieving MRR score of 0.500. In terms of Hits@10 ST-NewDE outperformed the other state of the art models by scoring 0.598 Hits@10. ST-DyHE and ST-NewDE also

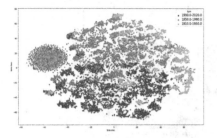

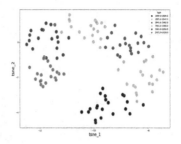

Fig. 4. Entity clustering w.r.t time.

Fig. 5. Time clustering w.r.t year.

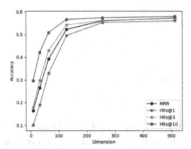

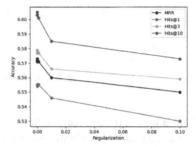

Fig. 6. Effect of d on ST-NewDE.

Fig. 7. Effect of R on ST-NewDE.

outperformed other models by achieving better accuracy in terms of MRR, Hits@1 and Hits@10. In case of our models, the best performing accuracy are marked in **bold**. In case of other models we have highlighted the better scores by underlining. Overall, in most of the cases, considering temporal part improves the performance of the triple-based models. In other words, for the queries of the form $(h, r, ?), (?, r, t)$, adding temporal information leads to a more accurate prediction. Similarly, adding the spatial and temporal information simultaneously, improves the accuracy of head or tail prediction even higher than other results (in most of the cases). In addition, this shows the way that time and location information is formulated in a model, has a direct impact on the results. Embedding temporal and spatial information in Dihedron space obtains a higher performance than Quaternion space, although both spaces are 4D. For example, on WikiData53K, ST-DyHE outperformed ST-QuatE (e.g. 0.515 vs 0.568 Hits@1). Moreover, in Dihedron space, ST-NewDE outperforms ST-DyHE in most of the cases. This is due to a more elegant formulation in the same space (Dihedron). While combining various transformations such as translation/rotation in Dihedron space (DyHE) obtains a higher performance than other models and spaces, our main formulation (ST-NewDE) obtains the highest accuracy in most cases. This confirms that the inclusion of all four parts namely entity (head and tail), relation, location and time provides orthogonal bases of the Dihedron space and computes the scores based on Dihedron product. This further reinforces the hypothesis that not only Spatio-temporal transformations are important, but also the *Spatio-temporal geometric representation* has a high impact on efficiently exploiting the location/time information. This design of the KGE models

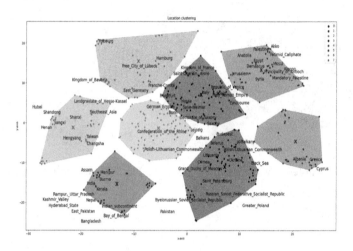

Fig. 8. Location clustering on YAGO3K.

leads to a more efficient formulation of the score function that works for such spatio-temporal data and outperforms other models. Such formulation has strong mathematical and geometric interpretation (discussed in method section). In addition to entity completion, we report the results of our model ST-NewDE and the best performing competitor ST-TransE. As shown in the Table 4, for completing the queries of the form $(h, r, t, l, ?)$ and $(h, r, t, ?, \tau)$, our model outperformed ST-TransE in all the used metrics. Such observation shows the efficiency of our model in head, tail, location, and time predictions. Note that like most of existing KGE models, our models can perform predictions on entities, relations, locations, and times that are seen during the training process. We consider inductive Spatio-temporal KGEs for the prediction of elements that are not given during training as future work. Regarding the number of parameters, it is noteworthy that ST-NEwDE has slightly fewer parameters than ST-TransE, and ST-DyHE has slightly more parameters than ST-TransE. Overall, the models have a very close number of parameters and their differences mainly come from the varying number of relation parameters which is significantly lower than the number of entity parameters.

Clustering Figure 4 depicts an ablation study on the clustering ability of our model. The quintuples are divided into three time categories of old (1910 - 1950) in orange, medium-old (1959 - 1990) in green, and recent (1990 - 2020) in blue. The results are generally "reasonable" and distinct while the overlapping parts belong to the cases with same head or tail entities appearing in different time periods. The dense cluster in the left side is caused by a high number of overlaps in the head or tail of the triples that belong to all the time categories.

The Fig. 5 illustrates the clustering ability of ST-NewDE over temporal part of the quintuples. For visualization purposes, each block of subsequent 20 years are grouped in one interval cluster. By using t-SNE, we visualize time embeddings in 2D space. First, our model puts all time points of the same 20-years block (1900–1920) in the same cluster and embeds the neighboring temporal intervals closely while the temporal order is preserved. Therefore, together with high performance in accuracy and the distinguished clustering results, we conclude that ST-NewDE efficiently captures the similarity of temporal part. As of ST-NewDE, it captured the location information or spatio-temporal part. As shown in Fig. 8, our model mainly embeds neighboring locations closely e.g., India, Nepal, Pakistan as well as cities in Germany (Freiburg, Hamburg).

Ablation Study on Dimension Here, we analyze the effect of dimension on the performance of ST-NewDE. In Table 3, we set $d = 100$ for all models. To have a fair comparison, we divided the used dimension relative to the space in which the models were designed. For Dihedron- and Quaternion-based models (4D algebra), we divided the dimension by four i.e. combination of all four parts will be 100 dimension. Similarly, for the models designed in complex vector space (e.g., ComplEx), we divided the dimension by two. For the ones in real space (e.g., TransE), we divided the dimension by one. Therefore, the performance improvement is assured not to be affected by other factors but only the core formulation. We experimented ST-NewDE with $d = \{8, 32, 64, 128, 256, 512\}$, and the other hyperparameters have been fixed. Figure 6 shows that by increasing the dimension on Wikidata53K, the performance improves quickly and then converges.

Ablation Study Regularization As can be seen in Fig. 7, we provided a study on the effect of regularization on the performance of ST-NewDE in Wikidata53K. According to Table 5, we noticed that the choice of regularization is important for improving the model performance. For some of the models such as DyHE, using a high value of regularization, improved the performance. In term of other models such as ST-NewDE, choice of a smaller regularization led to a better performance. We experimented the model with the following values for regularization $[0.1, 0.01, 0.001, 0.0001, 0.00001, 0.000001, 0.0000001]$. As shown, smaller regularization values obtain better performance than higher values. This shows that our model works well even without the N3 regularization.

6 Conclusion

In this paper, we address the problem of current KGE models performing on spatio-temporal KGs. We specifically proposed a family of embedding models which take advantage of the Dihedron algebra. The models were analysed in terms of the mathematical and *geometric interpretations*. We showed that our model facilitates *Spatio-temporal coordinate* representation and captures *Spatio-temporal and relational dependencies*. With these characteristics, the Dihedron-based KGE approach is capable of efficiently embedding spatio-temporal information into a rich geometric space. We additionally adapted the already exiting models to be able to encode spatio-temporal data. Our experiments on the subset of three public knowledge bases YAGO3K, DBpedia34K and WikiData53K showed that our approach usually achieves significantly higher performance than the extended state-of-the-art KGE models with time- or location-specific information. While our models predict entities, location, and time points which have been seen during training, we consider the prediction of unseen elements as future work.

Acknowledgement. We acknowledge the support of the following projects: SPEAKER (BMWi FKZ 01MK20011A), JOSEPH (Fraunhofer Zukunftsstiftung), the EU projects Cleopatra (GA 812997), PLATOON(GA 872592), TAILOR(EU GA 952215), CALLISTO(101004152), the BMBF projects MLwin(01IS18050) and the BMBF excellence clusters ML2R (BmBF FKZ 01 15 18038 A/B/C) and ScaDS.AI (IS18026A-F).

References

1. Balažević, I., Allen, C., Hospedales, T.M.: Tucker: Tensor factorization for knowledge graph completion. arXiv preprint (2019). arXiv:1901.09590
2. Bordes, A., Usunier, N., Garcia-Duran, A., Weston, J., Yakhnenko, O.: Translating embeddings for modeling multi-relational data. In: Neural Information Processing Systems (NIPS), pp. 1–9 (2013)
3. Chami, I., Wolf, A., Juan, D.-C., Sala, F., Ravi, S., Ré, C.: Low-dimensional hyperbolic knowledge graph embeddings. In: Proceedings of the 58th Annual Meeting of the Association for Computational Linguistics, pp. 6901–6914 (2020)
4. Dasgupta, S.S., Ray, S.N., Talukdar, P.: Hyte: hyperplane-based temporally aware knowledge graph embedding. In: Proceedings of the 2018 Conference on Empirical Methods in Natural Language Processing, pp. 2001–2011 (2018)
5. Dettmers, T., Minervini, P., Stenetorp, P., Riedel, S.: Convolutional 2d knowledge graph embeddings. In: Proceedings of the AAAI Conference on Artificial Intelligence, vol. 32 (2018)
6. García-Durán, A., Dumančić, S., Niepert, M.: Learning sequence encoders for temporal knowledge graph completion. arXiv preprint (2018). arXiv:1809.03202
7. Hobbs, J., Blythe, J., Chalupsky, H., Russ, T.A.: A survey of geospatial resources, representation and reasoning. Public Distribution of the University of Southern California (2006)
8. Ji, G., He, S., Xu, L., Liu, K., Zhao, J.: Knowledge graph embedding via dynamic mapping matrix. In: Proceedings of the 53rd Annual Meeting of the Association for Computational Linguistics and the 7th International Joint Conference on Natural Language Processing (vol. 1: Long Papers), pp. 687–696 (2015)
9. Ji, S., Pan, S., Cambria, E., Marttinen, P., Philip, S.Y.: A survey on knowledge graphs: representation, acquisition, and applications. IEEE Trans. Neural Netw. Learn. Syst. (2021)
10. Jin, W., Zhang, C., Szekely, P., Ren, X.: Recurrent event network for reasoning over temporal knowledge graphs. arXiv preprint (2019). arXiv:1904.05530
11. Kingma, D.P., Adam, J.Ba.: A method for stochastic optimization. arXiv preprint (2014). arXiv:1412.6980
12. Lacroix, T., Obozinski, G., Usunier, N.: Tensor decompositions for temporal knowledge base completion. arXiv preprint (2020). arXiv:2004.04926
13. Lacroix, T., Usunier, N., Obozinski, G.: Canonical tensor decomposition for knowledge base completion. In: International Conference on Machine Learning, pp. 2863–2872. PMLR (2018)
14. Leblay, J., Chekol, M.W.: Deriving validity time in knowledge graph. In: Companion Proceedings of the Web Conference 2018, pp. 1771–1776 (2018)
15. Lehmann, J., et al.: Dbpedia-a large-scale, multilingual knowledge base extracted from wikipedia, vol. 6, pp. 167–195 (2015, IOS Press)
16. Leidner, J.L.: A survey of textual data & geospatial technology. In: Werner, M. (ed.) Handbook of Big Geospatial Data, pp. 429–457. Springer, Cham (2021). https://doi.org/10.1007/978-3-030-55462-0_16
17. Li, Z., Liu, H., Zhang, Z., Liu, T., Xiong, N.N.: Learning knowledge graph embedding with heterogeneous relation attention networks. IEEE Trans. Neural Netw. Learn. Syst. (2021)

18. Lin, Y., Liu, Z., Sun, M., Liu, Y., Zhu, X.: Learning entity and relation embeddings for knowledge graph completion. In: Proceedings of the AAAI Conference on Artificial Intelligence, vol. 29 (2015)
19. Ma, Y., Tresp, V., Daxberger, E.A.: Embedding models for episodic knowledge graphs. J. Web Semant. **59**, 100490 (2019)
20. Mahdisoltani, F., Biega, J., Suchanek, F.: Yago3: a knowledge base from multilingual wikipedias. In: 7th Biennial Conference on Innovative Data Systems Research. CIDR Conference (2014)
21. Nayyeri, M., Vahdati, S., Aykul, C., Lehmann, J.: 5* knowledge graph embeddings with projective transformations. arXiv preprint (2020). arXiv:2006.04986
22. Qian, T., Liu, B., Nguyen, Q.V.H., Yin, H.: Spatiotemporal representation learning for translation-based poi recommendation. ACM Trans. Inf. Syst. (TOIS) **37**(2), 1–24 (2019)
23. Rotman, J.J.: Advanced Modern Algebra, vol. 114. American Mathematical Soc, New York (2010)
24. Sun, Z., Deng, Z.-H., Nie, J.-Y., Tang, J.: Rotate: knowledge graph embedding by relational rotation in complex space. arXiv preprint (2019). arXiv:1902.10197
25. Pellissier Tanon, T., Weikum, G., Suchanek, F.: YAGO 4: a reason-able knowledge base. In: Harth, A. (ed.) ESWC 2020. LNCS, vol. 12123, pp. 583–596. Springer, Cham (2020). https://doi.org/10.1007/978-3-030-49461-2_34
26. Trouillon, T., Welbl, J., Riedel, S., Gaussier, É., Bouchard, G.: Complex embeddings for simple link prediction. In: International Conference on Machine Learning, pp. 2071–2080. PMLR (2016)
27. Vrandečić, D., Krötzsch, M.: Wikidata: a free collaborative knowledgebase. Commun. ACM **57**(10), 78–85 (2014)
28. Wang, Z., Zhang, J., Feng, J., Chen, Z.: Knowledge graph embedding by translating on hyperplanes. In: Proceedings of the AAAI Conference on Artificial Intelligence, vol. 28 (2014)
29. Wildberger, N.J.: The geometry of the dihedrons (and quaternions). https://www.youtube.com/watch?v=qcmH0iKRF2w&t=1569s. Accessed 11 June 2021
30. Xu, C., Li, R.: Relation embedding with dihedral group in knowledge graph. arXiv preprint (2019). arXiv:1906.00687
31. Xu, C., Nayyeri, M., Alkhoury, F., Yazdi, H.S., Lehmann, J.: A time-aware knowledge graph embedding via temporal rotation. arXiv preprint (2020). arXiv:2010.01029
32. Yang, B., Yih, W.-T., He, X., Gao, J., Deng, L.: Embedding entities and relations for learning and inference in knowledge bases. arXiv preprint (2014). arXiv:1412.6575
33. Zhang, S., Tay, Y., Yao, L., Liu, Q.: Quaternion knowledge graph embeddings. arXiv preprint (2019). arXiv:1904.10281
34. Zhao, L., Deng, H., Qiu, L., Li, S., Hou, Z., Sun, H., Chen, Y.: Urban multi-source spatio-temporal data analysis aware knowledge graph embedding. Symmetry **12**(2), 199 (2020)

Hierarchical Topic Modelling
for Knowledge Graphs

Yujia Zhang[1]([⊠]) [iD], Marcin Pietrasik[1] [iD], Wenjie Xu[1] [iD],
and Marek Reformat[1,2] [iD]

[1] University of Alberta, 9211-116 Street, Edmonton, Canada
{yujia10,pietrasi,wx4,reformat}@ualberta.ca
[2] University of Social Sciences, 90-113 Łódź, Poland

Abstract. Recent years have demonstrated the rise of knowledge graphs as a powerful medium for storing data, showing their utility in academia and industry alike. This in turn has motivated substantial effort into modelling knowledge graphs in ways that reveal latent structures contained within them. In this paper, we propose a non-parametric hierarchical generative model for knowledge graphs that draws inspiration from probabilistic methods used in topic modelling. Our model discovers the latent probability distributions of a knowledge graph and organizes its elements in a tree of abstract topics. In doing so, it provides a hierarchical clustering of knowledge graph subjects as well as membership distributions of predicates and entities to topics. The main draw of such an approach is that it does not require any a priori assumptions about the structure of the tree other than its depth. In addition to presenting the generative model, we introduce an efficient Gibbs sampling scheme which leverages the Multinomial-Dirichlet conjugacy to integrate out latent variables, making the posterior inference process adaptable to large datasets. We quantitatively evaluate our model on three common datasets and show that it is comparable to existing hierarchical clustering techniques. Furthermore, we present a qualitative assessment of the induced hierarchy and topics.

Keywords: Knowledge graphs · Hierarchical clustering ·
Non-parametric model · Generative model

1 Introduction

Knowledge bases have received considerable research attention in recent years, demonstrating their utility in areas ranging from question answering [8,11] to knowledge generation [9,12,29] to recommender systems [4]. These knowledge bases are underpinned by graph structures called knowledge graphs which describe facts as a collection of triples that relate two entities via a predicate. Advances in artificial intelligence have spurred on the need to find representations of knowledge graphs which can be easily and accurately reasoned with

© The Author(s), under exclusive license to Springer Nature Switzerland AG 2022
P. Groth et al. (Eds.): ESWC 2022, LNCS 13261, pp. 270–286, 2022.
https://doi.org/10.1007/978-3-031-06981-9_16

by machines. One aspect of this is the increased research attention devoted to generative models for knowledge graphs which learn the latent probability distributions of a graph. These models work by decomposing the knowledge graph to a set of probability distributions that, when sampled together, generate its relations. The learning process, therefore, amounts to inferring the posterior distribution conditioned on the data.

Probabilistic topic models are types of generative models that have received considerable attention in the field of natural language processing. The aim of these models is to build abstract word topics from a corpus of documents and their words. In this sense, topics may be viewed as clusters of words. Most topic models operate under the intuition that words which co-occur in the same documents are likely to have similar semantics and therefore belong to the same topics. Hierarchical topic models extend this principle and organize the induced topics into a topic hierarchy whereby each ancestor topic represents a conceptually coarser version of its descendant topics.

In this paper, we present a model for generating a topic hierarchy from knowledge graphs which extends on existing topic models. In our model, topics are collections of entities and predicates, and are organized hierarchically in the form of a rooted tree. In generating these topics, our model also implicitly hierarchically clusters subjects by sampling a corresponding tree path. Furthermore, we employ a non-parametric prior over the tree, allowing our model to be free of any a priori assumptions about its structure other than its depth. We present an efficient Gibbs sampling scheme for posterior inference of our model. The approach leverages the Multinomial-Dirichlet conjugacy to integrate out parameters for faster inference. Our evaluation demonstrates our model's ability to induce a coherent topic hierarchy as well as hierarchical subject clustering.

2   Related Works

We divide the discussion of related works into two subsections, each of which our model shares a degree of similarity with: tag hierarchy induction models; and embeddings and clustering algorithms.

2.1   Tag Hierarchy Induction Methods

In the subsequent section, we introduce the concept of knowledge graph tags and how they can be leveraged to construct a topic hierarchy. Such a formulation is similar to that used in tag hierarchy induction methods which construct a hierarchy of tags based on the documents they annotate. One such method, described by Heymann and Garcia-Molina [16], uses the cosine distance to calculate tag similarity and generality. Tags are then added greedily, starting with the most general tag, as the child of the tag already in the hierarchy they are most similar to. Schmitz [30] proposed a method which uses subsumption rules to identify the relations between parents and children in the hierarchy. These rules form a directed graph which is then pruned to create a tree. Recently,

SMICT [25] applied principles from the aforementioned methods to knowledge graphs to induce a class taxonomy. This approach was extended in [26] to generate cluster hierarchies of knowledge graph subjects, yielding a result similar to our model. Frequency-based methods, like the ones mentioned above, often suffer from a problem where tags that appear more frequently are assumed to be more general. In an attempt to solve this, [2] introduce domain knowledge to the algorithm in [5] and verify the directionality of relations by searching for lexico-syntactic patterns on Wikipedia. This approach improves the quality of the induced hierarchy when compared to the original model. [15] and [34] both use a two phase approach in which a tag hierarchy is first induced using a strictly frequency-based approach and then optimized using domain knowledge in the form of an existing hierarchy.

2.2 Embeddings and Clustering Algorithms

Knowledge graph embedding methods map knowledge graphs from the discrete graph space to a continuous vector space. Such a representation is useful as it allows knowledge graphs to be easily integrated with common machine learning and deep learning methods. In the context of our work, knowledge graph embeddings may be used in conjunction with hierarchical clustering methods, allowing for a benchmark comparison. Perhaps the most canonical of embedding methods, TransE [9], applies the intuition that subject embeddings should be near object embeddings when translated by valid corresponding predicates. Such a formulation provides an objective function which is then optimized via stochastic gradient descent to learn the embeddings. In a related approach, RDF2Vec [28] uses breadth-first graph walks on the skip-gram language model [22] to generate embeddings. Factorization models such as RESCAL [24] and DistMult [35] learn embeddings by factorizing the knowledge graph adjacency tensor into the product of entity embeddings and relation specific translation matrices. ComplEx embeddings [33] extend DistMult to the complex domain to better handle asymmetry in the knowledge graph. ConvE [12] leverages the convolution operator in a neural framework by stacking embeddings as a martix and convolving them in two dimensions.

Having mapped a knowledge graph to a continuous space via embeddings, clustering is trivial since distances between embeddings may be easily calculated. The process is merely choosing the clustering algorithm best suited for the data. K-means [20] is perhaps the most common clustering algorithm used today and works by assigning entities to the cluster with the smallest centre distance before recalculating cluster centre based on the updated memberships. Another common approach, OPTICS [3], uses a density based approach which expands clusters so long as density criteria are being met. Spectral clustering encompasses a wide range of algorithms which operate on the eigenvalues of the input entities. To generate hierarchical clusters, agglomerate clustering builds a hierarchy bottom-up by joining clusters at higher levels in the hierarchy based on linkage criteria. We use the these clustering methods in conjunction with the aforementioned knowledge graph embeddings during our evaluation procedure.

This is similar to ExCut [13] which first generates embeddings before iteratively refining them using rule mining approaches to generate entity clusters.

3 Proposed Model

In this section, we describe our model by positioning it in the context of existing probabilistic topic models from which it draws inspiration. Specifically, we first introduce readers to Latent Dirichlet Allocation (LDA) [7] and Hierarchical Latent Dirichlet Allocation (hLDA) [6] before formalizing our model.

3.1 Problem Formulation

We define a knowledge graph as a collection of triples, $\mathcal{K}$, such that each triple relates a subject entity, s, to an object entity, o, via a predicate, p. Formally, $\mathcal{K} = \{\langle s, p, o \rangle \in \mathbf{S} \times \mathbf{P} \times \mathbf{O}\}$ where $\langle s, p, o \rangle$ is a triple, and $\mathbf{S}$, $\mathbf{P}$, and $\mathbf{O}$ are the sets of subjects, predicates, and objects in $\mathcal{K}$, respectively. We note that knowledge graphs are rarely bipartite in terms of $\mathbf{S}$ and $\mathbf{O}$. In other words, entities can take on the role of both subjects and objects in $\mathcal{K}$, thus $\mathbf{S} \cap \mathbf{O} \neq \emptyset$. Our goal is to find a representation of the knowledge graph in which entities and predicates are hierarchically organized such that entities representing coarse concepts subsume their fine grained counterparts. For instance, the concept Person is a coarser concept than Artist since it encompasses all persons, including artists and non-artists. A natural representation of this paradigm is a directed tree wherein coarse concepts occupy nodes closer to the root node. Nodes are then collections of entities and predicates which share similar semantics. Paths in the tree capture the progressive granularization of a concept.

3.2 Probabilistic Topic Models

Given a collection of documents and their words, $\mathbf{D}$, topic models generate abstract topics on the intuition that words belonging to the same topic are likely to occur in the same documents. Latent Dirichlet Allocation (LDA) [7] is a canonical example of the topic models used today. In this approach, each document, $d_i \in \mathbf{D}$, is a mixture of topics and each topic is a distribution of words. To generate a document, the number of document words, W_i, the document's topic mixture, θ_i, and each topic's word distributions, β_k, are sampled. For each document word, $w_{i,j}$, first a topic indicator $z_{i,j}$ is sampled according to θ_i then the word is generated from z_j's word distribution, β_{z_j}. This generative procedure is formally defined as follows:

- for each document; $d_i \in \mathbf{D}$
  - $W_i \sim \text{Poisson}(\xi)$
  - $\theta_i \sim \text{Dirichlet}(\alpha)$
- for each topic; $k \in 1, 2, ..., K$
  - $\beta_k \sim \text{Dirichlet}(\eta)$

– for each document; $d_i \in \mathbf{D}$
  • for each word in document; $w_{i,j} \in d_i$
    * $z_{i,j} \sim \text{Multinomial}(\theta_i)$
    * $w_{i,j} \sim \text{Multinomial}(\beta_{z_{i,j}})$

Learning the distributions which generate the documents amounts to inferring the posterior distribution. Although this problem is intractable for exact inference, it can be approximated with algorithms such as Variational Bayes [7] or Collapsed Gibbs Sampling [14]. We refer readers to the original papers for the full inference procedure.

LDA has been extended to generate a hierarchy of topics in Hierarchical Latent Dirichlet Allocation (hLDA) [6]. The foundation of hLDA is the nested Chinese restaurant process (nCRP) which is an extension of the Chinese restaurant process (CRP) [1]. The CRP is a recursively defined stochastic process which gets its name from the analogy of seating patrons at a Chinese restaurant. In this restaurant, there are an infinite number of tables and each table can seat an infinite number of guests. When a guest enters, the probability of him being seated at a table is proportional to the number of patrons already seated at the table. Formally, when seating guest g_i at a restaurant that has M non-empty tables, the probability of seating the guest at table m is:

$$
P(g_i = m | g_{i-1}, ..., g_1) = \begin{cases} \dfrac{|n_m^i|}{i - 1 + \gamma} & m \leq M \\ \dfrac{\gamma}{i - 1 + \gamma} & m = M + 1 \\ 0 & M + 1 < m \end{cases}
$$

where $|n_m^i|$ is the number of patrons sitting at table m when guest g_i arrives and γ is a hyperparameter which controls the probability that an incoming guest will be seated at an empty table.

The nCRP is used in hLDA as an infinitely deep and infinitely branching prior over a tree structure. In this process, a tree is generated by sampling a path, c_i, at each level in the tree via the CRP. Each node in a tree, $n_k \in \mathbf{N}$, has its own CRP and being seated at a table is analogous to taking a specific branch in the path down the tree. As before, the probability of taking a path is proportional to the amount of times the path has been taken before. When arriving at a node n_k with children $\mathbf{M}_k$ on the $(l-1)^{\text{th}}$ level in the tree, the probability of selecting an existing branch, $c_i[l] \in \mathbf{M}_l$ or creating a new branch, $c_i[l] = M_k^*$, is:

$$
P(c_i[l] = m | c_{i-1:1}, c_i[l-1:1]) = \begin{cases} \dfrac{|n_m^i|}{|n_k^i| + \gamma} & m \in \mathbf{M}_k \\ \dfrac{\gamma}{|n_k^i| + \gamma} & m = M_k^* \end{cases}
$$

where $\mathbf{c}_i[l]$ is the node on the path of d_i at level l, $M_k^* = \min(\mathbb{Z}^+ \setminus \mathbf{M}_k)$ is the smallest positive integer not in $\mathbf{M}_k$, and $|n_k^i|$ is the number of entities that have gone through node n_k when entity i arrived, $|n_k^i| = |\{j \in \mathbb{Z}^+ : j < i \wedge \mathbf{c}_j[l] = n_k\}|$.

Putting everything together, hLDA uses the nCRP to generate a tree of topics. The tree is bounded to a maximum depth of L and each node in the tree is associated with a topic β_k. Each document d_i samples a path through L nodes in the tree, c_i, and a topic distribution over levels in the tree analogous to the topic mixture in LDA, θ_i. For each word $w_{i,j}$ in d_i, a topic $z_{i,j}$ is sampled from θ_i and a word is generated from that topic. The generative process is summarized as follows:

- for each node in the tree; $n_k \in \mathbf{N}$
  - $\beta_k \sim \mathrm{Dirichlet}(\eta)$
- for each document; $d_i \in \mathbf{D}$
  - $c_i \sim \mathrm{nCRP}(\gamma)$
  - $\theta_i \sim \mathrm{GEM}(\rho, \pi)$
  - for each word in document; $w_{i,j} \in d_i$
    * $z_{i,j} \sim \mathrm{Multinomial}(\theta_i)$
    * $w_{i,j} \sim \beta_{c_i[z_{i,j}]}$

where $\mathrm{GEM}(\rho, \pi)$ stands for the stick-breaking process [27] and functions as the prior for topic levels. As with LDA, we refer the readers to the original papers for model inference.

3.3 Model Description

We present our model as an extension of hLDA which has been adapted to knowledge graphs. As such, we adopt the previously introduced concepts and notation, and focus on highlighting the differences.

The first difference is the departure from the domain of documents and words to that of subjects, predicates, and objects. We can think of a predicate-object pair as a *tag* which describes a subject in a way that is analogous to how a word describes a document. In this view, a tag, t, is defined as $\langle p, o \rangle$ and belongs to a subject such that $t_{i,j} \in \mathbf{T}_i$ denotes that tag $t_{i,j}$ belongs to subject s_i. This formulation is leveraged in our model by assigning a tag topic distribution, β^t, for each node in the tree. Furthermore, to capture the distributions of predicates in each cluster, we mix in a predicate specific topic, β^p. Predicates share their level indicators, $z_{i,j}$, with their corresponding tags. As such, the number of predicates belonging to a subject has to equal its tag count. We define the multiset of predicates which belong to subject s_i as $p_{i,j} \in \mathbf{P}_i$ such that $|\mathbf{P}_i| = |\mathbf{T}_i|$. Thus, each node is a collection of two topics whose elements span the domain of $\mathbf{T} \cup \mathbf{P}$.

Each subject s_i samples a path, c_i, through the tree using the nCRP as well as a level distribution, θ_i. A further departure from the original hLDA model is the replacement of the stick-breaking process as the prior of the level distribution with the Dirichlet distribution. This formulation is a return to the prior used in LDA and was chosen for two reasons. The first is that the Dirichlet distribution introduces only one hyperparameter in contrast to the stick-breaking process' two. This makes our model easier to apply a priori since hyperparameter sensitivity and selection present challenges in non-parametric models. The second is that the inference scheme is simpler when using the Dirichlet prior. Finally, the theoretical benefits of the stick-breaking prior are not justified in a practical context since the infinite distribution would get bounded in our model by the tree depth, L.

As mentioned previously, level indicators, $z_{i,j}$, are shared among corresponding predicates and tags. Thus, we sample one level indicator for each tag analogously to hLDA. This indicator is used in conjunction with the subject path to determine the node whose topics will be sampled from. Unlike hLDA which only samples words, our model samples predicates and tags from the selected node's predicate and tag topic distributions, $\beta^p[c_i[z_{i,j}]]$ and $\beta^t[c_i[z_{i,j}]]$, respectively. We use the notation $\beta^p[c_i[z_{i,j}]]$ and $\beta^t[c_i[z_{i,j}]]$ to denote the predicate and tag topic distributions of the node at level $z_{i,j}$ on path c_i. The generative process is defined as follows:

- for each node in the tree; $n_k \in \mathbf{N}$
  - $\beta^p \sim \text{Dirichlet}(\eta_p)$
  - $\beta^t \sim \text{Dirichlet}(\eta_t)$
- for each subject; $s_i \in \mathbf{S}$
  - $c_i \sim \text{nCRP}(\gamma)$
  - $\theta_i \sim \text{Dirichlet}(\alpha)$
  - for each tag in subject; $t_{i,j} \in \mathbf{T}_i$
    * $z_{i,j} \sim \text{Multinomial}(\theta_i)$
  - for each predicate in subject; $p_{i,j} \in \mathbf{P}_i$
    * $p_{i,j} \sim \text{Multinomial}(\beta^p[c_i[z_{i,j}]])$
  - for each tag in subject; $t_{i,j} \in \mathbf{T}_i$
    * $t_{i,j} \sim \text{Multinomial}(\beta^t[c_i[z_{i,j}]])$

η_p and η_t are hyperparameters of our model which control the sparsity of the topics such that lower η values result in sparser topics which are more dissimilar from one another. Furthermore, the ratio between η_p and η_t controls the relative importance of predicates to tags when calculating the likelihood functions. γ is a hyperparameter of the nCRP and controls the probability of creating a new path in the tree such that higher γ values will generate trees with a higher average branching factor. Finally, α is the topic level hyperparameter. We provide a graphical representation of our model using plate notation in Fig. 1.

3.4   Inference

Our model is intractable for exact inference, thus we approximate it using collapsed Gibbs sampling for posterior inference. The goal of the sampling scheme is to generate the subject paths, $\mathbf{c}$, and level indicators, $\mathbf{z}$, by inferring the latent parameters. For faster mixing, we integrate out the topic distributions, β^p and β^t, as well as the level distributions, θ, by leveraging the Multinomial-Dirichlet conjugacy. This reduces our inference scheme to simply sampling paths and levels alternately until the parameters of the model are learned, at which point we can collect samples to estimate the true posterior.

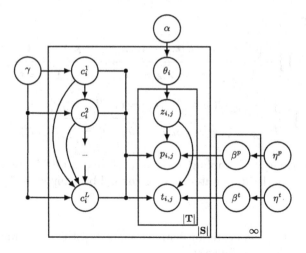

Fig. 1. Plate diagram for our model.

Sampling Paths. The posterior distribution of c_i, the path for subject s_i, conditioned on all other variables is:

$$\mathbb{P}(c_i|\mathbf{c}_{-i}, \mathbf{z}_i, \mathbf{P}_i, \mathbf{T}_i, \gamma, \eta_p, \eta_t) \propto \mathbb{P}(c_i|\mathbf{c}_{-i}, \gamma)\mathbb{P}(\mathbf{P}_i|c_i, \mathbf{P}_{-i}, \mathbf{z}_i, \eta_p)$$
$$\mathbb{P}(\mathbf{T}_i|c_i, \mathbf{T}_{-i}, \mathbf{z}_i, \eta_t) \qquad (1)$$

where $\mathbf{c}_{-i}$ denotes all paths in the tree excluding the path taken by subject s_i. Likewise, $\mathbf{P}_{-i}$ and $\mathbf{T}_{-i}$ denote the predicates and tags on the tree leaving out those belonging to to subject s_i. This expression is merely an application of Bayes' theorem which states the posterior is proportional to the likelihood times the prior. The first term, $\mathbb{P}(c_i|\mathbf{c}_{-i}, \gamma)$, is the nCRP prior and is calculated as outlined earlier in the paper. The second term, $\mathbb{P}(\mathbf{P}_i|c_i, \mathbf{P}_{-i}, \mathbf{z}_i, \eta_p)$, is the predicate likelihood given the choice of paths. In other words, it is the probability of observing the predicate data if subject s_i were to take path c_i. The calculation of this term is defined as follows:

$$\mathbb{P}(\mathbf{P}_i|c_i, \mathbf{P}_{-i}, \mathbf{z}_i, \eta_p)$$
$$= \prod_{l=1}^{L} \frac{\Gamma\left(\sum_{p_{i,j} \in \mathbf{P}_{-i}} \#[\mathbf{z}_{-i} = l, \mathbf{c}_{-i,l} = c_{i,l}, \mathbf{P}_{-i} = p_{i,j}] + \eta_p|\mathbf{P}|\right)}{\sum_{p_{i,j} \in \mathbf{P}_{-i}} \Gamma\left(\#[\mathbf{z}_{-i} = l, \mathbf{c}_{-i,l} = c_{i,l}, \mathbf{P}_{-i} = p_{i,j}] + \eta_p\right)}$$
$$\prod_{l=1}^{L} \frac{\prod_{p_{i,j} \in \mathbf{P}_i} \Gamma\left(\#[\mathbf{z}_i = l, \mathbf{c}_{i,l} = c_{i,l}, \mathbf{P}_i = p_{i,j}] + \eta_p\right)}{\Gamma\left(\prod_{p_{i,j} \in \mathbf{P}_i} \#[\mathbf{z}_i = l, \mathbf{c}_{i,l} = c_{i,l}, \mathbf{P}_i = p_{i,j}] + \eta_p|\mathbf{P}|\right)} \qquad (2)$$

where $\Gamma(.)$ is the gamma function and $\#[.]$ indicates the number of elements that satisfy the given conditions. Finally, the third term, $\mathbb{P}(\mathbf{T}_i|c_i, \mathbf{T}_{-i}, \mathbf{z}_i, \eta_t)$, is the tag likelihood given the choice of paths and is calculated analogously to the predicate likelihood:

$$\mathbb{P}(\mathbf{T}_i|c_i,\mathbf{T}_{-i},\mathbf{z}_i,\eta_t)$$

$$= \prod_{l=1}^{L} \frac{\Gamma\left(\sum_{t_{i,j}\in\mathbf{T}_{-i}} \#[\mathbf{z}_{-i} = l, \mathbf{c}_{-i,l} = c_{i,l}, \mathbf{T}_{-i} = t_{i,j}] + \eta_t|\mathbf{T}|\right)}{\prod_{t_{i,j}\in\mathbf{T}_{-i}} \Gamma\left(\#[\mathbf{z}_{-i} = l, \mathbf{c}_{-i,l} = c_{i,l}, \mathbf{T}_{-i} = t_{i,j}] + \eta_t\right)}$$

$$\prod_{l=1}^{L} \frac{\prod_{t_{i,j}\in\mathbf{T}_i} \Gamma\left(\#[\mathbf{z}_i = l, \mathbf{c}_{i,l} = c_{i,l}, \mathbf{T}_i = t_{i,j}] + \eta_t\right)}{\Gamma\left(\sum_{t_{i,j}\in\mathbf{T}_i} \#[\mathbf{z}_i = l, \mathbf{c}_{i,l} = c_{i,l}, \mathbf{T}_i = t_{i,j}] + \eta_t|\mathbf{T}|\right)} \tag{3}$$

The time complexity of sampling a single path, c_i, is $\mathcal{O}(|\mathbf{N}|(|\mathbf{S}| + |\mathbf{T}|))$, thus sampling all the paths in one iteration of the Gibbs sampler is $\mathcal{O}(|\mathbf{S}||\mathbf{N}|(|\mathbf{S}| + |\mathbf{T}|))$.

Sampling Levels. The posterior distribution of $z_{i,j}$, the level indicator for the jth tag in subject s_i is as follows:

$$\mathbb{P}(z_{i,j}|\mathbf{z}_{i,-j},\mathbf{P}_{i,-j}, i,-j, \mathbf{c}, \eta_p, \eta_t, \alpha) \propto \mathbb{P}(z_{i,j}|\mathbf{z}_{i,-j},\alpha)\mathbb{P}(p_{i,j}|\mathbf{P}_{i,-j}, \mathbf{c}, \mathbf{z}_i, \eta_p)$$
$$\mathbb{P}(t_{i,j}|\mathbf{T}_{i,-j}, \mathbf{c}, \mathbf{z}_i, \eta_t) \tag{4}$$

where $\mathbf{z}_{i,-j}$ are all the level indicators in subject s_i excluding $z_{i,j}$, the indicator for tag $t_{i,j}$. The prior for level indicators, $\mathbb{P}(z_{i,-j}|\mathbf{z}_{i,-j},\alpha)$, is obtained by integrating out the Multinomial distribution via the Multinomial-Dirichlet conjugacy and calculating the Dirichlet prior as follows:

$$\mathbb{P}(z_{i,j}|\mathbf{z}_{i,-j},\alpha) = \mathbb{E}(z_{i,j}|\mathbf{z}_{i,-j},\alpha)$$
$$= \mathbb{E}\left(\mathbb{E}(z_{i,j} = l)|\theta_1, \theta_2, ..., \theta_L, \mathbf{z}_{i,-j}, \alpha\right)$$
$$\propto \#[\mathbf{z}_{i,-j} = l] + \alpha \tag{5}$$

The predicate likelihood, $\mathbb{P}(p_{i,j}|\mathbf{P}_{i,-j}, c_i, \mathbf{z}_i, \eta_p)$, is calculated by counting the total number of predicates at the node specified by $z_{i,j}$ on path c_i that are the same as $p_{i,j}$:

$$\mathbb{P}(p_{i,j}|\mathbf{P}_{i,-j}, c_i, \mathbf{z}_i, \eta_p) = \mathbb{E}(p_{i,j}|\mathbf{z}_i, c_i, \eta_p)$$
$$\propto \#[\mathbf{z}_{-(i,j)} = z_{i,j}, \mathbf{c}_{z_{i,j}} = c_{i,z_{i,j}}, \mathbf{P}_{-(i,j)} = p_{i,j}] + \eta_p \tag{6}$$

The tag likelihood, $\mathbb{P}(t_{i,j}|\mathbf{T}_{i,-j}, \mathbf{c}, \mathbf{z}_i, \eta_t)$, is calculated analogously:

$$\mathbb{P}(t_{i,j}|\mathbf{T}_{i,-j}, c_i, \mathbf{z}_i, \eta_t) = \mathbb{E}(p_{i,j}|\mathbf{z}_i, c_i, \eta_t)$$
$$\propto \#[\mathbf{z}_{-(i,j)} = z_{i,j}, \mathbf{c}_{z_{i,j}} = c_{i,z_{i,j}}, \mathbf{T}_{-(i,j)} = t_{i,j}] + \eta_t \tag{7}$$

The time complexity of sampling a single topic, $z_{i,j}$, is $\mathcal{O}(L)$ and meaning that sampling all levels is $\mathcal{O}(|\mathbf{S}||\mathbf{T}|L)$.

Collapsed Gibbs Sampling. As mentioned previously, the collapsed Gibbs sampling process samples paths and levels alternately, as summarized in Algorithm 1. This approach creates a Markov chain which iteratively approaches its stationary distribution. As such, it is necessary to burn-in a fixed number of samples before samples approximating the posterior distribution may be obtained. Although Gibbs sampling is guaranteed to converge in the infinite case, the speed with which it does so is highly variable and difficult to predict a priori. Monitoring the likelihood of the model is therefore important in determining whether sufficient training has taken place. Furthermore, due to the non-parametric nature of our model, the selection of hyperparameters is critically important. Recall, for instance, that the tree's structure and size changes every time it is sampled. Thus, high γ values may induce trees with branching factors too high to feasibly perform inference on.

Algorithm 1. Gibbs Sampling Procedure

Input: Knowledge graph, $\mathcal{K}$; nCRP hyperparmeter, γ; topic hyperparameters, η^p and η^t; level hyperparameter α; Number of iterations, $iters$
Output: Hierarchical topic model for $\mathcal{K}$ defined by $\mathbf{c}$ and $\mathbf{z}$

1: Obtain $\mathbf{S}$, $\mathbf{P}$, and $\mathbf{T}$ from $\mathcal{K}$
2: **for** $iter = \{1, 2, ..., iters\}$ **do**
3:    **for** $i \in \{1, 2, ..., |\mathbf{S}|\}$ **do**
4:       Sample c_i using Equation 1
5:       **for** $j \in \{1, 2, ..., |\mathbf{T}|\}$ **do**
6:          Sample $z_{i,j}$ using Equation 4
7:       **end for**
8:    **end for**
9: **end for**

---

4 Evaluation

We split the evaluation of our model into two parts: quantitative and qualitative. In our quantitative evaluation, we train our model to obtain a hierarchical clustering of subject entities. This clustering is then evaluated by comparing against ground truth labels and calculating metrics of clustering performance. This gives insight into the quality of induced tree and allocation of subjects to leaf nodes. To assess the quality of the inferred topic clusters, we perform a qualitative evaluation by analyzing the membership distributions of predicates and tags to selected topics. What follows is a summary of our evaluation procedure and discussion of the results. The source code for our model along with the datasets used may be found on GitHub[1].

---

[1] https://github.com/yujia0223/hkg.

Table 1. Summary of ground truth classes used to derive clustering evaluation datasets.

	FB15k-237	YAGO3-10	DBpedia
Level 1	Person, Organization, Location, Event	Person, Organization, Body of Water	Person, Place
Level 2	Artist, Politician, Scientist, Officeholder, Writer, Musical Organization, Party, Enterprise, Nongovernmental Organization, County, Town, City, Mountain, Movie, Entertainment, Game, Contest	Artist, Politician, Scientist, Officeholder, Writer, Musical Organization, Party, Enterprise, Nongovernmental Organization, Stream, Lake, Ocean, Bay, Sea	Artist, Athlete, PopulatedPlace, NaturalPlace
Level 3	-	-	Actor, MusicalArtist, Painter, SoccerPlayer, GridironFootballPlayer, WinterSportPlayer, Swimmer, BodyOfWater, Mountain, Settlement, Island, Country
Level 4	-	-	AmericanFootballPlayer, IceHockeyPlayer, Lake, City, Town

4.1 Datasets

We use three real-world datasets in our evaluation: FB15k-237, YAGO3-10, and DBpedia. The datasets were chosen based on their ubiquity in existing literature and to highlight the scalability of our sampling scheme on large datasets. What follows is a brief description of each dataset.

FB15k-237. The FB15k-237 dataset [32] was constructed from the FB15k dataset [9] by removing redundant and inverse triples. It contains data queried from a version of Freebase that existed around 2013. Specifically, it is comprised of 272115 triples, 14541 entities, and 237 predicates. For our hierarchical clustering analysis, we followed a similar approach to generating a ground truth subset of the data as [18]. Namely, we first mapped entities to the WordNet taxonomy [23] through the *sameAs* predicate, which relates Freebase entities to YAGO entities. We then extracted triples containing subjects with labels on second level in the taxonomy from the sets provided in Table 1. This process yielded a dataset with 5301 subjects, 103550 triples, 10018 entities, and 190 predicates.

Table 2. Method results (mean ± standard deviation) on the FB15k-237, YAGO3-10, and DBpedia datasets. Underscore denotes significance at alpha value of 0.05 compared against our model as per t-test.

Method	FB15k-237 ARI	NMI	YAGO3-10 ARI	NMI	DBpedia ARI	NMI
RDF2VEC						
K-means	.308 ± .012	.567 ± .007	.070 ± .019	.199 ± .017	.223 ± .005	.416 ± .005
OPTICS	.087 ± .000	.283 ± .000	.009 ± .000	.172 ± .000	.001 ± .000	.311 ± .000
Agglom	.455 ± .000	.601 ± .000	.038 ± .000	.174 ± .000	.236 ± .000	.414 ± .000
Spectral	.539 ± .000	.678 ± .000	.071 ± .000	.218 ± .000	.218 ± .000	.410 ± .000
TransE						
K-means	.405 ± .049	.632 ± .009	.263 ± .009	.367 ± .003	.247 ± .029	.389 ± .024
OPTICS	.031 ± .000	.253 ± .000	.049 ± .000	.150 ± .000	.001 ± .000	.198 ± .000
Agglom	.491 ± .000	.599 ± .000	.226 ± .000	.337 ± .000	.198 ± .000	.383 ± .000
Spectral	**.658** ± .000	.684 ± .000	**.270** ± .000	.345 ± .000	.057 ± .000	.321 ± .000
DistMult						
K-means	.269 ± .011	.559 ± .013	.174 ± .012	.326 ± .015	.400 ± .008	.587 ± .010
OPTICS	.016 ± .000	.189 ± .000	.029 ± .000	.175 ± .000	.002 ± .000	.184 ± .000
Agglom	.379 ± .000	.621 ± .000	.202 ± .000	**.382** ± .000	.389 ± .000	.594 ± .000
Spectral	.505 ± .000	.600 ± .000	.035 ± .000	.124 ± .000	.150 ± .000	.478 ± .000
ComplEx						
K-means	.271 ± .020	.562 ± .016	.137 ± .012	.342 ± .009	.462 ± .013	.630 ± .015
OPTICS	.019 ± .000	.202 ± .000	.017 ± .000	.152 ± .000	.002 ± .000	.235 ± .000
Agglom	.385 ± .000	.630 ± .000	.181 ± .000	.299 ± .000	.442 ± .000	.628 ± .000
Spectral	.563 ± .000	.613 ± .000	.016 ± .000	.204 ± .000	.203 ± .000	.550 ± .000
ConvE						
K-means	.332 ± .031	.619 ± .013	.004 ± .003	.004 ± .001	**.474** ± .019	.612 ± .013
OPTICS	.040 ± .000	.254 ± .000	.012 ± .000	.088 ± .000	.002 ± .000	.238 ± .000
Agglom	.384 ± .000	.630 ± .000	.003 ± .000	.005 ± .000	.458 ± .000	.614 ± .000
Spectral	.556 ± .000	**.703** ± .000	.002 ± .000	.006 ± .000	.439 ± .000	**.639** ± .000
ExCut	.343 ± .011	.651 ± .002	.130 ± .007	.322 ± .011	.380 ± .016	.595 ± .005
Our Method	.656 ± .005	.669 ± .021	.044 ± .006	.218 ± .002	.406 ± .042	.582 ± .022

YAGO3-10. The YAGO3-10 dataset was derived from the YAGO3 database [21] which is a knowledge graph derived from Wikipedia and follows the hierarchical class structure of WordNet. As with FB15k-237, we mapped entities to the WordNet taxonomy before selecting the subset defined by classes in Table 1. This resulted in a dataset with 11954 subject, 84382 triples, 27572 entities, and 28 relations.

DBpedia. The DBpedia dataset was generated by querying DBpedia [19] for random entities belonging to classes on levels 4 and 5 as specified in Table 1. Specifically, 75 entities were extracted for each of these classes. Triples where these entities take on the subject role were then queried for, filtering out triples which indicate class membership. This process resulted in 908 subjects, 57191 triples, 31202 entities, and 345 predicates. The impetus for this dataset was to

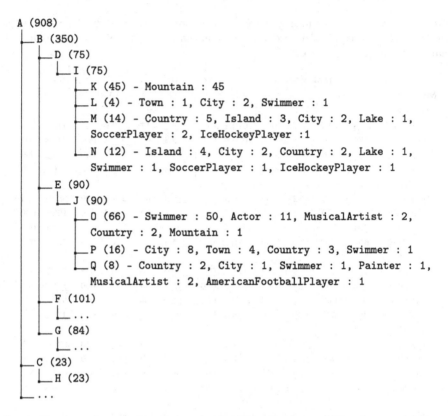

```
A (908)
 ├─B (350)
 │  ├─D (75)
 │  │  └─I (75)
 │  │     ├─K (45) - Mountain : 45
 │  │     ├─L (4) - Town : 1, City : 2, Swimmer : 1
 │  │     ├─M (14) - Country : 5, Island : 3, City : 2, Lake : 1,
 │  │     │   SoccerPlayer : 2, IceHockeyPlayer :1
 │  │     └─N (12) - Island : 4, City : 2, Country : 2, Lake : 1,
 │  │         Swimmer : 1, SoccerPlayer : 1, IceHockeyPlayer : 1
 │  ├─E (90)
 │  │  └─J (90)
 │  │     ├─O (66) - Swimmer : 50, Actor : 11, MusicalArtist : 2,
 │  │     │   Country : 2, Mountain : 1
 │  │     ├─P (16) - City : 8, Town : 4, Country : 3, Swimmer : 1
 │  │     └─Q (8) - Country : 2, City : 1, Swimmer : 1, Painter : 1,
 │  │         MusicalArtist : 2, AmericanFootballPlayer : 1
 │  ├─F (101)
 │  │  └─ ...
 │  └─G (84)
 │     └─ ...
 ├─C (23)
 │  └─H (23)
 └─ ...
```

Fig. 2. Excerpt of our induced tree on the DBpedia dataset. Numbers in brackets indicate number of subjects which visited the cluster on its path.

evaluate our model on a hierarchy not rooted in the WordNet taxonomy. The hierarchical relations between DBpedia classes were obtained from the DBpedia ontology mapping which may be found on the DBpeida website[2]. All querying to generate the dataset and ground truth clusters was performed in November of 2021.

4.2   Quantitative Evaluation

To quantitatively evaluate our model, we examined the hierarchical clustering of subjects in our induced topic hierarchy. This type of evaluation jointly assesses the quality of the tree structure as well as the allocation of paths along it. Specifically, we ran our model five times on each of the aforementioned datasets using 100 burn-in samples. We then sampled from our learned distributions to obtain a topic hierarchy. We evaluated the quality of the clustering using the Adjusted Rand Index (ARI) [17] and Normalized Mutual Information (NMI)

---

[2] http://mappings.dbpedia.org/server/ontology/classes/.

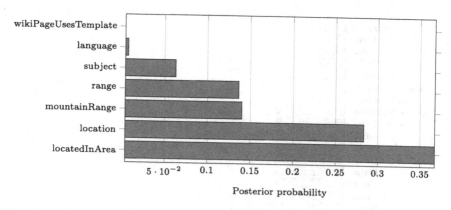

Fig. 3. Predicates and their posterior distribution for cluster K on the DBpedia tree as displayed in Fig. 2.

[31] as in previous works [18]. We compared our model against embedding based methods described in the related works section. Pretrained embeddings for these models were obtained from LibKGE[3] [10]. The mean and standard deviations of five runs are summarized in Table 2.

Our results indicate that our model is comparable with embedding based approaches. Indeed, the performance of all methods is highly variable with no method clearly outperforming the other. We note our model's underperformance on the YAGO3-10 dataset relative to other methods. We hypothesize that this is due to the high ratio of subjects to triples in this dataset. Such a characteristic results in a low amount of predicates and tags for each subject compared to other datasets. This in turn hinders our model's ability to approximate the true likelihood when calculating the posterior, resulting in lesser performance. Nevertheless, our model is still significantly better than many of the other methods as measured by a t-test. We conclude, therefore, that our model is capable of inducing coherent topic hierarchies on real world knowledge graphs.

4.3   Qualitative Evaluation

Cluster allocation is driven by the interaction of predicates and tags. Specifically, each cluster has predicate and tag membership distributions. This allows us to draw interesting observations in that we can describe a cluster by its predicate and tag distributions. This gives us insight into the composition of a cluster. Figure 2 provides an excerpt of our induced tree on the DBpedia dataset. On the other hand in Fig. 3, we provide an example of cluster K's predicate distribution from the DBpedia dataset. We note that this predicate distribution is consistent with the subjects whose path ends at this cluster. Namely, the predicates are consistent with these subjects, i.e., mountains. Furthermore, we can also analyze the distribution of objects to which the predicates are connected to. We highlight

---

[3] https://github.com/uma-pil/kge.

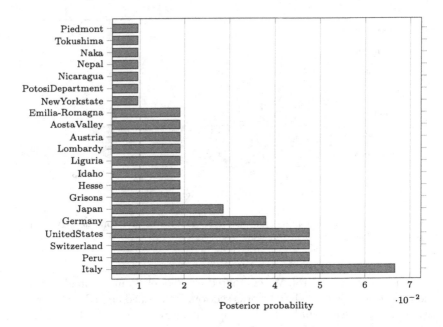

Fig. 4. Objects' posterior distribution for predicate locatedInArea

this in Fig. 4 which shows the object distribution for the predicate *locatedInArea* for cluster K. Based on the data that we used, the mountains in cluster K are most probably located in Italy, Peru, Switzerland, and United States.

5   Conclusion

In this paper we propose a model for discovering underlying hierarchical structures in knowledge graphs. For this purpose we adapt a hierarchical topic model used in natural language processing, namely hLDA, to the domain of knowledge graphs. Our model extends hLDA by introducing separate predicate and tag (predicate-object pair) topics, yielding a topic hierarchy consisting of predicate and tag distributions. Knowledge graph subjects take paths through this hierarchy which may be seen as an implicit hierarchical clustering of knowledge graph subjects. This formulation has the added benefit in that it is non-parametric, therefore does not require a priori assumptions about the tree structure other than its depth. To infer our model, we present an efficient Gibbs sampling scheme which leverages the Multinomial-Dirichlet conjugate to integrate out latent probability distributions allowing our model to scale to large datasets. We evaluate our model on three real world datasets and compare against benchmark methods. Our results demonstrate our model's ability to induce coherent topic hierarchies with high quality subject clusterings and explainable topic predicate and tag memberships.

References

1. Aldous, D.J.: Exchangeability and related topics. In: Hennequin, P.L. (ed.) École d'Été de Probabilités de Saint-Flour XIII — 1983. LNM, vol. 1117, pp. 1–198. Springer, Heidelberg (1985). https://doi.org/10.1007/BFb0099421

2. Almoqhim, F., Millard, D.E., Shadbolt, N.: Improving on popularity as a proxy for generality when building tag hierarchies from folksonomies. In: Aiello, L.M., McFarland, D. (eds.) SocInfo 2014. LNCS, vol. 8851, pp. 95–111. Springer, Cham (2014). https://doi.org/10.1007/978-3-319-13734-6_7

3. Ankerst, M., Breunig, M.M., Kriegel, H.P., Sander, J.: Optics: ordering points to identify the clustering structure. ACM Sigmod Rec. **28**(2), 49–60 (1999)

4. Bellini, V., Schiavone, A., Di Noia, T., Ragone, A., Di Sciascio, E.: Knowledge-aware autoencoders for explainable recommender systems. In: Proceedings of the 3rd Workshop on Deep Learning for Recommender Systems (2018)

5. Benz, D., Hotho, A., Stützer, S., Stumme, G.: Semantics made by you and me: self-emerging ontologies can capture the diversity of shared knowledge (2010)

6. Blei, D.M., Griffiths, T.L., Jordan, M.I.: The nested chinese restaurant process and bayesian nonparametric inference of topic hierarchies. J. ACM (JACM) **57**(2), 7 (2010)

7. Blei, D.M., Ng, A.Y., Jordan, M.I.: Latent dirichlet allocation. J. Mach. Learn. Res. **3**, 993–1022 (2003)

8. Bordes, A., Usunier, N., Chopra, S., Weston, J.: Large-scale simple question answering with memory networks. arXiv preprint (2015). arXiv:1506.02075

9. Bordes, A., Usunier, N., Garcia-Duran, A., Weston, J., Yakhnenko, O.: Translating embeddings for modeling multi-relational data. In: Advances in Neural Information Processing Systems 26 (2013)

10. Broscheit, S., Ruffinelli, D., Kochsiek, A., Betz, P., Gemulla, R.: Libkge-a knowledge graph embedding library for reproducible research. In: Proceedings of the 2020 Conference on Empirical Methods in Natural Language Processing: System Demonstrations, pp. 165–174 (2020)

11. Das, R., et al.: Go for a walk and arrive at the answer: reasoning over paths in knowledge bases using reinforcement learning. arXiv preprint (2017). arXiv:1711.05851

12. Dettmers, T., Minervini, P., Stenetorp, P., Riedel, S.: Convolutional 2d knowledge graph embeddings. In: Thirty-Second AAAI Conference on Artificial Intelligence (2018)

13. Gad-Elrab, M.H., Stepanova, D., Tran, T.-K., Adel, H., Weikum, G.: ExCut: explainable embedding-based clustering over knowledge graphs. In: Pan, J.Z. (ed.) ISWC 2020. LNCS, vol. 12506, pp. 218–237. Springer, Cham (2020). https://doi.org/10.1007/978-3-030-62419-4_13

14. Griffiths, T.L., Steyvers, M.: Finding scientific topics. Proc. Natl. Acad. Sci. **101**(suppl 1), 5228–5235 (2004)

15. Gu, C., Yin, G., Wang, T., Yang, C., Wang, H.: A supervised approach for tag hierarchy construction in open source communities. In: Proceedings of the 7th Asia-Pacific Symposium on Internetware, pp. 148–152. ACM (2015)

16. Heymann, P., Garcia-Molina, H.: Collaborative creation of communal hierarchical taxonomies in social tagging systems. Technical report (2006)

17. Hubert, L., Arabie, P.: Comparing partitions. J. Classif. **2**(1), 193–218 (1985). https://doi.org/10.1007/BF01908075

18. Jain, N., Kalo, J.-C., Balke, W.-T., Krestel, R.: Do embeddings actually capture knowledge graph semantics? In: Verborgh, R. (ed.) ESWC 2021. LNCS, vol. 12731, pp. 143–159. Springer, Cham (2021). https://doi.org/10.1007/978-3-030-77385-4_9

19. Lehmann, J., Isele, R., Jakob, M., Jentzsch, A., Kontokostas, D., Mendes, P.N., Hellmann, S., Morsey, M., Van Kleef, P., Auer, S., et al.: Dbpedia-a large-scale, multilingual knowledge base extracted from wikipedia. Semant. Web **6**(2), 167–195 (2015)

20. MacQueen, J., et al.: Some methods for classification and analysis of multivariate observations. In: Proceedings of the Fifth Berkeley Symposium on Mathematical Statistics and Probability, vol. 1, pp. 281–297. Oakland, CA, USA (1967)

21. Mahdisoltani, F., Biega, J., Suchanek, F.: Yago3: a knowledge base from multilingual wikipedias. In: 7th Biennial Conference on Innovative Data Systems Research. CIDR Conference (2014)

22. Mikolov, T., Sutskever, I., Chen, K., Corrado, G.S., Dean, J.: Distributed representations of words and phrases and their compositionality. Adv. Neural Inf. Process. Syst. **26**, 3111–3119 (2013)

23. Miller, G.A.: WordNet: an electronic lexical database (1998). MIT press

24. Nickel, M., Tresp, V., Kriegel, H.P.: A three-way model for collective learning on multi-relational data (2011)

25. Pietrasik, M., Reformat, M.: A simple method for inducing class taxonomies in knowledge graphs. In: Harth, A. (ed.) ESWC 2020. LNCS, vol. 12123, pp. 53–68. Springer, Cham (2020). https://doi.org/10.1007/978-3-030-49461-2_4

26. Pietrasik, M., Reformat, M.: Path based hierarchical clustering on knowledge graphs. arXiv preprint (2021). arXiv:2109.13178

27. Pitman, J.: Combinatorial stochastic processes. Technical report 621, Dept. Statistics, UC Berkeley, 2002. Lecture notes for St. Flour course, 2002 (2002)

28. Ristoski, P., Paulheim, H.: RDF2Vec: RDF graph embeddings for data mining. In: Groth, P. (ed.) ISWC 2016. LNCS, vol. 9981, pp. 498–514. Springer, Cham (2016). https://doi.org/10.1007/978-3-319-46523-4_30

29. Schlichtkrull, M., Kipf, T.N., Bloem, P., van den Berg, R., Titov, I., Welling, M.: Modeling relational data with graph convolutional networks. In: Gangemi, A. (ed.) ESWC 2018. LNCS, vol. 10843, pp. 593–607. Springer, Cham (2018). https://doi.org/10.1007/978-3-319-93417-4_38

30. Schmitz, P.: Inducing ontology from flickr tags. In: Collaborative Web Tagging Workshop at WWW2006, Edinburgh, Scotland, vol. 50, p. 39 (2006)

31. Shannon, C.E.: A mathematical theory of communication. Bell Syst. Techn. J. **27**(3), 379–423 (1948)

32. Toutanova, K., Chen, D.: Observed versus latent features for knowledge base and text inference. In: Proceedings of the 3rd Workshop on Continuous Vector Space Models and their Compositionality, pp. 57–66 (2015)

33. Trouillon, T., Welbl, J., Riedel, S., Gaussier, É., Bouchard, G.: Complex embeddings for simple link prediction. In: International Conference on Machine Learning, pp. 2071–2080. PMLR (2016)

34. Wang, S., Wang, T., Mao, X., Yin, G., Yu, Y.: A hybrid approach for tag hierarchy construction. In: Capilla, R., Gallina, B., Cetina, C. (eds.) ICSR 2018. LNCS, vol. 10826, pp. 59–75. Springer, Cham (2018). https://doi.org/10.1007/978-3-319-90421-4_4

35. Yang, B., Yih, W.T., He, X., Gao, J., Deng, L.: Embedding entities and relations for learning and inference in knowledge bases. arXiv preprint (2014). arXiv:1412.6575

Resources

Do Arduinos Dream of Efficient Reasoners?

Alexandre Bento[1]([✉]), Lionel Médini[1], Kamal Singh[2], and Frédérique Laforest[1]

[1] Univ Lyon, INSA Lyon, CNRS, UCBL, LIRIS, UMR5205,
69621 Villeurbanne, France
{alexandre.bento,lionel.medini,frederique.laforest}@liris.cnrs.fr
[2] Univ Lyon, UJM-Saint-Etienne, CNRS, Laboratoire Hubert Curien UMR 5516,
F-42023 Saint-Étienne, France
kamal.singh@univ-st-etienne.fr

Abstract. The Semantic Web of Things enhances the Internet of Things with Web technologies as well as Knowledge Graphs and reasoning. Traditional reasoners are too heavy in terms of memory footprint and/or processing time to be implementable on things. In this work, we present LiRoT, a lightweight incremental reasoner that can be embedded in constrained objects, so that reasoning on them in a fog architecture becomes possible. The focus of this work is to reduce drastically memory footprint while paying attention to processing time, hence usual optimization techniques are not fully adequate. We provide evaluations that (i) compare our system to the state of the art and (ii) show the effective benefits of the different optimizations we have implemented.

Keywords: Semantic web · Reasoning · Web of things · Embedded systems · Optimization

1 Introduction

Today, more and more applications require a connection to the physical world to capture information from the environment as well as to act on it. The Internet of Things provides answers to such needs by connecting sensors and actuators to computers; the Web of Things (WoT) intends to do it using the Web standards; and the Semantic Web of Things (SWoT) enhances the WoT by adding the expressive power of Knowledge Graphs as well as reasoning capabilities. At the same time, the huge number of things and the total volume of produced data has raised the need for distributed edge and fog architectures [22] where data are processed as close as possible to their production and consumption locations. Fog computing architectures can involve different types of nodes. Some of these fog nodes might have limited amounts of energy or/and bandwidth. Reasoning on devices with limited resources, such as microcontroller-based[1] ones, will be

---

[1] Microcontrollers are small processing units designed to run embedded applications, in contrast to more powerful microprocessors that can execute general purpose applications.

P. Groth et al. (Eds.): ESWC 2022, LNCS 13261, pp. 289–304, 2022.
https://doi.org/10.1007/978-3-031-06981-9_17

an important enabler for reasoning in such architectures. It is obvious that very small microcontrollers with a few kilobytes of RAM will not be able to process highly expressive reasoning tasks about very large datasets, and that enabling semantic reasoning on constrained devices is a matter of trade-offs. Nevertheless, significantly useful tasks such as classification using subsets of RDF-S[2] can be performed on devices with around 100 KB RAM and 100 MHz clock speed. However, modern state-of-the-art reasoners are optimized for speed and high data volumes and many of the optimizations on which they rely (e.g. exploiting multiple cores and highly parallel architectures) are not usable on such devices, so that finally none of them seems to fit for edge/fog reasoning.

Focusing on typical SWoT use cases on such architectures, the application, ruleset and ontology are usually known in advance. Hence, they can be flashed on the device (we herein assume there is enough room for that). Whilst running the application, sensor data that arrives periodically should be processed on the fly. Given our fog architecture assumption, we focus on a scenario where small data quantities issued by one or a few sensors require to update the reasoner's internal state. For example, in the CoSWoT[3] project, we target applications such as field watering or frost prevention that are based on temperature and humidity data provided for each field a couple of times each day, and that can rely on decisions made locally for each field (more details in our use case[4]). W3C WoT use cases also highlight connectivity and autonomy constraints in edge architectures[5], making reasoning on edge nodes relevant. We herein choose to process data incrementally [14] rather than as stream [4], to avoid the need for the reasoner to handle time or data windows.

In this work, we present LiRoT, a lightweight incremental reasoner that can be embedded in resource-constrained nodes of fog architectures. Starting from the well known RETE algorithm, we propose specific improvements that target the above SWoT use cases. The focus of our work is thus mainly on memory frugality, while also considering algorithmic optimization.

The paper is organized as follows. Section 2 reviews previous works on incremental reasoning optimizations and discusses their adequacy to SWoT; it also describes the classic RETE algorithm. Section 3 presents the proposed optimizations for LiRoT, a SWoT-compliant RETE-based reasoner. Section 4 presents two sets of evaluations, (i) to compare our reasoner to state of the art systems and (ii) to show the effects of the implemented optimizations. Section 5 discusses the results and describes optimizations that were experimented but did not improve performance. Section 6 concludes and sketches future directions for our work.

2   Related Work

In this section, we review the different works that can be applied to perform reasoning tasks with a focus on small devices. Given the strongly constrained

---

[2] https://www.w3.org/TR/rdf-mt/#rdfs_entailment.

[3] https://coswot.gitlab.io/.

[4] https://www.w3.org/TR/wot-usecases/#Agricultural-irrigation.

[5] https://www.w3.org/TR/wot-usecases/#edge-computing.

environments on which we intend to deploy our work, we herein focus on lightweight reasoning algorithms, namely rule-based ones. The OWL 2 RL profile[6] has been designed to foster the use of rule-based reasoners. The foundations of how and why to combine rules with ontologies for the Semantic Web are addressed in Description Logic Programs (DLP) [8], which bridge the gap between knowledge representation (KR) and in particular DL and LP[7]. Moreover, in order to allow their deployment on constrained environments, we herein restrict to subsets of the OWL 2 RL ruleset and the RDF-S entailment list[8].

2.1 The RETE Algorithm

A rule-based reasoner applies rules to the *facts*[9] contained in a knowledge base (KB) to produce new knowledge, and loops over the set of initial and produced knowledge until no new knowledge is issued. In the reasoning field, the facts that were already present in the KB at the beginning of this loop are called *explicit facts*, and those that have been inferred during the loop are called *implicit facts*.

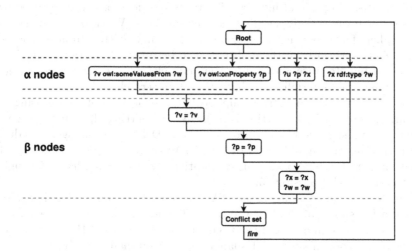

Fig. 1. Example of a RETE network for rule `cls-svf1` from the OWL 2 RL profile: (?v owl:someValuesFrom ?w) ∧ (?v owl:onProperty ?p) ∧ (?u ?p ?x) ∧ (?x rdf:type ?w) → (?u rdf:type ?v)

The RETE [6] algorithm is one of the most well-known algorithms to process rulesets over a KB. Rules are represented with a trie structure called the RETE network. A RETE network is composed of two main layers (see example on Fig. 1):

---

[6] https://www.w3.org/TR/owl2-profiles/#OWL_2_RL.

[7] https://en.wikipedia.org/wiki/Logic_programming.

[8] https://www.w3.org/TR/rdf11-mt/#entailment-rules-informative.

[9] As we herein consider the KB as being an ontology expressed in OWL 2 RL under RDF-based semantics, facts are RDF triples.

- The **alpha nodes**: each alpha node is associated with an atomic condition in a rule (e.g. ?x rdf:type foaf:Person), and performs a match operation over the whole knowledge base (i.e. explicit and implicit facts). Alpha nodes contain *alpha memories*, which store the facts that match the node's condition. An alpha node is related to a single rule, even if a same condition is shared by multiple rules. An alpha node has a single output edge, that leads to a beta node.
- The **beta nodes**: beta nodes perform join operations between two nodes. Beta nodes can be placed either after two alpha nodes, or after a beta and an alpha node if the associated rule has more than two conditions (then join operations are performed sequentially). Beta nodes contain *beta memories*, that store variable substitutions that are compatible with the node's parents. A beta node has exactly two input edges and one output edge.

2.2 Incremental Reasoning

SWoT use cases imply multiple data insertion and deletion steps to reflect the state of dynamically changing physical environments. When data evolves over time, explicit facts may need to be inserted in and deleted from a reasoner. Implicit facts derived from them also need to be updated. To tackle this issue, incremental maintenance allows insertion and deletion of explicit facts without re-performing the materialization operation from scratch.

The RETE algorithm natively supports incremental maintenance using two adaptations. Incremental insertion is managed by splitting alpha memories into two parts: new facts and already-processed facts. Deletion is managed by adding a data structure inside alpha and beta nodes to connect implicit facts with the explicit facts they come from. This is traditionally done using lists, although a tree structure is also possible [5].

Jena [3] is a reference rule-based OWL reasoner based on RETE that implements the DL subset of the first OWL specification. It provides an easy-to-use API for Java programmers and has been widely used as such. However, all constrained objects do not support this language, and especially a garbage-collecting function, so it is out of scope for the field of SWoT and embedded reasoning.

CLIPS[10] [12] is a widely used expert system tool. It uses a complete object-oriented language for writing expert systems. It employs the RETE algorithm for reasoning. However, RDF triples are not supported natively by CLIPS. Other tools, such as R-DEVICE [1], are needed to import RDF into CLIPS.

The Delete/Rederive (DRed) algorithm [10] handles deletion. It first over-deletes all implicit facts that depend on the deleted facts. Then it rederives the implicit facts that can be inferred another way. Rederivation is iteratively applied. Some other works use a variant of the DRed algorithm, for example in [19].

For supporting incremental reasoning, RDFox [13] uses a backward-forward algorithm. Unlike DRed that searches after over-deletion, the backward-forward

---

[10] http://www.clipsrules.net/.

algorithm uses an approach that first searches for the alternative derivations. This is done by using a combination of backward and forward chaining. The induced performance gain is particularly visible with implicit facts that are derived from numerous chained deductions (e.g. rdfs:subClassOf). GraphDB[11] uses the same approach. RDFox has a high memory footprint (especially due to a high number of indexes) and is optimized for architectures that allow parallel computing, hence it is not designed to be embedded in constrained devices.

HyLAR+ [17] is an incremental reasoner implemented in JavaScript and targeting Web applications, which has been improved with a so-called "tag-based" approach [18], that allows for fastly performing multiple fact insertion/deletion. This approach is inspired from that initiated by [7] on improving reasoning about evolving versions of ontologies. The general idea is to keep trace of previous reasoning computations originated by changes in the graph, in order to respond more quickly to similar changes in the future. The drawback of this approach is that it requires to store history as extra information, which is therefore not suitable for memory-constrained devices.

2.3   Embedded Reasoning

Many existing semantic reasoners are too resource-intensive to be directly ported on resource-constrained devices such as objects or sensors. Only a few works embed reasoning in constrained devices. Some of them are designed for mobile phones and not for more constrained devices. For example, [2] studied porting Description Logics (DL) reasoners on mobile Android-based devices. It is worth noting that smartphones have much higher computational capabilities as compared to the devices we target.

An OWL reasoner for embedded devices was proposed in [15], it is based on CLIPS. They considered OWL 2 RL. Their system was implemented and tested on Gumstix Verdex Pro which has 400 MHz CPU, 64 MB RAM, and 16 MB Flash. Note that in our work, we are targeting embedded devices that have several orders of magnitude less RAM (around 500 KB).

$RETE_{pool}$ [20] is a RETE-based reasoner that aims to reduce its memory footprint by reducing data duplication during rule based reasoning, while specifically considering OWL 2 RL. To do so, it uses one shared memory for all alpha nodes in the network. This way, duplicates are eliminated during insertion. In cases where a RDF store is used along the reasoner, another level of duplication is removed by using this store directly as the RETE memory, each alpha node having references to triples contained in the store. This saves memory, at the cost of speed degradation. Their experiments were conducted on smartphones and laptops.

Another work based on the RETE algorithm is called COROR [16]. It uses the following composition algorithms to reduce memory consumption. It selectively loads only the rules that are required, by creating a rule-construct dependencies

---

[11] https://graphdb.ontotext.com/documentation/free/reasoning.html#retraction-of-assertions.

set. Next it decomposes the RETE algorithm in two phases. The first phase does an initial matching. Then the next phase builds the next part of the RETE network using statistics collected from the first phase, that allow to reorder rules and conditions according to their selectivity (the most selective conditions are matched first; this optimization is well-known in database management systems, for ordering join operations). The first cycle is then completed by joining the facts obtained from the first phase. Results show that COROR reduces the memory footprint by 74% on average. COROR also uses rewriting for rules that are known to be resource–and time–consuming, such as rules involving owl:sameAs and wildcard conditions. COROR was experimented on a SunSPOT platform, which has a similar memory size to our target platforms but uses the Java language. These experiments showed very low speed: they reported 1561 s to process the WINE ontology that contains 1833 triples with the ρD* ruleset.

This state of the art has shown that the proposed reasoners of the literature are not well suited to be run on microcontroller-based platforms. Nevertheless, they include some optimization proposals that target memory footprint. The next section will present how we obtained a reasoner that can be run on platforms of the Arduino family, by extracting the best ideas of the literature and adapting them to this specific context.

3   LiRoT: Improving RETE for the SWoT

We propose optimizations over the traditional RETE algorithm. These optimizations are built with the objective to be embeddable on platforms like Arduino or ESP32 architectures. So they are focused on memory footprint, as it is a strict criterion for such platforms. We also pay attention to processing time, checking that it remains within acceptable bounds.

3.1   Term Indexing

To save memory, we use an index over the terms within the reasoner. Various implementations exist in the literature, relying on data structures such as linked lists, arrays or binary search trees. A hash table can also be used to do this in an efficient manner: indeed, on average, a hash table has a linear space complexity, and constant time complexity for insertion, search and deletion, which is better than the previously cited data structures. Hence we chose this option to make a term index.

3.2   Merging Alpha Memories

Alpha memories have already been the place for optimization of the RETE algorithm, like in RETE_{pool} where all alpha memories are merged into one common memory. Here, to optimize smartly the RETE algorithm memory footprint, we merge memories of alpha nodes that share a syntactically similar rule condition.

Let $c_1 = (s_1, p_1, o_1)$ and $c_2 = (s_2, p_2, o_2)$ be two conditions. c_1 and c_2 are similar if either:

- s_1 and s_2 (resp. p_1 and p_2, o_1 and o_2) are variable terms
- s_1 and s_2 (resp. p_1 and p_2, o_1 and o_2) are not variable terms and $s_1 = s_2$ (resp. $p_1 = p_2$, $o_1 = o_2$)

As we merge memories of all alpha nodes that share similar rule conditions, each rule condition is checked by only one node. This optimization saves both memory and time. This is particularly useful in rulesets like RDFS, where multiple rules use a wildcard condition (a condition that matches all facts in the KB, e.g. $?x\ ?p\ ?y$).

This optimization implies that each alpha memory is potentially connected to multiple beta nodes (one for each original alpha memory that has been merged), which is not the case in the original RETE network. It has an impact on the management of the alpha memories, as a fact remains new as long as one of its beta nodes has not yet processed it (see Sect. 2.2). Triggering the execution of the beta nodes connected to this alpha memory is thus required before pushing new facts in the already-processed part of the alpha memory.

3.3 Optimizing Incremental Maintenance

To handle incremental maintenance, we have used a similar philosophy as that of the backward-forward algorithm [13]. Although not initially designed to work with RETE, it is compatible with its network structure: each terminal beta node (i.e. the last beta node of a rule) stores a list of the implicit facts that it has produced. Every implicit fact contains the list of beta memories that led to its production. When an explicit fact is removed from the knowledge base, it is first removed from the memory of its matching alpha nodes, then the corresponding variable substitutions are removed from beta memories following the same route as when adding a new explicit fact. After this step, if an implicit fact has no more cause coming from the beta node that produced it, the algorithm first searches through all other terminal beta nodes if the same implicit fact was produced somewhere else. If not, it is deleted from the knowledge base. This avoids unnecessary deletions, in case an implicit fact was obtained in multiple ways.

3.4 Implementation Details

LiRoT is written in C: low-level languages are more suited for constrained platforms because they allow for more fine-grained memory management. Rust [11] could also have been an option, but it is relatively new and not yet available on most platforms.

LiRoT is composed of two modules:

- The core algorithm, based on RETE and including the proposed optimizations. It currently uses Sord[12], a C library providing a lightweight in-memory triplestore, to implement and store RDF triples.

---

[12] https://github.com/drobilla/sord.

– A wrapper to the core algorithm providing an RDF-JS-like API[13] to query the reasoner. It relies on Serd[14] to parse and serialize RDF triples in Turtle, N-Triples, N-Quads and TriG formats.

In the core algorithm, to implement the term index, we use the efficient hashtable implementation from the uthash[15] library. To compute terms hashes, we use the hash function provided by Sord.

LiRoT comes with two versions: a Linux version and an Arduino version. It was tested on Manjaro Linux, Arduino Due and ESP-32 platforms.

LiRoT source code is available at https://gitlab.com/coswot/lirot .

4   Evaluation and Results

4.1   Dataset and Evaluation Method

We used the LUBM benchmark [9] to generate 12 synthetic datasets of various sizes in the domain of universities. These datasets contain from 0 to 10,000 explicit triples representing assertions, in addition to the ontology itself (293 triples).

We have implemented the three rulesets RDFS-Simple, RDFS-Default and RDFS-Full (provided by Apache Jena[16]) for each tested reasoner. We provide the lists of rules in the LiRoT source code repository.

To compare our approach to other incremental reasoners and to assess the effectiveness of different versions of our algorithm, we have run four types of experiments on each of the 12 datasets:

– **full materialization.** The dataset is entirely loaded at once, then the reasoning is launched.
– **incremental insertions.** We randomly split each dataset into two equal parts. The first half is inserted into the reasoner. The second half is divided into five fragments, each representing 10% of the whole dataset, and are added sequentially.
– **incremental deletions.** The idea here is to delete multiple parts of a preloaded dataset. We first load the whole dataset, execute an initial reasoning task, then sequentially remove the same subsets as described before.
– **incremental insertions and deletions.** Here we sequentially add then delete parts of the dataset. The first half of the dataset is loaded, then each subset is first inserted then deleted. This is the closest test to the actual use

---

[13] https://rdf.js.org/.
[14] https://github.com/drobilla/serd.
[15] https://github.com/troydhanson/uthash.
[16] Rulesets are described at https://jena.apache.org/documentation/inference/#RDFSconfiguration.

cases that we performed using LUBM. In the use cases for which this reasoner is designed, the "same" triples will be added and removed (possibly with some variations).

We use two performance metrics: i) maximum resident set size[17] (maxRSS): in the context of constrained objects, memory size is a mandatory limitation, and ii) execution time.

Experimentation materials are available at https://gitlab.com/coswot/lirot-experiments-eswc-2022, where one can find datasets, scripts used to run all reasoners, raw results files and plots.

In the following sections, RETE denotes our baseline implementation of the RETE algorithm; RETE+alpha is RETE with the optimization on alpha memories as described in Sect. 3.2; RETE+terms denotes RETE with the index on terms described in Sect. 3.1; LiRoT is RETE with the optimizations on alpha memories and the index on terms. The optimization on incremental reasoning is present in all versions of RETE, RETE+alpha, RETE+terms and LiRoT.

4.2 Correctness Verification

To ensure that LiRoT produces correct results, we compared its output with that of Apache Jena, which is a well tested reasoner. We performed these tests both on the insertion and deletion algorithms, and got the same output in all cases.

4.3 Comparison with Other Reasoners

We compared our approach to two standard incremental reasoners: Apache Jena and RDFox. To do so, we ran each reasoner on the same desktop computer (MSI GF63 Thin 10SCXR-046FR Dragon Station, with an Intel Core i7-10750H CPU and 32 GB of DDR4 2666 MHz MHz RAM). We forced the execution of each reasoner on only one CPU thread, to mimic the behavior of more constrained devices. We ran each test 20 times, removed the most extreme values (5% lowest and 5% highest) and computed the average maxRSS and execution time for each reasoner and dataset.

Figure 2 compares LiRoT with Jena and RDFox for the easiest type of test (full materialization with the RDFS-Simple ruleset) and the most difficult one (successive insertions and deletions with the RDFS-Full ruleset).
All other results lie in between these two extreme cases; figures for the other tests are available online; they show similar trends to these figures.

With the RDFS-Simple ruleset and a full materialization, LiRoT uses 58–91% less memory than RDFox, and 92–98% less than Jena. With the RDFS-Full ruleset and incremental insertions and deletions, LiRoT uses 46–90% less memory than RDFox, and 91–98% less than Jena.

With RDFS-Simple and a full materialization, LiRoT is 72–93% faster than Jena. It is faster than RDFox for datasets under 1000 explicit facts and slower

---

[17] The maximum amount of RAM used by a program throughout its execution.

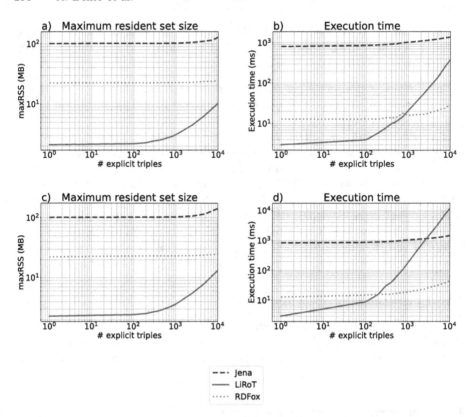

Fig. 2. maxRSS and execution time comparisons between LiRoT, Jena and RDFox. Figures a) and b) show a full materialization using the RDFS-Simple ruleset. Figures c) and d) show successive insertions and deletions using the RDFS-Full ruleset.

for larger datasets. With RDFS-Full and incremental insertions and deletions, LiRoT is faster than Jena (resp. RDFox) for datasets under 2500 (resp. 200) facts. Other configurations have maxRSS and execution time values in between these intervals.

4.4   Improvements of the RETE Algorithm

On the same desktop setup as previous evaluations, we ran each version of our RETE algorithm baseline and improvements.

Figure 3 shows the different maxRSS and execution times for all four RETE-based algorithms, and in the same configurations as above. With the RDFS-Simple ruleset (resp. RDFS-Full), we find that sharing similar alpha nodes across rules allows to save up to 24% (resp. 28%) memory compared to our baseline RETE implementation. The use of an index on terms saves up to 25% (resp. 29%) memory compared to baseline. The combination of both saves up to 32% (resp. 41%) memory.

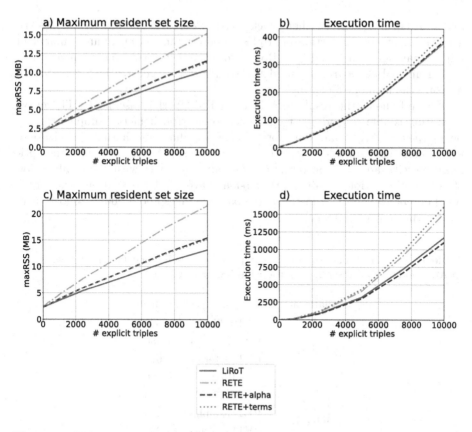

Fig. 3. maxRSS and execution time comparisons between different optimizations of the RETE algorithm. Figures a) and b) show a full materialization using the RDFS-Simple ruleset. Figures c) and d) show successive insertions and deletions using the RDFS-Full ruleset.

Figure 3 shows that for easier types of tests (full materialization with the RDFS-Simple ruleset), optimizations over the RETE baseline have a small impact on the execution time ($+1\%$ for the optimization on alpha nodes, $+7.6\%$ for the optimization on terms, -0.2% with both optimizations, for the largest dataset; the difference is even smaller for smaller datasets). For the most difficult type of tests (successive insertions and deletions with the RDFS-Full ruleset), optimizations have a more significant impact on the execution time (-37% for the optimization on alpha nodes, $+6.5\%$ for the optimization on terms, -23% for the combination of both). The differences among these optimizations are discussed in Sect. 5.2.

4.5 Improvements on Embedded Devices

We also provide tests for LiRoT on constrained devices, with the RDFS-Simple ruleset. We used an Arduino Due (clock speed of 84 MHz and 96 KB of SRAM) and an Adafruit ESP32 Feather (clock speed of 240 MHz and 520 KB of SRAM). We use incremental insertion of triples to determine the maximum number of triples that can be handled by LiRoT before the device runs out of memory.

On an ESP32 board, we used the standard ESP.getFreeHeap function to measure the memory footprint.

Figure 4 shows the results obtained on ESP32: the baseline RETE implementation is able to load 509 facts (ontology + explicit facts + implicit facts), before the board runs out of memory. With the alpha node-sharing optimization, this number goes up to 656 facts (+22% wrt baseline), 672 facts with the optimization with an index on terms (+24%) and 760 facts with the combination of both optimizations (+49%). All versions have similar execution time for an equal number of facts (for instance, the baseline takes about 200ms to load facts until it runs out of memory).

Table 1. Number of facts (ontology + explicit facts + implicit facts) after reasoning on Arduino Due and ESP32, with the RDFS-Simple ruleset

Platform	Version of RETE	# facts
ESP32	RETE	509
	RETE+alpha	656
	RETE+terms	672
	LiRoT	760
Arduino Due	RETE	240
	RETE+alpha	285
	RETE+terms	290
	LiRoT	349

The hardware architecture of the Arduino Due does not allow to dynamically measure memory usage with standard functions, as we did on the ESP32. To the best of our knowledge, no equivalent library allows to do it on this device. Hence, we measured the maximum number of facts that the reasoner is able to load and process before the board runs out of memory. Table 1 shows that the optimization on alpha nodes allows to load 19% more facts than the baseline implementation on Arduino Due; the use of an index on terms allows for 21% more facts, and the combination of both allows for 45% more facts, compared to baseline. All versions have similar execution time (about 550 ms for the baseline version).

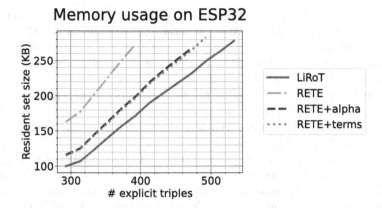

Fig. 4. Comparison of memory usage on a ESP32 board with different optimizations of the RETE algorithm, using the RDFS-Simple ruleset. The horizontal axis shows the number of explicit facts (ontology included).

5 Discussion

5.1 Memory Usage

On classic PC architectures, LiRoT significantly improves the memory footprint as compared to the literature. Its processing time is lower for small datasets in these settings, but increases rapidly when the number of facts grows. This shows that LiRoT can also be of help on intermediary devices that rely on the same architectures, but have less resources than even smaller computers (e.g. Raspberry Pi).

To the best of our knowledge, LiRoT is the only one able to be actually deployable on Arduino-based platforms[18].

These two points validate our hypothesis of LiRoT as designed for specifically targeting such devices. Processing time is of course to be taken seriously, but must be considered with respect to the findings of Sect. 4.5.

5.2 Processing Time

Considering the maximum number of facts that a small device can handle, LiRoT remains faster than other reasoners, given that these reasoners could be deployed on the devices. Our experiments also show that the time gained by the optimizations varies according to both sizes of the ruleset and the dataset. It is interesting to note that the LiRoT optimization is not the one that shows the best time

---

[18] Indeed, these devices have, in addition to a limited memory size for handling the application data (DRAM), the same kind of limitations for storing the program itself (IRAM). The reasoner should also be compiled specifically for the targeted platform, and use platform-specific available libraries.

performance among all optimizations presented in Sect. 4.4. However, as they do save memory space, we chose to keep all three optimizations in LiRoT.

Though we admit that restricting the operation range to processing very small amounts of data is not a common goal, it can still benefit to use cases where energy consumption is a critical issue, as well as to embed the reasoner on a relatively more powerful device that has to perform other tasks than reasoning. Indeed, in the CoSWoT project, we intend to include this reasoner as a WoT servient module, so that sensors and actuators can be run autonomously and locally process data as RDF triples.

Moreover, even if RDFox overpasses LiRoT when the number of facts grows, our trade-off assumption seems to remain valid for relatively higher number of facts than those that Arduino-based devices can handle. Our evaluations showed that there is less than 200 ms difference for a full materialization of 2000 facts for a whole run of insertions and deletions under the RDFS-Simple ruleset, which sounds to us as acceptable.

The optimizations we implemented for LiRoT come with a few trade-offs. The optimization on alpha nodes could cause concurrent access issues on alpha memories if it were implemented in a multi-threaded application (which LiRoT is not). Term indexing causes a slight loss in performance, as each term insertion requires to check if the term already exists; however, a hash table is a very efficient way to perform this task, so the performance loss is small, in comparison with the overall computation time. Finally, the optimized incremental maintenance algorithm requires to store intermediate reasoning results, which has an impact on memory footprint.

In order to improve the speed of our reasoner, we explored one dead-end that is worth mentioning, and are foreseeing two directions:

- One option we tried consists in using an efficient join algorithm in beta nodes. The naive approach, using a nested loop join, reviews all possible matches among two nodes. The sort-merge join algorithm can solve this issue by using sorted data structures in alpha and beta memories and limiting the number of merge steps to the minimum required, to the cost of nodes needing to maintain sorted data structures. However, the relevance of this optimization depends on the ruleset and the dataset sizes: it saves time when the two nodes to join have many elements in their respective memories. This is not the case in our experiments, which showed lower performance.
- The evaluations in which data were to be deleted show higher processing times than expected, while compared to RDFox for instance. Future works include improving the deletion algorithm by removing intermediary beta node memories, as done in [21].
- Parsing and serializing facts from and to a standard representation format is also time and memory-consuming. Even if Sord and Serd are optimized for this task, they offer many unneeded functionalities that could be removed from the reasoner. In order to diminish both memory usage and processing time, we are currently looking for a more global way to handle compressed RDF data without serialization/deserialization operations at the servient level, relying on binary representations such as CBOR-LD or HDT.

6   Conclusion and Future Works

We proposed LiRoT, a lightweight incremental reasoner, the first–to the best of our knowledge–that can be embedded on constrained devices such as Arduino Due and ESP32. LiRoT as a tool acts as an enabler for Semantic Web of Things by providing a reasoning capability to constrained devices. Our work also advances the existing state of the art in the fog computing paradigm.

LiRoT implements the RETE algorithm at its heart as the baseline algorithm. Additionally, we studied three optimization schemes to the baseline RETE that resulted in significant memory savings for incremental reasoning. We performed experiments on Linux as well as on embedded systems. We compared the performance of LiRoT with existing reasoners such as RDFox and Jena. As compared to these approaches, our experiments showed that for relatively modest numbers of facts (around 200 to 3000 facts depending on the complexity of the ruleset), which corresponds well to the paradigm of the Semantic Web of Things, LiRoT can do reasoning with lower computation times. LiRoT always had the lowest memory usage for performing reasoning on up to 10,000 facts (maximum number of facts tested). The memory usage was several orders of magnitude lower than RDFox and Jena.

On a desktop configuration, using the LUBM benchmark, our optimizations saved up to 32% of memory with the RDFS-Simple ruleset, up to 36% of memory with RDFS-Default and up to 41% with RDFS-Full. On embedded devices, with the RDFS-Simple ruleset, LiRoT was able to load up to 49% more facts than our baseline RETE implementation.

In the future, we would like to perform tests on other rulesets and to explore more optimization schemes as the ones presented in Sect. 5.2. We will also compare the energy consumption of different optimizations. We would also like to explore distributed and collaborative reasoning algorithms suited to SWoT and embedded environments.

Acknowledgment. This work is supported by grant ANR-19-CE23-0012 from the Agence Nationale de la Recherche, France, for the CoSWoT project.

References

1. Bassiliades, N., Vlahavas, I.: R-device: a deductive RDF rule language. In: Antoniou, G., Boley, H. (eds.) RuleML 2004. LNCS, vol. 3323, pp. 65–80. Springer, Heidelberg (2004). https://doi.org/10.1007/978-3-540-30504-0_6
2. Bobed, C., Yus, R., Bobillo, F., Mena, E.: Semantic reasoning on mobile devices: do androids dream of efficient reasoners? J. Web Seman. **35**, 167–183 (2015)
3. Carroll, J.J., Dickinson, I., Dollin, C., Reynolds, D., Seaborne, A., Wilkinson, K.: Jena: implementing the semantic web recommendations. In: Proceedings of the 13th International World Wide Web Conference on Alternate Track Papers & Posters, pp. 74–83 (2004)
4. Dell'Aglio, D., Della Valle, E., van Harmelen, F., Bernstein, A.: Stream reasoning: a survey and outlook. Data Sci. **1**(1–2), 59–83 (2017)

5. Doorenbos, R.B.: Production Matching for Large Learning Systems. Carnegie-Mellon Univ Pittsburgh PA Dept of Computer Science, Technical report (1995)

6. Forgy, C.L.: Rete: A fast algorithm for the many pattern/many object pattern match problem. In: Readings in Artificial Intelligence and Databases, pp. 547–559. Elsevier (1989)

7. Cuenca Grau, B., Halaschek-Wiener, C., Kazakov, Y.: History matters: incremental ontology reasoning using modules. In: Aberer, K. (ed.) ASWC/ISWC -2007. LNCS, vol. 4825, pp. 183–196. Springer, Heidelberg (2007). https://doi.org/10.1007/978-3-540-76298-0_14

8. Grosof, B.N., Horrocks, I., Volz, R., Decker, S.: Description logic programs: combining logic programs with description logic. In: Proceedings of the 12th International Conference on World Wide Web, pp. 48–57 (2003)

9. Guo, Y., Pan, Z., Heflin, J.: Lubm: a benchmark for owl knowledge base systems. J. Web Seman. 3(2–3), 158–182 (2005)

10. Gupta, A., Mumick, I.S., Subrahmanian, V.S.: Maintaining views incrementally. ACM SIGMOD Rec. 22(2), 157–166 (1993)

11. Matsakis, N.D., Klock, F.S.: The rust language. ACM SIGAda Ada Lett. 34(3), 103–104 (2014)

12. Meditskos, G., Bassiliades, N.: Clips-owl: a framework for providing object-oriented extensional ontology queries in a production rule engine. Data Knowl. Eng. 70(7), 661–681 (2011)

13. Nenov, Y., Piro, R., Motik, B., Horrocks, I., Wu, Z., Banerjee, J.: RDFox: a highly-scalable RDF store. In: Arenas, M. (ed.) ISWC 2015. LNCS, vol. 9367, pp. 3–20. Springer, Cham (2015). https://doi.org/10.1007/978-3-319-25010-6_1

14. Oliya, M., Pung, H.K.: Towards incremental reasoning for context aware systems. In: Abraham, A., Lloret Mauri, J., Buford, J.F., Suzuki, J., Thampi, S.M. (eds.) ACC 2011. CCIS, vol. 190, pp. 232–241. Springer, Heidelberg (2011). https://doi.org/10.1007/978-3-642-22709-7_24

15. Seitz, C., Schönfelder, R.: Rule-based OWL reasoning for specific embedded devices. In: Aroyo, L. (ed.) ISWC 2011. LNCS, vol. 7032, pp. 237–252. Springer, Heidelberg (2011). https://doi.org/10.1007/978-3-642-25093-4_16

16. Tai, W., Keeney, J., O'Sullivan, D.: Resource-constrained reasoning using a reasoner composition approach. Seman. Web 6(1), 35–59 (2015)

17. Terdjimi, M., Médini, L., Mrissa, M.: Hylar+ improving hybrid location-agnostic reasoning with incremental rule-based update. In: Proceedings of the 25th International Conference Companion on World Wide Web, pp. 259–262 (2016)

18. Terdjimi, M., Médini, L., Mrissa, M.: Web reasoning using fact tagging. In: Companion Proceedings of the The Web Conference 2018, pp. 1587–1594 (2018)

19. Urbani, J., Margara, A., Jacobs, C., van Harmelen, F., Bal, H.: DynamiTE: parallel materialization of dynamic RDF data. In: Alani, H. (ed.) ISWC 2013. LNCS, vol. 8218, pp. 657–672. Springer, Heidelberg (2013). https://doi.org/10.1007/978-3-642-41335-3_41

20. Van Woensel, W., Abidi, S.S.R.: Optimizing semantic reasoning on memory-constrained platforms using the RETE algorithm. In: Gangemi, A. (ed.) ESWC 2018. LNCS, vol. 10843, pp. 682–696. Springer, Cham (2018). https://doi.org/10.1007/978-3-319-93417-4_44

21. Wright, I., Marshall, J.A.: The execution kernel of rc++: Rete*, a faster rete with treat as a special case. Int. J. Intell. Game. Simul. 2(1), 36–48 (2003)

22. Yousefpour, A., et al.: All one needs to know about fog computing and related edge computing paradigms: a complete survey. J. Syst. Architect. 98, 289–330 (2019)

A Programming Interface for Creating Data According to the SPAR Ontologies and the OpenCitations Data Model

Simone Persiani[1], Marilena Daquino[2,3] (ID), and Silvio Peroni[2,3](✉) (ID)

[1] Department of Computer Science and Engineering, University of Bologna, Bologna, Italy
simone.persiani2@studio.unibo.it

[2] Research Centre for Open Scholarly Metadata, Department of Classical Philology and Italian Studies, University of Bologna, Bologna, Italy
{marilena.daquino2,silvio.peroni}@unibo.it

[3] Digital Humanities Advanced Research Centre (/DH.arc), Department of Classical Philology and Italian Studies, University of Bologna, Bologna, Italy

Abstract. The OpenCitations Data Model (OCDM) is a data model for bibliographic metadata and citations based on the SPAR Ontologies and developed by OpenCitations to expose all the data of its collections as sets of RDF statements compliant with an ontology named OpenCitations Ontology. In this paper, we introduce *oc_ocdm*, i.e. a Python library developed for creating OCDM-compliant RDF data even if the programmer has no expertise in Semantic Web technologies. After an introduction of the library and its main characteristics, we show a number of projects within the OpenCitations infrastructure that adopt it as their building block unit.

Keywords: OpenCitations · Python · Rdf · rdflib · citation data · bibliographic metadata · Spar Ontologies

1 Introduction

Data models are crucial artifacts that datasets suppliers should make available to document data and to enable users to understand and, thus, use appropriately suppliers' data. Sometimes, data models may be created (re)using terms defined in the same ontologies with different nuances, thereby generating diversity in data representation [7]. Of course, a data model can employ clearly defined ontological terms to ensure data consistency and facilitate integration tasks.

However, even if ambiguities are entirely avoided from a terminological perspective, creating datasets compliant with a particular data model can still be a challenge for people who are not experts in the related technologies, such as OWL and RDF. Further challenges can be due to data dynamics (e.g. extensions and modifications) [22] which must be performed accordingly to the data model

P. Groth et al. (Eds.): ESWC 2022, LNCS 13261, pp. 305–322, 2022.
https://doi.org/10.1007/978-3-031-06981-9_18

either to correct possible mistakes in an entity or to introduce new data. Additional complexities in data handling are introduced when the data model asks for tracking entities' provenance and changes every time an entity is modified.

To enable users (e.g. domain experts) to programmatically access the data organised according to a particular data model and to permit their modifications, applications (visual interfaces, web editors, etc.) must be developed to facilitate human-data interaction. However, an additional interface layer should be provided to permit programmers to develop such applications, since such programmers are experts in coding but not necessarily skilled in the technologies used by an underlying data model. Such an interface layer would enable creating and manipulating data transparently from the actual technologies used for their representation, such as RDF and, particularly, OWL ontologies.

The situation introduced above describes what happened in OpenCitations in the past few years. OpenCitations (http://opencitations.net) is an independent not-for-profit infrastructure organisation for open scholarship dedicated to the publication of open bibliographic and citation data by the use of Semantic Web technologies [27]. A few years ago, OpenCitations released the OpenCitations Data Model (OCDM) [7], a data model based on SPAR Ontologies [26], PROV-O [25], and other existing models, for describing all the entities in its collections, keeping track of their provenance and modifications in time. In addition of being reused by OpenCitations, the OCDM has also been recently adopted by other external projects dealing with bibliographic metadata and citations [7]. The more the OCDM is adopted, the more it is necessary to have a library to simplify the creation of applications dealing with OCDM-compliant data.

In this paper, we introduce a Python library, i.e. *oc_ocdm* [29], for enabling data owners and publishers to develop applications using OCDM-based data and provenance information. This library has already been used by OpenCitations in several components and projects, and it is the building block for all the future applications dealing with RDF data in OpenCitations' collections.

The rest of the paper is structured as follows. In Sect. 2, we introduce relevant existing libraries for simplifying the creation of RDF data compliant with data models and ontologies. In Sect. 3, we summarise the OpenCitations Data Model and list the requirements for the development of the library. In Sect. 4, we present the main characteristics of the library, including a discussion of its main modules and classes. In Sect. 5, we address the potential impact, adoption and community involved in the library development and reuse. Finally, in Sect. 6, we conclude the paper sketching out some future developments.

2   Related Works

Serving easy-to-access and effective instructions to reuse an ontology contributes to the recognition and validation of the quality of the ontology itself [13]. In recent years, several works have expanded on this aspect adapting FAIR (findable, accessible, interoperable, and reusable) principles to ontologies [12,14,31].

While ontology engineers have introduced best practices for documenting, versioning, and publishing Semantic Web artefacts, they rarely focus on the development of software to enable researchers to programmatically make use of ontologies in the early stages of their project pipelines (e.g. knowledge extraction and RDF data creation). To cope with potential data quality issues arisen by misleading interpretations of the ontologies, the development of SHACL [5] and ShEx [30] has enabled the validation of data conformance to a schema, which is defined in terms of the syntax and structure of a "shape". Nevertheless, human-readable documentation keeps being the primary way to correctly reuse ontologies, as it can more effectively convey the semantics and interpretation of ontology terms.

Only a few notable ontology providers provide effective solutions to systematically create and organise data according to an ontology. These include Python libraries, like GOATOOLS [21], which allows to reuse terms from Gene Ontology and perform data analysis, *pronto* [23], used to access specifications of the Open Biomedical Ontologies [33], or *motools* (https://github.com/motools), including a library for consuming terms of the Music Ontology. Alternatively, WYSIWYG tools, like *quickstatements* (https://quickstatements.toolforge.org) and *OntoRefine* (https://disc-semantic.uibk.ac.at/ontorefine) allow non-expert users to create and map data conforming to specific data models. It is worth noting that such efforts appear to be common in broad communities where diverse stakeholders, with more or less knowledge of Semantic Web technologies, must reuse the ontology to create data and conform to community standards. These solutions significantly prevent time-consuming quality checks, e.g. on crowdsourced data.

To the best of our knowledge, such aids are lacking in the publishing domain. Converters to transform bibliographic records into linked data according to RDA vocabularies [19] and other models exist, e.g. [9] and *bibtex2rdf* (http://www.l3s. de/~siberski/bibtex2rdf/). However, only records complying with library metadata standards are suitable for conversion and no programming interfaces are available for alternative formats, therefore excluding data produced by academic journals and venues. Similarly, well-known scholarly linked data providers [1,11,15] do not share interfaces for data creation according to their schemas [7]. In this work, we fill the gap providing a Python library for creating RDF data according to OCDM [7]. OCDM expands on several modules of the SPAR Ontologies [26], therefore allowing stakeholders in the publishing domain to easily create bibliographic and citation data regardless of their legacy formats – that could be stored according to other ontological models e.g. BIBO (https:// bibliontology.com).

3   Model and Requirements

The OpenCitations Data Model (OCDM) [7] includes terms to describe bibliographic and citation data of scholarly publications. Rather than being yet another ontology, OCDM addresses a broad selection of terms belonging to the SPAR Ontologies [26], which have been conveniently collected within the

OpenCitations Ontology (OCO, https://w3id.org/oc/ontology). Such guidelines, available as open access human-readable and machine-readable documentation, are adopted by several datasets that are either created and maintained by OpenCitations or by external ontology reusers.

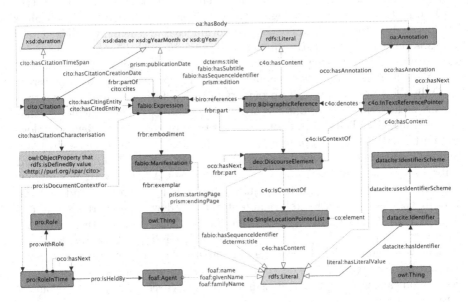

Fig. 1. The Graffoo diagram [10] of the OpenCitations Ontology. Yellow rectangles represent classes, green polygons represent datatypes, while blue and green arrows represent object properties and data properties, respectively. (Color figure online)

Specifically, the OCDM provides directives for recording dataset metadata, bibliographic entities metadata, identifiers, and provenance metadata (including versioning and provenance of changes in data). Dataset metadata include information on the distribution (e.g. a downloadable file) of the dataset. Bibliographic metadata (summarised in Fig. 1) include descriptions of bibliographic resources such as journals and articles (`fabio:Expression`), analog and digital editions of resources (`fabio:Manifestation`), in-text reference pointers (`c4o:InTextReferencePointer`), lists of pointers (`c4o:SingleLocation PointerList`), agents (`foaf:Agent`) and their roles (`pro:Role` linked to the agent via `pro:RoleInTime`), bibliographic references (`biro:Bibliographic Reference`), citations (`cito:Citation`), identifiers and their schemes (`datacite: Identifier` and `datacite:IdentifierScheme`).

Provenance metadata describe snapshots of data, which document the evolution of a particular entity as detailed in [28]. The provenance mechanism enforced by OpenCitations, summarised in Fig. 2, foresees an initial *creation* snapshot, potentially followed by operations like *modification, merge* and *deletion*, each corresponding to an additional snapshot.

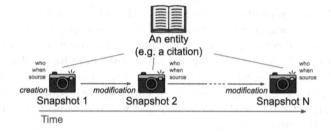

Fig. 2. The high-level description of the provenance layer of the OCDM to keep track of an entity's changes.

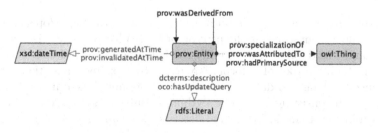

Fig. 3. The Graffoo diagram describing snapshots (`prov:Entity`) of an entity (linked via `prov:specializationOf`) and the related provenance information.

Every snapshot is linked to the described entity via `prov:specializationOf`, and to the previous snapshot via `prov:wasDerivedFrom` (see Fig. 3). Creation time (`prov:generatedAtTime`) and invalidation time (`prov:invalidatedAtTime`) of a snapshot are recorded along with the SPARQL Update query (`oco:hasUpdateQuery`) that encodes the changes applied with respect to the previous snapshot. The operation is also described with free text (`dcterms:description`), and the snapshot is linked to the source of metadata (`prov:hadPrimarySource`) and to the agent responsible for it (`prov:wasAttributedTo`).

The development of the *oc_ocdm* library was driven by the need of reengineering the existing OpenCitations' tools to make them modular. These tools should reuse basic software components, among which *oc_ocdm* has a central role. The library was developed considering the following requirements.

The first requirement was to adopt a development methodology that could make errors easier to spot at development time, thus ensuring better code quality even if with increased maintenance costs. We chose the Test Driven Development (TDD) method [2], which imposes a preliminary test design phase followed by an alternation of software development and testing.

The second requirement was the need to use a programming language compliant with the one that is used in other OpenCitations' applications, which led us to choose Python. To make the library easy to install for the final user, it was decided to package it, to manage its external dependencies and to release it on the PyPI online repository by making use of Poetry[1], a tool for dependency management and packaging in Python. In this way it is possible to manage the versioning of the library separately from that of the projects depending on it, making it simpler for users to follow the advancements in its development.

The final requirement was operational and concerned the design of a mechanism to consider the existing state of an entity defined somewhere (e.g. in a file or in a triplestore) in order to understand which modifications are applied to such an entity through the library.

4   Implementation

The Python library *oc_ocdm* (repository at https://github.com/opencitations/ oc_ocdm, documentation at https://oc-ocdm.readthedocs.io/) is based on rdflib[2] and was developed to simplify the handling of OCDM-compliant RDF graphs, including tasks of information extraction, shape validation, editing, provenance tracking and data serialisation. It's organised as a hierarchy of subpackages, each consisting of a set of Python modules. All the main subpackages are shown in Fig. 4 and in Fig. 5, and describe the main classes they define.

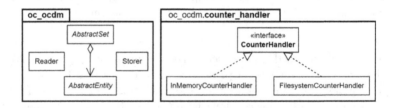

Fig. 4. The UML diagram of the main package and of the counter_handler subpackage.

The classes `Reader` and `Storer` are used for importing data from external sources and for either exporting data to a file or synchronising entities' status with an external triplestore.

Following OCDM's guidelines, all the entities are named using a URI that contains their local identifier, i.e. an incremental integer that uniquely identifies an entity among all entities of the same type. Thus, in order to enforce the uniqueness of the local identifiers for any given type of entity, the library provides a mechanism to correctly handle such counters. This functionality is provided by the abstract class `CounterHandler`, for which there currently exist only two implementations. On the one hand, the `InMemoryCounterHandler` temporarily stores counters via an in-memory data structure, with every progress being immediately lost when the instance of such class is destroyed. On the

---

[1] https://python-poetry.org/.
[2] https://github.com/RDFLib/rdflib.

other hand, `FilesystemCounterHandler` makes use of the file system to persistently read and write counters. Both `CounterHandler` implementations are in charge of keeping track of the last assigned integer number for each kind of entity, incrementing it by one unit when a new entity of the corresponding type is created.

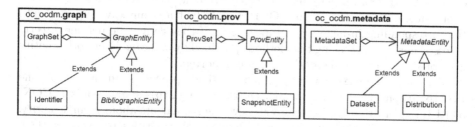

Fig. 5. The UML diagram of the `graph`, `prov` and `metadata` subpackages.

The *Set classes, shown in Fig. 5, are factories defining collections of entities (*Entity classes). Each *Set class contains a reference to a `CounterHandler` instance (see Fig. 6) for managing the assignments of unique URIs to the newly generated entities. The `AbstractSet` class is extended by three concrete classes: `GraphSet` for all kinds of bibliographic entities, `ProvSet` for entities' provenance snapshots, and `MetadataSet` for metadata about the dataset and its distributions. Various subclasses of `GraphEntity`, `ProvEntity` and `MetadataEntity` (all subclasses of the `AbstractEntity` class) were defined so as to represent all the possible types of entities described in the OCDM.

In *oc_ocdm*, all the subclasses of `GraphEntity` and `MetadataEntity` are able to internally track edits. It is worth mentioning that the library enables the generation of provenance information only for non-provenance entities. In addition, each *Entity internally holds a reference to the *Set in which it is contained, leading to the bidirectional containment relationships shown in Fig. 6.

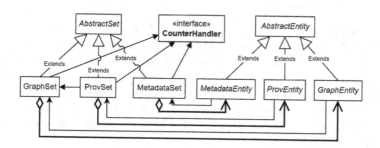

Fig. 6. A UML diagram showing the main relationships between classes in *oc_ocdm*.

4.1   Importing Data from a Persistent RDF Graph

The Reader class allows one to import entities from a persistent RDF graph and to parse them to produce their in-memory representations as a collection of Python objects. Entities with an rdf:type which is recognised by the library (i.e. the classes used the OCDM) are automatically converted into an instance of the corresponding *Entity class and collected inside a *Set. Additional statements about such entities that are not OCDM-compliant are, anyway, imported into the corresponding in-memory instances, even if they cannot be neither directly accessed nor modified using the methods provided by oc_ocdm. Every other entity is ignored and no statement about it are imported.

The import_entities_from_graph method processes instances of the rdflib.Graph class, while import_entity_from_triplestore sends a CON-STRUCT query to a triplestore to retrieve the statements about a single entity. Both methods enable importing only bibliographic entities (converting them into instances of a GraphEntity subclass) and they require an instance of the GraphSet class as input where to collect the imported entities.

Shape Validation. When using the methods import_entities_from_graph and import_entity_from_triplestore, the user can specify to perform shape validation on the imported graph, in order to filter out all the entities that do not respect the shape constraints described in the OCDM (constraints on a given property regarding its range datatype/class, the minimum/maximum amount of attributes associated to an entity, etc.). This operation is currently handled via the PyShEx library.

The shapes described in the OCDM were formalised into a proper ShExC file, that is the required input of PyShEx. Such a resource is included within the oc_ocdm package. ShEx was a design choice that we inherited from the initial phase of the development, which started a few years ago. We chose the ShExC format because of its simplicity and compactness, which makes it easy to be written and read also by non-expert users.

4.2   Data Manipulation

Oc_ocdm allows one to manipulate the content of OCDM-compliant entities only through the Python API exposed by their corresponding in-memory representations, while it does not offer a way to directly act on the persistent RDF graph. All the changes applied to a graph during a session are not made effective and persistent until the updated state of the in-memory objects is synchronised with the original dataset. The way in which the current OpenCitations' tools deal with such updates is to work with small chunks of the original dataset to reduce the impact on memory usage.

Once a *Set instance containing the imported entities has been obtained, it is possible to access each entity individually to read its content. Several getter methods are available for each *Entity class. For example, the title of a BibliographicResource instance can be obtained by calling the method get_title on it. For each getter method, the Python API provides also two corresponding setter methods: one for adding/modifying a value (e.g. has_title) and the other for removing predicate-object pairs (e.g. remove_title). It is worth mentioning that, with respect to methods naming, we decided to inherit the naming conventions used in the OCDM. In particular, setter methods recall the name of the ontology predicates to prevent misalignment between the OCDM and the library implementing it.

The method remove_every_triple modifies the *Entity by removing every triple from its in-memory representation, without deleting its persistent counterpart. This method could be used to clear the content of an *Entity instance to start writing on it from scratch.

In general, all the operations that can be performed on an *Entity allow:

- the creation of a new entity;
- the modification of an entity by adding, changing, or deleting its triples;
- the merging of an entity with another one (not applicable to instances of the ProvEntity subclasses);
- the deletion of the entity from the graph (not applicable to instances of the ProvEntity subclasses).

Creation of an Entity. The creation of a new entity can be done through one of the add_* methods made available by the GraphSet, ProvSet and MetadataSet classes. For example, a new Citation entity can be added to a GraphSet via the method add_ci. Each new instance is initialised with a triple stating its rdf:type and, optionally, with a user-provided rdfs:label.

Modification of an Entity. All the methods that apply changes to an entity (e.g. those that add or delete triples) perform preliminary checks to ensure compliance with the following constraints defined in the OCDM:

- the type of the object of a triple must comply with the one specified in the OCDM for the corresponding predicate (e.g. the value supplied to the method has_citing_entity must be of type BibliographicResource), otherwise an exception is thrown;
- predicates defined as functional properties (e.g. the method has_name of ResponsibleAgent) can be associated to at most one object. If called twice, the second value will override the first one.

Merge of Entities. The merge operation is motivated by user requirements, as it allows to manage deduplication and reconciliation of duplicated entities. Only two or more instances that belong to the same *Entity class can be merged together. The merge method is used on the *Entity to keep and specifies, as input, the other *Entity to merge. Hence, to merge n entities together (including the one to keep), the method must be called n − 1 times. Running the instruction A.merge(B) produces the following effects:

- for each predicate which is compliant with the OCDM for the particular type of the entities A and B, all corresponding objects from B are added to A (with overwriting in case of functional predicates). Every other statement from B which is not compliant with the OCDM gets ignored and it is not added to A;
- regarding the rdf:type predicate, the OCDM allows one to specify at most two values per entity. The first type value is mandatory and must be the same for both A and B, being in itself the requirement for enabling the merging operation. The second type value, if present in B, is added to A (or overwritten if a second value is already present in A);
- B is marked as to be deleted;
- all the other imported entities are scanned to replace B with A in all predicate-object pairs in which the object is B, thus redirecting all the references which still point to B.

Deletion of an Entity. An *Entity* can be marked as to be definitively removed from the persistent graph via the method mark_as_to_be_deleted. Invoking such method on an *Entity* E produces the following effects:

- E is marked as to be deleted;
- all the other imported entities are scanned to remove all predicate-object pairs having E as object, thus cleaning up dead links.

Additionally, it is possible to remove dead links also from other persistent entities that were not imported via the remove_dead_links_from_triplestore method from the GraphSet class. Such method is able to import from a triplestore all the entities that refer to at least one *Entity* which is currently marked as to be deleted and to remove the dead links from their in-memory graphs.

Marking an *Entity* as to be deleted cannot be undone. When we synchronise the deleted entity with the persistent graph, oc_ocdm automatically recognises that its persistent counterpart has to be completely deleted and it directly removes its persistent triples.

4.3   Change Tracking

Each *Entity* contains some private fields that are exploited by oc_ocdm to internally keep track of all the operations performed on it and to later reconstruct what happened. This does not apply to provenance entities, since there is no need to generate provenance information for them.

In addition to the rdflib.Graph that holds the current triples of the *Entity*, another rdflib.Graph named preexisting_graph is initialised during the construction of the Python instance. Such graph is intended to be read-only, as it represents the initial content of the entity at the time of importing it from the persistent graph. This allows, at any stage, to compute the changes introduced through oc_ocdm by making a comparison between the two in-memory graphs. There can be situations in which the initial content of an entity is unknown (e.g. when instantiating an *Entity identified by a URI that is already assigned to an existing entity, without importing its triples from the persistent graph). In this case the preexisting_graph remains empty, thus the library can only interpret new triples found inside the graph as to be added to the persistent graph.

For keeping track of merging operations, each *Entity internally sets a was_merged flag while also populating its merge_list with all the entities that were merged into it. Finally, the to_be_deleted flag is set when the *Entity is marked as to be deleted.

4.4   Provenance Generation

The OCDM envisions a provenance graph composed by trees of snapshot entities that describe the evolution of bibliographic entities and their external identifiers. In oc_ocdm, such entities are represented with instances of the GraphEntity subclasses, which are the only classes having a provenance graph associated.

The most recent snapshot of either a bibliographic entity or an external identifier (e.g. DOI, ISBN, ORCID, ...) represents its current persistent content. The provenance generation algorithm provided by oc_ocdm is responsible for iterating the imported GraphEntity instances and for generating a new snapshot exclusively for those whose content has been modified through the Python API. Usually, each new snapshot invalidates the previous one, and in turn becomes the most recent snapshot for a certain entity. However, in the event that an entity is removed from the persistent graph, the corresponding snapshot to be generated must invalidate both the previous one and also itself, and no other snapshots are linked to it afterwards.

Once the user is satisfied by the changes applied to the imported GraphEntity instances, it is possible to automatically generate a provenance snapshot (i.e. an instance of the SnapshotEntity class) for each involved *Entity. Such snapshot is intended to describe the changes produced on the related GraphEntity with respect to its preexisting_graph.

Multiple Operations on the Same Entity. The snapshot entity produced for a given GraphEntity can be of four different types which reflect the oc_ocdm's operations, namely: creation, modification, merging and deletion. When a composition of more than one of these operations is applied to an entity, it is necessary to define a scale of priority to be followed when choosing which of them should be associated with the new snapshot. By design, the highest priority operation is deletion, followed by merging, creation and, finally, modification. For instance, if an entity is modified and later it is deleted, then the particular details of the modification are lost (i.e. they are of no interest anymore), as the operation that correctly summarises the entire process is the deletion. In such a case, a new deletion snapshot would be produced.

The Special Case of the Merging Operation. When two or more entities are merged together, a merge snapshot must be generated only for the entity that the user chooses to keep in the persistent graph (i.e. the one to which the triples of the others are added). The merge snapshot needs to keep a reference to the latest snapshots of all the entities contained in the *merge_list* (see Sect. 4.3) by means of the predicate prov:wasDerivedFrom. It is mandatory that the deletion snapshots for the entities in the *merge_list* are generated only afterwards, as the merge snapshot must refer to the snapshots that represent the state of the involved entities prior to their deletion. This requirement imposes a strict ordering constraint in the production of snapshots.

The Provenance Generation Algorithm. The algorithm introduced in Listing 1 describes the rationale for handling the task described in this section. Comments have been inserted in particular branches of the pseudo-code to highlight all the possible scenarios that the algorithm needs to handle. In *oc_ocdm*, it is implemented by the generate_provenance method of the ProvSet class.

The algorithm iterates (line 2) over the entities A that have been effectively merged with any number of entities B_i (i.e. $A.merge(B_i)$), that is, all those entities that have been involved in a merging while not having been consequently marked as to be deleted. For these entities, the following scenarios must be addressed:

- **scenario A.1**: A does not exist in the persistent graph, hence a creation snapshot is generated;
- **scenario A.2**: A already exists in the persistent graph but each B_i in A's merge_list does not exist. Since it's not possible for a merge snapshot to refer to the snapshot of any B_i as they do not exist, such a merge operation can only be interpreted by *oc_ocdm* as a modification of A;
- **scenario A.3**: A already exists in the persistent graph and at least one B_i in A's merge_list exists as well. In this case, a merge snapshot is generated for A that references all the latest snapshots of such B_i entities.

Then, another iteration of the algorithm is performed (line 12) on all the remaining entities E. In this case, the following scenarios can occur:

- **scenario B.1**: E does not exist in the persistent graph and was not deleted, hence a creation snapshot is generated. Had it been deleted, no snapshot would have been generated, since the deletion of a non-existing entity does not produce any change to the persistent graph;
- **scenario B.2**: E already exists in the persistent graph and it was deleted (either explicitly or as a consequence of being involved in a merging operation), hence a deletion snapshot is generated;
- **scenario B.3**: E already exists in the persistent graph, it was not deleted but it was modified. In this last case, a modification snapshot is generated.

Algorithm 1: Pseudocode for provenance generation

---

1 *resultSet* ← *an empty set*
2 **foreach** *entity* **such that** (*wasMerged(entity)* **and not** *wasDeleted(entity)*) **do**
3   *latestSnapshot* ← *retrieveLatestSnapshot(entity)*
4   **if** *latestSnapshot* **is None then**
       // Scenario A.1 -> Creation
5   *resultSet.add(newCreationSnapshot(...))*
6   **else**
7   *snapshotsList* ← *getSnapshotsFromMergeList(entity.merge_list)*
8   **if** *wasModified(entity)* **and** *len(snapshotsList)* ≤ 0 **then**
          // Scenario A.2 -> Modification
9   *resultSet.add(newModificationSnapshot(...))*
10   **else if** *len(snapshotsList)* > 0 **then**
          // Scenario A.3 -> Merge
11   *resultSet.add(newMergeSnapshot(...))*

12 **foreach** *remaining entity* **do**
13   *latestSnapshot* ← *getLatestSnapshot(entity)*
14   **if** *latestSnapshot* **is None then**
15   **if not** *wasDeleted(entity)* **then**
          // Scenario B.1 -> Creation
16   *resultSet.add(newCreationSnapshot(...))*
17   **else**
18   **if** *wasDeleted(entity)* **then**
          // Scenario B.2 -> Deletion
19   *resultSet.add(newDeletionSnapshot(...))*
20   **else if** *wasModified(entity)* **then**
          // Scenario B.3 -> Modification
21   *resultSet.add(newModificationSnapshot(...))*

22 **return** *resultSet*

4.5   Data Synchronisation

Generally, the last step of a workflow that involves the manipulation of an OCDM-compliant dataset consists in the synchronisation of the in-memory content of the *Entity instances with a triplestore or a persistent RDF resource. All the relevant library operations are collected within the Storer class.

As far as the data serialisation task is concerned, the library supports three possible RDF file formats, namely: **N-Triples** for bibliographic entities and their external identifiers, **N-Quads** for provenance entities and **JSON-LD** (the default option for both kinds of entities).

The methods that the Storer class makes available enable one to work on the content of either a single *Entity or an entire *Set and permit considering the related export target (which can be either an RDF file or a SPARQL endpoint).

In particular, `store` and `store_all` methods are used to export respectively a single entity and an entire set of entities on the file system, while `upload` and `upload_all` are capable of generating batches of SPARQL 1.1 Update queries that are sequentially sent to a user-specified endpoint. Finally, the `upload_and_store` method combines the effects of `store_all` and `upload_all`.

Once the synchronisation task is executed, further modifications require the user to first call the `commit_changes` method either on single *Entity* instances or on an entire *Set*. Such method takes care of effectively destroying the Python objects of deleted entities and of resetting the internal state of the other ones (i.e. resetting their boolean flags and realigning the `preexisting_graph` with their updated persistent graph).

5   Potential Impact, Adoption and Community

In a previous article [7], we demonstrated the impact of OCDM with respect to a growing community, which includes a number of datasets and projects maintained by the OpenCitations infrastructure [16,18,27], a few OCDM adopters from diverse disciplines [4,20,24], a growing number of applications and services that rely on data served by OpenCitations (e.g., VOSViewer[3], CitationGecko[4], VisualBib[5], and OAHelper[6], DBLP[7] and Lens.org[8]), and data providers that align data to OpenCitations (e.g., OpenAIRE[9], MAKG, and WikiCite).

The library *oc_ocdm* has been tested and it is currently integrated into four applications and collaborative projects, namely: *Wikipedia Citations in Wikidata*[10], a project funded by the Wikimedia Foundation to extract citations from the English Wikipedia towards external bibliographic resources, transform data to RDF according to OCDM, and upload citations to Wikidata; *OpenCitations Meta*[11], a software to clean and transform tabular bibliographic metadata to RDF according to OCDM; *GraphEnricher*[12], a tool for identifiers discovery and data deduplication used to improve data quality of OpenCitations data; finally, *oc_ocdm* is used to define testing benchmarks of another Python library[13] used to perform time and provenance-aware queries on RDF datasets compliant with the OCDM.

Nevertheless, like other software developed by OpenCitations [6,8,17], also *oc_ocdm* has been developed with the aim of sharing a component that can

---

[3] https://www.vosviewer.com/.
[4] https://citationgecko.com/.
[5] https://visualbib.uniud.it/en/project/.
[6] https://www.otzberg.net/oahelper/.
[7] https://dblp.org.
[8] https://lens.org.
[9] https://www.openaire.eu/.
[10] https://meta.wikimedia.org/wiki/Wikicite/grant/Wikipedia_Citations_in_Wikidata.
[11] https://github.com/opencitations/meta.
[12] https://github.com/opencitations/oc_graphenricher.
[13] https://github.com/opencitations/time-agnostic-library.

be reused in different contexts. In particular, the broader community of SPAR Ontologies adopters can benefit of this programming interface, including current adopters for (1) data creation, (2) data analysis [3], (3) ontology-based data managements systems [32] and (4) academic journals [34].

Moreover, *oc_ocdm* can be reused by scholars in the Library and Information Science domain that want to produce bibliographic and citation data according to the SPAR Ontologies, which comply with most of the requirements for bibliographic ontologies (e.g. being based on FRBR conceptual model).

6   Conclusions

In this paper, we have introduced *oc_ocdm*, a Python library for enabling the development of applications using OCDM-based data and provenance information. After showing the main requirements for the development, we have introduced its organisation in terms of Python modules and classes and we have presented its current and future uses in the context of several components and projects related to OpenCitations, being the main building block for all the applications dealing with creating and modifying RDF data in OpenCitations' collections.

In the future, we aim at continuing the development of the library adding new features and reusing other existing components. For instance, since mid-February 2021 the development of PyShEx, that it is used to validate input data processed by *oc_ocdm*, has been slowed down. Therefore, we plan to convert the ShExC file into a compact SHACL file and use the pySHACL[14] library, which is actively maintained and better optimised.

Another aspect that deserves to be properly addressed is related to the parallel use of the library. Indeed, the current release of *oc_ocdm* is designed to work correctly if no more than one instance of *oc_ocdm* needs to access the indexes used to name new entities. If this condition is not met, episodes of race conditions could easily occur with the risk of assigning the same URI to more entities, therefore compromising the consistency and validity of all the produced data.

Finally, a last aspect that deserves to be addressed concerns the possibility of detaching from the library some aspects that can be applied to any RDF dataset, and not only to OCDM-compliant data. For instance, the way proposed by the OCDM to handle provenance and, in particular, change tracking is independent from the kinds of entities to track and could be devised, in the future, as a possible plugin for rdflib.

Acknowledgements. This work has been funded by the project "Open Biomedical Citations in Context Corpus" (Wellcome Trust, Grant n. 214471/Z/18/Z) and the project "Wikipedia Citations in Wikidata" (Wikimedia Foundation, https:// meta.wikimedia.org/wiki/Wikicite/grant/Wikipedia_Citations_in_Wikidata), and partially funded by the European Union's Horizon 2020 research and innovation program under grant agreement No 101017452 (OpenAIRE-Nexus). We would like to thank (in

---

[14] https://github.com/RDFLib/pySHACL.

alphabetic order) Fabio Mariani, Arcangelo Massari, and Gabriele Pisciotta for the constructive feedback.

References

1. Ammar, W., et al.: Construction of the literature graph in semantic scholar. In: Proceedings of the 2018 Conference of the North American Chapter of the Association for Computational Linguistics: Human Language Technologies, Vol. 3 (Industry Papers), pp. 84–91. Association for Computational Linguistics, New Orleans - Louisiana (2018). https://doi.org/10.18653/v1/N18-3011.https://aclanthology.org/N18-3011
2. Beck, K.: Test-Driven Development: By Example. The Addison-Wesley signature series, Addison-Wesley, Boston (2003)
3. Bertin, M., Atanassova, I., Sugimoto, C.R., Lariviere, V.: The linguistic patterns and rhetorical structure of citation context: an approach using n-grams. Scientometrics **109**(3), 1417–1434 (2016). https://doi.org/10.1007/s11192-016-2134-8
4. Colavizza, G., Romanello, M.: Citation mining of humanities journals: the progress to date and the challenges ahead. J. Eur. Periodical Stud. **4**(1), 36–53 (2019)
5. Corman, J., Reutter, J.L., Savković, O.: Semantics and validation of recursive SHACL. In: Vrandečić, D. (ed.) ISWC 2018. LNCS, vol. 11136, pp. 318–336. Springer, Cham (2018). https://doi.org/10.1007/978-3-030-00671-6_19
6. Daquino, M., Heibi, I., Peroni, S., Shotton, D.: Creating RESTful APIs over SPARQL endpoints using RAMOSE (2020). http://arxiv.org/abs/2007.16079
7. Daquino, M., et al.: The opencitations data model. In: Pan, J.Z., Tamma, V., d'Amato, C., Janowicz, K., Fu, B., Polleres, A., Seneviratne, O., Kagal, L. (eds.) ISWC 2020. LNCS, vol. 12507, pp. 447–463. Springer, Cham (2020). https://doi.org/10.1007/978-3-030-62466-8_28
8. Daquino, M., Tiddi, I., Peroni, S., Shotton, D.: Creating open citation data with BCite. In: Emerging Topics in Semantic Technologies, pp. 83–93. IOS Press, Amesterdam (2018)
9. Dunsire, G., Fritz, D., Fritz, R.: Instructions, interfaces, and interoperable data: the rimmf experience with RDA revisited. Cataloging Classif. Q. **58**(1), 44–58 (2020)
10. Falco, R., Gangemi, A., Peroni, S., Shotton, D., Vitali, F.: Modelling OWL ontologies with graffoo. In: Presutti, V., Blomqvist, E., Troncy, R., Sack, H., Papadakis, I., Tordai, A. (eds.) ESWC 2014. LNCS, vol. 8798, pp. 320–325. Springer, Cham (2014). https://doi.org/10.1007/978-3-319-11955-7_42
11. Färber, M.: The microsoft academic knowledge graph: a linked data source with 8 billion triples of scholarly data. In: Ghidini, C. (ed.) ISWC 2019. LNCS, vol. 11779, pp. 113–129. Springer, Cham (2019). https://doi.org/10.1007/978-3-030-30796-7_8
12. Franc, Y.L., Coen, G., Essen, J.P.V., Bonino, L., Lehväslaiho, H., Staiger, C.: D2.2 FAIR Semantics: First Recommendations (2020)
13. Gangemi, A., Catenacci, C., Ciaramita, M., Lehmann, J.: Modelling ontology evaluation and validation. In: Sure, Y., Domingue, J. (eds.) ESWC 2006. LNCS, vol. 4011, pp. 140–154. Springer, Heidelberg (2006). https://doi.org/10.1007/11762256_13
14. Garijo, D., Poveda-Villalón, M.: Best Practices for Implementing FAIR Vocabularies and Ontologies on the Web. Applications and practices in ontology design, extraction, and reasoning, vol. 49, p. 39 (2020)

15. Hammond, T., Pasin, M., Theodoridis, E.: Data integration and disintegration: managing springer nature SciGraph with SHACL and OWL. In: International Semantic Web Conference (Posters, Demos & Industry Tracks) (2017)
16. Heibi, I., Peroni, S., Shotton, D.: Crowdsourcing open citations with CROCI-An analysis of the current status of open citations, and a proposal (2019). arXiv preprint arXiv:1902.02534
17. Heibi, I., Peroni, S., Shotton, D.: Enabling text search on SPARQL endpoints through OSCAR. Data Sci. 2(1–2), 205–227 (2019)
18. Heibi, I., Peroni, S., Shotton, D.: Software review: COCI, the opencitations index of crossref open DOI-to-DOI citations. Scientometrics 121(2), 1213–1228 (2019). https://doi.org/10.1007/s11192-019-03217-6
19. Hillmann, D., Coyle, K., Phipps, J., Dunsire, G.: RDA vocabularies: process, outcome, use. D-Lib Mag. 16(1/2), 6 (2010)
20. Hosseini, A., Ghavimi, B., Boukhers, Z., Mayr, P.: EXCITE-A toolchain to extract, match and publish open literature references. In: 2019 ACM/IEEE Joint Conference on Digital Libraries (JCDL), pp. 432–433. IEEE (2019)
21. Klopfenstein, D., et al.: GOATOOLS: a python library for gene ontology analyses. Sci. Rep. 8(1), 1–17 (2018)
22. Käfer, T., Abdelrahman, A., Umbrich, J., O'Byrne, P., Hogan, A.: Observing linked data dynamics. In: Cimiano, P., Corcho, O., Presutti, V., Hollink, L., Rudolph, S. (eds.) ESWC 2013. LNCS, vol. 7882, pp. 213–227. Springer, Heidelberg (2013). https://doi.org/10.1007/978-3-642-38288-8_15
23. Larralde, M., Philipp, A., Henrie, A., Himmelstein, D., Mitchell, S., Sakaguchi, T.: althonos/pronto: 2.4.3 (2021). https://doi.org/10.5281/zenodo.5153400
24. Lauscher, A., et al.: Linked open citation database: enabling libraries to contribute to an open and interconnected citation graph. In: Proceedings of the 18th ACM/IEEE on Joint Conference on Digital Libraries, pp. 109–118 (2018)
25. Lebo, T., Sahoo, S., McGuinness, D.: PROV-O: the PROV ontology. W3C Recommendation 30 Apr 2013 (2013). http://www.w3.org/TR/2013/REC-prov-o-20130430/
26. Peroni, S., Shotton, D.: The SPAR ontologies. In: Vrandečić, D. (ed.) ISWC 2018. LNCS, vol. 11137, pp. 119–136. Springer, Cham (2018). https://doi.org/10.1007/978-3-030-00668-6_8
27. Peroni, S., Shotton, D.: OpenCitations, an infrastructure organization for open scholarship. Quant. Sci. Stud. 1(1), 428–444 (2020)
28. Peroni, S., Shotton, D., Vitali, F.: A document-inspired way for tracking changes of RDF data - the case of the OpenCitations Corpus. In: Hollink, L., Darányi, S., Meroño Peñuela, A., Kontopoulos, E. (eds.) Detection, Representation and Management of Concept Drift in Linked Open Data. CEUR Workshop Proceedings, vol. 1799, pp. 26–33. CEUR-WS, Aachen (2016). http://ceur-ws.org/Vol-1799/Drift-a-LOD2016_paper_4.pdf
29. Persiani, S.: opencitations/oc_ocdm (version 6.0.2) (2021). https://doi.org/10.5281/zenodo.5770647
30. Prud'hommeaux, E., Labra Gayo, J.E., Solbrig, H.: Shape expressions: an RDF validation and transformation language. In: Proceedings of the 10th International Conference on Semantic Systems, pp. 32–40. SEM 2014, Association for Computing Machinery, New York (2014). https://doi.org/10.1145/2660517.2660523
31. Riungu-Kalliosaari, L., Hooft, R., Kuijpers, S., Parland-von Essen, J., Tana, J.: D2.4 2nd report on FAIR requirements for persistence and interoperability (2020)

32. Senderov, V., et al.: OpenBiodiv-O: ontology of the openbiodiv knowledge management system. J. Biomed. Semant. **9**(1), 1–15 (2018). https://doi.org/10.1186/s13326-017-0174-5
33. Smith, B., et al.: The OBO foundry: coordinated evolution of ontologies to support biomedical data integration. Nat. biotech. **25**(11), 1251–1255 (2007)
34. Willighagen, E.: Adoption of the citation typing ontology by the journal of cheminformatics. J. Cheminformatics **12**(1), 1–3 (2020)

LD Connect: A Linked Data Portal for IOS Press Scientometrics

Zilong Liu[1]([✉])(iD), Meilin Shi[1](iD), Krzysztof Janowicz[1], Blake Regalia[1](iD), Stephanie Delbecque[2](iD), Gengchen Mai[1,3](iD), Rui Zhu[1](iD), and Pascal Hitzler[4](iD)

[1] STKO Lab, UC Santa Barbara, Santa Barbara, CA, USA
`{zilongliu,meilinshi,janowicz,regalia,gengchen_mai,ruizhu}@ucsb.edu`
[2] IOS Press, Amsterdam, The Netherlands
`s.delbecque@iospress.nl`
[3] Department of Computer Science, Stanford University, Stanford, CA, USA
`maigch@cs.stanford.edu`
[4] Data Semantics Lab, Kansas State University, Manhattan, KS, USA
`hitzler@k-state.edu`

Abstract. In this work, we describe a Linked Data portal, LD Connect, which operates on all bibliographic data produced by IOS Press over the past thirty-five years, including more than a hundred thousand papers, authors, affiliations, keywords, and so forth. However, LD Connect is more than just an RDF-based metadata set of bibliographic records. For example, all affiliations are georeferenced, and co-reference resolution has been performed on organizations and contributors including both authors and editors. The resulting knowledge graph serves as a public dataset, web portal, and query endpoint, and it acts as a data backbone for IOS Press and various bibliographic analytics. In addition to the metadata, LD Connect is also the first portal of its kind that publicly shares document embeddings computed from the full text of all papers and knowledge graph embeddings based on the graph structure, thereby enabling semantic search and automated IOS Press scientometrics. These scientometrics run directly on top of the graph and combine it with the learned embeddings to automatically generate data visualizations, such as author and paper similarity over all journals. By making the involved ontologies, embeddings, and scientometrics all publicly available, we aim to share LD Connect services with not only the Semantic Web community but also the broader public to facilitate research and applications based on this large-scale academic knowledge graph. Particularly, the presented scientometric system generalizes beyond IOS Press data and can be deployed on top of other bibliographic datasets as well.

Keywords: LD Connect · Knowledge graphs · Ontology engineering · Document embeddings · Knowledge graph embeddings · Scientometrics

---

Z. Liu and M. Shi—Both authors contributed equally to this work.

© The Author(s), under exclusive license to Springer Nature Switzerland AG 2022
P. Groth et al. (Eds.): ESWC 2022, LNCS 13261, pp. 323–337, 2022.
https://doi.org/10.1007/978-3-031-06981-9_19

1   Introduction

Knowledge graphs are playing an increasingly important role in academic search engines, and they serve as data backbones for data analytics at publishers and funding organizations. For example, Semantic Scholar provides a REST API[1] to facilitate author and paper lookup, conference peer review service, etc., based on its academic literature graph. SPECTER [3], a method for embedding scientific papers based on paper IDs, titles, and abstracts, has also been implemented in Semantic Scholar as a public endpoint[2] for retrieving embeddings computed for selected papers. However, these academic knowledge graphs suffer from several limitations. First, the access to these large-scale graphs, either via public endpoints or downloadable URLs, is limited. An example is AMiner, which consists of over 130M researchers and 320M publications in total (by the time of writing), has only released parts of its entire graph for the public to download[3] (i.e., 50K entities and 290M links). Second, few academic search engines share documentations about their ontologies, which hinders the semantic interoperability across different academic knowledge graphs. Third, spatial and temporal information is often unavailable in these graphs. For instance, while the data schema of Microsoft Academic Graph[4] contains affiliation information with geocoded outputs as Latitude and Longitude, no additional spatial contexts (e.g., the country of an affiliation) are provided nor visualization. Similarly, there is a lack of annotations about when a publication was received, reviewed, and accepted to support knowledge discovery during the entire publication process. Fourth, while pre-trained document embeddings such as SPECTER[5], are shared for academic knowledge graphs such as Semantic Scholar, these embeddings are learned from titles and abstracts instead of the full text of publications. Additionally, no knowledge graph embeddings trained on these graphs have been publicly available yet. Finally, the dataset and scientometric portal presented here are, while restricted to a single publisher, not merely a data export or service, but form the deployed data backbone of an academic publisher since several years, thus offering additional insights into the usage of bibliographic knowledge graphs in commercial practice.

With these limitations in mind, this paper presents LD connect, a Linked Data portal that serves, retrieves, visualizes, and analyzes IOS Press bibliographic data. More specifically, we introduce the construction of an academic knowledge graph using a newly designed (but aligned) ontology, the implementation of document embedding and knowledge graph embedding techniques, and the design of a scientometric system to support visualization and analysis of bibliographic data.

---

[1] https://www.semanticscholar.org/product/api.
[2] https://github.com/allenai/paper-embedding-public-apis#specter.
[3] https://www.aminer.cn/knowledge_graph.
[4] https://docs.microsoft.com/en-us/academic-services/graph/reference-data-schema.
[5] https://github.com/allenai/specter.

All resources in this paper, including a version of the datasets and pre-trained embeddings, underlying ontology, and scientometrics, are publicly available on GitHub[6] with detailed documentation.

The remainder of this paper is organized as follows. Section 2 provides an overview of LD Connect with the underlying ontology. Section 3 explains the need of embedding representation for both documents and knowledge graphs, and elaborates how they are generated. Section 4 demonstrates how IOS Press scientometrics are developed to answer identified competency questions for bibliographic analysis. Finally, Sect. 5 concludes the paper, and discusses future directions of improving and adopting LD Connect to other academic related datasets.

2   LD Connect

The ontology of LD Connect can be considered as an extension of the Bibliographic Ontology (BIBO)[7]. First, publications (`iospress:Publication`) are categorized as articles (`iospress:Article`) and chapters (`iospress:Chapter`), and contributors (`iospress:Contributor`) of a publication are categorized as authors (`iospress:Role.Author`) and editors (`iospress:Role.Editor`). Using list properties of container membership in RDF, the order of authorship in a paper is expressed as `rdf:_0`, `rdf:_1`, `rdf:_2`, etc. During bibliographic data triplification, since multiple Uniform Resource Identifiers (URIs) are assigned to a contributor for all contributed publications, co-reference resolution is performed to learn weights for matching contributors, and `owl:sameAs` relations are established among those URIs which indicate the same contributor based on whether their information, including first names, last names, and affiliations, is significantly similar. The same process is applied to one affiliation (`iospress:Organization`) shared by multiple contributors based on affiliation names and associated contributors. Figure 1(a) and Fig. 1(b) show ontology fragments of `iospress:Publication` and `iospress:Contributor`, respectively.

In addition, during the triplification process, spatial and temporal information is automatically generated and integrated into the knowledge graph. Affiliations are geocoded with Geocoding API provided by Google Map to fetch their geographic information. The ontology follows the OGC GeoSPARQL standard[8] to generate affiliation geometry. In addition, we provide spatial contexts about affiliations, including cities (`iospress-geocode:city`), countries (`iospress-geocode:country`), postal codes (`iospress-geocode:postalCode`), regions (`iospress-geocode:region`), and zones (`iospress-geocode:zone`). We also include rich temporal information about a publication such as its received date (`iospress:publicationReceivedDate`), accepted date (`iospress:publicationAcceptedDate`), preprint date (`iospress:publicationPreprintDate`), and publication date (`iospress:publicationDate`).

---

[6] https://github.com/stko-lab/LD-Connect.

[7] http://bibliontology.com/specification.html.

[8] https://www.ogc.org/standards/geosparql.

A statistical summary about these main classes is listed in Table 1. By the time of writing, there are over 100K articles, 530K contributors, and 60K geocoded locations in LD Connect linked by 11M relations, and these numbers are still growing as the dataset gets updated. It contains all papers published at IOS Press over the past 35 years.

Table 1. An overview of LD Connect as of 12/06/2021.

Class	Number of instances
iospress:Category	9
iospress:Series	44
iospress:Journal	133
iospress:Volume	2687
iospress:Issue	9732
iospress:Chapter	49874
iospress:Article	106131
iospress:Contributor	531126
iospress:Organization	547014
iospress:GeocodedLocation	60284

LD Connect is also made available via a SPARQL endpoint[9] for semantic search and more complex queries. The following SPARQL query shows an example of retrieving relevant information about papers whose first author is from affiliations located in China. The returned results include corresponding paper titles, associated keywords, publication years, journals, first authors, and their affiliations. At an academic publisher, such queries can be used to compare and potentially adjust the composition of editorial boards of journals to keep them geographically representative with respect to the locations of authors.

```
PREFIX rdf: <http://www.w3.org/1999/02/22-rdf-syntax-ns#>
PREFIX iospress: <http://ld.iospress.nl/rdf/ontology/>
PREFIX iospress-geocode: <http://ld.iospress.nl/rdf/geocode/>

select ?title (group_concat(?keyword; separator=',')
        as ?keywords) ?year ?journal ?first_author_name ?org_name
{
    ?paper iospress:publicationTitle ?title;
           iospress:publicationIncludesKeyword ?keyword;
           iospress:publicationDate ?date;
           iospress:articleInIssue/iospress:issueInVolume/
           iospress:volumeInJournal ?journal;
           iospress:publicationAuthorList ?author_list.
```

---

[9] http://ld.iospress.nl/sparql.

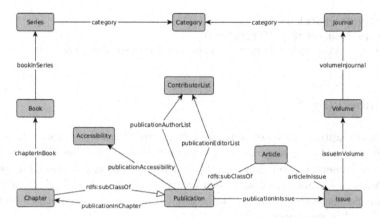

(a) Main classes and their relations to model `Publication`.

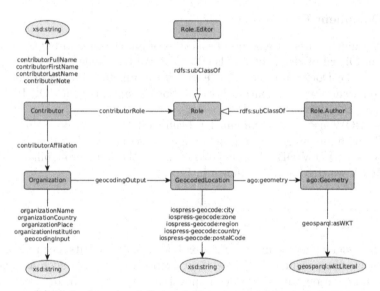

(b) Main classes and their relations to model `Contributor`.

Fig. 1. An overview of the ontology behind LD Connect. Edges with filled arrows are object/datatype properties, and edges with open arrow heads represent subclass relations. All classes and properties without any prefix are in the namespace `iospress`: <http://ld.iospress.nl/rdf/ontology/> .

```
?author_list rdf:_0 ?first_author.
?first_author iospress:contributorFullName ?first_author_name;
              iospress:contributorAffiliation ?org.
?org iospress:geocodingInput ?org_name ;
           iospress:geocodingOutput/
           iospress-geocode:country ?org_country.
```

```
bind(year(?date) as ?year)
values ?org_country {"China"@en}
} group by ?title ?year ?journal ?first_author_name ?org_name
```

3 Embeddings

In order to capture both semantic and structural knowledge about IOS Press publications, we take advantages of unsupervised embedding learning techniques, such as document embeddings [12,16] and knowledge graph embeddings [1,5] that encode each document and each entity in the graph as high dimensional embeddings, respectively. More details about the generation of both embeddings can be found in our paper [13].

3.1 Document Embeddings

To fill the gap of missing semantic properties of publications in LD Connect, we adopt the Distributed Bag of Words (PV-DBOW) [12] model, which is a specific version of the Doc2Vec model that encodes the full text bodies of documents (e.g., conference papers, journal articles, book chapters) from IOS Press into document embeddings.

PV-DBOW uses the maximum log likelihood (MLE) as its training objective. Given a document d_i represented as a sequence of words, i.e., $d_i = \{w_1, w_2, ..., w_T\}$, PV-DBOW aims at optimizing the joint probability distribution of each word given d_i, as shown in Eq. 1.

$$\sum_{t=1}^{T} \log p(w_t|d_i) \tag{1}$$

In the generation pipeline, the full text bodies of documents are first extracted from their corresponding PDF files. Text preprocessing steps such as tokenization, lemmatization, and stop word removal are carried out before texts are fed into the PV-DBOW model. Note that although there are more advanced text embedding techniques such as ELMo [17] and BERT [4], we selected PV-DBOW because 1) PV-DBOW is a rather simple but widely used neural network architecture that can be re-trained in a short amount of time, which is favored by LD Connect given its fast evolving nature and updates, and 2) implementation of PV-DBOW is highly reproducible for commercial production.

3.2 Knowledge Graph Embeddings

While document embeddings provide semantic knowledge about publications, structural knowledge is also needed to understand the relations among entities in LD Connect, such as journals, authors, and affiliations. Therefore, we utilize the knowledge graph embedding technique, TransE [1], to encode each entity and relation in LD Connect into a high dimensional vector space. Given one triple (h_i, r_i, t_i) in LD Connect, TransE encodes both entities and relations into the same embedding space - $\mathbf{h_i}, \mathbf{r_i}, \mathbf{t_i}$ - so that relation embedding $\mathbf{r_i}$ is treated as a translation operation from the head entity embedding $\mathbf{h_i}$ to the tail entity embedding $\mathbf{t_i}$. A plausibility scoring function $f_{TransE}(\cdot, \cdot, \cdot)$ is defined for each triple as shown in Eq. 2, where triples that exist in LD Connect receive lower plausibility scores, and those that do not exist receive higher scores.

$$f_{TransE}(h_i, r_i, t_i) = \| \mathbf{h_i} + \mathbf{r_i} - \mathbf{t_i} \| \qquad (2)$$

Similar to the reason why we use PV-DBOW, we choose to use TransE because it is more efficient to train, easier to interpret, and it has rather acceptable performance compared with other counterparts such as TransH [20], TransR [5], R-GCN [19], and TransGCN [2].

4   IOS Press Scientometrics

Scientometrics refers to the study of measuring and analyzing scholarly literature [8]. Research in scientometrics ranges from the study of growth and development in publications of a specific journal [11] to quantitatively characterizing the scientific output of a scholar [7], to designing a framework for measuring spatial and temporal citation patterns of both publications and researchers [6]. The large amount of structured bibliographic data provided by LD Connect enables knowledge discovery and data-driven analysis in studying the science of science. Our previous work have demonstrated prototypes of scientometric systems that use similar extended BIBO ontologies based on data from one journal published by IOS Press, or data enriched with research topics, expertise, and geographic information of institutions [9,10,15].

In this section, we present an overview of the current version of the scientometric system *IOS Press scientometrics*[10] built upon LD Connect which contains more enriched journal data and uses a more comprehensive ontology. Developed with JavaScript libraries such as D3.js[11] and Leaflet[12], IOS Press scientometrics consist of seven interactive modules for visual analysis, including *Home, Country Collaboration, Author Map, Author Similarity, Paper Similarity, Keyword Graph*, and *Streamgraph*. The Semantic Web journal is used as an example to explain how each module helps answer the competency questions listed below.

Q1: What is the spatial coverage of a journal based on the locations of author affiliations? The *Home* module provides an overview of the spatial coverage of the selected journal. A choropleth map displays the countries/regions of author affiliations included in the selected journal. Hovering the mouse over a country/region on the map displays the number of contributing authors. Figure 2

---

[10] http://stko-roy.geog.ucsb.edu:7200/iospress_scientometrics.
[11] https://d3js.org.
[12] https://leafletjs.com.

shows a total of 34 authors from Greece in the Semantic Web journal. The country/region is colored in proportion to the number of contributing authors. Countries/regions with higher than average authors are in darker shades of blue while those with fewer than average are shown in lighter shades.

Fig. 2. Spatial coverage of affiliations mentioned in the Semantic Web journal

Q2: What is the country collaboration pattern based on co-authorship? The *Country Collaboration* module uses a chord diagram to display collaboration patterns based on the co-authorship of the papers from the selected journal. The arc length represents the percentage of the total collaborative papers from each country/region. When hovering the mouse over the arc of a specific country/region, the total number of papers contributed by authors whose affiliations are from that country/region, and its percentage of the total collaborations is displayed. The probabilistic affinity between two countries/regions is shown when hovering the mouse over a specific chord. Figure 3(a) provides an overview of the collaboration pattern of the Semantic Web journal, and Fig. 3(b) highlights collaborations with the United States.

Q3: How are institutions of all authors geographically distributed on a global/local scale? The *Author Map* module allows users to drag, zoom in or out to see how institutions are clustered, and shows the count of each cluster. Users are able to observe at a local view and investigate further details about the institution of an author. For example, from Fig. 4 we can know that Ludger Jansen was working at Institute of Philosophy in the University of Rostock (when one of his papers was published in the Semantic Web journal), and the address of the institution is August-Bebel-Straße 28, 18051 Rostock, Germany.

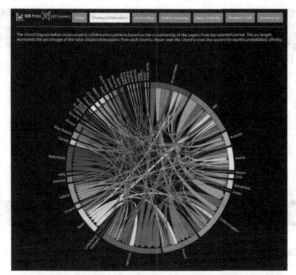

(a) Country Collaboration Overview

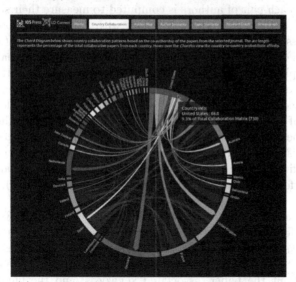

(b) Country Collaboration with the United States

Fig. 3. Country collaboration of the Semantic Web journal

Q4: Who are the most similar authors/papers to a selected author/paper? In *Author Similarity* module, similar authors across all journals are found based on the pre-trained knowledge graph embeddings discussed in Sect. 3.2. Cosine simi-

Fig. 4. Map visualization of clusters of author affiliations

larity between each pair of authors is computed to measure their similarity. The top 20 similar authors are retrieved, along with their institutions, addresses, and first 20 associated knowledge graph embeddings (see Fig. 5(a)). Hovering over stack bars of the selected author enables users to see the actual values of its knowledge graph embeddings. By clicking on one of the similar authors, a follow-your-nose author similarity search will be conducted with the selected author as a new input (see Fig. 5(b)). Similarly, the *Paper Similarity* module provides the functionality of searching for the most similar papers based on the pre-trained document embeddings discussed in Sect. 3.1. For each retrieved paper, its published year, a list of keywords, as well as first 20 corresponding document embedding are visualized (see Fig. 6(a)). Similarly, a follow-your-nose paper similarity search will perform once a similar paper is selected (see Fig. 6(b)).

Q5: How are the papers clustered based on similar keywords? The *Keyword Graph* module uses a force-directed graph to show the relationship among papers. Each paper from the selected journal is represented as a node. The nodes are clustered and linked together by shared keywords. Hovering the mouse over a node displays information about the paper, associated keywords, and the number of paper connections. Given the sheer number of keywords, the data have been split by years. In Fig. 7, an example node from the Semantic Web journal in 2016 is displayed.

Q6: What are the research topic trends of a journal across time? The *Stream-graph* module displays the trend of research topics/keywords in the selected journal over time. The top 20 keywords are selected and ranked according to the total number of papers containing the topic keywords. The streamgraph allows users to see the changes in the number of papers under certain topics. When

(a) Search results about *Pascal Hitzler*

(b) Search results about *Adila A. Krisnadhi*

Fig. 5. Information display of the selected author and similar authors, including the visualization of their first 20 knowledge graph embeddings.

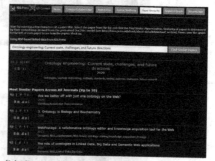

(a) Search results about *Are we better off with just one ontology on the Web?*

(b) Search results about *Ontology engineering: Current state, challenges, and future directions*

Fig. 6. Information display of the selected paper and similar papers, including the visualization of their first 20 document embeddings.

hovering the mouse over a specific keyword on the streamgraph, the information boxes will display author, paper, and year information associated with the selected keyword, as well as its count per year. Clicking on an author will link to its dereferencing interface developed on top of the Phuzzy.link framework [18]. Figure 8 shows an example of the keyword *Linked Open Data* with a count of 6 in 2016 from the Semantic Web journal.

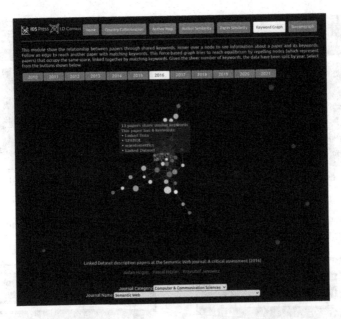

Fig. 7. Keyword graph visualization of the Semantic Web journal in 2016

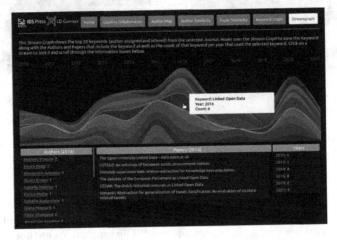

Fig. 8. Streamgraph visualization of top research topics in the Semantic Web journal

5   Conclusions and Future Work

In this work, we introduced a Linked Data-driven scientometric system on top of the LD Connect bibliographic knowledge graph that enables users to answer several competency questions by browsing and interacting with the system. The scientometrics showcase the potential to unveil the underlying characteristics of academic literature across space and time, as well as the ability to empower embedding-based similarity search on LD Connect. Being openly and freely available, the system is already in-use at IOS Press where it powers their data backbone, and will be publicly accessible[13] after transforming this prototype into production. At the same time, we aim to increase long-term availability and sustainability of our work in addition to the scientometrics. We have recently switched to an ongoing deployment where almost all steps described in this paper are automated, and therefore, time and costs can be reduced to a minimum. This deployment includes a pipeline that updates both the graph and embeddings when new data come in. We also plan to enrich our graph with more external information. For instance, we will associate contributors with their ORCIDs and include citation data for further bibliographic analysis.

It is worth noting that the scientometric system itself can be used and deployed by other researchers as a resource, such as for recommending reviewers and uncovering potential disparities between the geographic locations of authors versus journal editors. Also, the presented Linked Data-driven scientometrics are not restricted to LD Connect but can be deployed on other RDF-based datasets with minimal adjustments. While future work includes improving scalability of the scientometric system to support queries from journals with a larger volume, we also plan to develop new scientometrics to answer other interesting questions, such as how academic activity and collaboration change across space and time for both individuals and groups of scholars.

The ontology behind LD Connect can be aligned with other open bibliographic ontologies that are commonly in use for academic services, which will facilitate research in ontology alignment to improve semantic interoperability across academic knowledge graphs. Moreover, as a wide variety of spatial and temporal information is integrated with bibliographic data in the knowledge graph construction pipeline, we hope LD Connect highlights the importance of GeoEnrichment in ontology engineering for future research.

In addition to the ontology, a SPARQL endpoint, and open access to both the graph and the scientometrics, LD Connect is the first of its kind that shares both pre-trained document and knowledge graph embeddings, which overcomes copyright limitation to direct access to full text bodies of publications. In the future, we plan to incorporate spatial information in embedding generation by using techniques such as Space2Vec [14] to develop similarity search functions for discovering spatial similarity among entities in LD Connect. Furthermore, both shared embeddings serve as large-scale datasets with a wide diversity of research topics, author contribution, and their relations, opening up a great number of opportunities for research and applications in knowledge graphs, natural language processing, the Semantic Web, and beyond.

---

[13] http://ld.iospress.nl/scientometrics/.

References

1. Bordes, A., Usunier, N., Garcia-Duran, A., Weston, J., Yakhnenko, O.: Translating embeddings for modeling multi-relational data. Adv. Neural Inf. Process. Syst. **26**, 2787–2795 (2013)
2. Cai, L., Yan, B., Mai, G., Janowicz, K., Zhu, R.: TransGCN: coupling transformation assumptions with graph convolutional networks for link prediction. In: Proceedings of the 10th International Conference on Knowledge Capture, pp. 131–138 (2019)
3. Cohan, A., Feldman, S., Beltagy, I., Downey, D., Weld, D.S.: Specter: document-level representation learning using citation-informed transformers (2020). arXiv preprint arXiv:2004.07180
4. Devlin, J., Chang, M.W., Lee, K., Toutanova, K.: Bert: pre-training of deep bidirectional transformers for language understanding. In: NAACL-HLT (1) (2019)
5. Fan, M., Zhou, Q., Chang, E., Zheng, F.: Transition-based knowledge graph embedding with relational mapping properties. In: Proceedings of the 28th Pacific Asia Conference on Language, Information and Computing, pp. 328–337 (2014)
6. Gao, S., Hu, Y., Janowicz, K., McKenzie, G.: A spatiotemporal scientometrics framework for exploring the citation impact of publications and scientists. In: Proceedings of the 21st ACM SIGSPATIAL International Conference on Advances in Geographic Information Systems, pp. 204–213 (2013)
7. Hirsch, J.E.: An index to quantify an individual's scientific research output. Proc. Natl. Acad. Sci. **102**(46), 16569–16572 (2005)
8. Hood, W.W., Wilson, C.S.: The literature of bibliometrics, scientometrics, and informetrics. Scientometrics **52**(2), 291–314 (2001). https://doi.org/10.1023/A:1017919924342
9. Hu, Y., Janowicz, K., McKenzie, G., Sengupta, K., Hitzler, P.: A linked-data-driven and semantically-enabled journal portal for scientometrics. In: International Semantic Web Conference, pp. 114–129. Springer (2013)
10. Hu, Y., McKenzie, G., Yang, J.A., Gao, S., Abdalla, A., Janowicz, K.: A linked-data-driven web portal for learning analytics: data enrichment, interactive visualization, and knowledge discovery. In: LAK Workshops (2014)
11. Santha kumar, R., Kaliyaperumal, K.: A scientometric analysis of mobile technology publications. Scientometrics **105**(2), 921–939 (2015). https://doi.org/10.1007/s11192-015-1710-7
12. Le, Q., Mikolov, T.: Distributed representations of sentences and documents. In: International Conference on Machine Learning, pp. 1188–1196 (2014)
13. Mai, G., Janowicz, K., Yan, B.: Combining text embedding and knowledge graph embedding techniques for academic search engines. In: Semdeep/NLIWoD@ ISWC, pp. 77–88 (2018)
14. Mai, G., Janowicz, K., Yan, B., Zhu, R., Cai, L., Lao, N.: Multi-scale representation learning for spatial feature distributions using grid cells. In: International Conference on Learning Representations (2020)
15. McKenzie, G., Janowicz, K., Hu, Y., Sengupta, K., Hitzler, P.: Linked scientometrics: designing interactive scientometrics with linked data and semantic web reasoning (2013)
16. Mikolov, T., Sutskever, I., Chen, K., Corrado, G.S., Dean, J.: Distributed representations of words and phrases and their compositionality. Adv. Neural Inf. Process. Syst. **26**, 3111–3119 (2013)

17. Peters, M.E., et al.: Deep contextualized word representations. In: Proceedings of NAACL-HLT, pp. 2227–2237 (2018)
18. Regalia, B., Janowicz, K., Mai, G.: Phuzzy. link: a SPARQL-powered client-sided extensible semantic web browser. In: VOILA@ ISWC, pp. 34–44 (2017)
19. Schlichtkrull, M., et al.: Modeling relational data with graph convolutional networks. In: Gangemi, A., et al. (eds.) ESWC 2018. LNCS, vol. 10843, pp. 593–607. Springer, Cham (2018). https://doi.org/10.1007/978-3-319-93417-4_38
20. Wang, Z., Zhang, J., Feng, J., Chen, Z.: Knowledge graph embedding by translating on hyperplanes. In: Proceedings of the Twenty-Eighth AAAI Conference on Artificial Intelligence, pp. 1112–1119. AAAI Press (2014)

Chowlk: from UML-Based Ontology Conceptualizations to OWL

Serge Chávez-Feria(✉) [iD], Raúl García-Castro[iD], and María Poveda-Villalón[iD]

Ontology Engineering Group, Universidad Politécnica de Madrid, Madrid, Spain
serge.chavez.feria@upm.es, {rgarcia,mpoveda}@fi.upm.es

Abstract. Ontology conceptualization is an ontology development task that consists in generating a preliminary model based on the requirements that the ontology should represent. This activity is often carried out by generating the models as diagrams in a blackboard, paper or digital tools. The generated models drive the ontology implementation activity, where the model is formalized and completed using an implementation language. Normally, the ontology conceptualization output serves as guidance for the ontology implementation; however, ontology implementation is usually done from scratch using ontology editors. The goal of this work is to consider ontology conceptualizations as first-order artifacts in ontology development in order to boost the ontology implementation activity. For doing so we present Chowlk, a framework to transform digital machine-processable ontology conceptualization diagrams into OWL. Domain experts and ontologists benefit from this approach in several ways: 1) reduce time generating the first versions of the OWL file that can be invested on 2) focusing on the conceptualization diagrams that can be used both for 3) improving communication between ontology users and developers and 4) be reused during the ontology documentation stage.

Keywords: Ontology engineering · Ontology conceptualization · OWL

1 Introduction

Everyday more and more applications are being built on top of or in combination with semantic technologies. Ontologies play a crucial role in this development as they allow the representation of knowledge in a formal and structured way, being the OWL [4] language the default choice for their implementation because of its high level of expressiveness, reasoning capabilities and the fact that it has been designed for the web environment.

One of the first and most important steps in ontology development is the conceptualization one, during which the ontology development team defines a set of concepts and properties to represent the knowledge of a specific domain. Often,

This work has been supported by the BIMERR funded from the European Union's Horizon 2020 research and innovation programme under grant agreement no. 820621.

this conceptualization is materialized in a diagram that displays the relationships, attributes and axioms of the different concepts of an ontology. From this model, the ontology implementation is carried out normally using an ontology editor, such as Protégé [11], realizing the model into OWL code.

However, in this process the diagram is in most of the cases only used as a guideline to implement the ontology, translating the ontological elements and constructs to a formal syntax, being this process mostly manual and error-prone. Some tools have been proposed in the last years that allow the graphical creation or modification of ontologies following their respective visual notations [2,16].

In our case, rather than building a graphical ontology editor, the effort is driven towards the goal of allowing a smoother transition from the conceptualization activity to a first version of the actual implementation by taking the conceptualization output as a first order artifact in ontology development projects. For doing so, the Chowlk framework has been designed. The framework, shown in Fig. 1, consists of: 1) an UML-based visual notation; 2) a pair of diagrams.net templates implementing the visual notation; and 3) a converter from diagrams.net XML diagrams to OWL. It should be clarified that the resource presented in this paper is the converter that will be detailed in Sect. 3). However, for a better understanding of the converter, the visual notation is briefly presented in Sect. 2.

It should be clear at this stage that our goal is to fill the gap between the conceptualization and implementation of ontologies which is still a manual process, and as every manual procedure, it can be prone to errors. Even though, it is true that users can create ontologies directly in specialized editors such as Protégé [10] and avoid the creation of a diagram, our focus is on ontology users who follow developments where the conceptualization is the corner stone of the development process, and want to take full advantage of the effort made in the conceptualization step, for example to communicate and verify the model with users or clients as well as for documenting the ontology to publish or share it.

The validation of the Chowlk converter is described in Sect. 4 while a comparison with existing approaches is presented in Sect. 5. Future lines of work to evolve and improve the present work are proposed in Sect. 6.

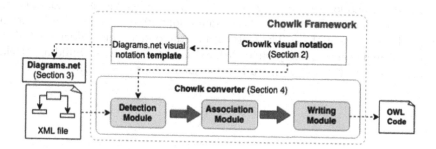

Fig. 1. Chowlk framework.

2   Visual Notation

The converter presented in this paper is based on the Chowlk visual notation that extends the UML_Ont profile [7]. It should be mentioned that while the original UML_Ont profile utilizes custom stereotypes and dependencies to cover OWL 1 constructs, the Chowlk notation binds the stereotypes used in the profile to OWL and RDF(S) constructs. Also, the visual notation used in this work proposes compact alternatives for representing property characteristics and axioms.

Due to the fact that the notation is considered an input for the converter instead of part of the resource presented in this paper, and for space matters, in this section only the main characteristics of the notation are included. While the notation has been partially published in [6], a more complete and updated version, including examples and alternative notation elements for those presented in this paper, is provided in the notation website[1].

Figure 2 provides an overview of the notation of the main OWL elements. Named **classes** are represented by labelled boxes. Unlabelled boxes or circles are used to represent anonymous classes and class intersections, unions, equivalences and disjoints. **Object properties** are represented by labelled arrows and **datatype properties** by labelled boxes attached to class boxes. Note that both types of properties can be represented by diamonds, notation needed in some cases, for example to represent equivalences or property hierarchies for datatype properties. For object properties, the relations between them (subproperty of, inverse or equivalent) can be represented both by arrows linking either the arrows representing the properties or the diamonds representing them.

Property characteristics (functional, inverse functional, transitive and symmetric) can be indicated before the property name or stating the characteristic construct in the diamond. **Class constraints** are represented between classes including the operator (universal, existential o cardinality) before the property over which the constraint is stated for subclass constraints. For equivalent class constraints or constraints in domains or ranges, unlabelled boxes are used in combination with the equivalent or domain/range indicator.

The Chowlk visual notation also allows to declare namespaces, for example to link entities from different ontology modules within a network or to indicate the reuse of other ontology elements. Finally, the notation includes a metadata block used not only for documenting the diagram but for ontology metadata generation during the conversion phase. The metadata is stated in a printed-document alike shape and makes use of the prefixes defined in the namespaces building block. Examples of namespaces and metadata blocks are shown in Fig. 3.

---

[1] https://chowlk.linkeddata.es/notation.html.

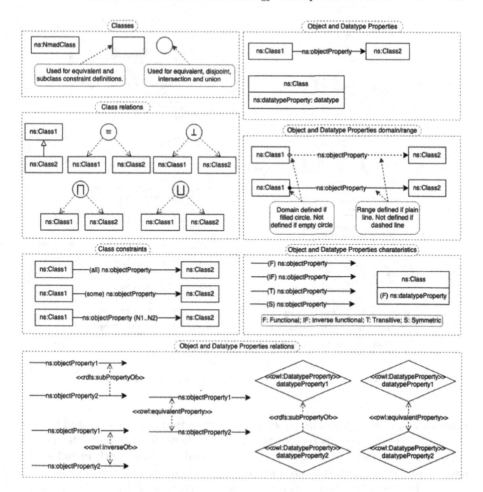

Fig. 2. Chowlk visual notation summary.

Figure 3 shows an excerpt of the BIMERR building ontology[2]. The figure shows basic elements such as classes, class hierarchies, object properties and datatype properties. Also, some more complex statements are represented as universal restrictions, for example between `building:Building` and `building:Storey` over the object property `bot:hasStorey`. Class cardinality constrains are shown for several datatype properties, for example the cardinality of the attribute `building:ifcIdentifier` for `building:Storey` is exactly 1.

Even though the presented visual notation is in some cases a one to one representation of the formalisms of the OWL language, it gives the freedom to develop lighter models. These less complex models can contain just boxes and plains arrows, without indicating restrictions or more complicated constructs,

---

[2] http://bimerr.iot.linkeddata.es/def/building#.

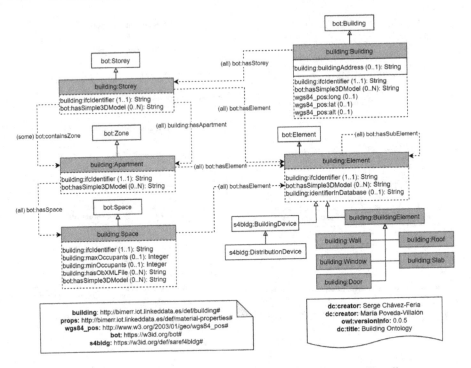

Fig. 3. Conceptualization example for an ontology using Chowlk.

almost like a conceptual map, which is easier to develop and understand by non ontology experts. For this reason, the notation allows for different alternatives for representing most of the OWL constructs and the framework includes two flavours of the notation that are implemented in two different templates.

The first template is a complete version containing all the building blocks described in the visual notation. This version was designed for ontology engineers who are knowledgeable about OWL. The second template is a lightweight version containing just a subset of the blocks, such as rectangles, arrows, and Boolean operators without more complicated constructs like restrictions. This second version was intended for users which are not familiar with OWL. Users can upload the templates and start making their conceptualizations by dragging and dropping the building blocks of the template into the diagramming layout of diagrams.net. This procedure reduces the entry barrier to start using the notation and avoids visual syntax errors when constructing the conceptualizations by providing already predefined combinations of the blocks in order to represent the OWL constructs.

3   The Chowlk Converter

Chowlk is a web application that takes as input an ontology conceptualization created with diagrams.net and generates the OWL implementation. The conceptualization is made following the Chowlk visual notation described in Sect. 2. The web application is available through its URL[3], and through its API[4]. The source code is shared in a GitHub repository[5] under the Apache 2.0 license. The software has a canonical citation using the DOI[6] provided by its Zenodo entry[7].

Figure 1 shows the modules in which the system is decomposed, namely: the detection module, the association module, and the writing module. The input to the system is a diagram representing the conceptual model of an ontology in XML format. After the conversion process, the tool outputs the ontology implementation in Turtle that can be downloaded to continue with the remaining ontology engineering process.

It is worth mentioning that even though the workflow shown in Fig. 1 has been defined within the Chowlk Framework, it can be reused to develop converters for other visual notations, just by adapting the detection stage which is in charge of detecting the underlying syntax of the blocks. Section 3.1 exposes the reasons to build the converter based on diagrams.net and the rest of the sub-sections cover in detail each of the modules in the transformation pipeline.

3.1   Selecting a Diagramming Tool

As already mentioned, the goal is not to produce a graphical ontology editor but to take advantages of conceptualizations that can be developed with a variety of diagramming tools. Indeed, the Chowlk notation is independent of the tool used to draw the diagram shapes or symbols and provides alternatives in case the diagramming tool does not support some symbols as the existential or universal operators. However, in order to use the converter to generate the OWL code from the conceptualization, diagrams.net should be used as the diagramming tool. The main reasons for choosing diagrams.net are:

1. It is flexible enough in terms of features and drawing options, so it allows to implement all the elements of the visual notation.
2. It supports synchronous collaborative diagram edition. In this sense, ontologists and domains experts, or other roles involved in ontology conceptualization, could be visualizing and/or editing the diagrams at the same time.
3. It is able to export diagrams in a structured format, such as an XML file. Figure 4 shows an example of the nested structure generated, where on the left side we have a very simple ontology excerpt composed by two classes and one object property, and on the right side the XML counterpart. Additionally,

---

[3] https://chowlk.linkeddata.es.
[4] https://chowlk.linkeddata.es/api.
[5] https://github.com/oeg-upm/Chowlk.
[6] https://doi.org/10.5281/zenodo.4312930.
[7] https://zenodo.org/record/4312930#.X9yNt9hKiUk.

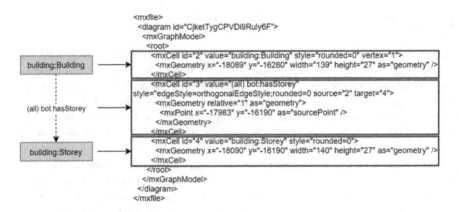

Fig. 4. Sample XML output of diagrams.net

each child element has a sequence of attributes that helps in the identification of each building block. Table 1 describes the fields used to describe the children elements. Some attributes apply to all the building blocks of the diagram such as the "id" field, while others only apply to specific shapes like the arrow blocks that should include a "target" and a "source" field.

4. It is a web-based open source platform. This feature lowers the barrier for its adoption, avoiding the process of having to download the software, install it and run it locally. The open source characteristic also opens the door to increase its functionalities whether through the extension of its source code or by means of plugins.

Table 1. XML diagrams.net data structure.

Block attribute	Block type	Definition
id	Classes, object properties, datatype properties	Unique identifier of the block in the diagram
value	Classes, object properties, datatype properties	Text content assigned to the block. Used to represent the URIs of the elements of the ontology
style	Classes, object properties, datatype properties	Allows to give style to the blocks and make a differentiation between the elements of the ontology
source	Object properties	Points to the block id that is connected to the source side of an arrow
target	Object properties	Points to the block id that is connected to the target side of an arrow

3.2   The Detection Module

Once the diagram is uploaded to the system, the transformation process triggers. The first step in the conversion procedure is performed by the detection module, where all the building blocks of the diagram are found. The detection of ontology elements is performed for all the building blocks represented in the diagram that follow the Chowlk visual notation, discarding any shape that does not correspond to the notation ones. This detection is done by analyzing the attributes of the XML data structure mentioned in Sect. 3.1. Specifically, the module searches for information in the "style" attribute of children elements to derive the type of shape it is dealing with. For instance, if the "style" attribute contains the keyword "edge", the module can interpret that the shape being analyzed is of type "arrow" that could represent an object property in the OWL language. Each element identified in the diagram populates a predefined data structure, where the fields change according to the type of ontology element. For example, in the case of an object property the data structure will store information regarding its prefix, the URI, if it is functional or symmetric, etc. These data structures facilitate later the querying of elements and searching for information during the subsequent stages.

In most of the cases the type of visual blocks used in the specification has a unique mapping to the OWL construct, like the namespace block. However, there are other situations in which the same type of building block is used to represent more than one OWL element. This is the case of concepts and attributes, where both use the rectangle definition, and it is needed to identify the geometry disposition of the blocks in the layout in order to disambiguate their meaning. In this particular example, if the algorithm detects a rectangle, then it searches for other rectangles above it in a close neighborhood. If they exist, the rectangle we are analyzing represents datatype properties, otherwise it represents a class.

In the current version of the converter, the source and target of arrows in a diagram must be anchored to other building blocks in order to identify the relationship. This characteristic in combination with the restriction that diagrams.net does not allow connections between arrows, impedes the creation of relationships between properties. This means that in order to represent rdfs:subPropertyOf relations between two object properties, the diamond option specified in the visual notation to represent object properties should be used. Diamond shapes can also be used as an optional alternative to state several other characteristics of the properties such as symmetry, functionality, range, domain, etc. If an object property is represented as an arrow in one part of the diagram and additional information is provided using the diamond shape, the definition of the property is generated by combining the information represented in both shapes.

Additionally, the converter is able to identify ontology metadata, and the namespaces and prefixes being used in the model, thanks to specific blocks dedicated to this type of information. Labels to each ontology element are added during the detection process.

Finally, the detector module also identifies any deviation from the visual notation and returns a report diagram indicating in which part of the diagram the ontology engineer is not following the correct syntax. For instance, if the ontology engineer attempts to instantiate a property without a prefix, or a prefix was detected in the ontology elements that was not included in the namespace declaration block, the module detects those errors and outputs: the id of the block involved, the label if available, and a generic explanation of the error. This example can be seen in Fig. 5.

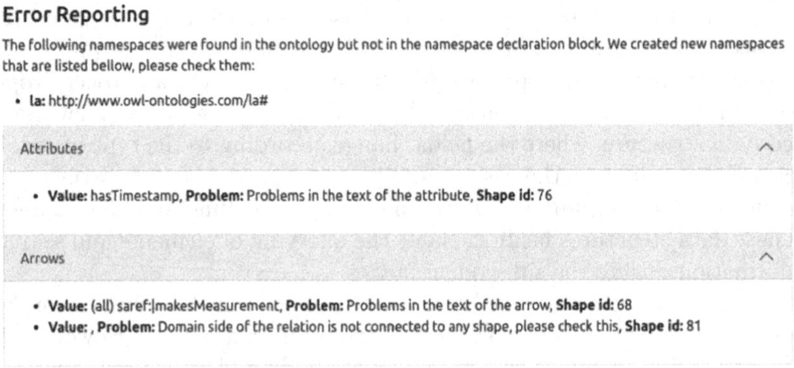

Error Reporting

The following namespaces were found in the ontology but not in the namespace declaration block. We created new namespaces that are listed bellow, please check them:

- **la:** http://www.owl-ontologies.com/la#

Attributes   ∧

  - **Value:** hasTimestamp, **Problem:** Problems in the text of the attribute, **Shape id:** 76

Arrows   ∧

  - **Value:** (all) saref:|makesMeasurement, **Problem:** Problems in the text of the arrow, **Shape id:** 68
  - **Value:** , **Problem:** Domain side of the relation is not connected to any shape, please check this, **Shape id:** 81

Fig. 5. Example of error report

3.3   The Association Module

The association module performs the connection between the classes, and the object and datatype properties instantiated in the diagram.

The correspondences are established following different procedures. In the case of associations between classes and object properties, the module checks if the identifier of the building block representing a class and the identifier in the "source" field of an object property is the same. For the case of association between classes and datatype properties, the module analyzes the location of the blocks representing them. If the datatype property block is below and close enough to a class block, it means those attributes are intended to be used with that class.

In a second step, the module analyzes if the object and datatype properties have a restriction with the class at hand. The module specifically searches for the following notation in the text of the properties: (some), (all), or (N1..N2), which indicates existential, universal, and cardinally restrictions respectively.

If the restrictions exist, the module maintains the connections previously created between the classes and the properties. Otherwise, the associations are eliminated because the properties have been diagrammed in that way only to give the potential user of the ontology an idea of how the properties are planned to be used. However, there is no formal restriction that states that it can only be used with that specific class.

Finally, the output of this module is an array that contains the concepts, objects properties and datatype properties associated through restrictions. This will facilitate the serialization of the restrictions in the final Turtle file.

3.4   The Writing Module

The writing module takes all the ontological elements detected in the previous steps and writes them one by one in an RDF file. The process starts by taking as basis a template that already incorporates common namespaces (rdfs, owl, rdf, xml, dcterms and vann) and their prefixes to avoid the user to indicate them. This default list is complemented with the namespaces and prefixes found in the metadata block. If there exist some prefixes detected on the elements of the ontology (e.g., concepts, relations, attributes) that were not declared on the namespace block, new namespaces invented by the tool are created automatically. Afterwards, the module writes high level information about the ontology declared in the ontology metadata block, such as title, authors, imports, etc. This information will be written in the `owl:Ontology` header.

Next, the module writes the definition of the object and datatype properties. The following information is included for both types of properties: English labels (which are automatically extracted from the URI), if they are functional, the domain, the range, and if they are sub-property or equivalent to another one. In the case of object properties additional characteristics are included if they are stated during the conceptualization: symmetric, transitive, inverse functional, or inverse of another property.

The process is similar for writing the classes, with the difference that in this case the module uses as input the output from the associations module. The writing module needs such data structure to know the type of restrictions that applies over each class with respect to the object and datatype properties. Relationships of the type `owl:disjointWith` and `owl:equivalentClass` with other concepts are also included. Instances and general axioms such as multiple disjoints between several classes are also added.

Once the writing process is finished, the converter provides the ontology in the Turtle format, and the file can be finally downloaded from the GUI of the web application.

Additionally, if the visual notation was not followed properly, a report is provided in the GUI of the application listing all the errors found during the parsing process. The report includes the id of the block containing the error, the label of the block for a rapid inspection in the diagram, and a generic explanation of the problem.

3.5   Current Limitations

The diagrams.net tool is a general purpose diagramming software, not specific for the development of ontologies, so the user has to be very careful when constructing the conceptualizations in order to avoid deviations from the visual notation being used. Even though the current version of the model can generate reports about the errors detected in model and the user can make the appropriate changes in the diagram, the process of identifying the blocks in the diagram manually can be very complex for very large conceptualizations.

Also, because diagrams.net does not allow to anchor the extremes of arrows to other arrows, the system cannot detect the `rdfs:subpropertyOf`, `owl:inverseOf` and `owl:equivalentOf` relationships. For that we need to use the diamond options of the Chowlk visual notation. For instance, to express that the object property "hasSpace" has a relationship of type `owl:inverseOf` with the property "isSpaceOf" we need to use the diamond shapes for the converter to be able to detect this kind of construct.

4   Validation

In the following section we provide a series of examples that prove the usage of the tool. Additionally, we verified the correctness of the results obtained by the converter by transforming the visual OWL constructs listed in the visual notation. Because of the simplicity of the tool, we do not include user experience evaluation in this first version, but it is something that we plan to do for the next iterations.

4.1   Adoption and Use

The service has been adopted in different projects from several institutions. For instance, Chowlk is being used as part of the ontology development pipeline in different H2020 European projects, such as BIMERR[8], and COGITO[9], within the research lab developing Chowlk, but also by external teams, for example in the BIM4EEB[10] and CosWot ANR projects[11].

Additionally, the system is being used to support the development of ontologies in different domains such as agriculture[12], public transport [14], time[13], ethics[14], material science[15], and ICT infrastructure [3]. Furthermore, some ontologies developed by international communities such as the W3C[16] has also being implemented using Chowlk, such as the WoT discovery ontology[17].

---

[8] https://bimerr.iot.linkeddata.es/.

[9] https://cogito.iot.linkeddata.es/.

[10] https://digitalconstruction.github.io/v/0.5/index.html.

[11] https://coswot.gitlab.io/.

[12] http://www.elzeard.co/ontologies/c3po/plant#.

[13] https://github.com/mnavasloro/ft3/blob/04c65c2b2ed2bd57f9ac6cfb32b7f4ebfda1f4c4/ft3.owl.

[14] https://krnlet.github.io/#.

[15] https://github.com/Mat-O-Lab/MSEO.

[16] https://www.w3.org/.

[17] https://github.com/w3c/wot-discovery/blob/24b2141e8e0cb74abd24cead0b4bbffb672e24c6/context/discovery-ontology.ttl.

Finally, the usage of the tool can be demonstrated by the issues, pull requests and forks made to the Github of the project. This demonstrate that Chowlk is being used not only to develop ontologies, but also being integrated in other ontology development softwares[18].

4.2   Validation Tests

The service has been tested against a set of 49 diagrams, where all the results obtained were valid ontologies. Each diagram contains a set of building blocks representing the OWL constructs defined in the Chowlk visual notation. The diagrams constructed with their corresponding OWL ontologies are available in the GitHub repository of the project[19] for its verification. Figure 6 shows an example of an input diagram and the ontology generated by the converter.

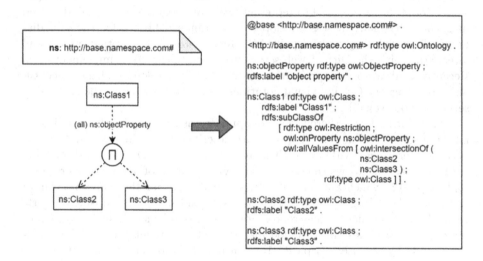

Fig. 6. Test conversion example.

As it was mentioned in Sect. 4.4, since it is not possible for the system to detect the `owl:inverseOf`, `owl:equivalentProperty` and `rdfs:subPropertyOf` axioms between object properties when they are represented using arrows, we tested those axioms representing the object properties with the diamond shapes.

5   Related Work

Several approaches have been proposed in the recent years with respect to visual ontology edition tools. The work developed in [5] presents a good review of the

---

[18] https://gitlab.com/kupferdigital/ontoflow.
[19] https://github.com/oeg-upm/Chowlk/tree/webservice/tests/inputs.

state of the art regarding tools with edition and visualization capabilities. From the spectrum of tools analyzed, only six include the visual edition of ontologies as a feature. It is important to remark that the following review only considers tools that are free for its usage.

On the one hand, there is a set of applications that are implemented as a web service. WebVOWL [16] is an application that has as a principal feature the visualization of OWL ontologies, which are displayed following the VOWL visual notation that has a graph representation. Among other capabilities it allows the customization of the visualization and the modification of the ontology by directly manipulating the elements of the graph. On the same line, OWLGrEd [2] is a framework that offers visualization capabilities following an UML based notation. The tool allows the visual edition of ontologies but only in its desktop version. Even though, the graphical edition of the ontologies is possible in both applications, neither of them allow collaborative work. Graffoo[20] is an open source tool that can be used to represent OWL ontologies as easy-to-understand diagrams. Originally, it was developed as a standard library for the yEd diagram editor including a set of pallets to create ontology conceptualizations and afterwards using the Ditto[21] web service to generate the OWL implementation. Recently, a library for diagrams.net was created to develop ontology conceptualization using the Graffoo visual notations; however, in this case the conversion service is not available.

On the other hand, there is a set of applications that require the local installation of software. The following tools are described based on their publications because there is no evidence of their availability. CMap Ontology Editor [8] is a set of tools that allows the creation and visualization of ontologies as conceptual maps [12], which are general artifacts that serve for the representation of any kind of knowledge. Ontotrack [10] is a standalone application that supports graph based and hierarchical representations of ontologies. It includes instant reasoning capabilities that provide instant feedback about the modeling decisions made by the user. Triple20 [15] is a manipulation and visualization tool developed using Prolog. Some of its characteristics include the representation of the ontology following a graph based and hierarchical view and the ability to handle large ontologies because all the data is stored in RAM memory. Finally, GrOWL [9] is a standalone Java application that, apart from the basic visualization and edition features, also makes use of shape, color and shade to encode properties in the nodes of the graph.

One common characteristic among the tools described previously is that all require the learning of a new development environment, the local installation of the software, or do not allow collaborative work.

The Chowlk converter eliminates the need for software installation by leveraging on existing popular diagramming tools that already provide collaborative edition features to generate the ontology conceptualizations. It could also be integrated with third party software. In addition, the proposed framework and

---

[20] https://essepuntato.it/graffoo/.
[21] https://essepuntato.it/ditto/.

converter are based on UML notation as it is commonly used in software engineering, and it is familiar to software engineers.

6   Conclusions and Future Work

This paper presents a system, Chowlk, to ease the ontology development process by leveraging the conceptualization activity outputs in order to transform the obtained diagrams into OWL code. Chowlk is implemented as a web application that allows the uploading of the diagram as an XML file and outputs the ontology in RDF/XML and Turtle formats speeding up the ontology developments.

The system was tested using a unit-test procedure with all the OWL constructs defined by the Chowlk visual notation, and also using it to generate the ontologies of the BIMERR ontology network.

We will explore the support for other visual notations for a broader adoption and testing of the tool. Additionally, further research should be carried out in order to support the updating of the conceptualizations. That is, how for a given ontology created by Chowlk, and then modified by an editor (Protégé), the changes can be appropriately represented in the diagram. The support for other standard formats such as SVG is also something to be explored in the next version. This could allow the converter to be independent of the diagramming tool to be used.

Finally, the sustainability plan for the system includes its continue use and evolution as part of current and future research projects and as part of the group ontology engineering tools suite roadmap. Some foreseen interactions exist between Chowlk and OnToology [1], by integrating the XML file as a resource in GitHub repositories from where OnToology can trigger Chowlk to generate the OWL code; and incorporating the pitfalls detection from OOPS! [13] within the conceptualization phase by the diagrams.net Chowlk plugin.

References

1. Alobaid, A., Garijo, D., Poveda-Villalón, M., Santana-Perez, I., Fernández-Izquierdo, A., Corcho, O.: Automating ontology engineering support activities with OnToology. J. Web Semant. **57**, 100472 (2018)
2. Barzdins, A., Barzdins, G., Cerans, K., Liepins, R., Sprogis, A.: UML style graphical notation and editor for OWL 2. In: Forbrig, P., Günther, H. (eds.) BIR 2010. LNBIP, vol. 64, pp. 102–114. Springer, Heidelberg (2010). https://doi.org/10.1007/978-3-642-16101-8_9
3. Corcho, O., et al.: A high-level ontology network for ICT infrastructures. In: Hotho, A., et al. (eds.) ISWC 2021. LNCS, vol. 12922, pp. 446–462. Springer, Cham (2021). https://doi.org/10.1007/978-3-030-88361-4_26
4. Dean, M., et al.: Owl web ontology language reference. W3C Recommendation, World Wide Web Consortium (2004). http://www.w3.org/TR/owl-ref/
5. Dudás, M., Lohmann, S., Svátek, V., Pavlov, D.: Ontology visualization methods and tools: a survey of the state of the art. Knowl. Eng. Rev. **33** (2018)

6. Garijo, D., Poveda-Villalón, M.: Best Practices for Implementing FAIR Vocabularies and Ontologies on the Web. IOS Press, Amsterdam (2020). https://doi.org/10.3233/SSW200034

7. Haase, P., Brockmans, S., Palma, R., Euzenat, J., d'Aquin, M.: D1.1.2 updated version of the networked ontology model. Technical report, Universität Karlsruhe (2009). NeOn Project. http://neon-project.org/

8. Hayes, P., Eskridge, T., Saavedra, R., Reichherzer, T., Mehrotra, M., Bobrovnikoff, D.: Collaborative knowledge capture in ontologies. In Proceedings of the 3rd International Conference on Knowledge Capture (2005)

9. Krivov, S., Williams, R., Villa, F.: Ontotrack: a semantic approach for ontology authoring. A tool for visualization and editing of OWL ontologies, GrOWL (2007)

10. Liebig, T., Noppens, O.: Ontotrack: a semantic approach for ontology authoring. Science, Services and Agents on the World Wide Web, Web Semantics (2005)

11. Musen, M.A.: The protégé project: a look back and a look forward. AI Matters 1(4), 4–12 (2015)

12. Novak, J., Cañas, A.: The theory underlying concept maps and how to construct and use them. Technical report (2006). Technical Report IHMC CmapTools 2006-01, Institute for Human and Machine Cognition (IHMC)

13. Poveda-Villalón, M., Gómez-Pérez, A., Suárez-Figueroa, M.C.: Oops! (ontology pitfall scanner!): an on-line tool for ontology evaluation. Int. J. Semant. Web Inf. Syst. 10(2), 7–34 (2014)

14. Ruckhaus, E., Anton-Bravo, A., Scrocca, M., Corcho, O.: Applying the lot methodology to a public bus transport ontology aligned with transmodel: challenges and results (2021). https://doi.org/10.3233/SW-210451. https://content.iospress.com/articles/semantic-web/sw210451

15. Wielemaker, J., Schreiber, G., Wielinga, B.: Using triples for implementation: the triple20 ontology-manipulation tool. In: Gil, Y., Motta, E., Benjamins, V.R., Musen, M.A. (eds.) ISWC 2005. LNCS, vol. 3729, pp. 773–785. Springer, Heidelberg (2005). https://doi.org/10.1007/11574620_55

16. Wiens, V., Lohmann, S., Auer, S.: Webvowl editor: device-independent visual ontology modeling. In: International Semantic Web Conference (2018)

QuoteKG: A Multilingual Knowledge Graph of Quotes

Tin Kuculo[1]([⊠])[iD], Simon Gottschalk[1][iD], and Elena Demidova[2][iD]

[1] L3S Research Center, Leibniz Universität Hannover, Hannover, Germany
{kuculo,gottschalk}@L3S.de
[2] Data Science and Intelligent Systems (DSIS), Universität Bonn, Bonn, Germany
demidova@cs.uni-bonn.de

Abstract. Quotes of public figures can mark turning points in history. A quote can explain its originator's actions, foreshadowing political or personal decisions and revealing character traits. Impactful quotes cross language barriers and influence the general population's reaction to specific stances, always facing the risk of being misattributed or taken out of context. The provision of a cross-lingual knowledge graph of quotes that establishes the authenticity of quotes and their contexts is of great importance to allow the exploration of the lives of important people as well as topics from the perspective of what was actually said. In this paper, we present QuoteKG, the first multilingual knowledge graph of quotes. We propose the QuoteKG creation pipeline that extracts quotes from Wikiquote, a free and collaboratively created collection of quotes in many languages, and aligns different mentions of the same quote. QuoteKG includes nearly one million quotes in 55 languages, said by more than 69,000 people of public interest across a wide range of topics. QuoteKG is publicly available and can be accessed via a SPARQL endpoint.

Keywords: Knowledge Graph · Quotes · Cross-lingual Alignment

1 Introduction

Quotes of public figures provide valuable information to understand their thoughts and attitudes, potentially leading to historically important actions, and thus serve as a crucial component in exploring world history [19]. Table 1 provides three examples of quotes, with the first one emphasising the relevance of historic quotes: in 1930, Winston Churchill recognised the value of reading them. The second example in Table 1 illustrates the relevance of quotes in world history: During a press conference in 2015, the German chancellor Angela Merkel said "Wir schaffen das" ("We can do this") when the European migrant crisis unfolded and Germany prepared for the reception of refugees from Northern

Resource DOI: 10.5281/zenodo.4702544
Permanent URL: https://quotekg.l3s.uni-hannover.de.

P. Groth et al. (Eds.): ESWC 2022, LNCS 13261, pp. 353–369, 2022.
https://doi.org/10.1007/978-3-031-06981-9_21

Africa and the Middle East. Since then, these three words defined Merkel's political course in the migrant crisis – and led both to a welcoming culture as well as the rise of nationalist protests and right-wing political parties [21,23].

Given this potential impact of words, it is of utmost importance to provide sources to quotes and to dismiss hoaxes [18,28]: The third example in Table 1 is a famous quote that has been attributed to different people, including Albert Einstein, Benjamin Franklin, and Mark Twain, but has not actually been said by any of them.[1] In general, a quote can be mentioned in different sources, and mentions can deviate. For example, "Wir schaffen das" might be mentioned as "We can do this" or "We will make it!" in English translations. Therefore, there is a need to align mentions to the same quote and to provide context information such as the source and description (e.g., "during a press conference").

Table 1. Three example quotes, together with their originators and dates. The last column gives examples of context that can be attributed to the mention of a quote, including source information, translations or validation of the quote's correctness.

Quote	By	Date	Selected context
It is a good thing for an uneducated man to read books of quotations. (en)	Winston Churchill	1930	*Source:* Roving Commission: My Early Life (1930) Chapter 9
Wir schaffen das. (de)	Angela Merkel	2015, Aug 31	*Translation:* We can do this (en)
The definition of insanity is doing the same thing over and over and expecting different results. (en)	Albert Einstein		Misattributed

In this paper, we introduce *QuoteKG* – a new knowledge graph that provides nearly one million quotes said by more than 69,000 persons of public interest in 55 languages. Quotes in *QuoteKG* come with detected sentiment and context such as their origin dates and sources. They are interlinked with their originators and other entities such as persons or events they refer to. Different mentions of the same quote are aligned across languages.

The creation of a knowledge graph covering quotes in many languages and their contexts faces several challenges detailed in the following.

- *Lack of context:* Most quote collections [10,24,34] lack context information and solely provide the quotes and their originators. To provide more context information in *QuoteKG*, we extract quotes from Wikiquote – a "free online compendium of sourced quotes from notable people"[2].

---

[1] Reasons for false attribution of quotes to persons include to appear educated or to lend authority from the person. [27].

[2] https://en.wikiquote.org/wiki/Main_Page.

– *Tedious extraction process:* Even though Wikiquote is a semi-structured resource, extraction of quotes and contexts is a tedious process. In particular, we must design an extraction pipeline that is flexible across languages and adopts their characteristics. For example, it is necessary to differentiate the quotes not said by a person, but said about a person (e.g., English: "Quotes about Albert Einstein", German: "Zitate mit Bezug auf Albert Einstein").
– *Missing alignment of quote mentions:* As quote mentions in Wikiquote are not linked across languages, another important step is cross-lingual quote alignment which we perform using a language-agnostic transformer model that we evaluate on a ground truth set of manually aligned quote clusters.

Our contributions are as follows: (i) We propose a schema to represent quotes and context information. (ii) We propose an extraction pipeline that extracts quotes, their mentions and context information from all Wikiquote language versions. (iii) We align quote mentions across languages using a cross-lingual language model. (iv) We make *QuoteKG* publicly available[3].

The remainder of this paper is structured as follows: First, we describe the impact of *QuoteKG* in the fields of Semantic Web, Natural Language Processing, Digital Humanities and others in more detail. Then, in Sect. 3, we describe the schema adopted for *QuoteKG*. In Sect. 4, we describe the *QuoteKG* creation pipeline. In Sect. 5, we provide statistics and examples of *QuoteKG*, followed by information about the availability and maintenance in Sect. 6. Section 7 gives an overview of related work. Finally, we provide a conclusion in Sect. 8.

2 Potential Impact

QuoteKG contains quotes, a new type of information that is, to the best of our knowledge, not yet present in existing knowledge graphs. Therefore, *Quote-KG* can potentially attract new audiences from several fields such as Digital Humanities and Natural Language Processing. While existing cross-domain or event-centric knowledge graphs such as Wikidata [36], DBpedia [2], and EventKG [12] target the representation of real-world entities, including persons of public interest and important events, they scarcely represent what people actually said – even though this information reflects persons' characteristics and can lead to an understanding of how particular events unfolded in the real world. Instead, facts about persons in knowledge graphs (e.g., properties representing birth dates, marriages, and awards received) without a doubt represent relevant facts in a person's life but typically do not reveal personal traits or surprising insights. Existing corpora of quotes like Quotebank [34] and the QUOTES500K dataset [10] provide large collections of English quotes. In contrast to these corpora, *QuoteKG* is a knowledge graph and provides societal relevant quotes and contexts in 55 languages, links them to other knowledge graphs, and aligns quote mentions across languages.

---

[3] https://www.quotekg.l3s.uni-hannover.de.

Potential applications of *QuoteKG* are manifold: (i) First and foremost, *QuoteKG* can add a new dimension to the exploration and investigation of the lives of public figures. While the creation of biography timelines from knowledge graphs has been studied in the past [1,13], such timelines do not consider the inclusion of quotes. *QuoteKG* can help to enrich such timelines with relevant quotes to make them more lively and informative. (ii) Similarly, the analysis of quotes related to a specific topic over time can support research in the fields of Digital Humanities, and can be used to gauge public opinion regarding specific events. For example, there have been analyses of how social movements and global events affect language [14] and how the words used by public persons carry political backgrounds [35]. (iii) Quotes also play an important role when observing information propagation [31] and the bias potentially caused by one-sided selection of quotes [25]. (iv) *QuoteKG* can also serve as an additional resource for machine translation, given that it contains 38,931 quotes with mentions in different languages. (v) *QuoteKG* can help answering questions such as "Who said 'yes, we can'?". (vi) *QuoteKG* contains 13,104 quotes labelled as misattributed or falsely claimed. They can be used as a resource for understanding the propagation of false or misleading information [28]. (vii) Finally, *QuoteKG* can take quote collections to a new level: There are plenty of websites available that provide collections of quotes[4], typically monolingual and primarily for entertainment purposes (e.g., images of inspirational quotes or quote mashup games[5]). The context information in *QuoteKG* can support the exploration and the search for quotes and provide important information surrounding a quote, and, as such, broaden the user's horizon.

3   *QuoteKG* Schema

The goal of the *QuoteKG* schema is to model quotes, their relationships with persons and other entities, as well as their different mentions, e.g. translations, typically in different contexts. To this end, *QuoteKG* is based on an extension of the *schema.org* vocabulary that provides a so:Quotation[6] class which is reused. According to the schema.org description, the so:Quotation class models quotes that are "Often but not necessarily from some written work" and can also refer to a "Quotation from an Event"[7]. Therefore, it fits well to our concept of a quote in *QuoteKG*. However, we extend the schema with a new class qkg:Mention which models the different mentions of a quote.

Figure 1 presents *QuoteKG*'s schema. Its classes are described in the following.

---

[4] https://www.brainyquote.com, https://www.goodreads.com/quotes, https://www.successories.com/iquote, ...

[5] http://natetyler.github.io/.

[6] so: https://schema.org/.

[7] https://schema.org/Quotation.

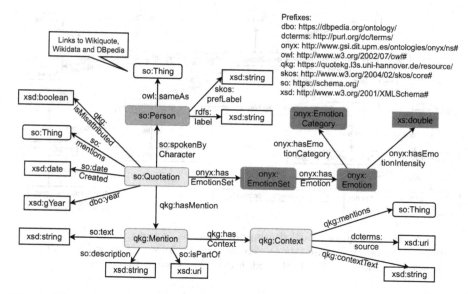

Fig. 1. The *QuoteKG* schema based on *schema.org*. Arrows visualize the rdfs: domain and rdfs:range restrictions on properties. Name spaces and prefixes are described in the top right corner. Orange classes are related to quotes and their mentions, blue ones to the sentiment of a quote and the green class is about the person. (Color figure online)

- **Person:** Each quote in *QuoteKG* is assigned to a person modeled as so: Person. For persons, *QuoteKG* provides additional type information (e.g., Politician) plus owl:sameAs relations to Wikidata and the different DBpedia and Wikiquote language editions.
- **Quote:** In *QuoteKG*, a resource typed as so:Quotation refers to the unique event of something being said by a person of public interest (so: spokenByCharacter) at a specific point in time (so:dateCreated). A quote may also refer to other entities (so:mentions) of any type.
- **Mention:** A quote can be mentioned in different contexts: For example, there may be translations of the quote in different languages, alternative records of the same quote, or different contexts that a quote is extracted from. Therefore, we introduce the class qkg:Mention. Mentions can be related to one or more qkg:Context objects.
- **Context:** The context of a mention provides additional attributes that come together with the specific mention. For example, its origin (e.g., a reference to a specific interview) and the original source (e.g., a link to a news website). To model context, we create the class qkg:Context.
- **Sentiment:** For each quote, we provide its sentiment using the Onyx ontology which is used for describing emotions [29]. A quote is assigned a score for a specific emotion category ("neutral", "negative" or "positive").

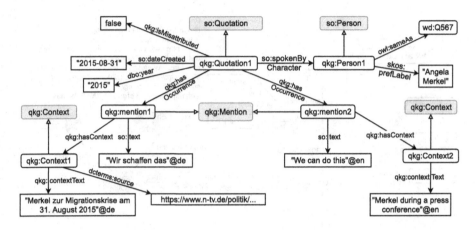

Fig. 2. An example quote modeled using the *QuoteKG* schema. ⇢ marks rdf:type relations. xsd:data type annotations were omitted for brevity. The prefixes and name spaces are the same as in Fig. 1, plus wd: https://www.wikidata.org/entity/.

Figure 2 shows an example instantiation of the *QuoteKG* schema. The quote "Wir schaffen das" introduced in Table 1 is connected to two instances of qkg: Mention, one representing a German mention, the other one an English one ("We can do this"). Both mentions come with additional context information.

4   Extraction and Alignment of Quotes

This section describes the input data and the implementation of the four main steps of the *QuoteKG* creation pipeline shown in Fig. 3.

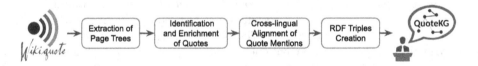

Fig. 3. Pipeline to create *QuoteKG* from Wikiquote.

4.1   Wikiquote

We base *QuoteKG* on Wikiquote – an online collection of quotes[8]. Wikiquote has a similar structure to Wikipedia: Independent versions of Wikiquote exist for different languages. Wikiquote contains pages, each of them about a given topic and divided into different sections and subsections. For *QuoteKG*, we focus on

---

[8] https://en.wikiquote.org/wiki/Main_Page.

```
*   '''Falling in love is not at all the most stupid thing that people do
|   but gravitation cannot be held responsible for it.'''
**  Jotted (in German) on the margins of a letter to him (1933), p. 56
```

Fig. 4. Example of a quote in the English Wikiquote, based on MediaWiki markup.

```
{{Citation|Tomber amoureux n'est pas du tout la chose la plus stupide
que font les gens | mais la gravitation ne peut en être tenue pour
responsable. |original=Falling in love is not at all the most stupid
thing that people do | but gravitation cannot be held responsible for
it.|langue=en}}
```

Fig. 5. Example of a quote in the French Wikiquote, using templates.

Wikiquote pages about persons that contain quotes attributed to them. Example pages are the English[9] and French page about Albert Einstein[10].

Each Wikiquote page is formatted using the MediaWiki markup[11] and contains semi-structured content that includes the person's description, sections with quotes, references and more. The quotes are given in one of the following representations: in the traditional MediaWiki markup as shown in Fig. 4 or using pre-defined *templates* that allow for a more structured definition of key-value pairs. For example, Fig. 5 shows the key-value pair (*key*: `Citation`, *value*: `Tomber amoureux...`).

While there are links between the pages describing persons in different languages, quote mentions are not linked across languages. Figure 4 and Fig. 5 show two mentions of the same quote by Albert Einstein. The first is from the English Wikiquote and shows an English quote, the second one from the French Wikiquote is given in French and English. The original German quote is not available in these two language versions.

In general, one can observe a large imbalance in Wikiquote regarding the covered persons and the number of quotes in different language versions. This imbalance can often be explained by the different sizes of Wikiquote language versions and the difference in the cultural significance of a person in one language community compared to another. For example, there exists a French page with 35 quotes and an Italian page with 2 quotes of the former footballer Michel Platini who used to play in Italy, but there is no English page. This imbalance also implies that there is no guarantee that Wikiquote will contain the original language version of a quote. *QuoteKG* can have multiple quote mentions of the same quote through cross-lingual quote mention alignment.

---

[9] https://en.wikiquote.org/wiki/Albert_Einstein.
[10] https://fr.wikiquote.org/wiki/Albert_Einstein.
[11] https://www.mediawiki.org/wiki/Help:Formatting.

Fig. 6. Excerpt of an example page tree from the English Wikiquote page about Albert Einstein. Section titles are underlined.

4.2 Extraction of Page Trees

In the beginning, our *QuoteKG* creation pipeline processes all Wikiquote language editions with at least 50 pages, excluding Simple English[12] and selects all pages about persons. From each Wikiquote page about a person, we create a *page tree*. The page tree consists of section titles plus quotes and contexts. An example page tree is presented in Fig. 6.

4.3 Identification and Enrichment of Quotes

In the second step of the *QuoteKG* creation pipeline, the page trees are transformed into a set of quotes with contextual information. To this end, we specify language-specific rules and enrich quotes and contexts with additional metadata.

To identify quotes, we first define a language-specific list of section titles denoting quotes (e.g., "Citations", "Zitate", "Citazioni") and contextual information (e.g., "útskýring", "Viitattu", "vydavatel"). In addition, we collect a list of template types representing quotes and consider all child nodes of section titles as quotes. From section titles and templates, we further gather the following:

- Dates: We identify the dates of quotes from a pre-defined list of template keys (e.g., "année d'origine" in French) for quotes extracted from templates. If such dates are not available or when dealing with quotes not extracted from templates, we extract dates from the section titles above the particular quote in the page tree and the contexts below the quote.
  We select the time expression with the highest level of precision (e.g., we select May 2020 over 2020). In case of conflicts, no date is chosen.
- Veracity: To reflect the authenticity of quotes and their contextual information, we capture whether a quote has been misattributed to the person. In Wikiquote, misattributed quotes are grouped into specified sections. We identify such sections with a manually created list of regular expressions (e.g., "Misattributed" (English) and "Fälschlich zugeschrieben" (German).
- Sources: Often, context contains links to websites where the quote was reported. We collect such external links from templates and from the markup.
- Linked entities: Quotes can be linked to entities such as other persons or organisations. We collect such links from templates and the markup.

---

[12] For more detailed statistics about Wikiquote language editions see: https://wikistats.wmcloud.org/display.php?t=wq.

- Language: While the Wikiquote pages are written in specific languages, their quotes can be written in their original language or translated. For this reason, we use language detection to designate the language of a quote and do not rely on the language of the page tree.
- Sentiment: We detect the sentiment of each quote mention (*positive, negative* or *neutral* with a score between 0 and 1) using XLM-RoBERTa-Twitter, an XLM-RoBERTa model trained on ~198 M multilingual tweets [3].
- Identity links: To establish `owl:sameAs` links between the QuoteKG entities, Wikidata and DBPedia, we use Wikidata's sitelinks[13].

For all persons and entities identified during this process, we extract additional information regarding their labels and types from DBpedia and Wikidata.

4.4 Cross-Lingual Alignment of Quote Mentions

After identifying and enriching quotes, we need to detect which of them represent mentions of the same quote said by a person of public interest. This task of cross-lingual alignment of quote mentions is treated as a clustering task at the end of which each cluster represents a quote with a set of mentions.

In detail, the clustering task is performed for each person in isolation. Given a person's quote mentions in a set of languages, we aim at creating clusters of highly similar mentions. To derive a similarity between two mentions, potentially from different languages, we compute the cosine similarity of sentence embeddings derived from the mentions' texts. As an embedding model, we use a language-agnostic transformer model pre-trained on millions of multilingual paraphrase examples in more than 30 languages, namely XLM-RoBERTa [8]. The ability of such models to adapt to previously unknown languages has been shown in [16]. Given such embeddings and the cosine similarity function, clustering is performed by detecting communities of quotes using a nearest-neighbour search. To do so, we chose UKPLab's Fast Clustering algorithm[14] that is optimised towards efficient similarity computations of our embeddings.

To aggregate the sentiments of all mentions in a cluster, we take the most frequent sentiment category and average over the scores of that category.

4.5 RDF Triples Creation

After identification of quotes and their contexts and cross-lingual alignment, we transform them into RDF triples following the schema presented in Fig. 1.

4.6 Implementation

We use the MWDumper[15] to process the Wikiquote XML dumps and parse the single pages given in the Wikipedia markup using the Bliki engine[16]. For lan-

---

[13] https://www.wikidata.org/wiki/Help:Sitelinks/en-gb.
[14] https://github.com/UKPLab/sentence-transformers/blob/master/examples/applications/clustering/.
[15] https://www.mediawiki.org/wiki/Manual:MWDumper.
[16] https://github.com/axkr/info.bliki.wikipedia_parser.

Table 2. Statistics of selected languages in *QuoteKG*.

Language	Persons	Quotes	Mentions	Mentions with contexts
English	19,073	267,740	271,541	193,848
Italian	18,803	145,235	146,103	48,107
German	3,461	16,012	16,441	4,330
Croatian	2,707	11,023	12,965	2,045
Welsh	239	461	508	247
All Languages	69,467	880,878	961,535	411,912

guage detection and time expression extraction, we use the langdetect[17] and dateparser[18] libraries. The Fast Clustering algorithm was run with a cosine similarity threshold of 0.8. The creation of knowledge graph triples and their serialisation is done via the RDFLib library[19]. The Java implementation of the dumper and the Python code for cross-lingual alignment and knowledge graph creation are publicly available on GitHub[20].

5 Statistics, Evaluation, Examples and Web Interface

In this section, we first provide general statistics of *QuoteKG*, evaluate the cross-lingual alignment and present example queries.

5.1 Statistics

In total, *QuoteKG* contains $880,878$ quotes with $961,535$ quote mentions. For $411,912$ mentions, context is available. Table 2 provides detailed statistics for selected languages. *QuoteKG* covers both high-resource languages such as English ($271,541$ quote mentions from $19,073$ persons) and Italian ($146,103$ quote mentions from $18,803$ persons), as well as low-resource languages such as Welsh (508 quote mentions from 239 persons).

5.2 Evaluation of the Cross-Lingual Alignment

We evaluate the quality of the cross-lingual alignment of quote mentions by comparing to a ground truth of correctly clustered mentions. Creating such a ground truth is a tedious process due to the large amount of possible clusterings

---

[17] https://pypi.org/project/langdetect/.
[18] https://github.com/scrapinghub/dateparser/.
[19] https://github.com/RDFLib/rdflib.
[20] https://github.com/tkuculo/QuoteKG.

and the number of pairwise comparisons[21]. We have selected eight persons with quotes in English, German and Italian and manually clustered their mentions. Ground truth clusters were then compared to the *QuoteKG* clusters by viewing the clustering process as a series of decisions for each of the pairs of mentions [30]. For example, we consider three positive pairs for a quote mentioned in three languages: (Mention$_1$, Mention$_2$), (Mention$_1$, Mention$_3$), (Mention$_2$, Mention$_3$).

Table 3 shows the results of this evaluation: Cross-language alignment in *QuoteKG* shows an average precision of 1.0 and an F_1 score of 0.99 for this ground truth data set. Following the imbalance of Wikiquote's coverage described in Sect. 4.1, there is a high number of true negatives, i.e., the majority of quotes are only mentioned once in all Wikiquote language versions. In total, there are only two mentions which are not clustered together but should have been. All the other clusters are correct.

Table 3. Evaluation of cross-lingual alignment for eight selected persons in English, German and Italian. TP: true positives (mention pairs that were correctly clustered together) TN: true negatives (mention pairs that were correctly not clustered together), FP: false positives, FN: false negatives, P: precision, R: recall, F_1: F_1 score.

Person	TP	TN	FP	FN	P	R	F_1
Alan Turing	10	935	0	1	1	0.91	0.95
Alexander the Great	5	491	0	0	1	1.0	1.0
Edward Snowden	6	697	0	0	1	1.0	1.0
Gustav Mahler	1	44	0	0	1	1.0	1.0
Jean-Claude Juncker	4	776	0	0	1	1.0	1.0
Marie Antoinette	4	347	0	0	1	1.0	1.0
Marie Curie	2	251	0	0	1	1.0	1.0
Tom Clancy	1	2849	0	0	1	1.0	1.0
Total	33	6390	0	1	1.0	0.99	0.99

Our ground truth set of manually aligned quote clusters is available on the *QuoteKG* website.

5.3   Example Queries

In this section, we present two example queries demonstrating how to use *Quote-KG* as a collection of quotes and as a resource to conduct research on the misattribution of quotes.

---

[21] When considering a person that has 10 quotes in 5 languages each, there are $\sum_i^{5-1} 10 \cdot i^2 = 1,000$ possible pairwise comparisons.

```
SELECT ?Person (COUNT(?quote) AS ?NumberOfQuotes) WHERE {
  ?quote a so:Quotation ;
    so:spokenByCharacter [
        skos:prefLabel ?Person ] .
} GROUP BY ?Person
ORDER BY DESC(COUNT(?quote))
```

Listing 1.1. SPARQL query: Persons with the most quotes.

QuoteKG as a Collection of Quotes and their Originators. Listing 1.1 shows a SPARQL query that returns the five persons with the most quotes in *QuoteKG*. Table 4 shows these persons together with the number of quotes. Without surprise, the persons with most quotes are philosophers and writers, including Friedrich Nietzsche and Oscar Wilde, plus Albert Einstein, known for many (misattributed) quotes [28].

Table 4. The first five results of the query in Listing 1.1.

?Person	?NumberOfQuotes
Friedrich Nietzsche	2,530
Oscar Wilde	1,786
Albert Einstein	1,627
Donald Trump	1,610
Johann Wolfgang von Goethe	1,537

Verification of Quotes. Misinformation on the Internet has become an increasingly important problem and requires methods that classify the veracity of information [33] and benefit from knowledge graphs such as ClaimsKG that provide annotated and erroneous facts [32]. While ClaimsKG provides wrong claims stated by persons extracted from fact-checking sites, *QuoteKG* has quotes labelled as wrongly attributed to persons, thus a different type of misinformation. The query shown in Listing 1.2 returns quotes of Albert Einstein that are marked as misattributed in *QuoteKG* (see Table 5), together with context information. Such context information can be a valuable resource for explaining misattribution in the case of quotes.

5.4   Web Interface

On the *QuoteKG* website, we offer a SPARQL endpoint and a demo Search & Demo interface where users can search for specific persons and display their quotes in selected languages. An example of this interface is shown in Fig. 7 which display Portuguese and English quotes of Johann Wolfgang von Goethe.

```
SELECT ?Text (SAMPLE(?contextText) AS ?contextTexts)
(SAMPLE(?source) AS ?Source) WHERE {
  ?quote so:spokenByCharacter [
    skos:prefLabel "Albert␣Einstein" ] ;
  qkg:isMisattributed true ; qkg:hasMention ?mention .

  ?mention so:text ?Text ;
    qkg:hasContext [
    qkg:contextText ?contextText ; so:source ?source
    ] .
} GROUP BY ?Text
```

Listing 1.2. SPARQL query: Quotes misattributed to Albert Einstein and their contexts.

Table 5. Two results of the query in Listing 1.2, returning quotes that were misattributed to Albert Einstein. Texts are shortened for brevity here.

?Text	?contextTexts	?Source
Everything is energy and that's all there is to it … It can be no other way. This is not philosophy. This is physics	There's no evidence that Einstein ever said this	http://quoteinvestigator.com/2012/05/16/everything-energy/
If the facts don't fit the theory, change the facts	The earliest published attribution of this quote to Einstein found on …, but no source to Einstein's original writings is given …	http://books.google.com/books?id=…

6   Availability

Availability: The *QuoteKG* website[22] provides access to a description of *QuoteKG* and its schema, to the SPARQL endpoint, to data downloads and will provide a canonical citation to this paper. *QuoteKG* is licensed under the Creative Commons Attribution Share Alike 4.0 International[23] license. Persistent access to the *QuoteKG* triple files is provided through an upload to the zenodo repository[24]. The code for the creation of *QuoteKG* is publicly available on GitHub[25] and is licensed under the MIT license[26].

Sustainability Plan: To account for updates in all Wikiquote language editions, we plan to release new versions of *QuoteKG* twice a year. To do so, we

---

[22] https://quotekg.l3s.uni-hannover.de.
[23] https://creativecommons.org/licenses/by-sa/4.0/legalcode.
[24] https://zenodo.org/record/4702544.
[25] https://github.com/tkuculo/QuoteKG.
[26] https://opensource.org/licenses/MIT.

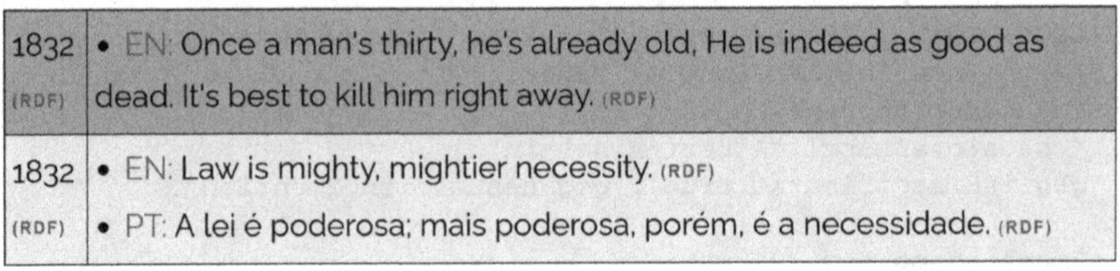

| 1832 | • EN: Once a man's thirty, he's already old, He is indeed as good as |
| (RDF) | dead. It's best to kill him right away. (RDF) |

| 1832 | • EN: Law is mighty, mightier necessity. (RDF) |
| (RDF) | • PT: A lei é poderosa; mais poderosa, porém, é a necessidade. (RDF) |

Fig. 7. An example of the Search & Demo interface showing two quotes of Johann Wolfgang Goethe which are available in Portuguese or English. The sentiment of quotes is indicated by colour (red: negative, green: positive). (Color figure online)

deploy a script that covers the entire pipeline depicted in Fig. 3, from the download of Wikiquote dumps in each language until the creation of triples.

Adherence to Standards: *QuoteKG* is modeled through the Resource Description Framework. Its schema is an extension of *schema.org*. We provide a machine-readable description of *QuoteKG* using the VoID vocabulary[27]. *QuoteKG* adheres to the Linked Data Principles: resources can be looked up through their URIs and they are interlinked with Wikidata and DBpedia.

7 Related Work

In this section, we give an overview of other corpora and knowledge graphs containing quotes, other usages of Wikiquote and about cross-lingual alignment.

Quote Corpora. Many collections of quotes have been created and maintained, mainly mono-lingual and without semantic annotations. Since the release of its first edition in 1941, The Oxford Dictionary of Quotations [20] aims at providing "the wit and wisdom of past and present" with a focus on the provenance of quotes. Provenance of quotes is also an indispensable criterion in the Book of Fake Quotes [4]. There are few machine-readable monolingual quote collections[28,29] [10,24,34]. These corpora are typically monolingual and extracted from news. Consequently, while they may have a large amount of quotes, they lack a mechanism to ensure societal relevance of quotes as in Wikiquote. As a knowledge graph, *QuoteKG* enables easy access to quotes and rich metadata.

Quotes in Knowledge Graphs. While DBQuote [26] allows user annotations of quotes extracted from Twitter and Wikiquote through an ontology, it only covers two languages (English and Korean) and has not been made available. To the best of our knowledge, *QuoteKG* is the first publicly available knowledge

---

[27] https://www.w3.org/TR/void/.
[28] https://www.kaggle.com/akmittal/quotes-dataset.
[29] https://github.com/JamesFT/Database-Quotes-JSON.

graph of quotes. Consequently, quotes have only been insufficiently covered in the Semantic Web: for example, Wikidata [36] contains less than 400 instances of the class "Phrase"[30] that are attributed to an author or creator – most of them only consisting of few words (e.g., "cogito ergo sum" and "covfefe"). Event-centric knowledge graphs such as EventKG [12] provide an understanding of the human history and world-shaking events. They do not include quotes that complement the deeds of public figures. Many applications based on knowledge graphs (e.g., for exploring the lives of persons of public interest [1,13]) could immediately profit from the inclusion of quotes.

Wikiquote. Until now, Wikiquote has rarely been used as a research corpus, presumably due to the required but tedious extraction process. One example is the work by Buscaldi et al. who manually tagged quotes of the Italian Wikiquote as humorous or not, and used their annotated corpus for training models for humour recognition [5]. Giammona et al. analysed the spread of ancient quotes in today's Web through Wikiquote [9] and Wikiquote was used for training the chatbot Poetwannabe [6]. With *QuoteKG*, we foresee to ease the access to quotes for a wide range of research questions.

Cross-Lingual Alignment. Several studies have shown that different languages share similar statistical properties that can be used to learn cross-lingual alignments between two languages, even without relying on any form of bilingual supervision [7]. While most works and datasets address bilingual alignment [11,15,17], there are only few works on cross-lingual alignment [22]. *QuoteKG* focuses on the specific task of cross-lingual alignment of quote mentions.

8 Conclusion

In this paper, we presented *QuoteKG* – a novel, multilingual knowledge graph of quotes. We have presented the *QuoteKG* schema based on schema.org as well as a pipeline that extracts quotes from the Wikiquote corpus and aligns them across languages. *QuoteKG* is publicly available and includes nearly one million quotes quotes in 55 languages, said by nearly $69,000$ people of public interest.

Acknowledgement. This work was partially funded by H2020-MSCA-ITN-2018-812997 under "Cleopatra".

References

1. Althoff, T., Dong, X.L., Murphy, K., Alai, S., Dang, V., Zhang, W.: TimeMachine: timeline generation for knowledge-base entities. In: Proceedings of the 21th ACM SIGKDD International Conference on Knowledge Discovery and Data Mining, pp. 19–28 (2015)

---

[30] https://www.wikidata.org/wiki/Q187931.

2. Auer, S., Bizer, C., Kobilarov, G., Lehmann, J., Cyganiak, R., Ives, Z.: DBpedia: a nucleus for a web of open data. In: Aberer, K., et al. (eds.) ASWC/ISWC -2007. LNCS, vol. 4825, pp. 722–735. Springer, Heidelberg (2007). https://doi.org/10.1007/978-3-540-76298-0_52

3. Barbieri, F., Anke, L.E., Camacho-Collados, J.: XLM-T: A Multilingual Language Model Toolkit for Twitter. arXiv preprint arXiv:2104.12250 (2021)

4. Boller, P.F., Jr., George, O.J., Jr., et al.: They Never Said It: A Book of Fake Quotes, Misquotes, and Misleading Attributions: A Book of Fake Quotes, Misquotes, and Misleading Attributions. Oxford University Press, Oxford (1989)

5. Buscaldi, D., Rosso, P.: Some experiments in humour recognition using the Italian Wikiquote collection. In: Masulli, F., Mitra, S., Pasi, G. (eds.) WILF 2007. LNCS (LNAI), vol. 4578, pp. 464–468. Springer, Heidelberg (2007). https://doi.org/10.1007/978-3-540-73400-0_58

6. Chorowski, J., Lancucki, A., Malik, S., Pawlikowski, M., Rychlikowski, P., Zykowski, P.: A talker ensemble: the University of Wroclaw's entry to the NIPS 2017 conversational intelligence challenge. In: Escalera, S., Weimer, M. (eds.) The NIPS '17 Competition: Building Intelligent Systems. TSSCML, pp. 59–77. Springer, Cham (2018). https://doi.org/10.1007/978-3-319-94042-7_4

7. Chung, Y.A., Weng, W.H., Tong, S., Glass, J.: Unsupervised Cross-Modal Alignment of Speech and Text Embedding Spaces. arXiv preprint arXiv:1805.07467 (2018)

8. Conneau, A., et al.: Unsupervised Cross-Lingual Representation Learning at Scale. arXiv preprint arXiv:1911.02116 (2019)

9. Giammona, C., Yanes, E.S.: From Print to Digital Texts, from Digital Texts to Print. Indirect Tradition of Latin Classics on the Web. Storie e Linguaggi 5(1) (2019)

10. Goel, S., Madhok, R., Garg, S.: Proposing contextually relevant quotes for images. In: Pasi, G., Piwowarski, B., Azzopardi, L., Hanbury, A. (eds.) ECIR 2018. LNCS, vol. 10772, pp. 591–597. Springer, Cham (2018). https://doi.org/10.1007/978-3-319-76941-7_49

11. Gottschalk, S., Demidova, E.: MultiWiki: interlingual text passage alignment in Wikipedia. ACM Trans. Web 11(1), 6:1–6:30 (2017)

12. Gottschalk, S., Demidova, E.: EventKG-the hub of event knowledge on the web-and biographical timeline generation. Semant. Web 10(6), 1039–1070 (2019)

13. Gottschalk, S., Demidova, E.: EventKG+BT: generation of interactive biography timelines from a knowledge graph. In: Harth, A., et al. (eds.) ESWC 2020. LNCS, vol. 12124, pp. 91–97. Springer, Cham (2020). https://doi.org/10.1007/978-3-030-62327-2_16

14. Haun, M.: How social movements and global events are changing language in 2020 (2020). Accessed 03 Dec 2021

15. Hieber, F.: WikiCLIR: A Cross-lingual Retrieval Dataset from Wikipedia. Universität (2014)

16. Hu, J., Ruder, S., Siddhant, A., Neubig, G., Firat, O., Johnson, M.: XTREME: a massively multilingual multi-task benchmark for evaluating cross-lingual generalisation. In: International Conference on Machine Learning, pp. 4411–4421. PMLR (2020)

17. Jing, Y., Xiong, D., Zhen, Y.: BiPaR: A Bilingual Parallel Dataset for Multilingual and Cross-Lingual Reading Comprehension on Novels. arXiv preprint arXiv:1910.05040 (2019)

18. Keyes, R.: The Quote Verifier: Who Said What, Where, and When. St. Martin's Griffin (2007)

19. Khurana, S.: These 4 Quotes Completely Changed the History of the World. https://www.thoughtco.com/quotes-that-changed-history-of-world-2831970. Accessed 01 Dec 2021

20. Knowles, E.: The Oxford Dictionary of Quotations. Oxford University Press, Oxford (2009)

21. Krämer, A.: Ein Satz mit Folgen (2021). https://www.tagesschau.de/inland/merkel-wir-schaffen-das-109.html. Accessed 01 Dec 2021

22. Liang, Y., et al.: XGLUE: A New Benchmark Dataset for Cross-Lingual Pretraining, Understanding and Generation. arXiv preprint arXiv:2004.01401 (2020)

23. Mushaben, J.M.: Wir schaffen das! Angela Merkel and the European refugee crisis. German Polit. **26**(4), 516–533 (2017)

24. Newell, C., Cowlishaw, T., Man, D.: Quote extraction and analysis for news. In: Proceedings of the Workshop on Data Science, Journalism and Media, KDD, pp. 1–6 (2018)

25. Niculae, V., Suen, C., Zhang, J., Danescu-Niculescu-Mizil, C., Leskovec, J.: QUOTUS: the structure of political media coverage as revealed by quoting patterns. In: Proceedings of the 24th International Conference on World Wide Web, pp. 798–808 (2015)

26. Piao, G., Breslin, J.G.: DBQuote: a social web based system for collecting and sharing wisdom quotes. In: Proceedings of the 5th Joint International Semantic Technology Conference, Poster and Demonstrations (2015)

27. Reucher, G.: Famos Quotes: Why are so many fake? (2021). Accessed 03 Dec 2021

28. Robinson, A.: Did Einstein really say that? Nature **557**(7703), 30–31 (2018)

29. Sánchez-Rada, J.F., Iglesias, C.A.: Onyx: a linked data approach to emotion representation. Inf. Process. Manag. **52**(1), 99–114 (2016)

30. Schütze, H., Manning, C.D., Raghavan, P.: Introduction to Information Retrieval, vol. 39. Cambridge University Press, Cambridge (2008)

31. Sims, M., Bamman, D.: Measuring information propagation in literary social networks. In: Proceedings of the 2020 Conference on Empirical Methods in Natural Language Processing (EMNLP) (2020)

32. Tchechmedjiev, A., et al.: ClaimsKG: a knowledge graph of fact-checked claims. In: Ghidini, C., et al. (eds.) ISWC 2019. LNCS, vol. 11779, pp. 309–324. Springer, Cham (2019). https://doi.org/10.1007/978-3-030-30796-7_20

33. Thorne, J., Vlachos, A.: Automated fact checking: task formulations, methods and future directions. In: Proceedings of the 27th International Conference on Computational Linguistics, pp. 3346–3359 (2018)

34. Vaucher, T., Spitz, A., Catasta, M., West, R.: Quotebank: a corpus of quotations from a decade of news. In: Proceedings of the 14th ACM International Conference on Web Search and Data Mining, pp. 328–336 (2021)

35. Viala-Gaudefroy, J., Lindaman, D.: Donald Trump's 'Chinese Virus': The Politics of Naming (2020). Accessed 03 Dec 2021

36. Vrandečić, D., Krötzsch, M.: Wikidata: a free collaborative knowledge base. Commun. ACM **57**(10), 78–85 (2014)

Stunning Doodle: A Tool for Joint Visualization and Analysis of Knowledge Graphs and Graph Embeddings

Antonia Ettorre(✉) [ID], Anna Bobasheva [ID], Franck Michel [ID], and Catherine Faron [ID]

Université Côte d'Azur, CNRS, Inria, I3S, Sophia Antipolis, France
{aettorre,bobasheva,fmichel,faron}@i3s.unice.fr

Abstract. In recent years, the growing application of Knowledge Graphs to new and diverse domains has created the need to make these resources accessible and understandable by users with increasingly diverse backgrounds. Visualization techniques have been widely employed as means to facilitate the exploration and comprehension of such data sources. Moreover, the emerging use of Knowledge Graph Embeddings as input features of Machine Learning methods has given even more visibility to this kind of representation, but raising the new issue of understandability and interpretability of such embeddings. In this paper, we show how visualization techniques can be used to jointly explore and interpret both Knowledge Graphs and Graph Embeddings. We present *Stunning Doodle*, a tool that enriches the classical visualization of Knowledge Graphs with additional information meant to enable the visual analysis and comprehension of Graph Embeddings. The idea is to help the user figure out the logical connection between (1) the information captured by the Graph Embeddings and (2) the structure and semantics of the Knowledge Graph from which they are generated. We detail the use of *Stunning Doodle* in a real-world scenario and we show how it has been helpful to interpret different Graph Embeddings and to choose the most suitable with respect to a specific final goal.

Keywords: Knowledge Graphs Visualization · Graph Embeddings

1 Introduction

During the last decade, the adoption of Knowledge Graphs (KGs) in multiple domains has increased steadily such that more and more projects rely on this kind of representation to store their data without compromising the semantics they bear. The fame of KGs has kept growing even more as they started to be used as information sources for many AI-powered applications in the most diverse fields, e.g. education [7,10], medicine [19] and finance [13]. One of the reasons for their increasing success is the possibility to easily employ them as input features for several Machine Learning methods by using a low-dimensional representation

© The Author(s), under exclusive license to Springer Nature Switzerland AG 2022
P. Groth et al. (Eds.): ESWC 2022, LNCS 13261, pp. 370–386, 2022.
https://doi.org/10.1007/978-3-031-06981-9_22

of such KGs, obtained through the graph embedding process. The diversity of circumstances in which Graph Embeddings (GEs), and therefore KGs, can be used opened the way to the exploitation of these data sources by users from various communities and with diverse backgrounds. In this context, two major needs for such users arise: *(i)* exploring and understanding the content and the structure of KGs and *(ii)* analyzing and interpreting the information captured by their GEs.

In recent years, the need for solutions to allow simple and straightforward exploration of KGs has stimulated the development of a plethora of visualization tools for such data sources. Indeed, visualization techniques are recognized as one of the main means to provide an immediate and simple understanding of complex concepts and structured data. Ideally, users could be able to gain in-depth understanding of the structure of a KG by analyzing the sheer ontologies and vocabularies on which the KG is based, and when it comes to RDF-based KGs, they could explore its content by means of SPARQL queries. Nevertheless, studying ontologies and vocabularies and writing SPARQL queries can be rather cumbersome tasks, especially when dealing with very large KGs. Moreover, these operations can only be carried out by practitioners of Semantic Web languages. Visualization techniques can help and simplify the exploration and understanding of KGs by non-expert users, offering easy-to-use interfaces, advanced statistics, and interaction functionalities. However, the visual exploration of KGs is a non-trivial task mainly due to the amount and heterogeneity of the information possibly described by KGs.

The comprehension of the information captured by GEs is an even more challenging process. Indeed, GEs are computed using "black-box" Machine Learning (ML) techniques that translate each element in the graph into a low-dimensional vector. Even though the algorithmic process to compute embeddings is well understood, a relation between the characteristics and role of the element in the graph and its vector representation in the embedding space cannot be established with certainty. In other words, multiple questions cannot be answered easily, such as:

- What do my embeddings represent?
- How are they related to the structure and semantics of my KG?
- How can I improve my embeddings to be better suited to my downstream application?

Recently, several research efforts have been made in this direction to start making sense of the information captured by GEs. Some approaches propose explainable models for computing GEs [12] or implement explanation strategies for specific embedding models [22]; while others propose methods to verify whether some specific piece of knowledge represented in a KG is actually encoded and captured by its GEs [9].

In this work, we aim to tackle this issue from a different point of view. We think that, as for KGs, the information borne by GEs could be explored and unveiled through the use of visualization techniques that would favor the discovery of the logical connection between the graph and its embeddings. Our goal is therefore to provide a visualization tool supporting the analysis and decoding of the information captured by KGEs by unveiling the relationship between, on one hand the structure and semantics of the KG, and on the other hand the KGEs generated from it.

To achieve this goal, in this paper we present *Stunning Doodle*, a tool designed for the visualization of RDF-based KGs and GEs. *Stunning Doodle* first provides a visualization of the graph to be analyzed, offering a rich overview of both the structure and the semantics of the data. We believe that this visualization should allow users to gain a general and immediate understanding of the displayed KG while presenting a complete and detailed view of each of its components. More interestingly, this visualization is enriched by connecting the nodes in the graph with the corresponding GEs to be analyzed. It enables to select a node in a KG and visualize its neighborhood in the embedding space, and conversely to pick any of its neighbors in the embedding space and visualize their links in the KG. We argue that the joint visualization of GEs and the KG from which they are generated, with its structure and semantics, will help users, may they be non-expert users, RDF experts or Machine Learning experts, to gain interesting insights into the information captured during the embedding process.

The remainder of this paper is organized as follows. In Sect. 2 we present our KGEs analysis tool, while in Sect. 3 we demonstrate how our tool can be used and useful in two real-world use-cases. In Sect. 4 we review related work. Finally, conclusions and future work are discussed in Sect. 5.

2   Stunning Doodle

Stunning Doodle has been developed to fill the gap in the field of visual analysis of KGEs. Its main goal is to provide users with an advanced visualization of RDF graphs, enriched with information extracted from the GEs generated from those graphs. To achieve this goal, *Stunning Doodle* offers two main functionalities: (1) the visualization and navigation of RDF graphs and, (2) for each node, the enrichment of such visualization with the addition of its neighborhood in the embedding space, w.r.t. a chosen similarity measure. Additionally, *Stunning Doodle* presents semantic-based filtering and customization functionalities for nodes and links which make the visualization clearly understandable even for users with no familiarity with RDF and SPARQL concepts. To improve the readability and comprehension of the displayed KG, *Stunning Doodle* implements a partial visualization of the graph and allows users to navigate and explore it incrementally.

2.1   Knowledge Graphs Visualization

The first function of *Stunning Doodle* is the dynamic visualization of a KG using a regular graph layout. The process starts with the upload of the RDF graph file in Turtle syntax. Once the file is uploaded, the graph will be displayed as shown in Fig. 1. On the left side of the page, under the "Upload Graph" menu, three menus provide relevant information about the graph: the list of the declared namespaces with their prefixes (1), and the legend, and advanced customization

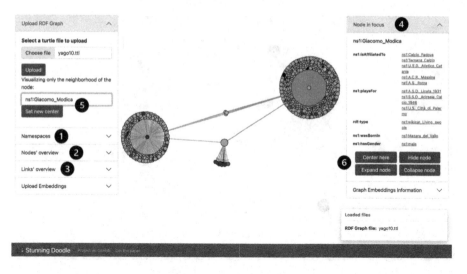

Fig. 1. Screenshot of the partial visualization of YAGO3 illustrating *Stunning Doodle'* exploration capabilities.

options for nodes (2) and links (3) that will be detailed later in this section. The visualization is interactive: nodes can be moved and zoom and spanning functionalities are available. Nodes can be selected by clicking on them, and the triples associated with the currently selected node are listed in the component "Node in focus" (4) on the right. In the example of Fig. 1, which shows an extract of YAGO3 [14], the selected node is the largest one at the bottom of the graph (ns1:Giacomo_Modica) and the triples for which this node is a subject are listed on the right. Each object in this list can be clicked to select the corresponding node in the graph.

Graph Exploration Interface. One of the main characteristics of *Stunning Doodle* is the ability to display nodes incrementally, thanks to the graph exploration system.

To deal with the large number of nodes possibly described in a KG, existing visualization tools employ different strategies: *(i)* relying on clusterized views which group similar (or close) nodes together [18], *(ii)* visualizing nodes incrementally based on the displayed area [3,4], or *(iii)* showing only the nodes that are considered to be more relevant based on diverse statistical metrics [20]. In *Stunning Doodle* we opted for a different approach: a user-guided incremental exploration. Indeed, through our tool, the user decides which nodes to visualize according to their interests and needs. This choice is motivated by the fact that different users can have different, possibly very specific requirements, e.g. analyzing only the facts related to individuals of a given type while disregarding all the other possible predicates; and these requirements do not always correspond to the standard clusterization criteria or relevance metrics. As Fig. 1 shows it,

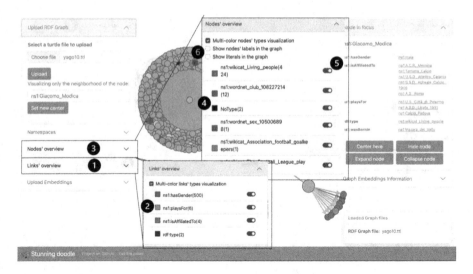

Fig. 2. *Stunning Doodle* interface with the available customization options. (Color figure online)

after the upload of an RDF graph, only one node, randomly selected among all the nodes in the graph, is displayed alongside its close neighborhood, i.e. nodes at one-hop distance from the main node. Users can change the displayed node by typing the URI of the node they want to visualize in the component "Upload Graph" and setting it as a new center (5). Users can also navigate the graph starting from a node of interest by using the buttons in the "Node in focus" component (6). Indeed, once a node is selected, users can decide to either *(a)* center the graph on that node, causing only the selected node and its neighborhood to be visualized, while hiding the other nodes and links; *(b)* hide the node, producing the removal of that node together with the nodes that are connected only to it; *(c)* expand the node, i.e. adding the direct neighbors of the node to the visualization; and *(d)* collapse the node, i.e. removing all the nodes that are only connected to that node while keeping the node itself. For example, the visualization in Fig. 1 has been obtained by centering the graph on the node ns1:male at first, and then expanding the selected node (ns1:Giacomo_Modica) to display its neighborhood.

Filtering and Customizing Nodes and Edges. Another helpful feature of *Stunning Doodle* is the possibility of filtering nodes based on their types (values of property rdf:type) and customizing nodes and links colors. These features help the user to easily recognize the nodes and links they want to analyze and to focus only on them while removing information that is irrelevant for their exploration needs.

In Fig. 2 we show an example customized visualization of an RDF graph and the settings of the visualization modes. These are optional and can be activated

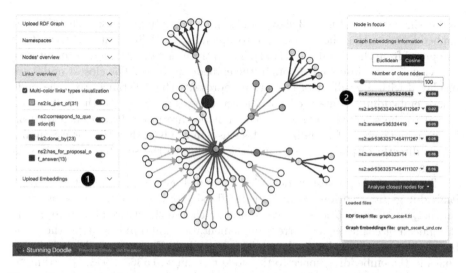

Fig. 3. Screenshot of *Stunning Doodle* showing closest nodes in the embedding space. (Color figure online)

by selecting the corresponding options in the "Nodes' overview" and "Links' overview" menus. In the "Links' overview" menu (1), all the possible predicates are listed, each one with the number of edges of that type currently displayed on the graph and the corresponding colors which can be customized by the user (2). Similarly, in the "Nodes' overview" menu (3), the types of all the displayed nodes are listed with the count of occurrences of every type (4). For each type, a filter is added to hide/show the corresponding nodes (5). In the same component, we find two additional options that allow the user to enable and disable labels and literals visualization (6). Once activated, the former will display labels, i.e. values of properties `rdfs:label` and `skos:prefLabel`, next to the nodes subject of these properties; while the latter will add literals as leaves in the graph.

2.2   Graph Embeddings Visual Analysis

In addition to the dynamic visualization of a KG, the key functionality provided by *Stunning Doodle* which represents a major step forward when compared with the state-of-the-art tools described in Fig. 4 is the possibility of having a first, simple visual analysis of the GEs computed from the visualized KG.

As explained in Fig. 1, the recent success of KGs is mainly due to the growing number of AI applications relying on KGs through the use of KGEs. Unfortunately, the main stumbling block to extensive utilization of such representation is its difficult interpretability.

With *Stunning Doodle*, we take a first step in the understanding of KGEs through the joint visualization of both KGEs and the KG from which they are generated. More precisely, *Stunning Doodle* enables to visualize, for each node, its closest nodes in the embedding space, according to both euclidean and cosine distance. Figure 3 shows an example of KGEs analysis. The "Upload Embeddings" menu allows the user to upload a CSV file containing the GEs computed from the visualized KG (1). Then, the user can select a node of interest and visualize its closest neighbors in the embedding space. The closest nodes are displayed with a gradient of color that represents their distance from the node whose embedding is analyzed, i.e. darker nodes are closer (in the embedding space) to the selected node while lighter nodes are more distant. If a relation between any couple of visualized nodes exists in the KG, then the corresponding link is directly displayed in the graph according to the selected customization settings.

The list of the closest nodes with their distance is shown in the "Graph Embeddings Information" menu on the right side of the page (2). Together with this list, additional options allow the user to choose the desired distance metric (Euclidean or cosine in the current implementation), and to customize the number of closest nodes to be displayed. The example in Fig. 3 shows the 100 closest nodes in the embedding space to the node ns1:answer536324943, according to cosine distance. The node currently selected (biggest node) is the node that we are analyzing and from which the distances are computed. This is evident by the fact that its URI is the first item in the list of nodes and it is also the darkest node in the graph, as the distance from itself is 0. The different shades of green of each node clearly highlight which nodes among the 100 visualized are closer, while the links show how they are connected in the KG.

While visualizing a node's neighbors in the embedding space, it is still possible to access the functionalities for navigating in the KG through the buttons in "Node in focus". Therefore, any displayed node can be expanded to show additional nodes linked to it in the KG even if they are not close in the embedding space. Naturally, any new node can be further expanded to visualize the desired portion of the KG. All the links and the displayed nodes whose embeddings are not close to the initial node will be visualized according to the selected customization options in "Nodes' overview" and "Links' overview". To switch back to the simple view of the KG it is sufficient to recenter the graph on any of the displayed nodes.

To sum up, *Stunning Doodle* enables the user to understand in a glance which nodes of a KG are considered to be similar in the embedding space, while keeping track of their connections in the KG. This permits to gain immediate insights on the information captured by KGEs, e.g. which predicates have the highest impact or what connectivity patterns are more taken into account during the embedding process.

2.3   Software Design and Limitations

For the sake of simplicity and flexibility of use, *Stunning Doodle* has been implemented as a lightweight web application relying on Python and Javascript, respectively for back-end and front-end. Its setup is fairly simple as it requires only to create a Python virtual environment and it allows the easy deployment on a server to be accessed remotely by multiple users through a web browser.

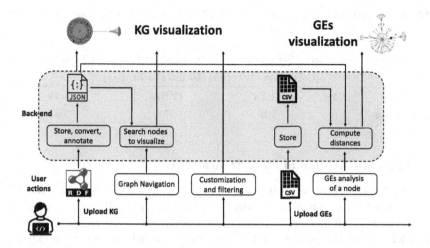

Fig. 4. Schema of the processing pipeline implemented in *Stunning Doodle*.

Stunning Doodle uses as input for the graph visualization a RDF Knowledge Graph stored in a file using Turtle syntax. For the analysis of the GEs, a CSV file containing the nodes' URIs associated with the corresponding embeddings is required. Figure 4 illustrates the processing pipeline of *Stunning Doodle* and shows how the interactions with users are handled. Firstly, users need to upload the RDF graph they want to visualize. This graph is parsed and converted into JSON, enriching each node description with information about its types and labels to enable on-the-fly customization of the visualization based on this information. When users navigate the graph by re-centering it or expanding new nodes, a remote request to the back-end is performed to compute and retrieve the new list of nodes to visualize. Since the actions of customization and filtering concern only the nodes and edges currently visualized, they do not require any remote request to the server. To enable the functionalities of analysis of GEs, a CSV file storing such embeddings must be uploaded by the user and stored on the server. At this point, users can analyze the embeddings of any displayed node, by visualizing its closest nodes in the embedding space. When this action is executed, the back-end is queried to compute the list of the closest nodes with their distances. After mapping the closest embeddings with their corresponding nodes and computing the edges connecting them, the results are returned to the front-end that displays the obtained subgraph.

The main limitation of *Stunning Doodle* is represented by the size of the KG to be analyzed. Indeed, the larger the KG and, consequently, the embeddings file are, the longer is the time needed to upload such files, to search for the nodes to be displayed and to compute the distances between embeddings. Table 1 reports the time needed to execute the main actions provided by *Stunning Doodle* on the graph shown in Fig. 1 and Fig. 2, which contains 920.435 triples describing 124.982 nodes. These timings were obtained running *Stunning Doodle* on a MacBook Pro with 2,8 GHz Quad-Core Intel Core i7 and 16 GB RAM.

Table 1. Execution time with number of displayed nodes when performing actions for the analysis of a subset of YAGO3.

	Load graph	Set center (ns1:male)	Expand node (ns1:AC_Sparta_Prague)	Closest nodes
Time	120 s	21,7 s	12,03 s	5,4 s
# nodes	–	502	271	100

2.4 Software Availability and Reusability

Stunning Doodle is made available as an open-source software under the MIT License. The tool is identified by means of a DOI provided through Zenodo [8] to improve its accessibility and citability. Moreover, *Stunning Doodle* source code is publicly accessible on the GitHub repository[1], which also includes extensive documentation and examples to guide the users. We plan on keeping improving *Stunning Doodle* with new functionalities and offering best-effort support to future users in case of bugs or issues of any kind. Furthermore, we welcome any feedback, idea or contribution by the community.

3 Use Cases: The OntoSIDES Scenario

To demonstrate the usefulness of *Stunning Doodle* and the added value of the functionalities it implements, we describe hereafter the two main use cases that motivated the development of such a tool. Both use cases stem from the *SIDES* real-world project in the field of e-education which aims to provide highly intelligent learning services to the medical students in France. The SIDES project involves the manipulation of OntoSIDES [17], a large KG describing the French higher education system for medical studies, with universities, students, learning material, and interactions between them. The graph includes in total more than 9.2 billion triples. To be able to exploit such an amount of information in an AI-powered downstream application, a first approach would be to rely on KGEs, as done in [10]. This scenario highlights two common needs for researchers and engineers working on the project:

1. understanding the OntoSIDES KG, its content and structure;
2. analyzing and comparing GEs generated from it.

In the following use cases, we considered a subgraph of OntoSIDES including only answers to questions of the *pediatrics* medical specialty. The extracted subgraph contains in total almost 650.000 triples.

3.1 Understanding a Knowledge Graph

We tested *Stunning Doodle* for the visualization and comprehension of the KG in the SIDES scenario. In this case, users are only aware of the high-level concepts

---

[1] https://github.com/antoniaettorre/stunning_doodle.

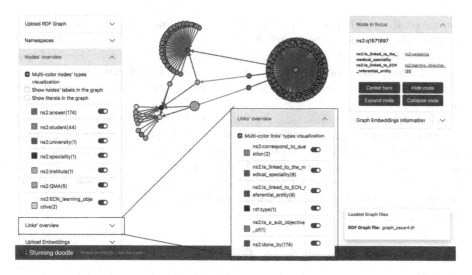

Fig. 5. Screenshot of *Stunning Doodle* showing the basic entities and relations in the OntoSIDES graph. (Color figure online)

that are supposedly defined in OntoSIDES and they need to understand what the described entities are, how they are modeled in the KG and how they are linked to each other through predicates. Expert users are normally able to gain this knowledge by running several prototypical SPARQL queries, but the use of *Stunning Doodle* facilitates this task by retrieving and displaying the same information only through a few clicks, and, at the same time, makes the KG accessible also to users with no expertise in SPARQL.

Once the graph file is uploaded, users can choose a node from which to start the graph exploration. This allows users to expand only the nodes that are relevant for them, displaying only the needed information. After expanding a few nodes, users should be able to visualize a small graph containing all the main elements of interest, i.e. questions, students, answers, institutes, and their links. Thanks to the "Nodes' overview" menu, users are able to distinguish the main entities in a glance, based on their colors. Considering the visualization (Fig. 5), it is evident that instances of *answers* (orange nodes) are directly linked to *questions* (spring green) and *students* (in gray). Moreover, it is possible to see that questions are associated with *specialties* (navy blue nodes) and *learning_objectives* (yellow). Taking a look at the "Links' overview" menu, predicates can also be easily discriminated in the graph thanks to their colors. Moreover, connections between distant nodes can be quickly identified to be possibly used as property paths in SPARQL queries. For example, to analyze which topics a student worked on, it is sufficient to follow the path going from a gray node (student) to a yellow node (*learning_objectives*), therefore, chaining the properties *done_by* (fuchsia line), *correspond_to_question* (pink), *is_linked_to_ECN_referential_entity* (brown). Additionally, few statistical observations on the number and types of entities can be done from the visualization. In the example in Fig. 5, the "expanded" student gave 174 answers, the displayed university has 44 students, all the questions are linked to the same medical specialty. Finally, each node can be analyzed in detail, by selecting the node and looking at the menu "Node in focus" which displays the triples associated with that node.

Thanks to its characteristics, *Stunning Doodle* proved to be not only useful but fundamental to gain a first, global understanding of the content and organization of the OntoSIDES KG. Its navigation systems, filtering functionalities, and customization capabilities allow the user to explore the graph step by step, focusing at each moment only on the pieces of information of interest for them. We claim that those features are very helpful for the understanding of Knowledge Graphs in any context.

3.2   Analyzing and Comparing GEs

The GEs analysis provided by *Stunning Doodle* allows to gain insights into the information encoded in the GEs and, therefore, to assess their meaningfulness for their final use. In particular, *Stunning Doodle* can be used for comparing GEs computed through different embedding models with several hyper-parameters settings, with the final goal of identifying and tuning the model generating the most meaningful embeddings.

In the framework of the SIDES project, GEs computed from OntoSIDES have been used as input features for a ML model designed to predict students' performance on medical questions [10]. In this context, (1) interpreting the information captured by GEs, (2) identifying the best model for their computation, and (3) tuning the hyper-parameters to obtain a meaningful representation of the nodes are crucial tasks.

In the following, we show how *Stunning Doodle* has been used to carry out these tasks, by analyzing the GEs generated from the OntoSIDES subgraph described in Sect. 3.1. We inspect and compare two sets of GEs, both computed using the *node2vec* model [11]. In the first case, the embeddings have been computed considering the graph directed (i.e. it can be traversed only going from subjects of triples to objects), while in the second case embeddings are obtained considering the graph undirected (i.e. links are bidirectional, it is possible to go from objects to subjects). In both cases, we study the embeddings of a student node (stu81235) to figure out which nodes are considered to be similar from the embeddings point of view and, therefore, what information embeddings can capture from the KG. After uploading the graph and embeddings' files, we select the node corresponding to ns2:stu81235 and we choose to visualize its closest nodes in the embedding space.

GEs from a Directed Graph. Figure 6 shows the result of visualizing the 100 nodes with the shortest euclidean distance (in the embedding space) from the node ns2:stu81235. The darkest node is the node ns2:stu81235 itself since its distance from itself is 0. Immediately, it occurs from the visualization that

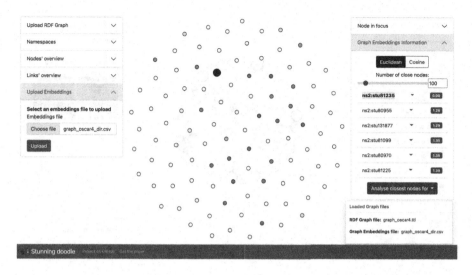

Fig. 6. Closest nodes in the embedding space to the node `ns2:stu81235` with GEs computed from a directed graph. (Color figure online)

its closest nodes are not directly linked between each other and, based on the gradient color, some of them are much closer than others. Looking at the nodes' list in the menu "Graph Embeddings Information", we can see that all the closest nodes are other instances of the student class. Therefore, we can assume that during the embedding process, nodes with the same type are recognized as similar, ending up close in the embedding space.

To investigate why some students are closer to `ns2:stu81235` than others, we can "expand" a few nodes to find connections among them. Figure 7 shows that the closest nodes (darkest color) are connected to the same university (fuchsia node) as `ns2:stu81235`. This highlights the fact that the link with the university has a high impact during the computation of the embeddings of a student's node. Nevertheless, we can notice that not all the students registered at the same university as `ns2:stu81235` are among its closest nodes (gray nodes in Fig. 7), meaning that there are other factors affecting the similarity between embeddings. Repeating the same analysis for other nodes of type `ns2:student` led to the discovery of similar patterns, i.e. all the analyzed nodes resulted to be similar to other students attending the same university. Thanks to these observations, we can conclude that, in this case, GEs are able to encode the information about the node type and the university for students' nodes.

GEs from an Undirected Graph. Figure 8 shows the closest nodes to `ns2:stu81235` for the embeddings computed from the undirected graph. The difference with the GEs computed in the previous case is immediately visible. Firstly, by looking at the list of the closest nodes we discover that for undirected embeddings the type of the node does not play an important role in the embedding process. Indeed, the closest nodes are of various types, including

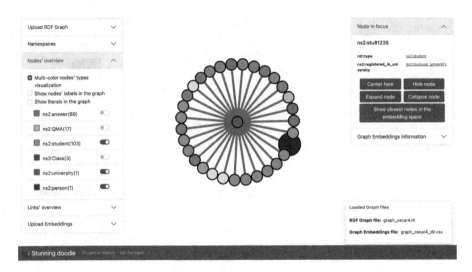

Fig. 7. Links between `ns2:stu81235` and its closest nodes in the embedding space. (Color figure online)

`ns2:answer` and `ns2:action_to_answer`. From the visualization, it is obvious that the connections between nodes assume a much more important role, as the closest nodes in the embedding space result to be the ones that are close in the KG as well (at 1-hop or 2-hops distance). On the other side, we notice that the opposite implication does not hold. Indeed, when we expand the node `stu81235` to display all the direct neighbors in the KG as done in Fig. 9, we can see that there are additional nodes which were not displayed before as they are not close in the embedding space (answers in orange). Therefore, we can assume that the distance in the KG, though important, is not the only parameter taken into account during the embedding process. Same observations can be done for other nodes of type `ns2:student`.

Finally, uniquely through the joint visualization of the distances in the embedding space and the KG structure, we can draw interesting conclusions on the meaning of GEs and, thus, identify the best settings to be used for the embedding computation. In our case, we discovered that GEs computed from directed graphs can capture semantic information from the KG, such as the nodes' type and the university associated with each student; while GEs obtained from undirected graphs tend to summarize the graph structure. As a consequence, if our final goal is the prediction of students' performance [10] the information about the attended university is likely relevant and GEs computed from the directed graph are more meaningful.

4   Related Work

Despite the growing attention dedicated to the issues of interpretability and understandability of KGEs, to the best of our knowledge, no work has been

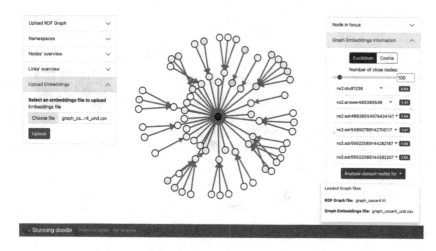

Fig. 8. Closest nodes in the embedding space to the node `ns2:stu81235` with GEs computed from an undirected graph.

done towards the visual analysis and comprehension of such representations. Yet, some solutions have been developed for the similar task of sentences and words embeddings visualization. *TensorFlow Projector*[2] allows users to upload and visualize their embeddings to help the comprehension of the information they summarize. Through this tool, it is possible to visualize the uploaded embeddings in a 3D space by using dimensionality reduction techniques. Moreover, the tool permits, for each element, to list its closest neighbors in the embedding space with the final goal of providing insights on the meaning of the computed embeddings. Although *TensorFlow Projector* is general enough to visualize any kind of embedding, i.e. word, sentence or graph, its usefulness in understanding KGEs is limited by the absence of KG-specific functionalities, such as displaying the graph structure or considering semantic information.

Concerning the visual analysis of KGs, multiple tools have been developed throughout the years. Recent research efforts [1,6] focused on listing, analyzing, and comparing existing Linked Data (LD) visualization tools able to deal both with KG schema and data. The majority of the available tools rely on a graph-based visualization and present several commonalities with respect to the provided functionalities and input mode. Many of these applications enable to visualize resources accessible through a public SPARQL endpoint [16,18], while only a few of them allow users to upload local RDF graphs [21]. Some of the tools complement graph visualization with advanced features, such as augmenting graph content with external information [16] and collecting statistics about the KG [20]. Different strategies are used to deal with large-scale KGs, e.g. [18] aggregates nodes in clusters, [2,21] try to identify and show firstly the most important concepts, while others, such as [5,15], rely on incremental exploration

---

[2] https://projector.tensorflow.org.

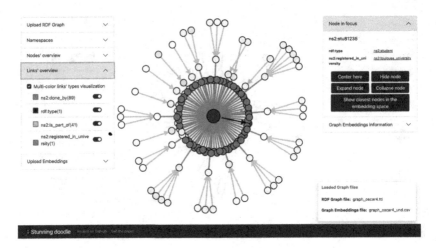

Fig. 9. ns2:stu81235 expanded to show its neighbors in the KG that are not close in the embedding space.

by the user. The conclusion drawn by [6] highlights the impossibility of identifying the best tool in the absolute and states that the research in this field is far from conclusion. Moreover, most of the existing applications are developed as proofs of concept or research tools, therefore they are often conceived for a very specific task and dataset and they can hardly be generalized, or they are rather cumbersome to set up. These limitations make them not suitable for widespread use by non-expert users.

Inspired by the GEs visualization implemented by Tensorflow, *Stunning Doodle* builds upon the functionalities offered by recent KGs visualization tools and extends them to enable a simple and straightforward visual analysis of the KGEs.

5   Conclusion and Future Work

Stunning Doodle is a first step to fill the gap in the field of visual analysis of KGEs. This visualization tool enables to build a link between the content and structure of any KG and its corresponding embeddings. We implemented a set of functionalities to facilitate the exploration and understanding of any KG and to analyze KGEs, connecting the two, and making sense of the information captured by the KGEs. We used *Stunning Doodle* to address two use cases requiring the study of a real-world KG and its embeddings. It has proven its usefulness in gathering meaningful insights for a more informed exploitation of KGEs.

As future work, we plan to implement new functionalities particularly useful for more expert users, such as the visualization of the result of SPARQL queries and the direct access to well-known SPARQL endpoints. Moreover, we aim to provide a deeper analysis of the uploaded GEs including advanced statistics on the closest nodes and additional similarity metrics. We also want to optimize the pre-processing pipeline to be able to display and analyze larger KGs. Our hope is that *Stunning Doodle* could build a large community of users and keep improving and growing throughout the years to satisfy their needs.

References

1. Antoniazzi, F., Viola, F.: RDF Graph Visualization Tools: A Survey, November 2018. https://doi.org/10.23919/FRUCT.2018.8588069
2. Asprino, L., Colonna, C., Mongiovì, M., Porena, M., Presutti, V.: Pattern-based visualization of knowledge graphs. arXiv preprint arXiv:2106.12857 (2021)
3. Bikakis, N., Liagouris, J., Kromida, M., Papastefanatos, G., Sellis, T.: Towards scalable visual exploration of very large RDF graphs. In: Gandon, F., Guéret, C., Villata, S., Breslin, J., Faron-Zucker, C., Zimmermann, A. (eds.) ESWC 2015. LNCS, vol. 9341, pp. 9–13. Springer, Cham (2015). https://doi.org/10.1007/978-3-319-25639-9_2
4. Bikakis, N., Liagouris, J., Krommyda, M., Papastefanatos, G., Sellis, T.: GraphVizdb: a scalable platform for interactive large graph visualization. In: 2016 IEEE 32nd International Conference on Data Engineering (ICDE), pp. 1342–1345. IEEE (2016)
5. Camarda, D.V., Mazzini, S., Antonuccio, A.: Lodlive, exploring the web of data. In: Proceedings of the 8th International Conference on Semantic Systems, pp. 197–200 (2012)
6. Desimoni, F., Po, L.: Empirical evaluation of linked data visualization tools. Future Gener. Comput. Syst. **112**, 258–282 (2020)
7. Ernst, P., Meng, C., Siu, A., Weikum, G.: KnowLife: a knowledge graph for health and life sciences. In: 2014 IEEE 30th International Conference on Data Engineering, pp. 1254–1257 (2014)
8. Ettorre, A.: antoniaettorre/stunning_doodle: First Version, December 2021. https://doi.org/10.5281/zenodo.5769192
9. Ettorre, A., Bobasheva, A., Faron, C., Michel, F.: A systematic approach to identify the information captured by knowledge graph embeddings. In: IEEE/WIC/ACM International Joint Conference on Web Intelligence and Intelligent Agent Technology (WI-IAT) (2021)
10. Ettorre, A., Rocha Rodríguez, O., Faron, C., Michel, F., Gandon, F.: A knowledge graph enhanced learner model to predict outcomes to questions in the medical field. In: Keet, C.M., Dumontier, M. (eds.) EKAW 2020. LNCS (LNAI), vol. 12387, pp. 237–251. Springer, Cham (2020). https://doi.org/10.1007/978-3-030-61244-3_17
11. Grover, A., Leskovec, J.: Node2Vec: scalable feature learning for networks. In: Proceedings of the 22nd ACM SIGKDD International Conference on Knowledge Discovery and Data Mining, pp. 855–864 (2016)
12. Kazemi, S.M., Poole, D.: Simple embedding for link prediction in knowledge graphs. In: Proceedings of the 32nd International Conference on Neural Information Processing Systems, pp. 4289–4300 (2018)
13. Liu, J., Lu, Z., Du, W.: Combining enterprise knowledge graph and news sentiment analysis for stock price prediction. In: Proceedings of the 52nd Hawaii International Conference on System Sciences (2019)
14. Mahdisoltani, F., Biega, J., Suchanek, F.: YAGO3: a knowledge base from multilingual Wikipedias. In: 7th Biennial Conference on Innovative Data Systems Research. CIDR Conference (2014)

15. Micsik, A., Turbucz, S., Györök, A.: LODmilla: a linked data browser for all (2014)
16. Nuzzolese, A.G., Presutti, V., Gangemi, A., Peroni, S., Ciancarini, P.: Aemoo: linked data exploration based on knowledge patterns. Semant. Web 8(1), 87–112 (2017)
17. Palombi, O., Jouanot, F., Nziengam, N., Omidvar-Tehrani, B., Rousset, M.C., Sanchez, A.: OntoSIDES: ontology-based student progress monitoring on the national evaluation system of French Medical Schools. Artif. Intell. Med. **96**, 59–67 (2019)
18. Po, L., Malvezzi, D.: High-level visualization over big linked data. In: International Semantic Web Conference (P&D/Industry/BlueSky) (2018)
19. Rotmensch, M., Halpern, Y., Tlimat, A., Horng, S., Sontag, D.: Learning a health knowledge graph from electronic medical records. Sci. Rep. **7**, 1–11 (2017)
20. Santana-Pérez, I.: Graphless: using statistical analysis and heuristics for visualizing large datasets. In: VOILA@ ISWC 2187, pp. 1–12 (2018)
21. Troullinou, G., Kondylakis, H., Stefanidis, K., Plexousakis, D.: RDFDigest+: a summary-driven system for KBs exploration. In: International Semantic Web Conference (P&D/Industry/BlueSky) (2018)
22. Ying, R., Bourgeois, D., You, J., Zitnik, M., Leskovec, J.: GNNExplainer: generating explanations for graph neural networks. Adv. Neural Inf. Process. Syst. **32**, 9240 (2019)

Capturing the Semantics of Smell: The Odeuropa Data Model for Olfactory Heritage Information

Pasquale Lisena[1]([✉]) [ID], Daniel Schwabe[2] [ID], Marieke van Erp[3] [ID],
Raphaël Troncy[1] [ID], William Tullett[4] [ID], Inger Leemans[3] [ID], Lizzie Marx[5] [ID],
and Sofia Colette Ehrich[3]

[1] EURECOM, Sophia Antipolis, France
{pasquale.lisena,raphael.troncy}@eurecom.fr
[2] Jožef Stefan Institute, Ljubljana, Slovenia
daniel.schwabe@ijs.si
[3] KNAW Humanities Cluster, Amsterdam, The Netherlands
marieke.van.erp@dh.huc.knaw.nl, {inger.leemans,sofia.ehrich}@huc.knaw.nl
[4] Anglia Ruskin University, Cambridge, UK
william.tullett@aru.ac.uk
[5] University of Cambridge, Cambridge, UK
esm38@cam.ac.uk

Abstract. Smells are a key sensory experience. They are part of a multi-billion euro industry and gaining traction in different research fields such as museology, art, history, and digital humanities. Until now, a semantic model for describing smells and their associated experiences was lacking. In this paper, we present the Odeuropa data model for olfactory heritage information. The model has been developed in collaboration with olfactory and art historians. Our model can express the various stages in a smell's lifetime – creation, being experienced, deodorisation – and their relation to locations, times and the agents that interact with them.

Keywords: Smell · Ontology · Cultural heritage · Vocabularies · Sensory mining

1 Introduction

Smells are a key sensory experience. As olfactory information goes straight from the nose, from the olfactory bulb, to the limbic system, the amygdala and hyppocampus, smells often evoke strong emotions and memories [32]. Throughout history, these emotive and mnemonic qualities of smelling have been recognised and described, for example in John Louis-Francois Ramond's *Travels in the Pyrenees* (French 1789; English translation 1813):

> There is a somewhat in **perfumes** which powerfully awakens the *memory of the past*. Nothing so soon recalls to the mind *a beloved spot, a regretted*

© The Author(s), under exclusive license to Springer Nature Switzerland AG 2022
P. Groth et al. (Eds.): ESWC 2022, LNCS 13261, pp. 387–405, 2022.
https://doi.org/10.1007/978-3-031-06981-9_23

situation, or *moments* whose passage has been deeply recorded in the heart, though lightly in the memory. The **fragrance** of a *violet* restores us to the *enjoyment* of many *springs*.

Senses such as vision and hearing are largely studied in signal processing and computer science, while others are underrepresented in scientific research. The sense of olfaction can be found in this latter group. However, the domain of smell, which is often perceived as a fringe one, is in fact quite broad and relevant by humanities and social science scholars [24,25,59]. A new interest in the odours of the past and how past odours are perceived in the present has stimulated research in cultural heritage, or more specific: into olfactory heritage [5] – where scents and smellscapes are understood as a form of both *Tangible* and *Intangible Heritage*[1] – and where scholars have become interested in past ways of smelling and historical smell scapes [28,47]. In addition to the perfume-making industry, we acknowledge the interest of GLAMs [23,62], urban design [22], tourism and environment preservation,[2] human-computer interfaces and 'computer nose' devices [8]. Emerging research has also been triggered by olfactory dysfunctions due to the COVID-19 pandemic, which make the preservation of past olfactory experience more urgent [40].

To capture information about historical smells, olfactory practices, and smell scapes, we can reach out to the rich digital heritage collections that have been developed over the last decades. References to smells and olfactory practices can be found in a large variety of digital texts and images: normative texts, medical texts and perfume handbooks, for instance, offer information about the production and usage of perfumes, or smell management [30,33]. Novels, poems and travel literature reveal connections between odours and identities, testifying to cultural sensitivities around smelling. They may also describe fragrant places, such as churches, parks, or sewers [29,59]. Olfactory clues, gestures and allegories can also be found in paintings, prints and other visual sources [56].

Cultural heritage data collections pose both an opportunity and a challenge. Up to now, most effort in olfactory mining has been put into mapping and classifying fragrances and malodours (specifically in the perfume and odour industries) and in computing the nose - the act of smelling and its effect on the body. Smells are notoriously hard to predict. Thusfar, olfactory informatics has been focused on computing what a molecule smells like based on its chemical structure [36,49,63]. Heritage texts and images however, provide a different type of information. They offer rich data about odour perception and valuation, about the cultural experience of smelling, including subjective interpretations of the perceived odours. To capture this information, different computer science technologies are required, such as image recognition, text mining, and semantic web technologies. Odeuropa[3] is the first major research project to combine these technologies to capture smell experiences in their historical and cultural con-

---

[1] It is relevant the inclusion of the perfumes of Grasse in the UNESCO list. Source: https://bit.ly/3opPRin. Last visited: 15/03/2022.

[2] Examples in Japan: https://bit.ly/3u4ySFD and in France: https://bit.ly/3rYpv7Q.

[3] https://www.odeuropa.eu.

text. Our goal is not so much to represent smells *per se* but rather to represent the historical and social aspects of smell perception and olfactory practices.

In this paper, we introduce the Odeuropa data model for representing odours and their experiences from a cultural heritage perspective. The data model reuses and extends established ontologies such as CIDOC CRM [14], to represent the relevant information as a set of interconnected events. The model is completed by a set of controlled vocabularies for representing crucial elements such as olfactory objects and gestures.

The ontology is developed using web technologies and is intended as a structure for realising an olfactory Knowledge Graph (KG) in the context of the Odeuropa project. The KG will include olfactory-related data extracted from text and images from the 17th to the early 20th century. This KG is intended to serve as a base for supporting heritage professionals, historians and scent designers in including Artificial Intelligence (AI) techniques in their daily practice [39].

This ontology is contributing to the domain in two ways:

- Offering a structure for representing smell-related information, a necessary step for the preservation of this intangible heritage. The ontology can potentially serve in all the previously mentioned areas of the olfactory domain;
- For the first time, closing a gap in the representation between objective observations and subjective experiences (in particular, sensory ones). This aspect is targeted also going beyond the olfactory domain, using a 2-level structure – with a first layer targeting senses in general and a second one focusing specifically on smells – enabling the description of sensorial experiences in fields such as history, literature, art, and cultural heritage.

The remainder of this paper is organised as follows. In Sect. 2, we describe related work on olfactory and semantic web modelling. In Sect. 3, we present our approach to designing the Odeuropa olfactory model with domain experts and the model requirements. In Sect. 4, we present our model, detailing the modelling decisions underneath, while in Sect. 5 are described the vocabularies that followed from the knowledge elicitation from domain experts. Section 6 reports about the evaluation of the data model. We showcase our model's use and expressivity in Sect. 7 with an example describing a smell experience. We conclude with a discussion and plans for future work in Sect. 8.

2   Related Work

The olfactory domain has typically been the purview of perfumers [27], psychologists [16], and sanitary scientists [61]. In the past decade, museologists, chemists, and historians have become interested in researching and preserving heritage smells [6] and curating smell archives, such as the Osmothèque in Versailles.[4] While there is a subfield of computer science that concerns itself with

---

[4] https://www.osmotheque.fr/en/the-collection/.

olfactory informatics, these studies are mostly focused on predicting or mapping olfactory characteristics of molecules [36,49,63], and semantic modelling of smells is as yet an under-researched area.

Previous work has shown that Knowledge Graphs are suitable to represent and exploit the domain information in cultural heritage [11,14,26], history [31], and art [1,13], as well as complex intangible domains such as event modelling [53, 60], biomedicine [54], ecological networks [58], and chemistry [41]. Whilst smells have a measurable component, namely their molecular composition, they remain a largely intangible and subjective concept as most smell discourse is based on personal observations. An olfactory model therefore needs to be able to deal with subjective observations that are anchored to a place and time.

In the past decade, (digital) humanities researchers have started working with semantic web researchers to develop ontologies and knowledge graphs for their domain such as [1,7,50,51]. For visual information, various ontologies and knowledge graphs are available such as [55]. Auditory information is well covered by for example [42], and a taste model was developed by [44]. The IoT community has been working on a digital senses model [12] and the concept of an artificial nose [19]. Recently, [52] combined odours, odorants, olfactory receptors and odorant-receptor interactions in a single MySQL database. However, this model focuses on the chemical compositions of odours, is not open data, and is less focused on the sensory impacts of olfactory experiences and heritage than Odeuropa. To the best of our knowledge, there is a lack of ontologies that specifically aim to represent sensory experiences.

3   Design Methodology and Model Requirements

As there is a large gap between the everyday practices of computer science and humanities researchers in which this model is to fit, we opted for a user-centred design methodology in which the olfactory and (art) historian experts were closely involved. In a series of meetings and hands-on exercises, the requirements of the model were elicited whereby the overall Odeuropa project goals were kept in mind as end-goal.[5] A core instrument in this process was the formulation of 74 competency questions for the model to answer.

With each step, the intermediate results were shared and progress on the design was measured according to the competency questions formulated and results of these were used to steer the next development iteration. A visual overview of our method is provided in Fig. 1.

The desired ontology will serve for storing together olfactory information from structured resources, as well as information from texts and images. Furthermore, the resulting knowledge graph is to be used to research smells through time and related to places. Historians are furthermore particularly interested in what people did with smells or how they created smells and what emotions these evoked. These requirements led to the following 7 categories of competency questions, all declinable in time and space:

---

[5] https://odeuropa.eu/objectives-timeline/.

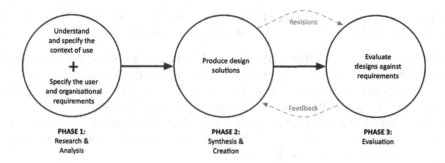

Fig. 1. User-centred design

Smells: About core properties of smell (source, carrier).
- *What are the most frequent smell sources in London in the 18th century?*
- *Which smells were perceived during spring?*

Noses and Gestures: Involving the actors perceiving the odours.
- *Which professions are more present in smelling experience descriptions?*
- *Which smelling gestures are described more frequently by tea-merchants?*

Identities: About the meaning of smells and their capability of being representative of something/someone.
- *Which flavours did people associate with femininity in Asia?*
- *What are the odours most associated with Ashkenazi Jewish practices?*

Emotions: Focusing on the interaction between olfaction and feelings.
- *What odours disgusted upper-class Europeans most?*
- *Which smell triggers memories of childhood?*

Practices: About smell-producing practices.
- *What types of cooking produce a bad smell?*
- *Which practice can reduce a smell intensity?*

Sites and contexts: About the presence of odours in particular places.
- *Which smells are associated with ships?*
- *Which smell could be perceived during the Crimean War?*

Texts and images: About how smells are represented in texts and images.
- *What are the adjectives used for orange aroma in the 15th century?*
- *Which smells can be found in paintings of the Rijksmuseum?*

4   The Odeuropa Data Model

This section describes the olfactory data model, highlighting our core modelling decisions and the main structure of the resulting ontology.

4.1   Extending Established Ontologies

Following best practices in the ontology development [10], we aim to re-use existing data models as base and extend them to represent domain-specific classes and properties. Given the lack of sensory-centered ontology (Sect. 2), we chose **CIDOC CRM** [14] as our core ontology for the following reasons:

- **It is a bridge to other cultural and heritage objects:** CIDOC CRM can be used to describe objects in museum and creative works [38], including paintings and textual resources. This makes it more natural to describe the relations between olfactory information and those elements;
- **It is already familiar to museums and digital libraries:** This can be an advantage when creating interlinking with existing collections and for eventual adoption by these institutions;
- **It is event-based:** [46] Due to the intangible nature of smells and the inevitable subjectivity in their usual descriptions, we decided to focus on the representation of olfactory events rather on odours themselves. In CIDOC CRM, events are the fundamental building blocks: the existence of anything is implying an event that generated it or made someone aware of it. An event can be described in relation to time, space, and involved participants can be linked to other events, including sub-events such as actions and gestures;
- **It is expressive and flexible:** The information to be represented may vary significantly, ranging from highly detailed olfactory experiences to brief mentions to a particular smell. The modularity of CIDOC CRM – itself made of events as building blocks which may be freely interconnected – provides the required flexibility in the representation. In particular, it allows to independently represent the event which generated (or transformed) the smell and the olfactory experience(s), giving the possibility of describing both or only one of them.

CIDOC CRM is extended by **CRMsci** [15], which adds properties about the scientific observation and description of natural phenomena. Here, the *observation* concept has to be understood in the broad sense of *experiencing something*, such that it can also be applied to sensory experiences beyond sight.

As a derivation of CIDOC CRM, the Odeuropa model follows the naming convention to prefix classes and property names with a number and a letter: CIDOC CRM uses E (for classes) and P (for properties); CRMSci uses S (classes) and O (properties); Odeuropa uses L (classes) and F (properties), taking two letters from "olfaction". In the text of this paper, we will omit these codes and letters for readability, while keeping them in the figures.

In addition, parts of the following ontologies are used:

- The READ-IT ontology, to represent emotions triggered by events [3];
- The PROV-O Ontology, for representing data provenance [34];
- The FOAF vocabulary, for describing people [18];
- Schema.org [20], e.g. to describe the genre and author of a text or painting.

4.2   A Three-Layered Model

Due to the complexity of the phenomena related to odours, we adopted a layered approach to construct the data model. We identified abstraction levels that

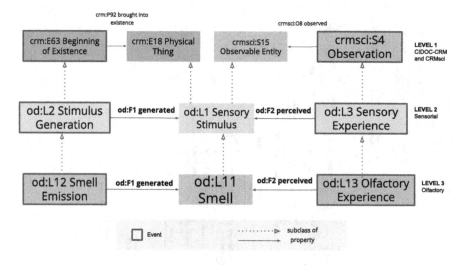

Fig. 2. The core of the Odeuropa data model

roughly correspond to the different aspects of interest. Accordingly, the Odeuropa Data model is organised in the following three levels:

- **Level 1** consists of the CIDOC CRM and CRMsci classes and properties that were used and/or extended. It represents an observation of a phenomenon;
- **Level 2** is an extension of Level 1 for representing sensorial experiences, not limited to olfaction. This level was developed because we identified commonalities shared by all senses and decided to provide more general classes and properties. This will help future extension of the model, including the representation of synaesthetic experiences;
- **Level 3** extends Level 2 by specifically targeting olfactory information.

The three levels are shown in Fig. 2, representing the core of the model. *Smell* (Sensory Stimulus) plays a central role, directly connected to two main types of events, namely *Smell Emission* (Stimulus Generation) and *Olfactory Experience* (Sensory Experience).

In this model, we consider a smell as a unique and non-repeatable entity, with defined time and space coordinates. By way of example, two roses have two distinct (but similar) smells, and the "smell of roses" exists only as a generalisation of the smells of all roses. A given smell can be generated by a unique *Smell Emission event*, but can be experienced multiple times, in distinct situations, by multiple people. This captures the fact that each person can perceive and describe the same smell differently [2].

Figure 3 reports all elements that are part of the data model. The information is organised around the three main events, directly linked to the Smell:

- The **Smell Emission** allows us to describe the smell generation from a smell source (e.g. tobacco) and the carrier of the smell (e.g. a pipe). These elements

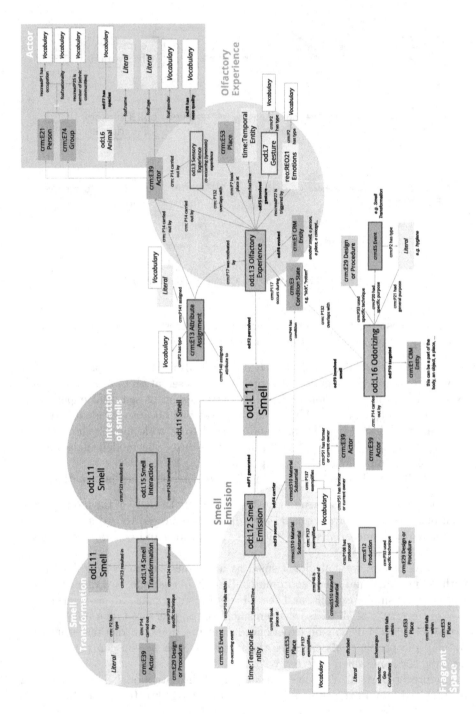

Fig. 3. The Odeuropa data model

can be further described through their components and/or the production process which creates them;

- The **Olfactory Experience** allows us to describe the perception of smell – who perceived the smell, their eventual emotions and gestures. In addition, it records the description that the perceiver makes of the smell, be it through adjectives (typed and linked to vocabularies using the Attribute Assignment class) or through the mention of (i.e., association with) evoked entities such as other smells, people, places, etc.;
- The **Odorizing** class allows to describe how a specific smell was used. For instance, it is possible to specify the purpose for which an odour was used – e.g. covering another smell, medical reason, etc. –, who was using it on what the smell is being used – e.g. a room, a part of the body, etc.

Both event classes inherit from CIDOC CRM some common properties to specify the time and space of the event and eventual co-occurring events. Given the subjective nature of the words used for describing smell, we preferred to model them as *Attribute Assignment* connecting the word (*assigned*) to the smell (*assigned attribute to*), with a direct link to the original person (*carried out by*) and the possibility to include the attribute (*has type*) in a category (e.g. hedonic, intensity, character, state, etc.).

Furthermore, the model includes also classes such as *Stimuli/Smell Transformation* – to represent events that modify a smell, e.g. opening a window – and *Stimuli/Smell Interaction* – to represent smells that are perceived as a combination of different smells, e.g. different foods in a dining room. Special care was devoted to model perceivers (i.e. the agents perceiving smells), by employing and extending the class *Actor* to represent people, groups and animals. Similarly, fragrant spaces are also represented, capturing those attributes that allow us to aggregate them – by type of place or by geographical contiguity.

4.3 Provenance Information

As we intend to trace smells through time, we need to keep track of the sources from which statements in our knowledge graph are derived and through what process. Furthermore, to anchor statements in time and place, we want to keep track of when they were published, and if possible who published them to – for example – map a debate on cultural differences with respect to a particular odour. To keep track of this in the KG, we apply the following strategy:

CIDOC CRM enables us to represent that a text (*Linguistic Object*) or image (*Visual Item*) contains a reference (*refers to*) to an entity, which is, in our domain, a Smell or an Olfactory Event. To include the information without drastically increasing the number of triples in the KG, these *refers to* links are instantiated on a subset of the graph, containing at least the core.

PROV-O [34] is used to record the ways this information was extracted from textual and visual sources, including the agent and/or software/algorithm which extracted the information and a confidence score in case automatic processes

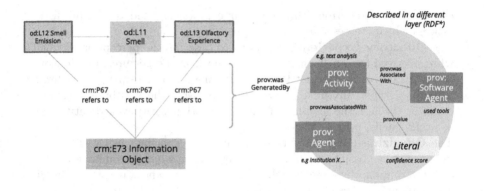

Fig. 4. The provenance of information represented in the Odeuropa data model

were involved. To keep the graph clean, we include this information in a second layer. This is realised by applying RDF* [21] and linking the provenance information to the relevant *refers to* properties, as shown in Fig. 4. In this way, the information and the meta-information are kept distinct, while it is always possible (when needed) to retrieve the provenance of a data excerpt.

5   Controlled Vocabularies

For the description of some fundamental olfactory-related concepts, a collection of controlled vocabularies was created. The use of vocabularies helps to better disambiguate entities, grouping synonyms and labels in different languages under a single identifier (URI). Our vocabularies are represented in SKOS [43], a format that allows us to define, for each concept, preferred and alternate labels, descriptions, broader, narrower and related terms. In this way, we can construct a hierarchy of terms, grouping the related ones and to instantiate bridges between concepts belonging to different vocabularies.

Following previous experiences in constructing controlled vocabularies in Digital Humanities [35,37], our collection is composed of previously-existing taxonomies – which are expressed using SKOS – and vocabularies built from scratch through the collaboration of domain experts and computer scientists.

The list of olfactory vocabularies converted in SKOS format is reported in Table 1 and consists of:

- The Fragrance Circle by Edward Drom, a smell wheel used in perfumery; [9]
- Michael Edwards' Fragrance Wheel, including 4 families and 14 subfamilies of olfactory groups used in modern perfumery;[6]
- The Odour wheel of historical books, for smell heritage preservation [5];
- The Nose-first classification of iconographies realised by Ehrich et al. for linking smells and their representation in art [17];

---

[6] https://en.wikipedia.org/wiki/Fragrance_wheel Last visited: 07/12/2021.

Table 1. Vocabularies converted in SKOS. Some classification systems have a second level which consists of smell sources rather than smell classes (reported as 1+1).

Name	Type	Levels	Top level concepts	Total concepts
Drom's fragrance circle	Odour wheel	2	16	77
Michael Edwards' scent wheel	Odour wheel	2	4	18
Odour wheel of historical books	Odour wheel	2	8	43
Nose-first classification of iconographies	Classification	1+1	25	168
Flavornet and human odour space	Classification	1+1	25	495
Zwaardemaker smell system	Classification	1+1	9	9

- The Flavornet odour space, the compilation of aroma compounds found in human odour space [4];
- The Linnaeus/Zwaardemaker smell system developed in 1895 [48].

These vocabularies were manually converted to a common format based on CSV and then processed and converted to SKOS. In addition, 3 multi-language vocabularies were developed in a collaboration between knowledge engineers and domain experts, representing:

- **Fragrant spaces**, listing interesting (from an olfactory point of view) such as churches, buildings, natural places, etc. These concepts are intended to be linked to instances of type E53 Place through *P137 exemplifies*;
- **Olfactory gestures**, simple actions which possibly occur during olfactory experiences, e.g. sniffing, covering the nose, etc. The included concepts are intended to be linked to instances of type L13 Olfactory Experience through *F5 involved gesture*;
- **Olfactory objects**, including entities (natural or human made) which are particularly relevant because emitting odours – e.g. a flower – or potentially carrying odour sources – e.g. a perfume bottle or a pomander. The included concepts are intended to be linked to instances of type L12 Smell Emission through *F3 source* or *F4 carrier*.[7]

The realisation of these vocabularies was carried out with synchronised spreadsheet tabs – one for each language – to collect the translations of each term. In addition, semantic relationships between terms inside the same vocabulary were instantiated – e.g. "Rose" *skos:broader* "Flower" or "Pipe" *skos:related* "Tobacco – and between vocabularies – e.g. "Library" *skos:related* "Book". An overview of the available languages is shown in Table 2. Please note that a given concept does not always have an appropriate translation in all languages.

---

[7] While some of these are clearly carriers (*wind, bottle*) and other smell sources (*jasmine, sulphur*), some specific elements can embody any of the two role depending on the context (*smoke*). For this reason, we decided to have a single vocabulary including all terms, reporting the preferred role when possible.

Table 2. Multi-language vocabularies for English (EN), German (DE), French (FR), Italian (IT), Dutch (NL), and Slovene (SL)

Name	Total concepts	EN	DE	FR	IT	NL	SL
Fragrant spaces	110	110	4	108	106	110	4
Olfactory gestures	35	35	0	33	32	16	0
Olfactory objects	417	400	172	378	381	390	402

6 Evaluation with Competency Questions

To guide the design of the data model and to provide a way to evaluate it, we used the set of 74 Competency Questions (CQ) [45] collected in Sect. 3 before the development of the model. These CQ were proposed by domain experts – historians and scholars with expertise in olfactory heritage – and are organised in 7 categories, reported in Table 3. These questions allowed the team to iteratively improve versions of the data model, in sequences of development and check. We considered the process of designing the ontology complete only when each CQ could be expressed with a proper SPARQL query, making sure that all the components and relations necessary to answer this question are in place.

In the final version of the model, we distinguish 4 different cases:

- The vast majority of questions can be answered with a SPARQL query. Example: *What are the most frequent smell sources in London in the 18th century?*
- A few questions cannot be answered by simple SPARQL queries, but require more AI methods to find a proper solution. Example: *Was muck perceived as more disgusting than smog?*
- 4 questions are answerable with SPARQL, but require external information that are outside the scope of the model – e.g. with the addition of knowledge bases such as WikiData. Example: *Which smell could be perceived during a war?*
- 1 question requires an extension of the model. Given that the challenging element of this question is not directly related to the olfactory/heritage information but to time representation, we decide to keep this issue open for future work. Example: *Which smells were perceived during morning?*

Apart from the last group, we consider the other cases satisfied by the model. Our results are summarised in Table 3.

Some of the CQs require additional AI techniques to be solved in a more exhaustive way. For example, when searching for bad smells, we are not only interested in the result for a query exactly matching the word *bad*, but we are interested in all kinds of *malodours*, smells described as *stinking, terrible, awful*, etc. We identified 2 possible strategies to address this situation:

Table 3. The number of competency question per category, together with the number of answerable one with the sole model (OK), in combination with AI techniques (AI), with the addition of external data (ExtData) and only with a further extension of the model (Extension)

Category	OK	AI	ExtData	Extension	Total
A. Smells	10	0	0	1	11
B. Noses and gestures	6	0	0	0	6
C. Identities	6	0	0	0	6
D. Emotions	6	0	0	0	6
E. Practices	8	5	0	0	13
F. Sites and contexts	9	0	2	0	11
G. Texts and images	19	0	2	0	21
TOTAL	64	5	4	1	74

- Sentiment detection on the words used for describing the smell (when we search for *good/bad* or *pleasant/unpleasant* smells);
- Rely on word embeddings and compute the similarity between the word in the graph (e.g. *reluctant, fetid*) and the searched one (e.g. *disgusting*).

7 Showcase: Modelling the Smell of a Location

To better understand and appreciate the expressivity and flexibility of the proposed data model, we showcase a modeling example. In Fig. 5, we model the olfactory information contained in a passage from Vita Sackville-West's *Knole and the Sackvilles* (1922). In this book, the author describes the house she grew up in but could not inherit due to aristocratic inheritance customs:

"They [**galleries** of **Knole**, ed.] have the **old, musty** smell which to **me**, whenever I met it, would bring back Knole. I suppose it is really the smell of all **old houses** - a mixture of **woodwork, pot-pourri, leather, tapestry**, and the little **camphor bags** which keep away the moth, and specifically about the pot pourri: **bowls of lavender** and **dried rose-leaves** stand on the window-sills; and if you **stir them** up you get the quintessence of the smell, a sort of **dusty** fragrance, **sweeter** in the under layers where it has held the damp of the spices."

The different olfactory sources mentioned are not physically combined together as in a recipe,[8] but they separately emit different smells which are combined (*Smell Interaction*) in the galleries of Knole. The author perceives this ensemble smell and describes this ensemble smell as *old* and *musty*. In the

---

[8] In that case, there will not be a Smell Interaction, but a single Smell Emission having as source the union of the different ingredients.

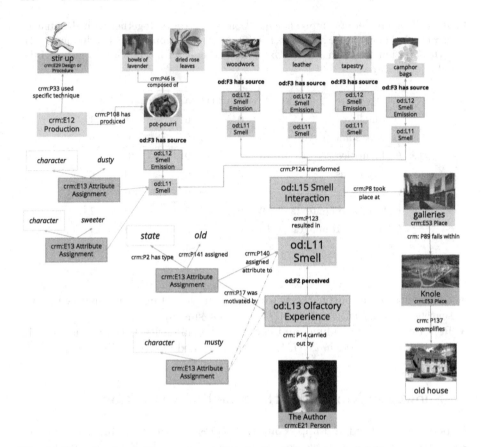

Fig. 5. An example of data representation in the Odeuropa model from a passage of Vita Sackille-West's *Knole and the Sackvilles* (1922)

text, one of the member smells, emitted by the pot-pourri is described, as *dusty* and *sweeter*, also mentioning the procedure of its realisation from lavender and rose leaves. Graph nodes can be interlinked with the controlled vocabularies of olfactory objects – e.g. leather, camphor bags – and fragrant spaces – e.g. old house. Further examples can be found in the Data Model presentation (see Table 5).

8   Conclusions and Future Work

In this paper, we introduced the Odeuropa Data Model, an olfactory extension of CIDOC CRM and CRMsci. The model can represent smell-related information, in particular describing the emission, use and experience of a given odour. The data model is accompanied by a set of multi-language controlled vocabularies for disambiguating of crucial olfactory information elements, such as the odour source or associated gestures. The model is implemented in OWL format and

Table 4. Re-used classes and properties in the Odeuropa data model

Ontologies	Reused classes	Reused properties
CIDOC CRM	10	16
CRMsci	1	0
FOAF	0	4
PROV-O	3	3
READ-IT	1	3
Schema.org	1	4
Time	1	1

Table 5. Resource table

Resource	URL
Data model complete presentation	https://bit.ly/3GuIHzL
Ontology (OWL)	https://github.com/Odeuropa/ontology
Ontology (Documentation)	http://data.odeuropa.eu/ontology/
Competency questions	https://bit.ly/odeuropa-cq
Vocabularies (RDF)	https://github.com/Odeuropa/vocabularies
Vocabularies (SKOSmos)	http://vocab.odeuropa.eu/
Vocabulary API	http://data.odeuropa.eu/api/vocabulary Doc: https://github.com/D2KLab/vocabulary-api
Odeuropa KG	http://data.odeuropa.eu/

published at http://data.odeuropa.eu/ontology under a Creative Commons 4.0 CC-BY License, along with its documentation. Odeuropa proposes 13 new classes and 10 new properties to capture olfactory information, defined as subclassed and subproperties of CIDOC CRM and CRMsci. To these, classes and properties from other models have been reused, as reported in Table 4.

Table 5 lists the pointers to all resources that we developed and published in the context of this work, available as resources to the whole community. In addition to the ontology and the competency questions, some olfactory controlled vocabularies are available via different access points in RDF (Turtle format) using SKOS, in a wide-public visualisation based on SKOSmos [57], through a HTTP API which can be used for interlinking.

The ontology and the vocabularies are part of the Odeuropa Knowledge Graph, hosted at http://data.odeuropa.eu/. At the time of writing, we are populating this graph with data extracted from text and images. This will constitute a multifaceted playground for the data model and for olfactory heritage research. Use of the knowledge graph may also inspire further extensions, validations and improvements of the Odeuropa Data Model.

In future work, we intend to further extend the data model. In particular, we aim to close the gap between the smell heritage domain and the perfume industry, for example by including the representation of chemical compounds and olfactory notes. In addition, we want to better investigate the capability of the model to represent synaesthetic experiences, i.e. the connections people perceive between different sensory experiences such as seeing colours when smelling fragrances. We also intend to extend the vocabularies by including new terms and translations and by adding new thesauri and classifications to our list.

Acknowledgements. This work has been partially supported by European Union's Horizon 2020 research and innovation programme within the Odeuropa project (grant agreement No. 101004469). Smells that helped get this paper out: citrus (to boost our energy levels), rosemary (to keep us alert) and the smell of hell (to keep us on our toes).

References

1. Achichi, M., Lisena, P., Todorov, K., Troncy, R., Delahousse, J.: DOREMUS: a graph of linked musical works. In: Vrandečić, D., et al. (eds.) The Semantic Web – ISWC 2018. LNCS, vol. 11137, pp. 3–19. Springer, Cham (2018). https://doi.org/10.1007/978-3-030-00668-6_1

2. Almagor, U.: Odors and private language: observations on the phenomenology of scent. Hum. Stud. **13**(3), 253–274 (1990). https://doi.org/10.1007/BF00142757

3. Antonini, A., et al.: Understanding the phenomenology of reading through modelling. Semant. Web **12**, 191–217 (2021). https://doi.org/10.3233/SW-200396

4. Arn, H., Acree, T.: Flavornet: a database of aroma compounds based on odor potency in natural products. Dev. Food Sci. **40**, 27–28 (1998)

5. Bembibre, C., Strlič, M.: Smell of heritage: a framework for the identification, analysis and archival of historic odours. Herit. Sci. **5**(1), 2 (2017). https://doi.org/10.1186/s40494-016-0114-1

6. Bembibre Jacobo, C.: Smell of Heritage. Ph.D. Thesis, UCL (University College London) (2020)

7. de Boer, V., van Doornik, J., Buitinck, L., Marx, M., Veken, T., Ribbens, K.: Linking the kingdom: enriched access to a historiographical text. In: Proceedings of the Seventh International Conference on Knowledge Capture, pp. 17–24 (2013)

8. Brooks, J., et al. (eds.): STT21: Smell, Taste, and Temperature Interfaces workshop. Yokohama, Japan (2021). https://stt21.plopes.org/

9. Brud, W.: Words versus odours, how perfumers communicate. Perfum. Flavorist **11**, 27–44 (1986)

10. Carriero, V.A., et al.: The landscape of ontology reuse approaches. In: Applications and Practices in Ontology Design, Extraction, and Reasoning, pp. 21–38. IOS Press (2020)

11. Carriero, V.A., et al.: ArCo: the Italian cultural heritage knowledge graph. In: Ghidini, C., et al. (eds.) The Semantic Web – ISWC 2019. LNCS, vol. 11779, pp. 36–52. Springer, Cham (2019). https://doi.org/10.1007/978-3-030-30796-7_3

12. Datta, S.K., Coughlin, T.: An IoT architecture enabling digital senses. In: 2016 IEEE 6th International Conference on Consumer Electronics-Berlin (ICCE-Berlin), pp. 67–68. IEEE (2016)

13. Dijkshoorn, C., et al.: The Rijksmuseum collection as linked data. Semant. Web J. **9**, 221–230 (2018)

14. Doerr, M.: The CIDOC conceptual reference module: an ontological approach to semantic interoperability of metadata. AI Mag. **24**(3), 75 (2003)
15. Doerr, M., Kritsotaki, A., Rousakis, Y., Hiebel, G., Theodoridou, M.: Definition of the CRMsci an extension of CIDOC-CRM to support scientific observation. Forth-Institution of Computer Science (2015)
16. Dravnieks, A.: Atlas of Odor Character Profiles. American Society for Testing and Materials, Philadelphia (1985)
17. Ehrich, S., Verbeek, C., Zinnen, M., Marx, L., Bembibre, C., Leemans, I.: Nose first. Towards an olfactory gaze for digital art history. In: First International Workshop on Multisensory Data and Knowledge (MDK), Zaragoza, Spain (2021)
18. Graves, M., Constabaris, A., Brickley, D.: FOAF: connecting people on the semantic web. Cat. Classif. Q. **43**(3–4), 191–202 (2007). https://doi.org/10.1300/J104v43n03_10
19. Guest, C., et al.: Feasibility of integrating canine olfaction with chemical and microbial profiling of urine to detect lethal prostate cancer. PLoS ONE **16**(2), e0245530 (2021)
20. Guha, R.V., Brickley, D., Macbeth, S.: Schema. org: evolution of structured data on the web. Commun. ACM **59**(2), 44–51 (2016)
21. Hartig, O.: The RDF* and SPARQL* approach to annotate statements in RDF and to reconcile RDF and property graphs. In: W3C Workshop on Web Standardization for Graph Data, Berlin, Germany (2019)
22. Henshaw, V.: Urban Smellscapes: Understanding and Designing City Smell Environments. Routledge/Taylor & Francis Group, New York (2014)
23. Howes, D.: Introduction to sensory museology. Senses Soc. **9**(3), 259–267 (2014). https://doi.org/10.2752/174589314X14023847039917
24. Howes, D.: Empire of The Senses: The Sensual Culture Reader. Routledge, London (2021)
25. Howes, D., Classen, C.: Ways of Sensing: Understanding the Senses in Society. Routledge, London (2013)
26. Isaac, A., Haslhofer, B.: Europeana linked open data - data.europeana.eu. Semant. Web J. **4**, 291–297 (2013)
27. Jaubert, J.N., Tapiero, C., Dore, J.: The field of odors: toward a universal language for odor relationships. Perfum. Flavorist **20**, 1 (1995)
28. Jenner, M.S.: Follow your nose? Smell, smelling, and their histories. Am. Hist. Rev. **116**(2), 335–351 (2011)
29. Kettler, A.: The Smell of Slavery: Olfactory Racism and the Atlantic World. Cambridge University Press, Cambridge (2020)
30. Kiechle, M.A.: Smell Detectives: An Olfactory History of Nineteenth-century Urban America. University of Washington Press, Seattle (2017)
31. Koho, M., Ikkala, E., Leskinen, P., Tamper, M., Tuominen, J., Hyvönen, E.: WarSampo knowledge graph: Finland in the second world war as linked open data. Semant. Web J. **12**, 265–278 (2021)
32. Krusemark, E.A., Novak, L.R., Gitelman, D.R., Li, W.: When the sense of smell meets emotion: anxiety-state-dependent olfactory processing and neural circuitry adaptation. J. Neurosci. **33**(39), 15324–15332 (2013)
33. Le Guérer, A.: Parfum. Le), Des origines à nos jours. Odile Jacob (2005)
34. Lebo, T., et al.: PROV-O: the PROV ontology. Technical report, World Wide Web Consortium (2013)

35. Leon, A., Gaitán, M., Insa, I., Sebastián, J., Alba, E.: SILKNOW. designing a thesaurus about historical silk for small and medium-sized textile museums. In: Ortiz Calderón, P., Pinto Puerto, Verhagen, P., Prieto, A. (eds.) Science and Digital Technology for Cultural Heritage. CRC Press, London (2020). https://doi.org/10.1201/9780429345470-34

36. Licon, C.C., et al.: Chemical features mining provides new descriptive structure-odor relationships. PLoS Comput. Biol. **15**(4), e1006945 (2019)

37. Lisena, P., et al.: Controlled vocabularies for music metadata. In: 19th International Society for Music Information Retrieval Conference (ISMIR), Paris, France (2018). http://ismir2018.ircam.fr/doc/pdfs/68_Paper.pdf

38. Lisena, P., Troncy, R.: Representing complex knowledge for exploration and recommendation: the case of classical music information. In: Cota, G., Daquino, M., Pozzato, G.L. (eds.) Applications and Practices in Ontology Design, Extraction, and Reasoning, Studies on the Semantic Web Series (SSWS), vol. 49, pp. 107–123. IOS Press (2020). https://doi.org/10.3233/SSW200038

39. Lisena, P., van Erp, M., Bembibre, C., Leemans, I.: Data mining and knowledge graphs as a backbone for advanced olfactory experiences. In: Brooks et al. [8] (2021). https://stt21.plopes.org/wp-content/uploads/2021/05/STT2021_Odeuropa.pdf

40. Mathis, S., et al.: Olfaction and anosmia: from ancient times to COVID-19. J. Neurol. Sci. **425**, 117433 (2021). https://doi.org/10.1016/j.jns.2021.117433, https://www.sciencedirect.com/science/article/pii/S0022510X21001271

41. de Matos, P., et al.: ChEBI: a chemistry ontology and database. J. Cheminform. **2**(1), 1 (2010). https://doi.org/10.1186/1758-2946-2-S1-P6

42. Meroño-Peñuela, A., et al.: The midi linked data cloud. In: International Semantic Web Conference, vol. 10588, pp. 156–164. Springer, Cham (2017). https://doi.org/10.1007/978-3-319-68204-4_16

43. Miles, A., Pérez-Agüera, J.R.: SKOS: simple knowledge organisation for the web. Cat. Classif. Q. **43**(3–4), 69–83 (2007)

44. Naravane, T., Lange, M.: Ontological framework for representation of tractable flavor: food phenotype, sensation, perception. In: ICBO (2018)

45. Noy, N.F., Hafner, C.D.: The state of the art in ontology design: a survey and comparative review. AI Mag. **18**, 53 (1997). https://doi.org/10.1609/aimag.v18i3.1306, https://ojs.aaai.org/index.php/aimagazine/article/view/1306

46. Pasin, M., Motta, E.: Ontological requirements for annotation and navigation of philosophical resources. Synthese **182**, 235–267 (2009). https://doi.org/10.1007/s11229-009-9660-3

47. Perkins, C., McLean, K.: Smell walking and mapping, chap. 10. Manchester University Press, Manchester (2020). https://doi.org/10.7765/9781526152732.00017, https://www.manchesterhive.com/view/9781526152732/9781526152732.00017.xml

48. Philpott, C.M., Bennett, A., Murty, G.E.: A brief history of olfaction and olfactometry. J. Laryngol. Otol. **122**(7), 657–662 (2008). https://doi.org/10.1017/S0022215107001314

49. Sanchez-Lengeling, B., Wei, J.N., Lee, B.K., Gerkin, R.C., Aspuru-Guzik, A., Wiltschko, A.B.: Machine learning for scent: learning generalizable perceptual representations of small molecules. arXiv preprint arXiv:1910.10685 (2019)

50. Schleider, T., et al.: The SILKNOW knowledge graph. Semant. Web 1–16 (2021)

51. Schouten, S., de Boer, V., Petram, L., van Erp, M.: The wind in our sails: developing a reusable and maintainable Dutch maritime history knowledge graph. In: Proceedings of the 11th on Knowledge Capture Conference, K-CAP 2021, pp. 97–104. Association for Computing Machinery, New York (2021). https://doi.org/10.1145/3460210.3493548

52. Sharma, A., Saha, B.K., Kumar, R., Varadwaj, P.K.: OlfactionBase: a repository to explore odors, odorants, olfactory receptors and odorant-receptor interactions. Nucleic Acids Res. (2021). https://doi.org/10.1093/nar/gkab763

53. Shaw, R., Troncy, R., Hardman, L.: LODE: linking open descriptions of events. In: Gómez-Pérez, A., Yu, Y., Ding, Y. (eds.) The Semantic Web, ASWC 2009. LNCS, vol. 5926, pp. 153–167. Springer, Heidelberg (2009). https://doi.org/10.1007/978-3-642-10871-6_11

54. Smith, B., et al.: The OBO foundry: coordinated evolution of ontologies to support biomedical data integration. Nat. Biotechnol. 25(11), 1251–1255 (2007)

55. Stamou, G., van Ossenbruggen, J., Pan, J.Z., Schreiber, G., Smith, J.R.: Multimedia annotations on the semantic web. IEEE Multimedia 13(1), 86–90 (2006)

56. van Suchtelen, A.: Fleeting Scents in Colour. Mauritshuis, Den Haag, the Netherlands (2021)

57. Suominen, O., et al.: Publishing SKOS vocabularies with Skosmos. Manuscript submitted for review (2015)

58. Torta, G., Ardissono, L., La Riccia, L., Savoca, A., Voghera, A.: Representing ecological network specifications with semantic web techniques. In: KEOD-International Conference on Knowledge Engineering and Ontology Development, vol. 2, pp. 86–97. SCITEPRESS-Science and Technology Publications, Lda. (2017)

59. Tullett, W.: Smell in Eighteenth-Century England: A Social Sense. Oxford University Press, Oxford (2019)

60. Van Hage, W.R., Malaisé, V., Segers, R., Hollink, L., Schreiber, G.: Design and use of the Simple Event Model (SEM). J. Web Seman. 9(2), 128–136 (2011)

61. Van Harreveld, A.P., Heeres, P., Harssema, H.: A review of 20 years of standardization of odor concentration measurement by dynamic olfactometry in Europe. J. Air Waste Manag. Assoc. 49(6), 705–715 (1999)

62. Verbeek, C., van Campen, C.: Inhaling memories. Senses Soc. 8(2), 133–148 (2013). https://doi.org/10.2752/174589313X13589681980696

63. Wu, D., Luo, D., Wong, K.Y., Hung, K.: POP-CNN: predicting odor pleasantness with convolutional neural network. IEEE Sens. J. 19(23), 11337–11345 (2019)

Stream Reasoning Playground

Patrik Schneider[1,2]($\boxtimes$), Daniel Alvarez-Coello[3,4], Anh Le-Tuan[5],
Manh Nguyen-Duc[5], and Danh Le-Phuoc[5]

[1] Vienna University of Technology, Vienna, Austria
[2] Siemens AG Österreich, Vienna, Austria
patrik@kr.tuwien.ac.at
[3] University of Oldenburg, Oldenburg, Germany
[4] BMW Technologies E/E Architecture, Wire Harness, Garching, Germany
[5] Technical University of Berlin, Berlin, Germany

Abstract. Stream Reasoning is a well established field not only in the
Semantic Web, but is also adapted in the knowledge representation and
reasoning and AI community in general. In the Semantic Web area, there
have been valuable efforts in building data generators and benchmarks,
however they are not well suited for evaluating more expressive stream
reasoning approaches, since the focus is on a graph-based data model
and more limited reasoning features, such as query answering. This paper
aims at filling the gap, so the different communities can compare, discuss,
and benchmark the various approaches for stream reasoning based on a
common playground. We will present the *stream reasoning playground*
that targets streaming reasoning as the first-class modelling and processing feature. Our playground includes an easy-to-extend platform for
data stream generation with pluggable data formatters, whereby different
data stream sources, and modelling problems for two interesting application scenarios, i.e., intelligent traffic management and vehicle stream
data analytics, are provided. Furthermore, we present a more generic
scenario for time-series data, where a workflow for streaming time-series
data from various datasets is facilitated by using mapping functions. To
illustrate a first application of the playground, we report on the usage
experience of well-known stream reasoner developers in the "model and
solve" Hackathon event of the annual Stream Reasoning workshop.

1 Introduction

Stream Reasoning (SR) is a well established field not only in the Semantic Web,
but also in the AI community in general and focuses on inference, *i.e.*, deducing or inducing implicit facts over data streams. SR has been actively evolving
for more than a decade now, and there exists a wide range of approaches to
reason over streams [9, 21]. Since approaches to stream reasoning can be considerably diverse, it has become desirable to have an agile and well-defined
playground accompanied by the corresponding scenarios, datasets, and (output)
tooling to compare and test the different approaches by fast cycles of iterative
tasks. Notably, the RDF Stream Processing (RSP) community has developed

© The Author(s), under exclusive license to Springer Nature Switzerland AG 2022
P. Groth et al. (Eds.): ESWC 2022, LNCS 13261, pp. 406–424, 2022.
https://doi.org/10.1007/978-3-031-06981-9_24

several successful platforms, e.g., TripleWave [22], RSPLab [32], RSP4J [31], and LSBench [18] for benchmarking and comparing different RSP approaches. However, these platforms usually require a graph-based data model, i.e., RDF [17], that can be queried by some extension of SPARQL [25], i.e., C-SPARQL [4] or CQELS [19].

This brings us to the core of the problem, the underlying data model and query language puts already (desired) restrictions on the scope of usage regarding: (a) the expressive power of an approach, (b) the underlying data model and syntax that can be consumed, and (c) the reasoning tasks that could be solved. Hence, RSP-based tools cannot be simply adapted and used for evaluating and benchmarking logic programming- or complex event detection-based approaches. In this paper, we present the SR Playground that is an initiative and underlying *open-source framework* for providing such resources to the SR community, which should eventually lead to a better understanding of the manifolds of (formal) languages, approaches/pipelines, and reasoners. This imposes the following requirements and the derived features to the framework:

- *(F1) Stream reasoning as a first-class use case*, where the prime focus of the playground is the evaluation of a wide range of stream reasoning approaches/pipelines, i.e., RSP-, logic programming-, and complex-event-based approaches (see Sect. 4).
- *(F2) Consumer agnosticism*, where streams for several consumer modelling languages and input formats should be generated, whether the consumer is graph-, rule-, or complex event-based (see Sect. 2).
- *(F3) Extensibility*, where a simple extension with new stream players and data sources is desired, whether the data source can be a simulation tool, web stream, or collections of (preprocessed) datasets (see Sect. 2/3).
- *(F4) Availability and agility*, where the playground should be easy to deploy and fast to update, in case changes or extensions occur, e.g., a syntax change in the generated streams (see Sect. 2).
- *(F5) Base scenarios*, where already challenging scenarios from relevant domains should be given as a starting point (see Sect. 3).
- *(F6) Multiple tasks*, where for each scenario a range of reasoning tasks is given as plain text, which should go beyond query answering, and include the use of a background knowledge base (see Sect. 3).

Alas, the above features cannot be always aligned, since for instance agnosticism makes the process of extensions harder. Taking the above considerations into account and aiming at a well-balanced framework, we present the SR Playground with the following contributions:

- An easy-to-configure and extendable platform for stream generation capable of producing streams of different scenarios based on stream players and pluggable data formatters. The platform is quickly deployable using the playground's Github repository and a Docker-container-based deployment.
- Two well-defined Intelligent Transport Systems (ITS) scenarios, that consist of (a) a traffic-simulation-generated vehicle flow, and (b) a driving trace from the perspective of an ego vehicle's camera moving in a city.

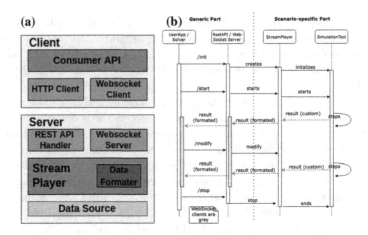

Fig. 1. (a) Overview of architecture and (b) interaction between the components, where the data source is a traffic simulation tool.

- A scenario with a workflow that indicates the steps for streaming time-series data from various datasets that is facilitated by using mapping functions.
- A case study, including the lessons learned from the first usage of the platform in the Stream Reasoning Hackathon 2021, where different *model and solve* tasks were given to participants.

The rest of this paper is organized as follows: Sect. 2 introduces the SR playground. Then, in Sect. 3, we describe two out-of-the-box scenarios that were built with the playground and suggest a workflow to reuse this work with a custom scenario. Then, a case study of a hackathon that tested the presented playground is presented in Sect. 4. Related work is covered in Sect. 5. Lastly, the final remarks are available in Sect. 6.

2   Platform - Stream Reasoning Playground

In this section, we introduce the *Stream Reasoning Playground* (SRP), an infrastructure to stream out semantically annotated data. The SRP's design is based on a client-server architecture that is illustrated in Fig. 1a. The *Server* is a publisher that generates and broadcasts data streams, whereas the *Client* is a data stream consumer that can be piped to the users' Stream Reasoning solver.

On the server-side, the *Stream Player* streams out annotated data streams based on a data source-specific implementation. For example, in the Scenario *A* (Sect. 3.1), the Stream Player embeds a microscopic traffic simulation tool [33] and forwards the simulation tool's states, e.g., vehicle positions and traffic light states. In its simplest form, the Stream Player reads preprocessed datasets and forwards their content to the Data Formatter. Since our architecture is designed to be easily extendable, new scenarios can be added with little effort.

The *Data Formatter* is embedded inside the Stream Player and allows data from the data source to be mapped into several output data formats (*i.e.,* Datalog, RDF, etc.) as the publisher desires. The data streams from the Stream Player are broadcasted via a *Websocket Server* to one or more clients. On the server-side, there is also a *REST API Handler* that allows users (clients, developers, or administrators) to operate and manipulate the behavior of the Stream Player using a given REST API. For example, via the REST API, users can request the background knowledge base (KB) used in their reasoning engine, start/stop streaming, and modify the streaming rate or stream output formats.

On the client-side, users can pipe their reasoner (*e.g.,* a SR solver) to the Playground via the *Consumer API*. This is illustrated in the sequence diagram of Fig. 1b, which describes the data flow of the Stream Playground for Scenario A (Sect. 3.1), where a user can send control commands to the Server as HTTP requests via a *HTTP client*, e.g., a web browser, and receives the data stream from the Server via a *Websocket client*. A user can set up, start, and stop the SR playground via the following commands:

- Initialise a stream player : `/init`
- Start playing a stream : `/start`
- Modify the behavior of a stream: `/modify`
- Stop streaming: `/stop`
- Get the background knowledge base: `/getkb`

At initialization a user can select the scenario and dataset using the `?streamtype` and `?streamid` arguments, as well as the output format using the `?templatetype` argument. All the possible initialization parameters are defined for a scenario in the configurations defined in the `config.yaml` file.

Example 1. HTTP requests to initialize and start of a SUMO traffic stream with JSON-LD as the output format:

- ⟨IP_ADDRESS⟩:⟨PORT⟩/init?streamtype=sumo&streamid=streamSumo1 &templatetype=traffic-json
- ⟨IP_ADDRESS⟩:⟨PORT⟩/start

Feature Coverage. Regarding extensibility (F3), we designed the architecture with two layers, where the generic layer is scenario-independent and allows a unified REST API and Websocket server facing the clients. The scenario-dependent part is implemented via stream players and the re-use of data formatters, where a new Stream Player inherits from an abstract player that defines the interfacing. Importantly, the Stream Player acts as a Python generator (introduced in PEP 255)[1] using the `yield` keyword to return stream messages.

Consumer agnosticism (F2) requires that different output formats, given by a modelling language, should be supported regardless of the scenario. We assume that the data sources provided are either relational (as with DB tables and log files) or tree-shaped (as with JSON files). Then, the input tuples are transformed by distinct data formatters, where we support the following types of formatters:

---

[1] https://www.python.org/dev/peps/pep-0255/.

- *Template-based* formatters are based on template files for stream messages given in the configuration. A set of variables is predefined and replaced on execution by the Stream Player, e.g., as shown in Scenario A (Sect. 3.1).
- With *in-line* formatters the transformations are hard-coded in the Stream Player, as shown for instance in Scenario B (Sect. 3.2).
- *Mapping-language-based* formatters are based on a standardized mapping language such as RML [10], which consists of mapping rules made of a logical source, a subject map and zero or more predicate-object maps.

The type of formatter can be chosen according to the complexity of the scenario, where for simpler scenarios template- or in-line-based and for complex scenarios mapping-language-based formatters can be used.

Regarding availability (F4), the SRP is published under the Apache 2.0 license. The source code and all the scenario data can be found in the Github repository: https://github.com/patrik999/stream-reasoning-challenge. For a fast deployment, it is dockerized and can be delivered as a Docker container.[2]

3   Scenarios

In this section, we outline a traffic management, a vehicle signal processing with object detection, and a custom time-series streaming scenario. The scenarios differ regarding the most complex reasoning tasks and background KB in Scenario A, the highest complexity of streams and novelty of the domain in Scenario B, and the easiness to extend with new sources in Scenario C.

3.1   Scenario A - Traffic Management

The first scenario is in the domain of urban traffic management and involves traffic management for Cooperative Intelligent Transportation Systems (C-ITS). Traffic is observed from a third-person, top-down perspective, and streams of vehicle movements and signal phases (states) of traffic lights in a given road network are generated. In this scenario, we have identified the following (possible) tasks to be tackled:

1. Gathering traffic statistics, *e.g.,* counting the number of vehicles passing;
2. Event detection, *e.g.,* detecting accidents or traffic jams;
3. Diagnosis, *e.g.,* finding the cause for a traffic jam;
4. Motion planning, *e.g.,* routing the vehicles optimally through the network.

Unexpected events could be triggered, *e.g.,* vehicle breakdowns, which lead to possible traffic disruptions. The data source in this scenario is a microscopic traffic simulation framework [33] called Simulation of Urban MObility (SUMO).[3]

---

[2] https://www.docker.com/.
[3] https://www.eclipse.org/sumo/.

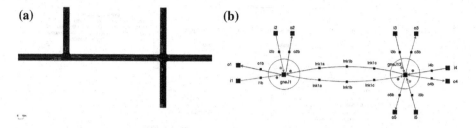

Fig. 2. (a) SUMO rendering of two intersections and (b) corresponding abstract flow model, where green/brown edges are the "we" or "ew" traffic orientation.

The data streams are generated on the fly by different simulation runs on SUMO, where the simulation design is taken from the experiments by Eiter et al. [11].

Data Generation. In Fig. 2a, we show the provided road networks of the scenario rendered in SUMO with two intersections that connect three roads (one horizontal and two vertical). Figure 2b is the graph representation of the road network of Fig. 2a including nodes for intersections, links, sources, and sinks. As shown in the figure, the street layout has two intersections and tree roads with two in/outgoing lanes each, different road segments between intersections, and each intersection with a traffic light controller generating traffic light states based on a static signal plan. We also provide different traffic scenarios to generate traffic streams of varying density, where the streamid parameter in the API is used for the initialization of the density:

- Light traffic with free flow (30 vehicles): &streamid=streamSumo1;
- Medium traffic with free flow (120 vehicles): &streamid=streamSumo2;
- Heavy traffic with traffic jam (180 vehicles): &streamid=streamSumo3.

Background KB. A static background KB adds additional immutable information to the data streams. In this scenario, it captures the SUMO graph representation, including simple traffic regulations, traffic signal plan constraints, and a simple vehicle taxonomy. The road network of the SUMO model was rendered into a graph representation encoded as Datalog facts and RDF triples. It is split into segments of the same length as shown in Fig. 2b, where the type *its:node* defines connection points between two edges, and a single edge is represented by an *its:link* property, where its subject is the source of the edge and the sink is given by the *its:linkedTo* property with the additional information on the traffic flow direction, i.e., "we" denotes west to east. Traffic regulations currently come with speed limits (in m/s) that are assigned to an edge using the *its:maxSpeed* property. Conflicts between traffic lights, i.e., lanes that are not allowed to be simultaneously red, are given for each intersection by the *its:conflictingTL* property, which are (hard) constraints regarding a signal plan. We also provide a simple vehicle taxonomy, which adds sub-types using *rdfs:subClassOf* axioms, where vehicles in the streams are relate to leaf types.

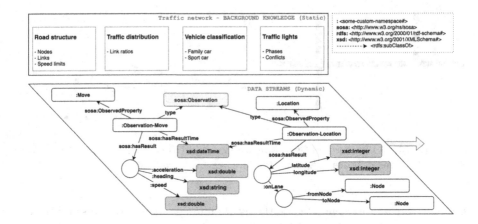

Fig. 3. Representation of the schema used for the streams of scenario A.

Streams. The given data streams are the means for processing and performing the intended evaluation, e.g., a hackathon task. The traffic streams are directly extracted from the SUMO simulation, where we distinguish between *vehicle* and traffic light *signal streams*. The generation of stream messages in this scenario is driven by each simulation step, whereby each step results in a single message for each vehicle and each traffic light signal. Figure 3 provides an overview of the model used for the traffic streams, where the background KB is static and the traffic streams capture moving objects, e.g., the vehicles, and their values as observations. Two observable properties have slightly different annotation patterns for their observations, where the first pattern describes movement-related attributes, such as speed, heading, and acceleration, and the second pattern describes position-related attributes, such as the GPS position or the active lane of the vehicle. Note that the properties *its:onLane* and the vehicle model in *rdf:type*, e.g., carA, refer to facts in the background KB. The attributes for *traffic signal* messages are not outlined here, but are simpler and include the intersection, the traffic light ID, and the signal state (green or red), as well as the message time as a time-point.

We also provide for each data stream a predefined template that allows to render the messages to RDF triples (encoded in JSON-LD) or Datalog facts, where the rendering can be set in the stream initialization by the template types of `traffic-json`, resp., `traffic-asp`. A given set of attributes can be used in the template to complete it on execution.[4] For example in vehicle streams, the variables `$VehicleID$`, `$Type$`, and `$Timestamp$` can be used in templates to add the respective SUMO-generated values.

---

[4] A full list of attributes is given in https://github.com/patrik999/stream-reasoning-challenge/blob/master/hackathon-2021/Hackaton_Overview.pdf.

Example 2. In the following, we give a (simplified) example message rendered in Datalog: `speed(vehicle:20, 20, 1001)`. `vehModel(vehicle:20, carA).`, stating that *:vehicle:20* of type *carA* moves at the speed of 20 at time-point 1001. The same message rendered as RDF triples is shown below:[5]

```
<vehicle:20> a sosa:Platform;
    a its:carA;
    sosa:madeObservation <obs:20_1001>.
<obs:20_1001> a sosa:Observation;
    sosa:hasResult [ its:speed "20"^^xsd:float ];
    sosa:resultTime "1001"^^xsd:int.
```

Example Tasks. We introduce possible example tasks that could be used for a model and solve hackathon. The tasks increase in difficulty, so the first tasks could be given as a starting point.

Tasks 1. Collection of network traffic statistics updated frequently:

1. Calculating the number of vehicles (NoV) and average speed of all vehicles on each edge;
2. Separated by vehicle super-type, calculate the NoV and their average speed;
3. Based on red traffic lights, detect any vehicles that do a red-light violation.

Tasks 2. Detection of legal/illegal behavior and driving patterns of individual vehicles:

1. Detect the vehicles that perform standard maneuvers: vehicles appear/ disappear in the network, turn left/right, or make a short stop;
2. Detect the vehicles that violate traffic rules: speeding, accident, or u-turn;
3. Detect the vehicles that must stop because of another vehicle's accident.

3.2 Scenario B - Vehicle Signals and Surrounding Objects

The second scenario introduces the challenge of automotive applications to SR modelers/developers, which was triggered by industry stakeholders (e.g., BMW, Bosch, and Siemens) to use SR tools and develope descriptive requirements via well-formulized/standardized data models (e.g., RDF) and query/reasoning tools. Different from the previous scenario, which focuses on a complex background-KB, this scenario will focus on facilitating the access to a large collection of open stream data sources of the automotive industry, provided along with a DNN processing pipeline. We believe that this application domain can foster interesting applications for SR community in years to come, which also distinguishes the playground from current RSP counter-parts.

---

[5] Additionally to the standard namespaces **rdf**, **rdfs**, and **xsd**, we have **sosa:** <http://www.w3.org/ns/sosa/>, vsso: ¡https://github.com/w3c/vsso#¿, semkg: <http://vision.semkg.org/onto/v0.1/>.

In this light, the second scenario consists of data streams produced from the perspective of vehicles. For each time step, data representing the driving context is generated. Although the driving context involves data from several domains (*e.g.*, traffic, weather, infrastructure, and others [16]), we focus on two specific stream sources: (1) the stream of the vehicle's location and movement *e.g.*, speed, acceleration, etc.; (2) the stream of objects that are detected from the images captured by the camera attached in front of the vehicle. In this scenario, we find the following tentative tasks to be solved:

1. Finding relevant behavioral patterns from the driving context, *e.g.*, the flow of traffic around.
2. Finding possible reasons for particular situations, *e.g.*, what was the reason for a particular maneuver?
3. Detection of complex events, *e.g.*, dangerous situations on the road.

Data Generation: The second scenario concerns with the object scene flow for autonomous vehicles. The data used in this scenario is based on a few traces from the KITTI dataset [13], a well-known dataset that has been extensively used for benchmark comparisons in tasks related to autonomous driving. The data consists of images captured from a camera attached to a car, its GPS location, and its speed and acceleration. The data is semantically annotated with SSN [14] and VisionKG[6] vocabulary [20] and is provided as a Linked Data stream using the streaming platform described in Sect. 2.

To annotate the location and movement of the car, we follow the schema as described in Scenario A (see Fig. 3). The result of each *IndividualMove* observation includes the speed (m/s), and the acceleration in X axis and Y axis of the vehicle. The *IndividualLocation* gives the GPS coordinates of the vehicle.

Figure 4 illustrates the semantic schema of the stream data of detected objects. To detect the objects from the images captured by the camera, we use an object detection algorithm (*e.g.*, FRCNN [27] or Yolo [26]) which is annotated as a *procedure* (*sosa:Procedure*). To perform object detection for our data, we use Yolo algorithm version 4 [5]. An object detection is a result of an *observation* (sosa:Observation) that is made by using Yolov4. The result of the object detection is 1:1 associated with a frame by the property *:observedFrame*. A detected object contains a box (*Box*) and a label that names the object (*e.g.*, car, van). X and Y are the coordinates of the center of the box in the image. The width and height values are the size of the box. Figure 5 illustrates an example of how a detected object from a frame is symbolically annotated (see Footnote 5). The box that is labelled with "van" is described as follows. Line 1 represents that the detected object 0 belongs to the detection 0. Line 2 links the detected object 0 with the label "van", and box 0. The coordination and the size of box 0 are annotated from lines 4 to 7.

Example Tasks.

- *Task 1:* Query (detect) other vehicles behavior in the stream of labelled objects collected by the ego-vehicle. Possible tasks are to detect:

---

[6] https://vision.semkg.org/.

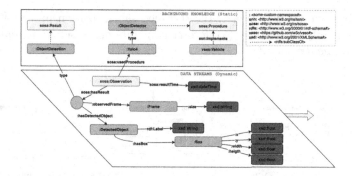

Fig. 4. Schema used for the semantic annotation of objects detected from the video frames of the KITTI dataset

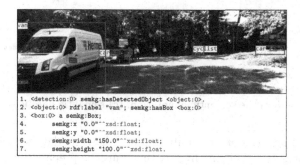

Fig. 5. Example of the semantic annotations of detected objects in RDF.

   1. Detect all oncoming traffic or all crossing traffic.
   2. Detect if one object (vehicle) is stationary or moving.
- *Task 2:* Driving scene understanding, find the explanations for certain observations, e.g., stopping because of pedestrians, traffic lights, or other cars.

3.3 Streaming a Custom Time-Series Scenario

While the previous subsections provided details about ready-to-use data streams produced from two distinct scenarios, the platform's users might be interested in streaming out a custom scenario. Hence, we present in this subsection a possible workflow to stream a custom time-series scenario, together with an example of using a public dataset.

Workflow: The workflow, illustrated in Fig. 6, shows how to reuse our platform with arbitrary time-series data sources and annotation schemes. Please note that it aims to show one way of reusing our platform. Alternatively, users might opt to design and implement their own solutions. We assume that a data source representing the scenario of interest is available. Such data source could be either a stored time-series dataset or directly a data stream (*e.g.,* readings of an actual sensor, simulation of transactions, among others). The workflow follows the general description of the so-called semantization process [29], and consists of mainly three steps: (a) *dataset preparation*, (b) *semantic annotation*, and (c) *message broadcast*.

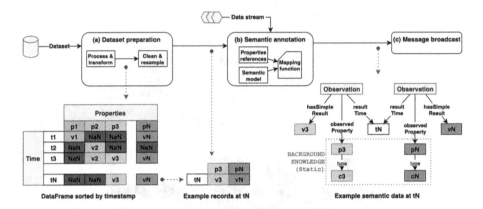

Fig. 6. Overview of the suggested workflow for streaming a custom scenario from a dataset or an actual data stream. (a) datasets must be prepared by the user. (b) a mapping function does the semantic annotation. (c) semantic data is streamed out.

The *dataset preparation* step only applies when the data source is a stored dataset. As datasets are highly diverse, they must be prepared by the user. The idea is to transform the dataset into one data frame, where columns refer to the values of the properties (*aka.*, features, variables) that will be streamed, and rows associate a set of values with the corresponding time. The data frame can then be read as a whole or iterated over its rows. Since values are sometimes unavailable for all columns at a specified time, null values can be cleaned up for simplicity. Alternatively, the user can choose to resample the series to impute or fill in the missing values to have all columns with values at each row.

The *semantic annotation* step annotates the input values with a schema defined from a semantic model. It has a mapping function that populates the schema with the source data values. The mapping itself could be performed in different ways. In the case of RDF data, we recommended using the RDF Mapping Language (RML) [10]. There are a few existing implementations of RML interpreters. For further reference see, for example, *RMLStreamer*.[7] Alternatively, one can also programmatically do the mapping.

Lastly, the *message broadcast* step takes the resulting semantic data at the current time and streams it out as a message by yielding it to the stream player. An example of the workflow is presented next.

An Example with the comma2k19 Dataset: To demonstrate the suggested workflow, we applied it on the *comma2k19*[8] dataset [28]. This dataset is publicly available and consists of multiple commute journeys (*aka.*, routes) that are split

---

[7] https://github.com/RMLio/RMLStreamer.
[8] https://github.com/commaai/comma2k19.

into one-minute segments. Data was collected from two vehicles driven mainly on a highway in California, United States. It has data properties available in three groups: Controlled Area Network (CAN) bus, Inertial Measurement Unit (IMU) and Global Navigation Satellite System (GNSS).

We have implemented[9] the proposed workflow with one of the dataset segments and with two dynamic vehicle properties: *Speed* and *SteeringWheelAngle*. However, if needed, the principle could be replicated with more segments or properties. Figure 7 shows an excerpt from the time series data before and after the *data preparation* step.

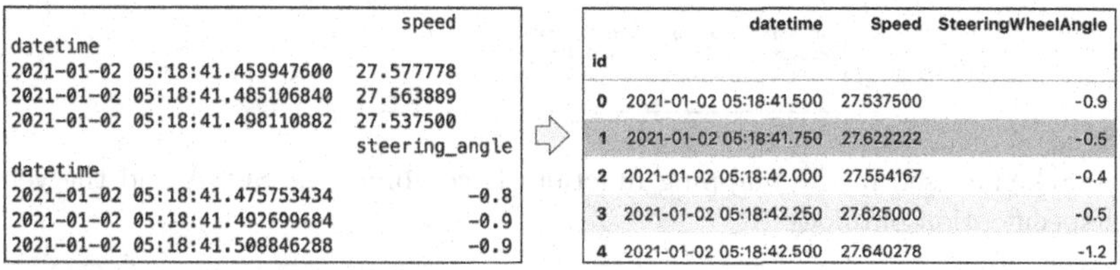

Fig. 7. Sample sequences of speed and steering wheel angle before and after the dataset preparation.

Regarding the *semantic annotation*, we defined a custom schema and a mapping function. The schema, shown in Fig. 8, is based on the combination of the ontologies Sensor, Observation, Sample, and Actuator (SOSA) [15] and the Vehicle Signal Specification (VSSo) [34] (See footnote 5). It is used as a template for the mapping function, implemented with RML rules. An excerpt of the rule that maps the *Speed* is shown below:

```
rr:subjectMap [
    rr:template "http://sr-challenge/vehicle/speed/observation/{id}" ;
    rr:class sosa:Observation];
rr:predicateObjectMap [
    rr:predicate sosa:hasSimpleResult;
    rr:objectMap [
        rml:reference "Speed" ;
        rr:datatype xsd:float ] ] ;
```

Consequently, the resulting semantic data is passed as a message that the player will stream out. For instance, the semantic data (in RDF turtle format) at a particular time looks like the following:

```
<http://sr-challenge/vehicle/speed/observation/1> a sosa:Observation;
    sosa:hasSimpleResult "27.622222222222224"^^xsd:float;
    sosa:observedProperty  indv:Speed;
    sosa:resultTime "2021-01-02 05:18:41.750"^^xsd:dateTime .
```

---

[9] https://github.com/patrik999/stream-reasoning-challenge/blob/master/example-custom-scenario/workflow.ipynb.

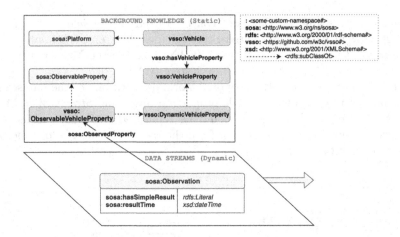

Fig. 8. Schema used in the mapping function. It combines the SOSA and the vehicle signal specification ontology.

```
<http://sr-challenge/vehicle/steering/observation/1> a sosa:Observation;
    sosa:hasSimpleResult "-0.5"^^xsd:float;
    sosa:observedProperty indv:SteeringWheelAngle;
    sosa:resultTime "2021-01-02 05:18:41.750"^^xsd:dateTime .
```

Please refer to the platform's Github repository for further details, such as the complete RML mapping rules, the Stream Player implementation, and explanations of the workflow.

4   Case Study and Lessons Learned: SR Hackathon 2021

The stream generation platform, together with the Scenarios A and B, was for the first time applied in the Stream Reasoning (SR) Hackathon 2021,[10] which was organized as part of the SR Workshop in Milan, Italy.[11] This hackathon allowed both onsite/remote participation and was designed as a "model and solve" challenge, where participants had the freedom to choose their own SR pipelines and reasoners. The principal milestones were: (1) Hackathon announcements prior to the competition, including the description of preliminary tasks to let the participants familiarize with the platform; (2) Introduction of participants, general overview of the platform, and detailed discussion of the tasks at the beginning of the event; (3) Intermediate short sync discussions between organizers and participants to clarify problems and fix problems in the stream players and data formatters; (4) Presentation of solutions by all teams and an online voting to determine the most interesting solutions at the end.

---

[10] http://streamreasoning.org/stream-reasoning-hackathon-2021.
[11] http://streamreasoning.org/events/srw2021.

4.1   Solutions of the Participants

In this section, we give a short overview of the participating teams and their suggested solutions, where the solutions of Oxford University and University of Calabria collected the most votes.

FAU Erlangen-Nürnberg. The team of FAU used *Stream Containers*, which are designed to map RDF streams to a RESTful architecture by the decomposition and decentralization of stream query evaluation. The decomposition is based on a stream-to-relation (S2R) operator for creating a snapshot of tuples based on a window function, and a relation-to-stream (R2S) operator for creating streams with newly time-stamped instances. *Example solution:* SELECT ?v ?s (AVG(?z) AS ?s) WHERE {?v :madeObservation ?y. ?y :hasResult ?x. ?x its:speed ?z } GROUP BY ?v, where the sliding window is managed by the embedding stream container.

NCSR Demokritos. NCSR's team applied the solver Wayeb [1] for complex event forecasting using a streaming extension of symbolic automata. Symbolic automata extend deterministic finite automata with Boolean algebra that can be defined over an infinite domain. Complex (event) patterns can be defined as regular expressions with concatenation as · , and Kleene-star as ∗. *Example solution:* R = (speed > 13)·(speed > 13), detecting that the event (speed > 13) occurs twice on the (vehicle) speed stream.

Oxford University. The team of Oxford suggested a hybrid approach to participate at the hackathon. According to the given tasks, a decision module selects between two solvers, namely RDFox [23] and MeTeoR. In RDFox, they applied standard Datalog rules including aggregation. MeTeoR was applied, where metric temporal logic (MTL) operators such as $\boxminus_{[0,5]}$ for simulating a window operator, e.g., of 5 time steps (ts), were needed. *Example solution:* avgSpeed[?Z,?T] :- AGGREGATE(onLane[?X, ?Z], speed[?X,?S] ON ?Z BIND AVG(?S) AS ?T).

University of Calabria. The team of UniCal competed with an approach that extends static Answer Set Programming (ASP) with streaming features. The solver is called I-DLV-sr [7] and combines an incremental grounding solver [6] with streaming data forwarded from Apache Flink. *Example solution:* accid(X) :- speed(X,0) always in [25]., where the last part of the rule states that speed(X,0) has to occur always in a 25 ts window.

TU Wien. The TU Wien team followed the spirit of a hackathon and extended the static ASP solver Clingo [12] to handle streaming data. This was achieved by a Python-based stream handler that emulates a window operator and generates time-dependent facts. *Example solution:* cNoV(N,X,Y,T) :- link(X,Y), time(T), N=#count{I:onLane(I,X,Y,T))}, N ≠ 0., where cNoV(N,X,Y,T) can then be chained in a new rule to sum up several time points.

4.2 Lessons Learned

The lessons learned include a user survey after the hackathon based on the introduced features, but also covers our own evaluation of organizational aspects.

Hackathon Survey. We conducted a survey after the hackathon, where six questions were asked to the participants.[12] The questions related directly to the platform features with the results shown in parenthesis: Overall suitability (4.2 out of 5 points), suitability for SR (3.4 out of 5 points), extensibility with new features (4.4 out of 5 points), preferred usability aspects (deployment, API, and fast error fixing), and difficulty of the "model and solve" tasks (well balanced). The last question regarding platform improvements is discussed below.

Own Evaluation. Using Github and providing an easy-to-follow installation to replicate the environment via Docker were positive decisions and allowed participants to quickly deploy after we made changes. The release of the initial hackathon's tasks a few weeks before the competition helped participants get familiar with the platform and set up their working environments on time, where the use of the messaging platform Slack was essential for the rapid communication between (remote) teams and organizers. The increasing difficulty of the tasks was an appropriate way to keep the engagement of participants. Tasks that were not fully solved are a clear indicator of possible future work. Giving the participants the freedom to "model and solve" the solutions resulted (as expected) in different ways of solving the tasks, which is an excellent way to cope with the diverse techniques and approaches existent in the SR area.

What Could Be Improved? From the participants perspective, the following suggestions for improvements were given: (1) For Scenario A, provide a larger set of streams, (2) give a clearer definition of the tasks and some example solutions, (3) provide all formats for all scenarios including plain JSON, (4) add benchmarking features for automated measurements, (5) extend the protocols so other communication methods such as Apache Kafka Producer API could be used. From our perspective, we released the platform/documentation only a few weeks in advance, hence the timeline was tight to spot, and issues needed to be fixed on-the-fly. Therefore, the earlier the competition details and tools can be released, the better.

Based on the evaluation, we conclude that the reuse of the proposed platform is recommended as it constitutes a strong foundation for the preparation of future competitions, where the focus could be on the other scenarios presented or on entirely new scenarios taken from different domains such as robotics.

---

[12] The questions and results are provided in https://github.com/patrik999/stream-reasoning-challenge/blob/master/hackathon-2021/Survey.pdf.

5   Related Work

There have been many benchmarks and data generators for RDF stream data processors. The earliest ones are SRBench [35], LSBench [18], and CityBench [2], which focus on the query features of C-SPARQL and CQELS-QL. SRBench uses data from three sources, i.e., LinkedSensorData, Geonames and DBpedia to create data streams. LSBench provides social network stream data via its simulated data generator. CityBench provides stream data from a Smart City application for the city of Aarhus, Denmark. Recently, integrated tools and benchmarks, such as TripleWave [22] and RSPLab [32] aim to reduce the effort required to design and execute reproducible experiments as well as share their results. RSPLab integrates two existing RSP benchmarks (LSBench and CityBench) and two RSP engines (C-SPARQL engine and CQELS). It provides a programmatic environment to deploy in the cloud RDF Streams and RSP engines. While LSBench's and CityBench proposed two test-drivers that push RDF Stream to the RSP engines subject of the evaluation, RSP Lab is developed to be benchmark-independent. The most common processing features of this line of work are based on SPARQL such as C-SPARQL and CQELS-QL. In some cases, complex event query patterns such as [3] and [8] are also introduced.

There have been some extensions of the above RSP-based data generators to accommodate reasoning features. However, such extensions only cover some small set of reasoning features such as RDFS or a fragment of OWL-DL. In SRP, reasoning is the first-class feature by design which is motivated the emergent application scenarios around V2X and autonomous driving. Our playground has been encouraged by the developers of stream reasoners participating at the annual stream reasoning workshops (see Footnote 11). For instances, the ones listed in Sect. 4 are among the well-known reasoners. Moreover, with the extensibility of the playground presented above, the mentioned benchmarking systems can be reused in our players for their systems.

6   Conclusion

This work is sparked by the diversity of approaches in the field of SR, making it challenging to compare formal languages, approaches/pipelines, and reasoners. Existing works from the RSP community have constituted a good starting point, but they are too restrictive regarding the data model and the reasoning tasks (i.e., RDF and query answering). To overcome these limitations, we presented the Stream Reasoning Playground (SRP), which is an open source platform and treats SR as a first-class use case. As indicated by a user survey, it provides a satisfying level of extensibility (F3), consumer agnosticism (F2), availability/agility (F4), and a base scenarios with reasoning tasks that were considered as "well balanced" (F5/F6). The SRP comes with an easy-to-configure and extensible platform to generate streams for different scenarios based on stream players and pluggable data formatters. Notably, besides two well-defined ready-to-use ITS scenarios, we have described a workflow for streaming custom time-series data.

The features were evaluated in a case study based on the SR Hackathon 2021, where we reported on developed solutions, a user survey, and lessons learned.

We consider that the following aspects deserve the attention of the future development of the SRP: (i) defining and including benchmarking features and the corresponding metrics to validate and compare the performance of different solutions to a standard set of tasks; (ii) extending SRP to support streams that include probabilities and addition of probabilistic reasoning features to integrate deep learning models and common-sense reasoning, such as [24,30]; (iii) organizing a repository for community-contributed scenarios and data sets; and (iv) improving the semantization step by adding new formatters covering other languages and a tighter integration of RML.

Acknowledgements. This work was funded by the German Research Foundation under grant nr. 453130567 (COSMO), the German Ministry for Education and Research BIFOLD grant nr. 01IS18025A and 01IS18037A, the German Academic Exchange Service grant nr. 57440921, and the EU Horizon 2020 Research and Innovation program under grant nr. 779852 (IoTCrawler).

References

1. Alevizos, E., Artikis, A., Paliouras, G.: Wayeb: a tool for complex event forecasting. CoRR abs/1901.01826 arXiv:1901.01826 (2019)
2. Ali, M.I., Gao, F., Mileo, A.: Citybench: a configurable benchmark to evaluate RSP engines using smart city datasets. In: Arenas, M., et al. (eds.) The Semantic Web - ISWC 2015. Lecture Notes in Computer Science, vol. 9367, pp. 374–389. Springer, Cham (2015). https://doi.org/10.1007/978-3-319-25010-6_25
3. Anicic, D., Fodor, P., Rudolph, S., Stojanovic, N.: EP-SPARQL: a unified language for event processing and stream reasoning. In: Srinivasan, S., Ramamritham, K., Kumar, A., Ravindra, M.P., Bertino, E., Kumar, R. (eds.) Proceedings of the 20th International Conference on World Wide Web, WWW 2011, pp. 635–644. ACM (2011). https://doi.org/10.1145/1963405.1963495
4. Barbieri, D.F., Braga, D., Ceri, S., Della Valle, E., Grossniklaus, M.: C-SPARQL: SPARQL for continuous querying. In: Proceedings of the 18th International Conference on World Wide Web, pp. 1061–1062 (2009)
5. Bochkovskiy, A., Wang, C.Y., Liao, H.Y.M.: Yolov4: optimal speed and accuracy of object detection. arXiv preprint arXiv:2004.10934 (2020)
6. Calimeri, F., Ianni, G., Pacenza, F., Perri, S., Zangari, J.: Incremental answer set programming with overgrounding. Theory Pract. Log. Program. **19**(5–6), 957–973 (2019). https://doi.org/10.1017/S1471068419000292
7. Calimeri, F., Manna, M., Mastria, E., Morelli, M.C., Perri, S., Zangari, J.: I-DLV-sr: a stream reasoning system based on I-DLV. Theory Pract. Log. Program. **21**(5), 610–628 (2021). https://doi.org/10.1017/S147106842100034X
8. Dell'Aglio, D., Dao-Tran, M., Calbimonte, J., Phuoc, D.L., Valle, E.D.: A query model to capture event pattern matching in RDF stream processing query languages. In: Blomqvist, E., Ciancarini, P., Poggi, F., Vitali, F. (eds.) Knowledge Engineering and Knowledge Management, vol. 10024, pp. 145–162 (2016). https://doi.org/10.1007/978-3-319-49004-5_10

9. Dell'Aglio, D., Della Valle, E., van Harmelen, F., Bernstein, A.: Stream reasoning: a survey and outlook: a summary of ten years of research and a vision for the next decade. Data Sci. **1**(1–2), 59–83 (2017). https://doi.org/10.3233/DS-170006

10. Dimou, A., Vander Sande, M., Colpaert, P., Verborgh, R., Mannens, E., Van de Walle, R.: RML: a generic language for integrated RDF mappings of heterogeneous data. In: Proceedings of the 7th Workshop on Linked Data on the Web, April 2014

11. Eiter, T., Falkner, A.A., Schneider, P., Schüller, P.: ASP-based signal plan adjustments for traffic flow optimization. In: Giacomo, G.D., et al. (eds.) ECAI 2020–24th European Conference on Artificial Intelligence, Including 10th Conference on Prestigious Applications of Artificial Intelligence (PAIS 2020). Frontiers in Artificial Intelligence and Applications, vol. 325, pp. 3026–3033. IOS Press (2020)

12. Gebser, M., Kaufmann, B., Kaminski, R., Ostrowski, M., Schaub, T., Schneider, M.T.: Potassco: the potsdam answer set solving collection. AI Commun. **24**(2), 107–124 (2011)

13. Geiger, A., Lenz, P., Stiller, C., Urtasun, R.: Vision meets robotics: the kitti dataset. Int. J. Robot. Res. **32**(11), 1231–1237 (2013)

14. Haller, A., et al.: The modular SSN ontology: a joint W3C and OGC standard specifying the semantics of sensors, observations, sampling, and actuation. Semant. Web **10**(1), 9–32 (2019). https://doi.org/10.3233/SW-180320

15. Janowicz, K., Haller, A., Cox, S.J., Le Phuoc, D., LefrançSois, M.: SOSA: a lightweight ontology for sensors, observations, samples, and actuators. J. Web Semant. **56**, 1–10 (2019). https://doi.org/10.1016/j.websem.2018.06.003

16. Klotz, B., Troncy, R., Wilms, D., Bonnet, C.: A driving context ontology for making sense of cross-domain driving data (2018). https://www.researchgate.net/publication/331991645_A_driving_context_ontology_for_making_sense_of_cross-domain_driving_data

17. Klyne, G., Carroll, J.J.: Resource description framework (RDF): concepts and abstract syntax. W3C Recommendation (2004). http://www.w3.org/TR/2004/REC-rdf-concepts-20040210/

18. Le-Phuoc, D., Dao-Tran, M., Pham, M.-D., Boncz, P., Eiter, T., Fink, M.: Linked stream data processing engines: facts and figures. In: Cudré-Mauroux, P., et al. (eds.) The Semantic Web – ISWC 2012. LNCS, vol. 7650, pp. 300–312. Springer, Heidelberg (2012). https://doi.org/10.1007/978-3-642-35173-0_20

19. Le-Phuoc, D., Dao-Tran, M., Xavier Parreira, J., Hauswirth, M.: A native and adaptive approach for unified processing of linked streams and linked data. In: Aroyo, L., et al. (eds.) The Semantic Web – ISWC 2011. LNCS, vol. 7031, pp. 370–388. Springer, Heidelberg (2011). https://doi.org/10.1007/978-3-642-25073-6_24

20. Le-Tuan, A., Kien-Tran, T., Nguyen-Duc, M., Yuan, J., Hauswirth, M., Yuan, J.: VisionKG: towards a unified vision knowledge graph. In: Proceedings of the ISWC 2021 Posters and Demonstrations Track. CEUR Workshop Proceedings (2021)

21. Margara, A., Urbani, J., van Harmelen, F., Bal, H.: Streaming the web: reasoning over dynamic data. J. Web Seman. **25**, 24–44 (2014). https://doi.org/10.1016/j.websem.2014.02.001

22. Mauri, A., et al.: TripleWave: spreading RDF streams on the web. In: Groth, P., et al. (eds.) The Semantic Web – ISWC 2016. LNCS, vol. 9982, pp. 140–149. Springer, Cham (2016). https://doi.org/10.1007/978-3-319-46547-0_15

23. Nenov, Y., Piro, R., Motik, B., Horrocks, I., Wu, Z., Banerjee, J.: RDFox: a highly-scalable RDF store. In: Arenas, M., et al. (eds.) The Semantic Web - ISWC 2015. Lecture Notes in Computer Science, vol. 9367, pp. 3–20. Springer (2015). https://doi.org/10.1007/978-3-319-25010-6_1

24. Phuoc, D.L., Eiter, T., Lê Tuán, A.: A scalable reasoning and learning approach for neural-symbolic stream fusion. In: Thirty-Fifth AAAI Conference on Artificial Intelligence, AAAI 2021, Thirty-Third Conference on Innovative Applications of Artificial Intelligence, IAAI 2021, The Eleventh Symposium on Educational Advances in Artificial Intelligence, EAAI 2021, pp. 4996–5005. AAAI Press (2021). https://ojs.aaai.org/index.php/AAAI/article/view/16633
25. Prud'hommeaux, E., Seaborne, A.: SPARQL query language for RDF. W3C Recommendation, January 2008. http://www.w3.org/TR/rdf-sparql-query/
26. Redmon, J., Divvala, S., Girshick, R., Farhadi, A.: You only look once: unified, real-time object detection. In: Proceedings of the IEEE Conference on Computer Vision and Pattern Recognition, pp. 779–788 (2016)
27. Ren, S., He, K., Girshick, R., Sun, J.: Faster R-CNN: towards real-time object detection with region proposal networks. Adv. Neural Inf. Process. Syst. **28**, 91–99 (2015)
28. Schafer, H., Santana, E., Haden, A., Biasini, R.: A commute in data: the comma2k19 dataset. arXiv:1812.05752 (2018)
29. Shi, F., Li, Q., Zhu, T., Ning, H.: A survey of data semantization in Internet of Things. Sensors **18**(2), 313 (2018). https://doi.org/10.3390/s18010313
30. Suchan, J., Bhatt, M., Varadarajan, S.: Commonsense visual sensemaking for autonomous driving - on generalised neurosymbolic online abduction integrating vision and semantics. Artif. Intell. **299**, 103522 (2021). https://doi.org/10.1016/j.artint.2021.103522
31. Tommasini, R., Bonte, P., Ongenae, F., Della Valle, E.: RSP4J: an API for RDF stream processing. In: Verborgh, R., et al. (eds.) The Semantic Web, ESWC 2021. LNCS, vol. 12731, pp. 565–581. Springer, Cham (2021). https://doi.org/10.1007/978-3-030-77385-4_34
32. Tommasini, R., Della Valle, E., Mauri, A., Brambilla, M.: RSPLab: RDF stream processing benchmarking made easy. In: d'Amato, C., et al. (eds.) The Semantic Web – ISWC 2017. LNCS, vol. 10588, pp. 202–209. Springer, Cham (2017). https://doi.org/10.1007/978-3-319-68204-4_21
33. Treiber, M., Kesting, A.: Traffic Flow Dynamics: Data, Models and Simulation. Springer, Heidelberg (2013). https://doi.org/10.1007/978-3-642-32460-4
34. Wilms, D., Alvarez-Coello, D., Bekan, A.: An evolving ontology for vehicle signals. In: 2021 IEEE 93rd Vehicular Technology Conference (VTC2021-Spring), pp. 1–5. IEEE, Helsinki (2021). https://ieeexplore.ieee.org/document/9448884/, https://doi.org/10.1109/VTC2021-Spring51267.2021.9448884
35. Zhang, Y., Duc, P.M., Corcho, O., Calbimonte, J.-P.: SRBench: a streaming RDF/SPARQL benchmark. In: Cudré-Mauroux, P., et al. (eds.) The Semantic Web – ISWC 2012. LNCS, vol. 7649, pp. 641–657. Springer, Heidelberg (2012). https://doi.org/10.1007/978-3-642-35176-1_40

In-Use Track

The Dow Jones Knowledge Graph

Ian Horrocks[1(✉)], Jordi Olivares[2], Valerio Cocchi[3], Boris Motik[1],
and Dylan Roy[2]

[1] University of Oxford, Oxford, UK
Ian.Horrocks@cs.ox.ac.uk
[2] Dow Jones, New York, USA
[3] Oxford Semantic Technologies, Oxford, UK

Abstract. Dow Jones is a leading provider of market, industry and portfolio intelligence serving a wide range of financial applications including asset management, trading, analysis and bankruptcy/restructuring. The information needed to provide such intelligence comes from a variety of heterogeneous data sources. Integrating this information and answering complex queries over it presents both conceptual and computational challenges. In order to address these challenges Dow Jones have used the RDFox system to integrate the various sources in a large RDF knowledge graph. The knowledge graph is being used to power an expanding range of internal processes and market intelligence products.

1 Background and Motivation

Dow Jones is a leading provider of market, industry and portfolio intelligence serving a wide range of financial applications including asset management, trading, analysis and bankruptcy/restructuring.[1] Dow Jones supports businesses, governments and financial institutions with award-winning journalism, deep content archiving and indexing, robust data sets, and flexible information tools; it provides a portfolio of information solutions covering diverse customer needs including uncovering market advantage, integrating data into workflows and managing risk. The goal of Dow Jones is to deliver trusted news and data that can help businesses and society to make better decisions.

High quality market intelligence is critical to corporate decision making. For example, decision makers in a given company may need to be alerted to news items about competitor companies that operate in a related sector to themselves or one of their subsidiaries. The information needed to answer such questions can come from a wide range of heterogeneous sources, including structured sources such as company data and financial data, and unstructured sources such as news feeds. This data needs to be integrated so as to allow for suitable queries to be formulated across multiple sources. This can be very challenging: even if the data from all sources is loaded into a single database, the resulting schema can be very complex, and formulating suitable queries can be very difficult, requiring a combination of knowledge and expertise in the domain, the data sources

---

[1] https://www.dowjones.com/.

P. Groth et al. (Eds.): ESWC 2022, LNCS 13261, pp. 427–443, 2022.
https://doi.org/10.1007/978-3-031-06981-9_25

and the query language. Moreover, the resulting queries can be computationally challenging for a typical database system.

The solution adopted at Dow Jones is to use relevant information from multiple sources to construct a large RDF knowledge graph. This is achieved by using the standard direct mapping to transform structured sources into RDF triples [3], and by using an NLP process to extract relevant facts from news feed articles and transform them into triples. Fast loading and updating of the graph is critical to the feasibility of this approach: the graph currently consists of approximately 2.6 billion triples, and while some of these are derived from relatively static data sources (such as company data) others come from rapidly changing sources (such as news feeds).

The knowledge graph will power a wide range of internal processes and market intelligence products, with knowledge from the graph being accessed via SPARQL queries [6]. For example, Dow Jones researchers maintain data about competitor relationships between companies, and to support this they use an in-house tool to explore information about companies and their relationships retrieved from the graph via back-end SPARQL queries. Using SPARQL queries over the knowledge graph provides applications with both power and flexibility, but it means that fast SPARQL query answering is critical in order to provide acceptable response times in applications.

In order to meet these requirements, Dow Jones has chosen to use the RDFox knowledge graph system. RDFox provides fast parallelised data loading and is able to load all 2.6 billion triples in approximately 26 min using only a relatively modest 4 vCPU server; it also supports incremental updates, and can add/delete several thousand triples per second. RDFox also provides a highly optimised SPARQL engine which not only exploits novel in-memory data structures but also employs sideways information passing to optimise complex SPARQL queries; as a result, typical queries can be answered in milliseconds, and even hard "stress-test" queries can be answered in only a few seconds.

In the remainder of the paper we will provide more details about the construction of the knowledge graph and how it is used in applications (Sect. 2); review the relevant features of RDFox, and in particular the data loading and query answering capabilities that are critical in this setting (Sect. 3); present some data on system performance (Sect. 4); and discuss future plans for extending the system and its application (Sect. 5).

2   Knowledge Graph Construction and Applications

2.1   Knowledge Graph Construction

The knowledge graph integrates data from a wide range of sources that are maintained by and hosted in various different parts of the company. The majority of the data comes from the following sources:

- *Basic information about companies.* This is stored in a relational database and consists of basic information about more than 70 million companies including name, address, normalized code for the region of the address, industry codes (NAICS, SIC, NACE), other identifiers (such as DUNS or LEI), and other name aliases that might be found in news feeds.

- *Company hierarchy information.* This is stored in a relational database and consists of information that links a DUNS coded company to its parent company, forming a company hierarchy graph in relational form.
- *Executives.* This is stored in a relational database and consists of information about more than 140 million company executives including their name, the companies that they are associated with and the roles that they play in these companies.
- *Stock information about companies.* This is stored in a relational database and consists of information about approximately 100,000 company stock market listings including their stock ticker (a unique identifier assigned to each security traded on a particular market), whether this is the main listing or not, and in which stock exchanges they are listed.
- *Stock exchanges.* This is stored in a CSV file and consists of information about stock exchanges including their name, location and relationship to other exchanges. The data is publicly available and can be accessed from https://www.iso20022.org/market-identifier-codes.
- *Geonames.* This is a public domain geographical names database derived from official public sources, and extended and improved via crowdsourcing. It consists of information about more than 25 million locations, including name variants, latitude, longitude, elevation, population, etc. The data is already available as RDF and can be accessed via https://www.geonames.org/.

The relational data sources are transformed into RDF triples via the standard Direct Mapping of Relational Data to RDF [3]. A similar process is used to transform the CSV data into RDF. The Geonames data is already in RDF form. Integration of this data is relatively straightforward as the structured sources are well curated, and include industry standard identifiers such as ticker symbols, DUNS numbers, and NAICS codes. Some cleanup of "messy" identifiers may be required in the future if other data sources are added, but this is not currently an issue for Dow Jones.

The above sources yield a total of approximately 2.3 billion triples, and constitute about 90% of the triples in the knowledge graph. These sources are continuously curated and updated, but the rate of change is relatively low, and in order to simplify the system architecture the whole ETL process is simply repeated once per month.

In addition to this relatively static data, the graph also includes data extracted from financial news articles from several sources:

- Articles from Dow Jones publications including the Wall Street Journal,[2] Market Watch[3] and Baron's Magazine.[4] Approximately 7–10 thousand such

---

articles are available at any time, and this set is constantly changing as old articles are deleted, new articles are added and existing articles are edited.
- Articles from the Dow Jones Factiva feed.[5] Approximately 150–250 thousand such articles are available at any time, and like the Dow Jones articles the set of available articles is constantly changing.

Each available Dow Jones article is represented by an entity in the knowledge graph, along with meta-data such as its title, news topics, and companies and regions mentioned. Some of this meta-data is available directly, but some, such as companies and regions mentioned, must be extracted from the text. This is done using a custom NLP process that extracts not only this kind of meta-data, but also so called *signals* that indicate relevant events such as earnings announcements, initial public offerings (IPOs), acquisitions, mergers and Chap. 11 bankruptcy filings. Each such signal is also represented by an entity in the graph. The data extraction process exploits domain knowledge stored in the graph and uses it to identify target entities (such as companies and regions), and is designed so as to be easily adaptable to capture any kind of entity or signal that might be of interest to Dow Jones customers, and that might help them to identify relevant news content. Article and signal meta-data is stored as triples in the knowledge graph associated with the relevant article and signal entities; additional triples link signal entities to relevant articles, companies, regions, etc.

Articles from the Factiva feed are processed in the same way, but due to the very large number of such articles they are only stored in the graph if they are found to contain relevant entities or signals.

The above process typically yields in the range of 4–5 thousand new signals each day, amounting to approximately 30–40 thousand triples. These are added to the knowledge graph incrementally, which takes only a few seconds. At the same time, triples relating to older articles that have been deleted from the relevant news-feeds are removed from the knowledge graph; this is again done incrementally, and again requires only a few seconds.

When signals are first added to the knowledge graph they are marked as "potential" by adding a suitable triple to the signal entity. Potential signals are checked and curated by human experts, and if confirmed the "potential" triple is deleted; otherwise the whole signal is deleted. These deletions are again achieved via incremental updates; such updates involve deleting only a small number of triples, which typically requires only a few milliseconds.

Finally, Dow Jones also maintains data about competitor relationships between companies. This data is actively curated on a continuous basis using an in-house tool that exploits knowledge graph queries to identify and analyse possible competitors. The resulting competitor relationship data is stored back into the knowledge graph. This is again realised via incremental updates; as in the case of signal curation, the number of triples involved in each update is relatively small and such updates can be performed in only a few milliseconds.

When all these sources are loaded into RDFox the resulting knowledge graph contains approximately 2.6 billion triples.

---

[5] https://factiva.com/.

The resulting graph structure is very simple. The Direct Mapping of relational sources produces a structure that directly mimics the source tables; articles and signals are represented by single entities, with attached (meta-) data triples; and triples are used to link signals, articles and other entities in the graph. Dow Jones have chosen to use the W3C Shape Expressions Language (ShEx) to describe this structure [18]. This could in principle be used for data validation, but it is used at Dow Jones simply to document the graph structure. Dow Jones application developers and knowledge engineers use the ShEx schema to help them to write queries, and they chose ShEx over SHACL [19] for this purpose because they find ShEx syntax to be more natural and easier to understand. For example, the following extract specifies the graph structure of stock listings:

```
cande-shex:StockListing {
 a [ cande:StockListing ] ;
 cande:lists_company IRI // orm:continuation cande-shex:Company ;
 cande:has_ticker_symbol xsd:string ;
 cco:designated_by @cande-shex:StockListingIdentifier * ;
 cande:listed_in IRI // orm:continuation cande-shex:StockExchange ;
 cande:is_primary_listing xsd:boolean ;
}


cande-shex:StockListingIdentifier {
 a [ cande:SEDOL cande:ISIN cande:CUSIP ] ;
 common:id_literal xsd:string ;
}
```

From this the developers and engineers can quickly identify the relevant predicates for accessing information about stock listings, e.g., they can access the ticker symbol via the `cande:has_ticker_symbol` predicate, and for navigating to other entities, e.g., they can navigate to the relevant company entity via the `cande:lists_company` predicate; moreover, they can see that the structure of companies is specified by the shape expression `cande-shex:Company`.

As mentioned above, ShEx could in principle be used for data validation, but it is not supported by RDFox. However, it would be an easy matter to translate ShEx into SHACL, which is supported by RDFox, if data validation were required.

2.2   Knowledge Graph Applications

The knowledge graph can be used to answer questions that would be difficult or impossible to answer without integrating multiple data sources. For example, given a company C specified by `<companyIri>`, the following query Q_1 retrieves competitor companies that are listed in the stock exchange and are in the same or related sector as C or that are in the exact same sector as one of C's direct subsidiaries:

```
SELECT DISTINCT ?competitor ?industryCode ?industryCodeType
WHERE {
  BIND(?company AS <companyIri>)
  {
    ?company cande:has_industry_code/skos:relatedMatch/^skos:relatedMatch
      ?industry .
  } UNION {
    [] cco:is_subsidiary_of ?company ;
       cande:has_industry_code ?industry .
  }

  ?industry a ?industryCodeType ;
            cande:has_id ?industryCode .
  FILTER(?industryCodeType IN (djid:DJIDCode, djn:DJNCode, naics2017:NAICSCode))
  ?competitor cande:has_industry_code ?industry .
  [] cande:lists_company ?competitor ;
}
```

Answering this query requires integrating basic company data, company hierarchy data, competitor relationships data and stock listings data. Such queries can be relatively easily constructed by consulting the ShEx specification outlined in Sect. 2.1 above.

The knowledge graph can power a wide range of internal processes and market intelligence products. One such internal process is the construction of the knowledge graph itself, and in particular the extraction of signals from news articles. Here the knowledge graph is used to support validation and disambiguation; for example, if the we find a potential signal of the form A is buying B, then A and B should both be companies, and should be identified with specific companies represented in the knowledge graph.

Another example is the identification of competitor relationships between companies. As already mentioned in Sect. 2.1, data about competitor relationships is stored in the knowledge graph and is presented to customers in "quote pages" which provide detailed information about given companies. Dow Jones researchers continuously curate this competitor data using a tool that supports identifying and exploring possible competitors. Users can specify a range of different search parameters and filters, and these are converted into SPARQL queries over the knowledge graph which return (details about) relevant companies; see, for example, query Q_1 above. Queries are constructed using templates whose slots can be filled with values derived from the user-specified search and filter parameters; in the case of Q_1, the company of interest can be specified in <companyIri>. The system is designed so that it is easy to add new parameters, filters and query templates as needed to meet user requirements.

An example of a product in which the knowledge graph will be used is the Wall Street Journal (WSJ) Bankruptcy Pro.[6] This product provides a searchable archive of relevant articles, and supplements articles with important data such as competitor analyses, risk factor identification, capital structure, credit ratings and recent filings. Users can specify a range of different search parameters and

---

[6] https://wsjpro.com/.

filters, and these will be converted into SPARQL queries over the knowledge graph which return (pointers to) relevant articles and data. Queries will be constructed using templates in the same way as for the competitor research tool described above.

The knowledge graph will also enable a range of new and more powerful applications that are currently under development including, e.g., personalised recommendations for customers, including recommending relevant authors and news articles, analysis of investment risk factors, and checks on regulatory compliance. Many of these applications will involve heavy use of RDFox's reasoning capabilities.

3 RDFox

As we have seen in Sects. 2.1 and 2.2, construction and maintenance of the knowledge graph depends on fast loading and updating of triples, while applications of the knowledge graph depend on fast responses to SPARQL queries. These were the main considerations that led Dow Jones to select the RDFox system.

RDFox is a high performance knowledge graph and semantic reasoning engine. Originally the result of research at the University of Oxford [14], RDFox is now developed and marketed by Oxford Semantic Technologies.[7] RDFox exploits a patented in-memory architecture and parallelised computation to provide high performance for data loading, reasoning and query answering. Key features of RDFox include:[8]

- RDF triples, rules, and OWL 2 [17] and SWRL [8] axioms can be imported either programmatically or from files in a range of formats including turtle, datalog and OWL. RDF data can also be validated using the SHACL constraint language.
- Information can be accessed directly from external data sources, such as CSV files, relational databases, and Apache Solr.[9]
- Triples, rules and axioms can be exported into a number of different formats, and the contents of the system can also be (incrementally) saved into a binary file, which can later be used to restore the system's state.
- Multi-user support with ACID transactional updates [5].
- Access control allows for individual information elements in the system to be assigned different access permissions for different users.
- Full support for SPARQL 1.1, and functionality for monitoring query answering and accessing query plans.
- Materialization-based reasoning, where all triples that logically follow from the triples and rules in the system are materialized as new triples.
- Incremental update of materialized graphs: reasoning does not need to be performed from scratch when the information in the system is updated.

---

[7] https://www.oxfordsemantic.tech/.

[8] See https://arxiv.org/pdf/2102.13027.pdf for a survey of RDF stores and their features.

[9] https://solr.apache.org/.

– Explanation of reasoning results: RDFox is able to return a proof for any new
  fact added to the store through materialization.

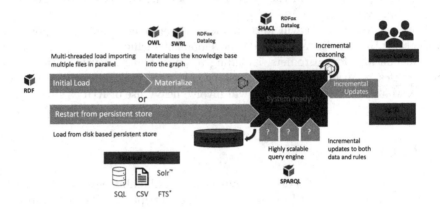

Fig. 1. RDFox architecture

Figure 1 illustrates RDFox's basic features and functionality. At startup,
RDFox can load data, rules, axioms and constraints in a range of different formats as described above. It is also possible to import data directly from external
legacy sources including relational databases, CSV files and Apache Solr. Alternatively, the system can be restored from a previous state saved in a binary file.
An important feature in the Dow Jones application is that RDFox can import
multiple sources in parallel, and we will discuss this in more detail below.

After loading, RDFox performs materialization-based reasoning and constraint
validation using a parallelized variant of the seminaïve algorithm [1,13] (see
Sect. 3.1). Once the initial materialization process is complete the store is ready for
subsequent operations including querying and incremental updating. Access control and ACID transactions allow for control over user access to data and ensure
predictable behaviour when multiple users are updating the store. The state of the
system can also be saved in a binary file for subsequent reloading.

Incremental updates can include deletion and addition of data, and also
deletion and addition of rules, axioms and constraints. RDFox deals with such
updates using FBF, a novel extension of the delete and rederive (DRed) view
maintenance algorithm that avoids excessive overdeletion [11,12]. Like data loading, incremental updates are parallelized for improved performance. RDFox uses
a highly optimised SPARQL engine with sideways information passing; this is
another important feature in the Dow Jones application that we will discuss
in more detail below. Each query is evaluated on a single thread, but multiple
queries can be evaluated in parallel using multiple threads.

3.1 Parallelized Materialization

As already mentioned, RDFox materializes all implied triples using a parallelized
variant of the seminaïve algorithm [1,13]. The triples that make up the RDF

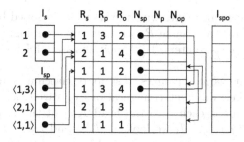

Fig. 2. Data Structure for Storing RDF Triples

graph are stored in a table. The triples are considered one at a time and matched to the rules, with parallelization being achieved by assigning triples to available threads. For example, given the following rules

$$\langle ?x, C, ?y\rangle \wedge \langle ?y, E, ?z\rangle \rightarrow \langle ?x, D, ?z\rangle \tag{R1}$$

$$\langle ?x, D, ?y\rangle \wedge \langle ?y, E, ?z\rangle \rightarrow \langle ?x, C, ?z\rangle \tag{R2}$$

and a triple $\langle a, E, b\rangle$, a thread will match it to the triple $E(?y, ?z)$ in rule (R1) and evaluate subquery $\langle ?x, C, a\rangle$ to derive triples of the form $\langle ?x, D, b\rangle$, and it will handle rule (R2) analogously. We thus obtain independent subqueries, each of which is evaluated on a distinct thread. The difference in subquery evaluation times is irrelevant because of the large number of queries (i.e., proportional to the number of triples) so threads are fully loaded.

A naïve application of this idea would be inefficient: if we have triples $\langle a, C, b\rangle$ and $\langle b, E, c\rangle$, then we would derive the triple $\langle a, D, c\rangle$ twice—that is, we would consider the same rule instance twice. To address this source of inefficiency, the seminaïve algorithm evaluates subqueries only over the triples that appear *before* the triple being processed. For example, if $\langle a, C, b\rangle$ is processed first, then $\langle b, E, c\rangle$ will not be visible to the subquery and $\langle a, D, c\rangle$ will not be derived; however, when $\langle b, E, c\rangle$ is processed, $\langle a, C, b\rangle$ will be visible to the subquery and $\langle a, D, c\rangle$ will be derived.

To support this idea in practice, RDFox uses patented data structures that support both efficient evaluation of subqueries and efficient parallel updates [9,13]. Like systems such as Hexastore [20] and RDF-3X [16], RDFox maintains indexes over stored triples to support efficient (sub)query evaluation; RDFox, however, uses hash-based indexes that allow for efficient 'mostly' *lock-free* parallel updates [7]: most of the time, at least one thread is guaranteed to make progress regardless of the remaining threads.

RDFox stores triples in a six-column *triple table* as shown in Fig. 2. As usual in RDF systems, resources are encoded as integer IDs using a dictionary, with IDs produced by a counting sequence so they can be used as array indexes. Columns R_s, R_p, and R_o contain the integer encodings of the subject, predicate, and object of each triple. Each triple participates in three linked lists: an *sp*-list connects all triples with the same R_s grouped (but not necessarily sorted) by

R_p, an op-list connects all triples with the same R_o grouped by R_p, and a p-list connects all triples with the same R_p without any grouping; columns N_{sp}, N_{op}, and N_p contain the next-pointers. Triple pointers are implemented as offsets into the triple table.

RDFox maintains various indexes to support matching triples with different binding patterns (i.e., different configurations of variables in the triple). For example, index I_s maps each s to the head $I_s[s]$ of the respective sp-list; to match a triple $\langle s, ?y, ?z \rangle$ in I, we look up $I_s[s]$ and traverse the sp-list to its end; if $?y = ?z$, we skip triples with $R_p \neq R_o$. Index I_{sp} maps each s and p to the first triple $I_{sp}[s, p]$ in an sp-list with $R_s = s$ and $R_p = p$; to match a triple $\langle s, p, ?z \rangle$ in I, we look up $I_{sp}[s, p]$ and traverse the sp-list to its end or until we encounter a triple with $R_p \neq p$. Index I_{spo} contains each triple in the table, and so it can match fully specified triples $\langle s, p, o \rangle$. Other indexes include I_p, and I_o and I_{op}. Indexes I_s, I_p, and I_o are realised as arrays indexed by resource IDs. Indexes I_{sp}, I_{op}, and I_{spo} are realised as open addressing hash tables storing triple pointers.

Lock-freedom is achieved using compare-and-set (CAS) instructions: CAS(loc, exp, new) loads the value stored at location loc into a temporary variable old, stores the value of new into loc if old = exp, and returns old; hardware ensures that all steps are atomic (i.e., without interference). CAS can be used directly to update the linked lists in the triple table. For example, if thread T^1 has added a triple $\langle 1, 3, 6 \rangle$ to the table and is trying to add it to the N_{sp} list after the triple $\langle 1, 3, 2 \rangle$, then T^1 will set the N_{sp} pointer of the $\langle 1, 3, 6 \rangle$ entry to point to the N_{sp} pointer from the $\langle 1, 3, 2 \rangle$ entry and will use a CAS instructions to try to set the N_{sp} pointer from the $\langle 1, 3, 2 \rangle$ entry to point to the $\langle 1, 3, 6 \rangle$ entry; if the CAS instruction fails, then some other thread must have changed the N_{sp} pointer, in which case T^1 repeats the insertion procedure.

The process of adding a new triple to the table is more complex as one must atomically query I_{spo} (to check for duplicates), add the triple to the table, and update I_{spo}. To do this, RDFox implements a form of localised locking: if a thread does not find the new triple in I_{spo}, then it identifies a suitable empty bucket and tries to lock it by using a CAS instruction to store a special marker in the bucket. If this fails then some other thread may have already inserted the same triple, and so the whole operation is repeated beginning with the query to I_{spo}. If the CAS instruction succeeds, then we can add the new triple to the table, store it in the bucket (effectively releasing the localised lock), and then update all remaining indexes. In the meantime, we make sure that other threads do not skip over the bucket until the marker is removed.

3.2 Parallelized Data Loading

Although originally designed to support parallelized materialization, the lock-free data structures described in Sect. 3.1 also allow for the parallelization of data loading. This can be achieved simply by assigning a thread to each data source to be loaded. Each thread can then add triples to the triple table in the same way as the multiple threads used for materialization.

Additionally, when data is being loaded from files containing RDF triples in turtle format, each file can use one thread for parsing and multiple threads for adding parsed triples to the triple table. Parsing is single threaded because the syntax of IRIs makes it difficult to parallelize, and in any case parsing is typically much faster than adding triples to the data structures, so a single parser thread can keep several data addition threads fully occupied. If the data is split into multiple files, then these can be loaded in parallel using multiple threads.

3.3   SPARQL Query Answering

The indexed triple table described in Sect. 3.1 is designed to support efficient (sub)query evaluation during materialisation and so already supports efficient join evaluation in SPARQL query answering. However, SPARQL queries can be (heavily) nested; i.e., the outer level query can have sub-queries as components. A simple example is a query $Q = Q1$ MINUS $Q2$. In this case query Q is made up of two sub-queries $Q1$ and $Q2$, with the answer to Q being the answer to $Q1$ minus the answer to $Q2$. Note that $Q1$ and $Q2$ could themselves contain sub-queries and that this nesting of queries can continue to arbitrary depth. In order to make query answering be more efficient and to use less memory we want to evaluate the query "top-down", that is, starting with the outer level queries and working inwards. In our example, a naïve "bottom-up" method would compute the answers to $Q1$ and $Q2$, and then subtract the answer to $Q2$ from the answer to $Q1$; however, this would require computing and storing the full answers to both sub-queries. In our "top-down" method we would iterate through the answers to $Q1$, and for each such answer we would check if it is also an answer to $Q2$, retaining it as an answer to Q only if it is not an answer to $Q2$. This requires very little storage, and only requires us to check $Q2$ for tuples that we already identified as answers to $Q1$. This technique is known in the literature as Sideways Information Passing (SIP) [1]; in our example, information about answers to $Q1$ is passed "sideways" to $Q2$.

The above example is relatively simple, but SPARQL is a large language containing many operators for modifying and combining queries (Filter, Bind, And, Union, Minus, Distinct, Project, etc.) as well as a large number of built-in functions for manipulating values including, e.g., arithmetic functions (plus, minus, etc.), aggregation (sum, max, min, etc.) and string manipulation (concatenate, sub-string, etc.). It is extremely challenging to design a SPARQL query evaluation algorithm that maximises the efficiency benefits of SIP while at the same time guarantees to conform to the SPARQL semantics, i.e., to compute the same answers as would be computed by a naïve bottom-up method. Neumann and Weikum presented a SIP algorithm for basic SPARQL pattern matching queries [15], but this doesn't consider nested queries using some or all of the above mentioned features. RDFox uses a patented algorithm that extends SIP optimisation to arbitrary queries by compiling the query into a tree and introducing variable normalisation and expansion nodes as needed to ensure safe application of SIP [10]. The combination of SIP and the optimised data structures discussed in Sect. 3.1 allow for extremely efficient evaluation of SPARQL queries: most queries used in applications of the Dow Jones knowledge graph can be answered in only a few milliseconds, and even the most complex queries require only a few seconds (see Sect. 4).

4 Performance

In this section we present some performance data for RDFox using both standard benchmarks and the Dow Jones knowledge graph.

4.1 Test Data and Environments

For standard benchmarks we used both LUBM and WatDiv [2, 4]. We used a version of LUBM with 10,000 universities (LUBM-10k), which comprises approximately 1.3 billion asserted triples, with a further approximately 0.5 billion triples added via materialisation of (rules derived from) the LUBM ontology; the graph for query answering therefore comprises approximately 1.8 billion triples. We used WatDiv 100M, which comprises approximately 150 million asserted triples; WatDiv does not have an ontology. Each benchmark comes with a standard set of test queries. These tests used RDFox 5.4.0 running on a c5.18xlarge AWS instance with 3.0 GHz Intel Xeon processors, 72 vCPUs and 144 GiB of RAM.

For the knowledge graph tests we used the Dow Jones Knowledge Graph (DJKG) described in Sect. 2.1, which comprises approximately 2.6 billion triples, and a set of three test queries:

Q1 retrieves all the signals and their properties that were derived from an English language article that was published between 2020-05-24 and 2020-05-26, and that talks about either Africa or North America.
Q2 retrieves all the signals and their properties that were derived from an English language article that was published between 2020-01-01 and 2020-09-28, and that talks about a company with a given identifier.
Q3 retrieves the number of different companies in the knowledge graph grouped by identifier type, industry, and country.

Q1 and Q2 are typical application queries; Q3 is not a realistic application query but is designed to stress-test SPARQL query engines. The SPARQL for these queries is too verbose to be given here, but they are available at https://bit. ly/3qGJS9I along with all non-confidential data. These tests used RDFox 4.0.0 running on a Google Cloud N1 with 4 vCPUs and 125 GB of RAM.

4.2 Data Loading

The data was split into multiple files to facilitate parallel loading. In the case of WatDiv and LUBM the data was split into 72 files and loaded using 72 threads; in the case of DJKG the data was split into 4 files and loaded using 4 threads. Table 1 shows the loading time for the three data sets (Time) as well as the number of threads (Threads), the loading rate in triples per second (T/s), the

Table 1. Data loading times

Dataset	Time	Threads	T/s	T/T/s	Speedup
DJKB	1,560.0	4	1,666,667	416,667	—
LUBM	273.0	72	4,761,905	66,138	—
WatDiv	272.3	1	400,285	400,285	1.00
WatDiv	85.7	4	1,271,852	317,963	3.18
WatDiv	26.6	16	4,097,658	256,104	10.24
WatDiv	17.5	32	6,242,710	195,085	15.56
WatDiv	17.5	64	6,228,441	97,319	15.56
WatDiv	16.8	72	6,487,959	90,111	16.21

relativised loading rate in triples per thread per second (T/T/s), and the speedup relative to a single thread (Speedup) in the WatDiv case.

The same DJKB loading test was repeated using several other knowledge graph systems. RDFox was at least an order of magnitude faster than any of these other systems; unfortunately the licence conditions of these systems mean that we are not able to present their results here.

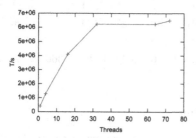

Fig. 3. Number of threads vs. loading speed (triples per second)

Even the relativised (T/T/s) loading rates are not directly comparable across different datasets as there may be a large difference in, e.g., the cost of parsing, which can depend on many factors (such as the structure of URIs). In order to give a clearer idea of the effectiveness of parallel loading we therefore repeated the WatDiv loading test using different numbers of threads; these results are also presented in Table 1, and we have additionally plotted T/s against the number of threads in Fig. 3. As we can see, the speedup from 1–32 threads is relatively consistent, with 32 threads giving a nearly 16 times speedup, but there is little or no additional speedup after that. The reasons for this are not fully understood, and are difficult to investigate in a cloud computing environment; however, we believe that there are only 36 physical cores, with the 72 vCPUs coming from hyper-threading, so significant speedup beyond 36 times is not necessarily to be expected.

Table 2. Results on WatDiv 100M benchmark (times in ms)

Query	#ans	R1	R2	R3	R4	R5	Avg	ms/ans
L1	2	1	1	1	1	1	1	<1
L2	595	16	210	24	60	7	63	<1
L3	24	1	1	1	1	1	1	<1
L4	603	20	11	9	8	7	11	<1
L5	958	16	41	5	12	1	15	<1
S1	6	1	1	1	1	1	1	<1
S2	249	10	12	19	50	5	19	<1
S3	0	29	26	33	30	29	29	–
S4	13	113	333	233	15	30	145	11
S5	68	15	13	11	17	10	13	<1
S6	81	4	15	11	1	6	7	<1
S7	0	1	1	1	1	1	1	–
F1	7	7	6	11	11	4	8	1
F2	58	1	6	9	4	1	4	<1
F3	128	1	7	12	5	1	5	<1
F4	382	6	5	4	5	6	5	<1
F5	43	1	1	1	1	1	1	<1
C1	201	30	23	34	23	20	26	<1
C2	22	140	65	75	62	59	80	4
C3	4,244,261	1,830	1,640	1,380	1,360	1,450	1,532	<1

4.3   Query Answering

The results on the WatDiv queries are presented in Table 2. For each query we give the number of answers (#ans), the time to return all answers in 5 separate runs (R1–R5), and the average time (Avg); we also give the average time per answer (ms/ans). All times are in milliseconds. As can be seen, RDFox answers most queries in only a few milliseconds; query C3 takes an average of 1,532ms, but this is mainly due to the time taken to return over 4 million answers. The average time per answer is less than 1ms in most cases, and never more than 11 ms.

Table 3. Results on LUBM 10k benchmark (times in ms)

Query	#ans	t-1	t-10	t-100	t-all	ms/ans
q1	4	2	1	1	1	0.21
q2	2,528	1,440	2,210	17,800	459,000	181.57
q3	6	1	1	1	1	0.11
q4	34	1	1	1	1	0.01
q5	719	1	1	1	1	0.00
q6	104,403,077	1	1	1	68,933	0.00
q7	67	1	1	1	1	0.01
q8	7,790	1	1	1	39	0.01
q9	2,721,773	1	1	5	128,000	0.05
q10	4	1	1	1	1	0.13
q11	224	1	1	1	1	0.00
q12	15	2	1	1	1	0.03
q13	46,366	1	1	1	303	0.01
q14	79,211,095	1	1	1	37,733	0.00

The results on the LUBM queries are presented in Table 3. For each query we give the number of answers (#ans), the time to return the first answer (t-1), the first 10 answers (t-10), the first 100 answers (t-100) and all answers (t-all); we also give the average time per answer (ms/ans). All times are in milliseconds. Most of the queries are relatively easy for RDFox, with all answers being returned within 1s, and in most cases in less than 1ms. Queries q6, q9 and q14 take several seconds to fully evaluate, but this is only because of the very large numbers of answers, ranging from 2.7 million up to more than 104 million; the times to return the first 100 answers, and the times per answer, are still in the (sub) millisecond range. Query q2 is the only query that can be considered non-trivial; this is a "triangle" query, where there is no query plan that can avoid computing a very large intermediate result that is subsequently pruned by other query atoms. Even on this query, RDFox returns the first answer in only 1.4 s, and returns subsequent answers at a rate of approximately one every 180 ms.

For the three DJKB test queries, the average execution time for RDFox was 300 ms for Q1, 12 ms for Q2 and 10,700 ms for Q3. As mentioned in Sect. 4.1, Q3 is not a realistic query but has been designed as a stress test. The same test was again repeated with several other knowledge graph systems; RDFox was always at least an order of magnitude faster and in some cases several orders of magnitude faster.

5   Discussion and Future Directions

Using a knowledge graph at Dow Jones has had many advantages: it facilitates the integration of data from multiple heterogeneous sources, SPARQL queries provide a powerful and flexible mechanism for accessing information, and this can be used to power a wide range of internal processes and user facing products.

Constructing and maintaining a large knowledge graph can be computationally challenging, as can answering SPARQL queries over the graph. However, RDFox boasts several features that help it to perform well on these tasks, in particular lock-free data structures, parallelised data loading, incremental data updates and a highly optimised SPARQL engine. As a result it can load the entire 2.6 billion triple data set in only 26 min and can answer typical application queries in only a few milliseconds.

Currently, the majority of the data in the knowledge graph is kept up to date by simply reloading it on a regular basis (once per month). This is feasible given RDFox's fast loading time, but is clearly not ideal. Dow Jones developers are therefore working an a rearchitected system in which RDFox is connected directly to data sources (a feature already supported by RDFox) and the graph is automatically updated whenever the source data changes.

So far the knowledge graph has mainly been used as part of internal processes such as the extraction of signals from news feed articles and the maintenance of competitor relationships data. Work is underway to integrate the knowledge graph into a wider range of internal processes, for example to support the Risk and Compliance team, and into existing customer facing products. It is also planned to develop several new and more powerful applications that were previously infeasible due to data integration issues and/or query performance issues. One specific goal is to increase customer engagement by providing user specific recommendations for relevant articles in news feeds.

References

1. Abiteboul, S., Hull, R., Vianu, V.: Foundations of Databases. Addison Wesley Publ. Co., Reading (1995)
2. Aluç, G., Hartig, O., Özsu, M.T., Daudjee, K.: Diversified stress testing of RDF data management systems. In: Mika, P., et al. (eds.) ISWC 2014. LNCS, vol. 8796, pp. 197–212. Springer, Cham (2014). https://doi.org/10.1007/978-3-319-11964-9_13
3. Arenas, M., Bertails, A., Prud'hommeaux, E., Sequeda, J.: A direct mapping of relational data to RDF. W3C Recommendation (2012). http://www.w3.org/TR/rdb-direct-mapping/
4. Guo, Y., Pan, Z., Heflin, J.: LUBM: a benchmark for OWL knowledge base systems. J. Web Semant. 3(2–3), 158–182 (2005)
5. Härder, T., Reuter, A.: Principles of transaction-oriented database recovery. ACM Comput. Surv. 15(4), 287–317 (1983)
6. Harris, S., Seaborne, A.: SPARQL 1.1 Query Language. W3C Recommendation (2013). https://www.w3.org/TR/sparql11-query/
7. Herlihy, M., Shavit, N.: The Art of Multiprocessor Programming. Morgan Kaufmann, Boston (2008)
8. Horrocks, I., Patel-Schneider, P.F., Boley, H., Tabet, S., Grosof, B., Dean, M.: SWRL: a semantic web rule language combining OWL and RuleML. W3C Member Submission (2004). http://www.w3.org/Submission/SWRL/

9. Motik, B., Nenov, Y., Horrocks, I.: Parallel materialisation of a set of logical rules on a logical database (US Patent 10817467) (2020)
10. Motik, B., Nenov, Y., Horrocks, I.: Complex query evaluation using sideways information passing (US Patent 11216456) (2022)
11. Motik, B., Nenov, Y., Piro, R., Horrocks, I.: Incremental update of datalog materialisation: the backward/forward algorithm. In: Proceedings of the 29th National Conference on Artificial Intelligence (AAAI 15), pp. 1560–1568. AAAI Press (2015)
12. Motik, B., Nenov, Y., Piro, R., Horrocks, I.: Maintenance of datalog materialisations revisited. Artif. Intell. **269**, 76–136 (2019)
13. Motik, B., Nenov, Y., Piro, R., Horrocks, I., Olteanu, D.: Parallel materialisation of Datalog programs in centralised, main-memory RDF systems. In: Proceedings of the 28th National Conference on Artificial Intelligence (AAAI 14), pp. 129–137. AAAI Press (2014)
14. Nenov, Y., Piro, R., Motik, B., Horrocks, I., Wu, Z., Banerjee, J.: RDFox: a highly-scalable RDF store. In: Arenas, M., et al. (eds.) ISWC 2015. LNCS, vol. 9367, pp. 3–20. Springer, Cham (2015). https://doi.org/10.1007/978-3-319-25010-6_1
15. Neumann, T., Weikum, G.: Scalable join processing on very large RDF graphs. In: SIGMOD Conference, pp. 627–640. ACM (2009)
16. Neumann, T., Weikum, G.: The RDF-3X engine for scalable management of RDF data. VLDB J. **19**(1), 91–113 (2010)
17. OWL 2 Web Ontology Language Overview (Second Edition). W3C Recommendation (2012). http://www.w3.org/TR/owl2-overview/
18. Prud'hommeaux, E., Boneva, I., Labra Gayo, J.E., Kellogg, G.: Shape expressions language 2.1. W3C Community Group Report (2019). http://shex.io/shex-semantics/
19. Shapes Constraint Language (SHACL). W3C Recommendation (2017). https://www.w3.org/TR/shacl/
20. Weiss, C., Karras, P., Bernstein, A.: Hexastore: sextuple indexing for semantic web data management. PVLDB **1**(1), 1008–1019 (2008)

CONSTRUCT Queries Performance on a Spark-Based Big RDF Triplestore

Adam Sanchez-Ayte[(⊠)], Fabrice Jouanot, and Marie-Christine Rousset

Université Grenoble Alpes, Saint-Martin-d'Hères, France
{adam.sanchez,fabrice.jouanot,
marie-christine.rousset}@univ-grenoble-alpes.fr

Abstract. Despite their potential, CONSTRUCT queries have gained little attraction so far among data practitioners, vendors and researchers. In this paper, we first exhibit performance bottlenecks of existing triplestores for supporting CONSTRUCT queries over large knowledge graphs. Then, we describe a novel Spark-based architecture for big triplestores, called TESS, that we have designed and implemented to overcome the above limitations by using parallel computing. TESS ensures ACID properties that are required for a sound and complete implementation of CONSTRUCT-based forward-chaining rules reasoning.

Keywords: CONSTRUCT queries · Big knowledge graphs · Spark

1 Introduction

CONSTRUCT queries are SPARQL queries that enable ETL[1] data pipelines (to reduce large datasets to workable datasets), graph interoperability (to merge graphs from different sources) and are a key component in several W3C specifications (e.g., SPIN[2], and later SHACL[3]) for supporting rule-based inference.

However, despite their potential, CONSTRUCT queries have gained little attraction so far among data practitioners, vendors and researchers. In fact, CONSTRUCT queries are available since the first SPARQL specification (2008) but their usage has been very limited in public SPARQL endpoints. According to an analytical study of large SPARQL query logs conducted in [3], CONSTRUCT queries represented only 1.84% from a total of 90 millions of unique queries collected from 14 datasets between 2013 and 2017.

Query performance evaluation has been centered on SELECT queries and have been neglected for CONSTRUCT queries for which no performance benchmark on large datasets is available. For example, from 29 SPARQL queries proposed to measure performance of different types of queries in Task 2 of the MOCHA 2018 Challenge[4], none of them was a CONSTRUCT query.

---

[1] Extraction, Transformation, Load.
[2] https://spinrdf.org/spin.html.
[3] https://www.w3.org/TR/shacl-af/#rules.
[4] https://project-hobbit.eu/challenges/mighty-storage-challenge2018/.

© The Author(s), under exclusive license to Springer Nature Switzerland AG 2022
P. Groth et al. (Eds.): ESWC 2022, LNCS 13261, pp. 444–460, 2022.
https://doi.org/10.1007/978-3-031-06981-9_26

For current SPARQL implementations, the output size of CONSTRUCT queries is restricted. For example, in Virtuoso, while millions of rows can be fully streamed for SELECT queries, CONSTRUCT query results cannot be fully outputted beyond 1 million triples[5].

In this paper, we first exhibit (Sect. 3) performance bottlenecks of existing triplestores (namely Virtuoso and GraphDB) for supporting CONSTRUCT queries over large knowledge graphs, even when we decompose their computation into the evaluation of SELECT queries followed by the construction and the storage of the graph output. For this, in the absence of appropriate benchmarks, we have set up an experimental protocol (described in Sect. 3.2) on top of a big knowledge graph, called OntoSIDES [13], at the core of a learning management system used in medical studies in France.

Then, we describe (Sect. 4) a novel Spark-based architecture for big triplestores, called TESS, that we have designed and implemented to overcome the above limitations by using parallel computing. TESS ensures a part of ACID properties that are required for a sound and complete implementation of CONSTRUCT-based forward-chaining rules reasoning. We report in Sect. 4.2 the experimental results on the performance of TESS that we have obtained.

Beforehand, Sect. 2 provides the background of this work. Finally, Sect. 5 positions it w.r.t. the related work and Sect. 6 concludes the paper.

2 Background

Let I, L, B, and V be pairwise disjoint sets of IRIs, literals, blank nodes, and variables, respectively. An RDF graph is a set of RDF triples $(s, p, o) \in (I \cup B) \times I \times (I \cup L \cup B)$. A named graph is a pair consisting of an IRI and an RDF graph. An RDF dataset is a collection of RDF named graphs. The CONSTRUCT and SELECT queries that we consider are built on SPARQL 1.1 graph patterns [5].

Definition 1 (SPARQL 1.1 graph pattern).

– *A basic graph pattern is a set of triple patterns* $(s, p, o) \in (I \cup V) \times (I \cup V) \times (I \cup L \cup V)$.
– *A SPARQL 1.1 graph pattern is an expression P generated from the following grammar:*

$$P :: = BGP \mid (P_1 \ Union \ P_2) \mid (P_1 \ And \ P_2) \mid (P_1 \ Opt \ P_2) \mid P \ FILTER \ R$$
$$\mid Graph \ g \ P \mid FILTER \ NOT \ EXISTS \ P$$

where BGP is a basic graph pattern, $g \in V \cup I$ and R is a constraint expression over variables in P.

Definition 2 (SELECT queries). *By $\overline{x}$ we denote a vector of variables.*

---

[5] https://community.openlinksw.com/t/sparql-query-limiting-results-to-100000-triples/2131.

- *A simple SELECT query is of the form:*
  SELECT $\bar{x}$ WHERE { GP } where GP is a SPARQL 1.1 graph pattern including variables in $\bar{x}$. When evaluated over an RDF graph (or dataset) G, there are as many answers $\mu(\bar{x})$ as mappings μ allowing to match GP with a subgraph of G.
- *An aggregate SELECT query is of the form*

$$\text{SELECT } \bar{x}, \ f(\bar{y}) \ \text{WHERE} \{ \ GP \ \} \ \text{GROUP BY } \bar{x}$$

  where f is an aggregate function and GP a SPARQL 1.1 graph pattern including variables in $\bar{x} \cup \bar{y}$. When evaluated over G, there are as many groups as mappings allowing to match the tuple $\bar{x}$ with tuples of values $\bar{v}$ and as many answers $(\bar{v}, av)$ where av is computed by the aggregate function on the corresponding group.
- *A nested SELECT query is a SELECT query for which the WHERE clause is of the form { GP { SQ } } where GP is a SPARQL 1.1 graph pattern and SQ is a (simple or aggregate) SELECT query. The inner SELECT query is called a subquery and is evaluated first. The subquery result variable(s) can then be used in the outer SELECT query.*

Definition 3 (CONSTRUCT queries). *A CONSTRUCT query is written:*

$$\text{CONSTRUCT} \{ \ Template \ \} \ \text{WHERE} \ \{ \ GP \ [\{ \ SQ \ \}] \ \}$$

where GP is a SPARQL 1.1 graph pattern, Template is a basic graph pattern (possibly containing blank nodes) with variables appearing in GP, and SQ is an optional SELECT subquery.
The result of the evaluation over an RDF graph G is the union of graphs obtained by instantiating the variables x in Template with values $\mu(x)$ for each mapping μ satisfying the WHERE clause.
The induced SELECT query is: SELECT $\bar{x}$ WHERE $\{GP \ [\{ \ SQ \ \}] \ \}$
where $\bar{x}$ is made of all the variables in the graph pattern GP.

Based on Definitions 2 and 3, the computation of the result of a CONSTRUCT query can be decomposed into the evaluation of its induced SELECT query followed by the construction of the output graph as the union of the template instances obtained by replacing each variable by its corresponding value in the answer set of the SELECT query. Figure 1 shows an example of a CONSTRUCT query and of its induced SELECT query.

CONSTRUCT-Based Forward-Chaining Rules Reasoning
Like in SHACL and SPIN specifications, CONSTRUCT queries can be used to express rules that allow to derive inferred RDF triples from existing asserted triples. For example, the CONSTRUCT query in Fig. 1 participates to the rule-based definition of two properties for answers in the e-learning setting of Onto-SIDES (see Sect. 3.1): the numerical property has_for_result and the boolean property stronglyWrong.

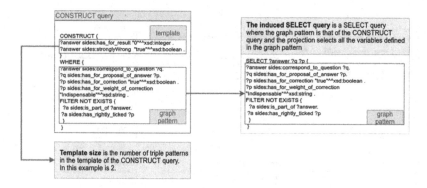

Fig. 1. Example of a CONSTRUCT query and of its induced SELECT query

A forward-chaining reasoner can thus be implemented on top of any RDF triplestore by iterating the triggering of the CONSTRUCT queries and the adding of their results in the triplestore. The termination is guaranteed when the rules are safe, i.e., when no new blank nodes appear in the the template of the corresponding CONSTRUCT queries. In addition, when the rules are non recursive, i.e., when the underlying dependency graph [8] is acyclic, they can be organized in independent reasoning layers that can be computed at compile time. Then, rule triggering can be ordered in a serial or parallel manner so that each corresponding CONSTRUCT query is evaluated only once.

3   Performance Evaluation of Virtuoso and GraphDB

In the absence of appropriate benchmarks (for CONSTRUCT queries or for SELECT queries on big knowledge graphs), we have chosen to conduct our performance evaluation on the OntoSIDES knowledge graph the size of which (12 Billions triples) is comparable to that of Wikidata (14 billions triples as of 2020) and DBpedia (21 billions triples as of 2021).

3.1   OntoSIDES Benchmark for CONSTRUCT Queries

OntoSIDES is a big knowledge graph at the core of an ontology-based learning management system used in medical studies in France, in which the educational content, the traces of students' activities and the correction of exams are described in RDF using a lightweight ontology [13]. Thanks to an automatic mapping-based data materialization and rule-based data saturation, OntoSIDES contains about 12 Billions triples to date, and describes training and assessments activities performed by more than 145,000 students over almost 6 years. Students activities are described at the granularity of time-stamped clicks of answers done by students for choosing among the proposals of answers associated to multiple choices questions.

Q1

```
CONSTRUCT
{?question sides:has_for_number_of_proposals
?np}
WHERE { select ?question (COUNT (?p) As
?np)
{?question
sides:has_for_proposal_of_answer?p}
group by ?question
}
```

Q2

```
CONSTRUCT {
?answer sides:has_for_number_of_wrong_tick
?nw }
WHERE {
select ?answer (COUNT (?a) As ?nw)
{?a sides:is_part_of ?answer.
?a sides:has_wrongly_ticked ?p}
group by ?answer
}
```

Q3

```
CONSTRUCT
{?answer
sides:has_for_number_of_missed_right_tick
?nm }
WHERE {select ?answer (COUNT(?p) As
?nm)
{?answer sides:correspond_to_question ?q.
?q sides:has_for_proposal_of_answer ?p.
?p sides:has_for_correction
"true"^^xsd:boolean.
FILTER NOT EXISTS {?a sides:is_part_of
?answer.
?a sides:has_rightly_ticked ?p}
}
group by ?answer
}
```

Q4

```
CONSTRUCT
{?answer:has_for_number_of_discordance
"0"^^xsd:integer}
WHERE {?answer a sides:answer.
FILTER NOT EXISTS {?answer
sides:has_for_number_of_wrong_tick ?nw}
FILTER NOT EXISTS {?answer
sides:has_for_number_of_missed_right_tick
?nm}
}
```

Q5

```
CONSTRUCT
{?answer:has_for_number_of_discordance
?count }
WHERE {
select ?answer (?nw + ?nm as ?count)
{
?answer sides:has_for_number_of_wrong_tick
?nw.
?answer
sides:has_for_number_of_missed_right_tick ?nm
}
}
```

Q6

```
CONSTRUCT
{?answer:has_for_number_of_discordance
?nw}
WHERE {?answer
sides:has_for_number_of_wrong_tick ?nw.
FILTER NOT EXISTS {?answer
sides:has_for_number_of_missed_right_tick
?nm}
```

Q7

```
CONSTRUCT
{?answer sides:has_for_number_of_discordance
?nm}
WHERE {?answer
sides:has_for_number_of_missed_right_tick ?nm.
FILTER NOT EXISTS {?answer
sides:has_for_number_of_wrong_tick ?nw. }
}
```

Q8

```
CONSTRUCT
{?answer sides:has_for_result 1}
WHERE {?answer
sides:has_for_number_of_discordance
"0"^^xsd:integer
}
```

Q9

```
CONSTRUCT
{ ?answer sides:has_for_result
"0"^^xsd:integer .
?answer sides:stronglyWrong
"true"^^xsd:boolean .}
WHERE {
?a sides:is_part_of ?answer.
?a sides:has_wrongly_ticked ?p.
?p sides:has_for_weight_of_correction
"Unacceptable"^^xsd:string
}
```

Q10

```
CONSTRUCT
{?answer sides:has_for_result "0"^^xsd:integer
.
?answer sides:stronglyWrong
"true"^^xsd:boolean .
}
WHERE {
?answer sides:correspond_to_question ?q.
?q sides:has_for_proposal_of_answer ?p.
?p sides:has_for_correction
"true"^^xsd:boolean .
?p sides:has_for_weight_of_correction
"Indispensable"^^xsd:string .
FILTER NOT EXISTS {
?a sides:is_part_of ?answer.
?a sides:has_rightly_ticked ?p }
}
```

Q11

```
CONSTRUCT {?answer sides:has_for_result
"0"^^xsd:integer .
?answer sides:stronglyWrong
"true"^^xsd:boolean .
}
WHERE {
?answer sides:correspond_to_question ?q.
?q rdf:type sides:QUA.
?answer sides:has_for_number_of_discordance
?d.
FILTER (?d > 0)
}
```

Q12

```
CONSTRUCT {?answer sides:has_for_result
0.5^^xsd:decimal }
WHERE {
?answer sides:has_for_number_of_discordance
"1"^^xsd:integer .
?answer sides:correspond_to_question ?q.
?q sides:has_for_number_of_proposals
"5"^^xsd:integer.
FILTER NOT EXISTS {?answer
sides:stronglyWrong "true"^^xsd:boolean }
}
```

Q13

```
CONSTRUCT {?answer sides:has_for_result
"0.2"^^xsd:decimal }
WHERE {?answer
sides:has_for_number_of_discordance
"2"^^xsd:integer .
?answer sides:correspond_to_question ?q.
?q sides:has_for_number_of_proposals
"5"^^xsd:integer .
FILTER NOT EXISTS {?answer
sides:stronglyWrong "true"^^xsd:boolean }
}
```

Q14

```
CONSTRUCT {?answer sides:has_for_result
"0.425"^^xsd:decimal }
WHERE {?answer
sides:has_for_number_of_discordance
"1"^^xsd:integer .
?answer sides:correspond_to_question ?q.
?q sides:has_for_number_of_proposals
"4"^^xsd:integer .
FILTER NOT EXISTS {?answer
sides:stronglyWrong "true"^^xsd:boolean }}
```

Q15

```
CONSTRUCT {?answer sides:has_for_result
"0.1"^^xsd:decimal
}
WHERE
{?answer sides:has_for_number_of_discordance
"2"^^xsd:integer .
?answer sides:correspond_to_question ?q.
?q sides:has_for_number_of_proposals
"4"^^xsd:integer .
FILTER NOT EXISTS {
?answer sides:stronglyWrong
"true"^^xsd:boolean }
}
```

Q16

```
CONSTRUCT
{?answer sides:has_for_result
"0"^^xsd:integer}
WHERE
{?answer sides:correspond_to_question ?q.
?q sides:has_for_number_of_proposals ?np.
?answer sides:has_for_number_of_discordance
?n.
FILTER (?np > 3 && ?np < 6 && ?n > 2).
}
```

Q17

```
CONSTRUCT
{?answer sides:has_for_result
"0.3"^^xsd:decimal}
WHERE {
?answer sides:has_for_number_of_discordance
"1"^^xsd:integer .
?answer sides:correspond_to_question ?q.
?q sides:has_for_number_of_proposals
"3"^^xsd:integer .
FILTER NOT EXISTS {
?answer sides:stronglyWrong
"true"^^xsd:boolean }
}
```

Q18

```
CONSTRUCT
{?answer sides:has_for_result "0"^^xsd:integer
}
WHERE
{?answer sides:has_for_number_of_discordance
?n.
?answer sides:correspond_to_question ?q.
?q sides:has_for_number_of_proposals
"3"^^xsd:integer.
FILTER (?n > 1)
}
```

Fig. 2. 18 CONSTRUCT queries over OntoSIDES knowledge graph

Table 1. Classification by category

Category	Queries
Simple (BGP)	Q5, Q8, Q9
Aggregate subquery	Q1, Q2, Q3
FILTER on terms (FRT)	Q11, Q16, Q18
FILTER NOT EXISTS on graph patterns (FGP)	Q3, Q4, Q6, Q7, Q10, Q12, Q13, Q14, Q15, Q17

Table 2. Classification by their template size

Template size	Queries
1	Q1, Q2, Q3, Q4, Q5, Q6, Q7, Q8, Q12, Q13, Q14, Q15, Q16, Q17, Q18
2	Q9, Q10, Q11

Among the 48 properties defined in the OntoSIDES ontology, 6 properties are defined by 18 rules expressed as CONSTRUCT queries provided in Fig. 2.

As summarized in Table 1 and Table 2, the considered CONSTRUCT queries cover a variety of SPARQL 1.1 features.

3.2 Experimental Protocol

The goal is to study how CONSTRUCT query evaluation performance is impacted by the growing size of the input RDF datasets. We first explain how we have built the different RDF datasets over which the 18 CONSTRUCT queries described above will be evaluated. Then, we describe the measures that we consider for evaluating the performance of each CONSTRUCT query in isolation as well as for the whole process of forward-chaining reasoning.

RDF Datasets. We have extracted *10 datasets* from the whole OntoSIDES knowledge graph (before saturation) which growing sizes ranging from *121 millions of triples* to *1.6 billion triples* as shown in Table 3.

Table 3. Ontosides datasets

Dataset	Size (millions triples)
D1	121
D2	194
D3	273
D4	380
D5	497
D6	633
D7	791
D8	977
D9	1209
D10	1604

For doing so, we have adapted the notion of traversal views introduced in [11] and we have structured the OntoSIDES knowledge graph (before saturation) as the union of named graphs whose IRI is a given student's IRI, so that each of these named graph contains the RDF description of all the answers done by the student and of all the corresponding questions. The 10 datasets have been obtained by grouping increasing numbers of students' named graphs (from 880 students' named graphs for the D1 dataset to 8845 students' named graphs for the D10 dataset). By doing so, each extracted dataset contains the required data for each of the 18 CONSTRUCT queries to produce a sound and complete result for the computation of the inferred properties on meaningful fragments of the full OntoSIDES knowledge graph.

Performance Measures

For each CONSTRUCT query, in addition to measuring its execution time that we denote the *construct execution time*, we will also measure:

- the *body execution time*, the time to evaluate its induced SELECT query
- the *template execution time*, the time to instantiate the template. Since triple-stores do not provide the *template execution time*, we will compute it as the difference between the *construct execution time* and the *body execution time*
- the *construct storing time*, the update time needed to add the output of a CONSTRUCT query to the triplestore
- the *inference time*, the sum of the *construct execution time* and the *construct storing time*, which estimates the cost of a CONSTRUCT query used as an update rule.

Given the set of the 18 CONSTRUCT queries in Fig. 2 used as rules, we will also evaluate the performance of both *serial* and *parallel* implementations of CONSTRUCT-based forward-chaining reasoning. Based on their dependency graph, the rules can be structured in 4-depth layers of reasoning:

- Layer 1 = {Q1, Q2, Q3, Q9, Q10}
- Layer 2 = {Q4, Q5, Q6, Q7}
- Layer 3 = {Q8, Q12, Q11}
- Layer 4 = {Q13, Q14, Q15, Q16, Q17, Q18}

The serial versus parallel implementations of CONSTRUCT-based forward-chaining reasoning differ in the sequential versus parallel execution of the CONSTRUCT queries within each layer. The layers are themselves handled by increasing depth. We will measure and compare:

- the *serial forward-chaining reasoning time*, as the sum of *inference times* of all the queries applied sequentially in the order induced by the different layers,
- the *parallel forward-chaining reasoning time*, as the sum of the parallel execution and update times for each of the 4 reasoning layers.

Hardware. The server used in our experiments has the following characteristics:

- Processor: 32 cores, Intel(R) Xeon(R) Gold 6144 CPU @ 3.50Ghz.
- Disk: 7 disks, 2 Terabytes size each.
- Memory: 566 Gigabytes RAM.

3.3   Limitations of Virtuoso and GraphDB

Virtuoso is a column-store triplestore where SPARQL queries are translated into SQL to be executed. GraphDB is a native triplestore where SPARQL queries are executed directly on data. In our experiments, we used Virtuoso 07.20.3229 Community Edition, GraphDB 9.0.0 Enterprise Edition and Docker 19.03.8. Both have been configured for optimal parallelization and memory usage according their online documentation. Blazegraph was not considered because it was outperformed by Virtuoso in 3 of 4 tasks in Mocha 2018 (RDF data ingestion, data storage, versioning).

Each dataset from Table 3 was stored in a Virtuoso and a GraphDB triplestore. Each triplestore run on top of a Docker container configured for 32 CPU cores and 128 GB RAM.

Figure 3a shows how the forward-chaining reasoning time (y axis) evolves in function of the sizes (x axis) of the 10 datasets reported in Table 3:

- for GraphDB, the CONSTRUCT-based forward-chaining rules reasoning can be completed in a reasonable time for D1, D2 and D3 datasets only.
- Virtuoso does not show up at all because the output of each of the 18 CONSTRUCT queries was greater than 1 million triples which is the maximum limit for a CONSTRUCT query output in Virtuoso.

In Fig. 3b, the y axis corresponds to the sum of the execution times of the SELECT queries induced by the 18 CONSTRUCT queries. Virtuoso does not suffer of the above limitations on output size for SELECT queries. However, we have discovered that for the datasets greater than D4 (380 million triples), Virtuoso does not compute the correct answers for *aggregate* queries like Q3 (and others aggregate queries outside the strict setting of our experiment). We have used PostgreSQL as reference to validate the correctness of the results after transforming the SELECT queries into SQL queries.

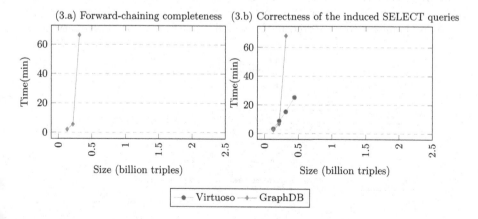

Fig. 3. Forward chaining reasoning time performance. Best viewed in color

These experiments show the limitation of Virtuoso for outputting CON-STRUCT results of more than 1 million triples, and to compute correct answers to *aggregate* SELECT queries over datasets of size greater than 380 millions triples. They also show the limitation of GraphDB to compute SELECT or CONSTRUCT queries in a reasonable time over datasets of size greater than 275 millions triples. This motivates the needs for an architecture supporting parallel computing.

4   TESS Architecture and Performance

In this section, we describe TESS, a novel architecture for big RDF triplestores, and we provide experimental results on its performance. TESS has been designed to support CONSTRUCT queries. Since CONSTRUCT queries can be used to update triplestores, it is important to guarantee data integrity during this updating step. This is particularly important when CONSTRUCT queries are used for supporting rule-based reasoning. For this reason, we have included in the TESS architecture a transactional management module that enforces atomicity property.

4.1   TESS Architecture

TESS is based on a modular architecture that supports log-based transactions for data updates. Transactions in TESS are highly scalable and enables key data management features like query point-in-time and rollback operations.

For the implementation, we chose Spark [18], (the leading platform for large-scale SQL and batch processing as of today [9]), as base technology to select the proper software for each architecture component.

Figure 4 shows the 5 layers of the TESS architecture with the selected tech-nologies for each component (on the right side) and two inputs types supported by TESS: at the top, the Spark Application for CONSTRUCT-based forward-chaining reasoning and, on the left side, a SPARQL query. However, only the SPARQL query has an external output since the outcome of the forward chaining reasoning is meant to be stored in the distributed storage for later querying.

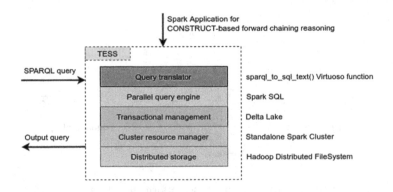

Fig. 4. TESS triplestore architecture

We now describe each component. The modular architecture makes possible to disable some of them, like the transactions manager or the cluster resource manager if they are not useful, for instance if the CONSTRUCT queries are not used for updates or if parallel computation is not necessary.

Distributed Storage. We define a schema of 4 columns (i.e. <s, p, o, g>) to store RDF data. One-table layout is very efficient for updates because an update does not need to be normalized into the many tables of other layouts. In addition, ACID properties provided by Delta Lake only supports transactions on one table at a time [22]. This structure, using a column storage format for performance purpose (a collection of versioned Parquet [19] files), is referenced as the ACID table. The storage is based on Hadoop Distributed FileSystem (HDFS) cluster [20] which operates in a fully distributed mode. It comprises a namenode (master) and a datanode (slave) servers.

Cluster Resource Management. This component allows to execute a query plan using high parallel computing based on standalone Spark cluster. It comprises a master and workers nodes, usually as many workers as queries/rules to manage with. The master receives Spark applications and schedules worker resources to be run among them. A Spark application is organized around jobs, the top level work unit. By default, Spark jobs within an application are executed serially, but they can also be run in parallel if concurrency is enabled at application level.

Transactions Management. This layer is in charge of the reliability and the correctness of RDF data with update transactions. Based on Delta Lake [21] that adds ACID service to Spark: a) keep track of all the commits made to the ACID table and b) use time travel for loading the ACID table at a given version or timestamp [10].

Parallel Query Engine. A query optimization/execution component based on Spark SQL which starts from the logical query plan to generate an optimized physical query plan. Then, the optimized query plan is used to generate efficient code to exploit modern compilers and CPUs.

Query Translator. This component is needed to rewrite SPARQL queries into SQL in order to use Spark SQL, a Spark module for dataframe-based structured data processing. We retained sparql_to_sql_text() Virtuoso [12] function to generate self-joins queries for the ACID table.

4.2 TESS Performance Evaluation

For our experiments, we used Spark 3.1.1, Delta Lake 0.8.0 and HDFS 2.9.2. Each dataset from Table 3 was stored in the HDFS as an ACID table.

Serial Forward-Chaining Reasoning Time

For this experiment, TESS run on top of a network of 5 Docker containers: 2 containers for the Hadoop namenode and datanode. 2 containers for the Spark Standalone cluster (1 for the master node, 1 for the worker node) and 1 container for sending the Spark Application to the Spark Standalone Cluster in client mode. We assign 128GB RAM and 32 CPU cores to each container member of the Spark Standalone cluster.

Figure 5 shows that TESS completes the forward-chaining reasoning for all datasets in reasonable time and that the time grows linearly w.r.t the size of the input datasets. (see black curve CONSTRUCT with square shaped dots). It also makes explicit how the *construct execution time* is split into the *body execution time* (see blue curve with triangle shaped dots) and the *template execution time* (see red curve with circle shaped dots). We observe that the impact of *body execution time* is much greater than the *template execution time* for CONSTRUCT-based forward-chaining reasoning.

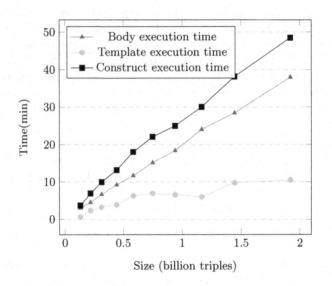

Fig. 5. Evaluation of serial forward-chaining reasoning time. Best viewed in color.

Figure 6 shows the individual performance of each of the 18 queries. We observe that for most of the queries, the coefficient of the linear progression of time in function of dataset size is very small (for Q1, Q14, Q15, Q17 and Q18) or small (for Q5, Q6, Q7, Q9, Q11, Q12, Q13 and Q16). The same figure shows that the difference between *construct execution time* and *body execution time* may be important when the graph output of the CONSTRUCT queries is not restricted to a single triple pattern, like in the queries Q9, Q10 and Q11.

In Fig. 7, we focus on the 5 most expensive queries, namely Q2, Q3,Q4, Q8 and Q10, and we show the correlation between the query output size and the

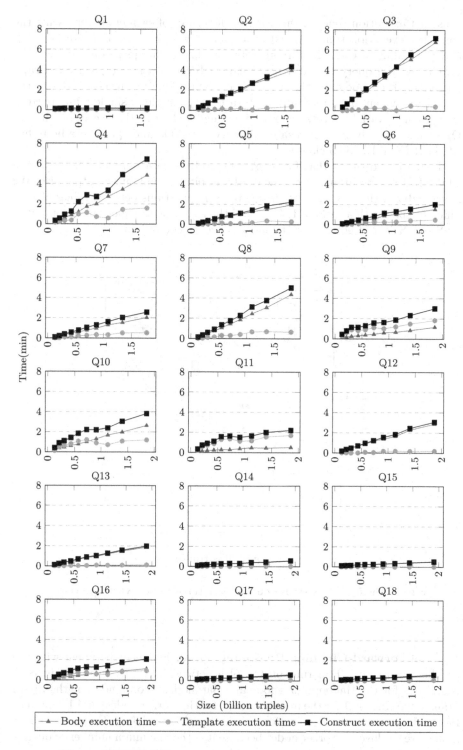

Fig. 6. CONSTRUCT queries performance. Best viewed in color.

construct execution time. In this Figure, the center of each circle represents the construct execution time (y axis) of a query for a given dataset size (x axis), and the radius of each circle represents the size of the query output size. For Q2,Q3,Q4, the query cost can be explained both by the complexity of graph patterns in their body and the size of their output. For Q8, the cost is due to the size of its output since its graph pattern is very simple: a single triple pattern with a single variable. Yet, its execution time is close to the execution time of Q2 (whose body has an aggregate subquery) or of Q3 and Q4 (whose body has FILTER NOT EXISTS clauses).

Figure 7 also shows that for queries like Q10 with a template size > 1, the CONSTRUCT performance can be costly despite a small query output size. We have analyzed that its high cost is due to join operations between tables of very different size (with a ratio of 1/453), and the fact that the query plan computed by Spark SQL did not choose the most efficient type of joins.

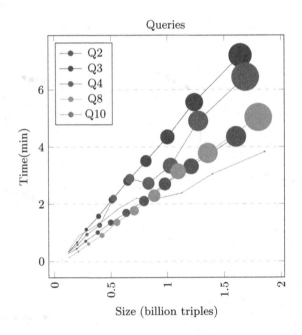

Fig. 7. The 5 most expensive queries performance. Best viewed in color.

Parallel Forward-Chaining Reasoning Time. For the experiment with the parallel algorithm, TESS runs on top of a network of 10 Docker containers: 2 containers for the Hadoop namenode and datanode. 7 containers for the Spark Standalone cluster (1 for the master node, 6 for the worker nodes) and 1 container for sending the Spark Application to the Spark Standalone Cluster in client mode. We deployed 6 worker nodes because it is the maximum number of queries in a layer of reasoning. The Spark Application executes a query per worker node. We reduce the number of workers from 6 to 1 for setting up the experiment with the serial algorithm for comparison purposes. Furthermore, we assign 64 GB RAM and 4 CPU cores to each container member of the Spark Standalone cluster.

Figure 8 shows how the TESS implementation of the parallel forward-chaining algorithm outperforms the serial algorithm for all the datasets. The execution time seems to grow linearly but with a much smaller coefficient than for the serial case.

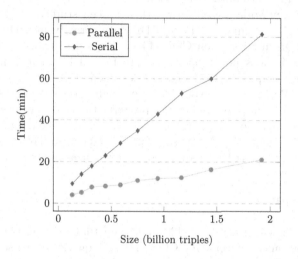

Fig. 8. Parallel vs Serial performance. Best viewed in color.

The source code of the Spark cluster and CONSTRUCT-based forward chaining implementation along with the 18 CONSTRUCT queries used in the experiments are available in https://github.com/asanchez75/ontosides-bpe.

5   Related Work

To the best of our knowledge, CONSTRUCT queries performance has been barely covered in two benchmarks: the Berlin SPARQL Benchmark (BSBM) [2] and the Featured-Based SPARQL Benchmark Generation Framework (FEA-SIBLE) [15]. However, it was restricted to *centralized triplestores* (e.g. Virtuoso, Sesame, Jena Fuseki and OWLIM-SE) and the size of the biggest dataset was 232.5 millions triples.

SELECT queries performance has received more attention. In [16], an extensive analysis of eleven SPARQL benchmarks has been carried out on *centralized triplestores* (e.g. Virtuoso and Fuseki). Despite one of the benchmarks (BowlognaBench [6]) included aggregate queries, they were not considered and the survey only covered SELECT queries without aggregate. The size of the biggest dataset in the survey was 232 millions triples.

An exhaustive evaluation of Big RDF frameworks is performed in [4]. The survey shows how distributed storage and parallel query processing for RDF data have evolved over time. For SPARQL parallel query processing, MapReduce has been replaced gradually by Spark, whereas for distributed storage, Hive and HBase has been superseeded by HDFS and Parquet. Although some of the frameworks considered aggregate queries in their performance studies, none of them dealt with FILTER NOT EXISTS queries. None of the approaches provided ACID support for RDF data transactions.

Big RDF frameworks based on Spark has been studied in [1]. In contrast with SANSA [17], Bellman [7] loads RDF data directly into Dataframes and execute SPARQL queries (even CONSTRUCT queries) translated into Spark-SQL. However, it does not support full SPARQL 1.1 and it is not clear if it supports aggregate queries due to lack of documentation.

Regarding RDF storage layouts, even though the experiments reported in [14] show that vertical partitioning and property tables outperform single table layout for some scenarios, single table layout remains as the dominant layout in real-world deployments (e.g. Virtuoso). The implementation of the other layouts is a time consuming task that requires data normalization and query rewriting.

6   Conclusion

We have first shown in a real-world application that existing triplestores have intrinsic limitations for supporting CONSTRUCT queries at big scale. Then, we have described TESS, a novel modular Spark-based infrastructure for big RDF triplestores that we have designed and implemented based on modern technologies for distributed computing over big data. We have built on components offered by the growing ecosystem of Big Data SQL management tools.

A distinguishing point of TESS is that it implements part of ACID properties, namely atomicity, which is required to reliably support CONSTRUCT-based updates of triplestores. This is particularly crucial when CONSTRUCT queries are used to implement forward-chaining rules reasoning.

Our experiments have demonstrated that TESS triplestores can manage full SPARQL 1.1 CONSTRUCT queries on large datasets. We have also shown the performance gain when we exploit TESS components to implement parallel CONSTRUCT-based forward-reasoning. As future work, we plan to conduct a query performance comparison between CPU-based and GPU-based TESS architecture, and a performance study of workload (supporting thousands of queries per second) for CONSTRUCT-based ontology modularization.

Acknowledgements. This work has been supported by the the French National Research Agency with projects *LabEx* PERSYVAL Lab (11-LABX-0025-01), *DUNE* SIDES 3.0 (ANR-16-DUNE -0002-02), *P3IA* MIAI@Grenoble Alpes (ANR-19-P3IA-0003) and *CE23* CQFD (ANR-18-CE23-0003).

References

1. Agathangelos, G., Troullinou, G., Kondylakis, H., Stefanidis, K., Plexousakis, D.: Rdf query answering using apache spark: review and assessment. In: 2018 IEEE 34th International Conference on Data Engineering Workshops (ICDEW), pp. 54–59 (2018). https://doi.org/10.1109/ICDEW.2018.00016
2. Bizer, C., Schultz, A.: The berlin sparql benchmark. Int. J. Semantic Web Inf. Syst. **5**, 1–24 (2009). https://doi.org/10.4018/jswis.2009040101
3. Bonifati, A., Martens, W., Timm, T.: An analytical study of large SPARQL query logs. VLDB J. **29**(2–3), 655–679 (2020) https://doi.org/10.1007/s00778-019-00558-9, https://hal.archives-ouvertes.fr/hal-03118422
4. Chawla, T., Singh, G., Pilli, E.S., Govil, M.: Storage, partitioning, indexing and retrieval in big rdf frameworks: a survey. Comput. Sci. Revi. **38**, 100309 (2020). https://doi.org/10.1016/j.cosrev.2020.100309, https://www.sciencedirect.com/science/article/pii/S1574013720304093
5. Chen, Y., Kokar, M., Moskal, J.: Sparql query generator (SQG). J. Data Semant. **10**, 1–17 (2021). https://doi.org/10.1007/s13740-021-00133-y
6. Demartini, G., Enchev, I., Wylot, M., Gapany, J., Cudre-Mauroux, P.: Bowlognabench-benchmarking RDF analytics, vol. 116 (2012). https://doi.org/10.1007/978-3-642-34044-4_5
7. (GSK), G.: Project bellman. https://gsk-aiops.github.io/bellman/, Accessed 27 Nov 2021
8. Hassanpour, S., O'Connor, M.J., Das, A.K.: Visualizing logical dependencies in SWRL rule bases. In: Dean, M., Hall, J., Rotolo, A., Tabet, S. (eds.) RuleML 2010. LNCS, vol. 6403, pp. 259–272. Springer, Heidelberg (2010). https://doi.org/10.1007/978-3-642-16289-3_22
9. Pointer, I.: Infoword. What is apache spark? the big data platform that crushed hadoop. https://www.infoworld.com/article/3236869/what-is-apache-spark-the-big-data-platform-that-crushed-hadoop.html, Accessed 05 Dec 2021
10. Laskowski, J.: The internals of delta lake. https://books.japila.pl/delta-lake-internals/, Accessed 05 Dec 2021
11. Noy, N.F., Musen, M.A.: Specifying ontology views by traversal. In: McIlraith, S.A., Plexousakis, D., van Harmelen, F. (eds.) ISWC 2004. LNCS, vol. 3298, pp. 713–725. Springer, Heidelberg (2004). https://doi.org/10.1007/978-3-540-30475-3_49
12. OpenLink Software: Virtuoso universal server. https://virtuoso.openlinksw.com/, Accessed 05 Dec 2021
13. Palombi, O., Jouanot, F., Nziengam, N., Omidvar-Tehrani, B., Rousset, M.C., Sanchez, A.: Ontosides: ontology-based student progress monitoring on the national evaluation system of French medical schools. Artif. Intell. Med. **96**, 59–67 (2019)
14. Ragab, M., Sakr, S., Tommasini, R.: Benchmarking spark-SQL under alliterative rdf relational storage backends (2019)
15. Saleem, M., Mehmood, Q., Ngonga Ngomo, A.-C.: FEASIBLE: a feature-based SPARQL benchmark generation framework. In: Arenas, M., et al. (eds.) ISWC 2015. LNCS, vol. 9366, pp. 52–69. Springer, Cham (2015). https://doi.org/10.1007/978-3-319-25007-6_4

16. Saleem, M., Szárnyas, G., Conrads, F., Bukhari, S.A.C., Mehmood, Q., Ngonga Ngomo, A.C.: How representative is a sparql benchmark? an analysis of rdf triplestore benchmarks. In: The World Wide Web Conference, WWW 2019, pp. 1623–1633. Association for Computing Machinery, New York (2019). https://doi.org/10.1145/3308558.3313556

17. Stadler, C., Sejdiu, G., Graux, D., 0001, J.L.: Querying large-scale RDF datasets using the sansa framework. In: Suárez-Figueroa, M.C., Cheng, G., Gentile, A.L., Guéret, C., Keet, C.M., Bernstein, A. (eds.) Proceedings of the ISWC 2019 Satellite Tracks (Posters & Demonstrations, Industry, and Outrageous Ideas) co-located with 18th International Semantic Web Conference (ISWC 2019), Auckland, New Zealand, 26–30 October 2019. CEUR Workshop Proceedings, vol. 2456, pp. 285–288. CEUR-WS.org (2019). http://ceur-ws.org/Vol-2456/paper74.pdf

18. The Apache Software Foundation: Apache spark. https://spark.apache.org/, Accessed 05 Dec 2021

19. The Apache Software Foundation: Apache parquet. https://parquet.apache.org/, Accessed 05 Dec 2021

20. The Apache Software Foundation: Hadoop cluster setup. https://hadoop.apache.org/docs/stable/hadoop-project-dist/hadoop-common/ClusterSetup.html, Accessed 05 Dec 2021

21. The Linux Foundation: Delta lake documentation. https://delta.io/, Accessed 05 Dec 2021

22. Zaharia, M., Ghodsi, A., Xin, R., Armbrust, M.: Lakehouse: a new generation of open platforms that unify data warehousing and advanced analytics. In: 11th Conference on Innovative Data Systems Research, CIDR 2021, Virtual Event, Online Proceedings, 11–15 January 2021 (2021). www.cidrdb.org, http://cidrdb.org/cidr2021/papers/cidr2021_paper17.pdf

Matching Multiple Ontologies to Build a Knowledge Graph for Personalized Medicine

Marta Contreiras Silva[(✉)], Daniel Faria, and Catia Pesquita

LASIGE, Faculdade de Ciências da Universidade de Lisboa, Lisbon, Portugal
mcdsilva@fc.ul.pt

Abstract. A rich biomedical knowledge graph can support the multi-domain data integration necessary for the application of Artificial Intelligence models in personalised medicine. Constructing such a knowledge graph from already available biomedical ontologies relies on ontology matching, however, current ontology matching systems are geared towards the alignment of pairs of ontologies of the same domain one at a time. This approach, when applied to a multi-domain problem such as personalised medicine in an all vs. all fashion, poses scalability issues while also ignoring the particularities of the multi-domain aspect.

In this work we evaluate a state-of-the-art ontology matching system, AgreementMakerLight, in the task of building a network of 28 integrated ontologies to construct a knowledge graph for Explainable AI in personalised oncology, highlighting its shortcomings. To address them, we have developed a novel holistic ontology alignment strategy building on AgreementMakerLight that clusters ontologies based on their semantic overlap measured by fast matching techniques with a high degree of confidence, and then applies more sophisticated matching techniques within each cluster. We implemented two within cluster alignment strategies, one based on pairwise alignment and another on incremental alignment.

The within-cluster incremental alignment reduced alignment time by 80% when compared with within-cluster pairwise alignment, achieving 88% coverage of its mappings. Compared to an all vs. all pairwise approach, holistic approaches reduce total running time by up to 60%.

Keywords: Ontology matching · Holistic ontology matching · Biomedical ontologies · Knowledge graphs

1 Introduction

Data-centric approaches like personalized medicine have taken the forefront in biomedical research, driven by the increasing availability of biomedical data. Artificial Intelligence (AI) is positioned as a promising solution to handle these large heterogeneous datasets composed of various types of data (e.g. genomic, clinical and image data). However, the evolution of AI has favored black-box approaches

P. Groth et al. (Eds.): ESWC 2022, LNCS 13261, pp. 461–477, 2022.
https://doi.org/10.1007/978-3-031-06981-9_27

that, while effective, do not foster user trust or understanding—aspects which are critical in personalized medicine, as it often involves life-or-death decisions.

To address this limitation of black-box AI approaches, there have been renewed efforts towards developing explanatory mechanisms for AI [4]. Frequent among the approaches proposed for explainable AI (XAI) is the use of Knowledge Graphs (KG), which comprehensively encode the knowledge in a domain, and can be leveraged to support user-friendly explanations when used in concert with AI methods [3,16]. The challenge is that, the more complex the domain, the more complex and comprehensive will be the KG needed to support XAI in that domain, and few domains approach the complexity of personalized medicine.

The area of personalized medicine deals with knowledge stemming from many specific subdomains that interact in various ways, ranging from molecules (e.g. chemical compounds, genes, and proteins) to clinical and demographic factors[10]. Accordingly, ontological representations of these domains have been the subject of intense investigation. In BioPortal [19], an online repository of biomedical ontologies, there are currently more than 800 ontologies (totalling almost 9 million classes). Some of these ontologies are designed and developed as community efforts that function as community-approved representations of reality, while others are developed by single research teams and serve a more specific and localized purpose. Thus, rather than build a KG from the ground up, we posit that one can harness the wealth of publicly available knowledge through ontology matching to build the ontological layer of a KG for personalized medicine [23].

The KATY project[1] aims to develop an AI-empowered personalized medicine system to assist medical professionals and researchers in diagnosing patients more accurately, making predictions about their future health, and recommending better treatments. KATY will tackle the challenge of translating AI-based suggestions into practical decision-making processes and treatment strategies that clinicians can understand and trust by combining high performing black-box machine learning approaches with a comprehensive knowledge graph. The KG will serve as input to AI methods (e.g. directly, through embeddings) as well as encode the AI outcomes themselves to create a shared semantic space for data, scientific context and predictions capable of supporting explanation methods [31].

In a preliminary step, a careful selection of ontologies that span the domain of interest was conducted, as the goal is to reconcile the ontologies into a single cohesive knowledge model, through ontology matching techniques, to form the backbone of the knowledge graph. This resulted in a catalogue of relevant ontologies and controlled vocabularies which comprises 78 ontologies, of which 16 are referenced directly from the public data resources, and the remaining 62 were selected from our survey of BioPortal. Of these 78 ontologies, 28 were considered core to the KATY project, and the remaining 50 were considered potentially relevant, to be used only if the coverage of the 28 core ontologies is found insufficient when integrating datasets into the KATY knowledge graph.

This paper describes the process of matching the 28 core ontologies to build an integrated semantic backbone for the knowledge graph, focusing on finding

---

[1] http://katy-project.eu/.

simple equivalence mappings between pairs of entities belonging to the set of ontologies. We further detail the requirements for ontology matching in this application, discuss the challenges found when applying a state of the art ontology matching system, and present a novel approach for holistic ontology matching that builds on an existing system, AML [9], addressing the requirements and challenges in biomedical ontology holistic matching. We performed a series of experiments to demonstrate the impact of the holistic approach and measure improvements over the baseline state of the art system.

2   Challenges in Holistic Biomedical Ontology Matching

Ontology matching (or alignment) is the process of establishing mappings (or correspondences) relating the entities (classes, properties or individuals) of two ontologies with overlapping domains. A mapping is usually represented as a tuple $< e_1, e_2, r, c >$ where e_1 and e_2 are entities of the two ontologies, r is the semantic relation between them (e.g. $\equiv, \geq, \leq, \perp$) and optionally c is a confidence score indicating how certain about the mapping is the person or algorithm who produced it [7]. A collection of mappings between two ontologies is called an alignment, and is typically stored in a file external to the ontologies, in the Alignment RDF format[2] that is the de facto standard in the field.

Matching biomedical ontologies is a challenging task on its own [8], both in terms of computational resources (as they are typically quite large) and in terms of the richness and complexity of the information available to match them, including a substantial lexical component where homonyms and synonyms abound [25], the presence of cross-references that establish correspondences but with no formal semantics, and the presence of logical definitions which correspond to complex ontology mappings [15,20].

Holistic ontology matching is an extension of the pairwise ontology matching process for a set $\Omega = \{O_1, ..., O_N\}$ of ontologies with $N \geq 2$, where a final alignment A is produced between all of them [18]. The basic approach to do this consists of uniting the alignments between all pairwise combinations of the ontologies to align, which is evidently a sub-optimal strategy computationally as in implies performing a quadratic number of ontology alignment steps.

This holistic matching challenge has been recognized by the ontology matching, schema matching and linked data communities [22,27], with strategies to address it being usually based on exploring two different concepts: partitioning the search space in groups within which pairwise alignment is employed [11] or applying incremental matching according to a predefined order [13,30]. Gruetze et al. [11] proposed grouping of linked data concepts by topic using Wikipedia and then running an alignments only between concepts in the same group. Saleem et al. [30] developed a method to incrementally create an integrated schema encompassing all input schema trees, by first clustering the nodes based on linguistic label similarity and then applying a tree mining technique. Hertling et al. [13] analyzed the impact of the ordering of ontologies in linear executions

---

[2] https://moex.gitlabpages.inria.fr/alignapi/format.html.

of alignments to produce an alignment of multiple ontologies and demonstrated that near-optimal results can be achieved with linear efforts. Orthogonally to these works, Megdiche *et al.* [18] developed an approach based on linear programming that is able to find a stable alignment between multiple ontologies independently of the order of alignment tasks.

Building an integrated KG containing multiple ontologies requires, not only holistic matching to produce an external alignment between them, but actually merging the ontologies. Osman *et al.* [21] categorized ontology merging works according to whether they are applied after all pairwise alignments are found [1, 5] or integrated into an incremental matching approach starting from a seed ontology [2,32].

Thus, aligning and integrating the 28 selected KATY ontologies to form the backbone of a KG for precision oncology requires tackling challenges at these three levels: biomedical ontology matching, holistic ontology matching, and holistic ontology integration. Moreover, it also requires addressing requirements in terms of quality, coverage and scalability.

Ensuring **high quality mappings** between the ontologies is a strong requirement for a system that must work with a minimal human involvement due to the size of the task, but has a high-stakes target application in healthcare. Alignments need to achieve both high precision and high recall, since both types of errors can compromise XAI approaches, either by proposing wrong explanations or not finding suitable ones.

Achieving a sufficient **coverage of all domains** in personalized oncology is mandatory to make sure that all data required to train the AI models is well-described according to domain ontologies in a way that supports building explanations. The integration of molecular and clinical data is the key to personalized medicine, which seeks to understand the play between genotype, phenotype and environment and how it bears on the effectiveness of treatments or the prognosis of diseases. This aspect requires that not only the KG covers multiple domains but that it also includes sufficient granularity.

Finally, **scalability** must also be considered. Matching 28 ontologies means that there are nearly 1.2 million classes plus their associated properties and individuals that need to be processed. Moreover, since ontologies evolve and new relevant data may be added to the KG, the time required to build the network of ontologies should not be a limiting factor in updating the system.

To build a high quality network of biomedical ontologies we need to strike a balance between quality, coverage and scalability. Filtering out lower quality mappings may result in lower coverage, higher coverage requires the ability to align more ontologies, but more sophisticated ontology matching algorithms that are able to produce higher quality and higher coverage alignments are harder to scale.

3   Enhancing AML for Holistic Ontology Matching

3.1   AgreementMakerLight

AgreementMakerLight (AML) is an automated ontology matching system predicated on the design principles of scalability and extensibility [9]. It has been one of the best performing systems in the yearly Ontology Alignment Evaluation Initiative (OAEI) for the past eight years, excelling particularly in tracks involving biomedical ontologies [26]. This is thanks to features such as the weighting system it uses to differentiate labels and synonyms enabling fine-grained lexical matching, or its use of cross-references and logical definitions (which are singular to the biomedical domain) [8]. Given the stellar performance of AML's matching algorithms in biomedical ontology matching, using it as the baseline matching system is an added guarantee of the quality of the produced mappings.

Of note, AML's lexical matching, cross-reference matching and logical-definition matching algorithms are all implemented using a hash-search strategy that means they run in linear time [8], and therefore can be used for profiling the suite of ontologies to match with regard to their overlap, in a holistic matching scenario.

However, like all matching systems participating in the OAEI, AML is only prepared to perform pairwise matching of ontologies, and produces ontology alignments that are external to the ontologies, in the Alignment RDF format. It doesn't include the functionality of integrating ontologies through their alignment, which would be required to build a KG automatically through ontology matching. Indeed, building a KG automatically through the alignment of multiple ontologies is beyond the state of the art in ontology matching evidenced by the OAEI.

3.2   Extensions to AML

Setting aside the possibility of matching multiple ontologies simultaneously, and contemplating only the scenarios of pairwise matching or incremental matching, the only core functionality missing from AML for holistic ontology matching is the ability to merge two ontologies through their alignment, with a *simple merge*, as defined by [21].

We extended AML by implementing the functionality of converting an RDF alignment into an OWL ontology that imports the aligned ontologies and adds the axioms corresponding to the mappings: *equivalentClass* or *subClass* for equivalence or subsumption mappings between classes; *equivalentProperty* or *subProperty* for equivalence or subsumption mappings between properties; and *sameIndividual* axioms for equivalence mappings between individuals. To enable the pairwise strategy, we also implemented the functionality of merging two or more ontologies (or OWL alignments) into a single ontology, which will be necessary to combine multiple pairwise alignments into a single KG. Furthermore, for both the pairwise and the incremental strategies, we implemented the functionality of merging an ontology with all its imports, as it would be unwieldy to have a knowledge graph with OWL import statements for several local OWL ontology and/or alignment files.

3.3   Implementing Holistic Matching Strategies Using AML

Using AML and the extensions detailed above, we implemented two distinct holistic matching strategies: pairwise and incremental, which can be preceded by a clustering step and applied within-cluster, or applied globally to the full suite of ontologies to match. Since the global algorithms are the same as the within-clustering algorithms in the particular case where the number of clusters is 1, we present only the more general within-cluster algorithms. The use of clustering is motivated by the fact that there are multiple near-orthogonal sub-domains in the biomedical domain, and we can isolate groups of ontologies from each sub-domain, for which performing sophisticated ontology matching against ontologies of other sub-domains would likely produce more erroneous than correct mappings.

To enable clustering, we perform an initial anchoring step for all pairwise combinations of ontologies using linear-time matching algorithms, whereby we calculate the fraction of classes of the smallest ontology of each pair that have the same URI, direct cross-references, shared cross-references, overlapping logical definitions, or equivalent labels or synonyms to classes in the largest ontology of the pair. This anchoring is substantially quicker than performing a full pairwise matching strategy, and has the objective of determining the overlap between all ontologies with a high degree of confidence. From the anchoring results, we build an affinity matrix indicating the semantic overlap between each pair of ontologies, which we use as input for spectral clustering, to define groups of ontologies with a higher level of overlap and therefore likely within the same sub-domain.

In the within-Cluster Pairwise Alignment (CPA) strategy, each pairwise combination of the ontologies in each cluster is matched then merged, then all the merged ontology pairs of a cluster are combined and merged into a KG[3], and finally the KGs of each cluster are merged into a final single KG using the anchoring algorithm, as detailed in Algorithm 1.

In the within-Cluster Incremental Alignment (CIA) strategy, the pair of ontologies within each cluster that has the greatest overlap is matched and merged, then the resulting merged ontology is matched against the next ontology in the cluster, and so on until all ontologies in the cluster have been matched. Then, anchoring is performed also incrementally between the KG produced from each cluster, to produce a final single KG. The algorithm is detailed in Algorithm 2.

In both strategies, we used AML's automatic matching with default configurations, but with no ontologies used as background knowledge, and with the alignment repair step switched off. Using ontologies as background knowledge would be nonsensical in this setting, as any ontology that could be used effectively as background knowledge source should be, in principle, included in the suite of ontologies to match (as the goal is to build a comprehensive knowledge graph) and therefore will be merged with the other ontologies which is effectively

---

[3] We use KG to denote the integrated network of ontologies which constitute the semantic backbone of the full fledged KG.

Algorithm 1. Within-cluster pairwise alignment (CPA)

```
input:  C->O (map of clusters to ontologies)
init:  CM->OM = new map of cluster to ontologies
init:  KG = new list of ontologies
init:  OK = new list of ontologies
for  Cᵢ in  C:
     Oᵢ = C.get(Cᵢ)
     for  j = 0 to  Oᵢ.length-1:
          for  k in j+1 to  Oᵢ.length:
               A = AML.match(Oᵢ[j],Oᵢ[k])
               O = merge(convert(A))
               OMᵢ.add(O)
     CM->OM.put(Cᵢ,OMᵢ)
     KG[i] = OMᵢ[0]
     for  j = 1 to  OMᵢ.length:
          merge(KG[i],OMᵢ[j])
for  i = 0 to  KG.length-1:
     for  j = i+1 to  KG.length:
          A = AML.anchor(KG[0],KG[0])
          O = merge(convert(A))
          OK.add(O)
init:  KGF = OK[0]
for  i = 1 to  OK.length:
     merge(KGF,OK[i])
output:  KGF
```

Algorithm 2. Within-cluster incremental alignment (CIA)

```
input:  C->O (map of clusters to sorted ontologies)
init:  KG = new list of ontologies
for  Cᵢ in  C:
     Oᵢ = C.get(Cᵢ)
     KG[i] = Oᵢ[0]
     for  j = 1 to  Oᵢ.length:
          A = AML.match(KG[i],Oᵢ[j])
          KG[i] = merge(convert(A))
init:  KGF = KG[0]
for  i = 1 to  KG.length:
     A = AML.anchor(KGF,KG[i])
     KGF = merge(convert(A))
output:  KGF
```

equivalent to having it as a source of background knowledge. As for the choice not to perform repair, it is predicated on our desire for completeness of the alignment over coherence [24]. Furthermore, alignment repair algorithms take arbitrary choices when faced with conflicting mappings to remove, so while it is critical to ensure the final KG is coherent, this should involve human revision to ensure the mappings removed or edited are indeed inaccurate.

4 Integrating Biomedical Ontologies in a Personalized Oncology KG

4.1 Ontologies

The goal of our study is the integration of the 28 ontologies selected to cover the personalized oncology domain into a single KG[4]. These ontologies are listed in Table 1, together with the biomedical sub-domains they cover, which total 19, from molecular biology to drug-side effects. Taken together, these ontologies contain 1,191,785 classes, 2,634 properties and 397,535 individuals.

4.2 Alignment Strategies

To integrate the 28 KATY ontologies, we compared the two holistic matching strategies, CPA and CIA. Additionally, as a reference point, we also tested the global pairwise alignment (GPA) strategy, which corresponds to a naive use of the state of the art in ontology matching (and algorithmically, as detailed in Sect. 3, is the same as CPA when the number of clusters is 1) (Fig. 1).

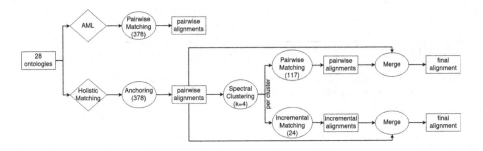

Fig. 1. Overview of the alignment strategies

4.3 Results

The global pairwise alignment (GPA) of the 28 ontologies translated into 378 alignment runs resulting in 378 pairwise alignments with a total of more than half a million mappings. The duration of the loading and matching processes[5] and total number of mappings found are presented in Table 3.

---

[4] Although the UMLS provides mappings between some of our ontologies, its usage license does not allow public reuse.

[5] Experiments were run in a machine with 100Gb of available RAM.

Table 1. Ontologies used and their domains

Acronym	Ontology	Domains	Classes
ACGT-MO	Cancer Research and Management ACGT Master Ontology	Clinical feature, sample status	1769
ATC	Anatomical Therapeutic Chemical Classification	Drug	6567
CCTOO	Cancer Care: Treatment Outcome Ontology	Response to treatment, drug screening	1133
ChEBI	Chemical Entities of Biological Interest Ontology	Metabolic, drug	171058
CL	Cell Ontology	Cellular	10984
CLO	Cell Line Ontology	Cell line	44873
CMO	Clinical Measurement Ontology	Clinical feature, sample status	3054
DCM	DICOM Controlled Terminology	Histological images	4561
DOID	Human Disease Ontology	clinical feature	17642
DTO	Drug Target Ontology	drug target interaction	10075
EFO	Experimental Factor Ontology	Experimental	28816
FMA	Foundational Model of Anatomy	Anatomical data	78977
GENO	Genotype Ontology	Genomic	425
GO	Gene Ontology	Genomic, biological pathway	50713
HCPCS	Healthcare Common Procedure Coding System	Clinical feature, drug sampling	7094
HGNC	HUGO Gene Nomenclature	Genomic	32917
HP	Human Phenotype Ontology	Biological feature	27482
ICDO	International Classification of Diseases Ontology	Clinical feature	1313
LOINC	Logical Observation Identifier Names and Codes	Clinical feature	268552
MONDO	Mondo Disease Ontology	Clinical feature	43735
NCIT	National Cancer Institute Thesaurus	Biological feature, clinical feature	166884
OAE	Ontology of Adverse Events	Drug side effect, response to treatment	5762
OMIM	Online Mendelian Inheritance in Man	Biological feature	97261
OPMI	Ontology of Precision Medicine and Investigation	Clinical feature, clinical trial	2939
ORDO	Orphanet Rare Disease Ontology	Clinical feature	14886
PDQ	Physician Data Query	Clinical feature, drug screening	13452
PMAPP-PMO	PMO Precision Medicine Ontology	Genomic, clinical feature, clinical trial, sampling	76154
SO	Sequence Ontology	Genomic, transcriptomic	2707

The two clustering-based approaches, CPA and CIA, require ontologies to be clustered, which involves an initial step of anchoring followed by spectral clustering, as detailed in Sect. 3. The anchoring step also translated into 378 (lightweight) alignment runs resulting in a set of 378 pairwise alignments, as well as in an affinity matrix computed based on these alignments. The duration and total mappings found by the anchoring step are also presented in Table 3.

Figure 2 presents a heatmap representation of the semantic overlap computed by the anchoring step. Individual heatmaps for each component of the anchoring process are available as supplementary materials[6]. A few ontologies have a high number of direct cross-references between them or reuse classes from each other extensively. Logical definitions are less relevant to establish the semantic overlap between ontologies, since the majority of ontologies used does not declare them. The Lexical Matcher is the method that is able to find more correspondences for more ontology pairs.

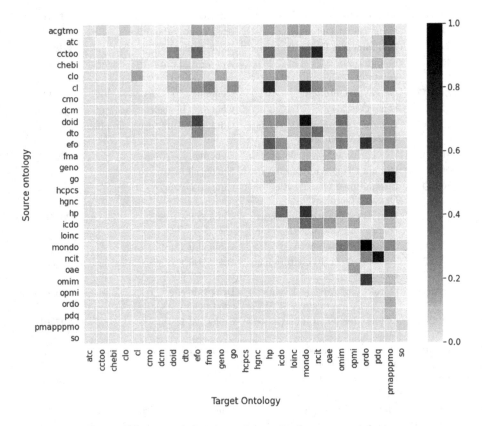

Fig. 2. Heatmap of the semantic overlap between ontologies

---

[6] https://github.com/liseda-lab/holistic-matching-aml.

The affinity matrix was then used as input to clustering with spectral clustering. We tested cluster numbers between 3 and 6, and empirically selected 4 clusters, with the aid of the constructed heatmaps, which are shown in Table 2.

Table 2. Ontologies organized by cluster

Cluster	Ontologies	Classes
C1	NCIT, PDQ, LOINC, ChEBI, CCTOO	621079
C2	PMAPP-POM, GO, HP, ATC, FMA, CL, CLO, OAE, ACGT-MO, ICDO, SO, HCPCS, GENO	314820
C3	CMO, OPMI	5993
C4	MONDO, ORDO, DOID, OMIM, EFO, HGNC, DTO, DCM	249893

We applied the CPA and CIA strategies on the four clusters. In the CPA, the alignment tasks are run between all pairwise combinations of ontologies within each cluster, in the same manner as the GPA strategy. This translated into 117 pairwise alignment tasks and output alignments, which were merged within each cluster to produce 4 intermediate cluster KGs. In the CIA, only n-1 alignment tasks are required to integrate the n ontologies in each cluster incrementally, so 24 alignment tasks were necessary in total to produce 4 intermediate cluster KGs. The final step of each strategy was the merging of the cluster KGs through the anchoring algorithm to create a fully integrated KG. Again, the statistics of the alignment processes are summarized in Table 3.

Since we employed the GPA strategy only as a reference point for the state of the art, we did not perform the merging of the pairwise alignments into a single KG (as it would be beyond the state of the art). Thus, the GPA runtime is directly comparable to the sum of anchoring and within-cluster alignment runtimes for the CPA and CIA strategies. We note that, while the GPA takes more than 31 h to complete, anchoring+CPA takes less than 24 h and anchoring+CIA less than 16 h. It is obvious that although matching times are greatly reduced in CPA (by nearly 12 h) and CIA (by nearly 19 h) when compared with GPA, the process of loading the ontologies using the OWL API is responsible a considerable portion of the time spent in running the alignment processes.

Table 4 presents the final alignment sizes produced by each strategy[7], where for CPA and CIA, the final alignment size results in combining the within-cluster mappings with between-cluster anchoring mappings, where the latter contributed over 200,000 mappings for both strategies. CPA produced a final KG that is 60% smaller than the total GPA mappings, while CIA produced one that is 65% smaller. We note that CPA strategy led to a greater number of mappings than the original global anchoring, whereas the CIA strategy led to a number of mappings just under the global anchoring. While this might

---

[7] Individual statistics available in the supplementary materials.

Table 3. Alignment results

Strategy	Runtime (hh:mm)			Alignment	
	Load	Match	Total	Mappings	Tasks
GPA	11:47	19:51	31:37	554547	378
Anchoring	11:47	01:59	13:46	427300	378
CPA	02:25	07:42	10:07	219021	117
CIA	01:05	01:05	02:10	193503	24

GPA: global pairwise alignment. CPA: within-cluster pairwise alignment. CIA: within cluster incremental alignment.

suggest that the CIA strategy is losing relevant mappings, it is in fact a natural consequence of the incremental strategy, due to the fact that AML is configured to produce (mostly) 1 to 1 alignments. Thus, if in the pairwise strategy we have 3 mappings between equivalent classes $c_1 A$, $c_1 B$ and $c_1 C$ of ontologies A, B and C, in the incremental strategy we would have only 2 mappings, since once ontologies A and B are combined into AB in a first iteration, AML will generally produce only 1 mapping between each class of AB and C, so $c_1 C$ would be mapped to either $c_1 A$ or $c_1 B$, but not both. We note, however, that the third mapping would be semantically redundant, as it is implied by other two. Thus, the CIA strategy is expected to capture less mappings than the CPA strategy, but most of the missing mappings will be semantically redundant. A comparison of the alignments produced by CPA and CIA revealed that all CIA mappings are contained in the CPA alignment, with CIA covering 88% of the mappings found by CPA.

Table 4. Merged alignments results

Strategy	Total Mappings
GPA	554547
CPA+anchoring	442649
CIA+anchoring	417131

GPA: global pairwise alignment. CPA: within-cluster pairwise alignment. CIA: within cluster incremental alignment.

4.4 Discussion

The holistic alignment of real world ontologies is a challenge that state of the art ontology matching systems that compete in the OAEI have yet to address. The very good performance of systems such as AML[17] and LogMap[14] in the biomedical tracks at OAEI[12] is impressive, but pales in comparison to the challenges of matching ontologies that have not stood the scrutiny applied to benchmark ontologies in organized challenges. In the course of this work, we encountered several hurdles due to syntactical issues in the ontologies or unexpected uses of some properties that had to be solved to ensure adequate coverage. As an example we highlight the case of the Experimental Factor Ontology (EFO) ontology that establishes cross-references between a single class and 77 classes in the Human Disease Ontology (DOID). A cross-reference is usually interpreted as an equivalent or closely related class, and this is explored by AML to produce equivalence mappings, but in this case the underlying relation between the one class and the 77 is one of subsumption. Addressing such cases correctly, will require adapting AML.

A recognized challenge in holistic matching is that the order of the matching tasks can impact the quality of the final alignment [13]. To circumvent this issue, [18]) developed a method that performs simultaneous matching of ontologies but unfortunately results in substantial losses in performance when compared to pairwise methods. The cost of determining the order for incremental matching is not considered by other works (e.g. [13]), however we argue that it must be considered a part of the alignment process. Moreover, employing simply lexical similarity is less than ideal in the biomedical domain where there is a high level of synonymy that is not always captured by the lexical component of the ontologies. In this work, we employ the same method to determine cluster affiliation and matching order, which is based on the semantic overlap between ontologies as measured by very high lexical similarity but also based on cross-references and logical definitions, which are particular to the biomedical domain. While in other works, clustering or tree mining is employed to determine the order of matching, we chose to apply clustering to actually partition the search space. This not only allowed a reduction of the matching tasks, but since clusters are based on semantic overlap and group together ontologies of the same domain, it can also mitigate the problem of false positives caused by homonyms. Let's take the example of the class *Gingiva* in the Foundational Model of Anatomy (FMA) ontology and the class *Gum* in the National Cancer Institute Thesaurus (NCIT). While 'gum' and 'gingiva' are synonymous words, in this case *Gum* actually refers to a type of chemical. However, since NCIT and FMA were actually placed in different clusters, the impact of these type of mappings can be minimized.

Although it is not possible to directly measure the quality of the resulting alignments short of a manual evaluation (as no reference alignments exist for these ontologies), an analysis of the number of mappings obtained can shed light on some interesting aspects. The GPA represents an upper bound on the number of mappings. It finds 120 thousand more mappings that anchoring, which we hypothesize to have a lower precision but increased recall, since the extra method employed by the full AML pipeline compared to anchoring are mostly methods that were designed to increase recall, assuming that the performance of AML in these ontologies is comparable to its performance in the OAEI biomedical benchmarks. One advantage of the clustering-based approaches is that they have the potential to increase the precision of mappings between clusters, by only establishing mappings based on the high precision and lower recall anchoring strategy, while increasing recall within the clusters, by employing more sophisticated alignment methods. Moreover, as detailed in Sect. 4.2, the CIA strategy is expected to find less mappings than the CPA strategy, but these will be mostly semantically redundant mappings.

5 Conclusions

The rich panorama of both publicly available data and ontologies in the biomedical domain represents an opportunity for developing explainable knowledge-enabled systems. In multi-domain areas, such as personalized medicine, this requires the integration of multiple data sources and ontologies. Holistic ontology matching and integration holds the promise to scale semantic data integration to multiple sources [28], however holistic ontology matching in the biomedical domain is still an open challenge.

We have developed a novel approach for holistic ontology matching that builds on an existing system, AML[9], addressing the requirements and challenges of the biomedical domain. We demonstrated that the straightforward application of the pairwise alignment approach to all ontology pairs takes up to 100% more time than the novel clustering-based approaches. We further demonstrated that the within-cluster incremental alignment approach is five times faster than the within-cluster pairwise alignment approach. All approaches were able to generate a fully integrated KG, meaning that all ontologies have mappings to one or more of the other ontologies, effectively responding to the coverage requirement. The quality assessment of the resulting alignment is not straightforward, since there are no holistic reference alignments within the biomedical domains, and out of the 378 pairwise alignments, only one pair is covered by an existing reference (FMA-NCI) but it employs an outdated version of the ontologies and was produced semi-automatically.

The proposed approach can be extended with further refinements. To increase the coverage and semantic richness of the KG, complex mappings can be applied to more accurately capture the relations between their entities. While the KG construction will be mostly automated, expert feedback will be paramount to ensure an accurate KG that can support explanations. To make the most efficient use of feedback, we will develop algorithms to identify potentially doubtful mappings that require user validation, and algorithms that propagate the user feedback automatically [6].

The experience of applying a state of the art ontology matching system to a large set of real world biomedical ontologies for holistic matching and integration resulted in lessons learnt for future endeavours. One of the identified challenges was the comparative evaluation of the alignment quality produced by pairwise and holistic approaches. One future opportunity is to build upon the set of reference alignments made available by the OAEI to create a holistic reference alignment following the approach described by Roussille et al.[29]. Another lesson was the fact that the ontology loading times slow down the alignment process substantially, also this was partly due to the fact that AML still employs an older version (3.4) of the OWL API. Preliminary testing showed that a new version of the OWL API (5.1) speeds up the loading by a factor of 2. Perhaps the biggest challenge was in handling the varying degrees of quality of the ontologies, with formatting issues and non-standard uses of the cross-reference property that required *ad hoc* solutions to circumvent, and will require further extensions to AML to handle more adequately.

Building on decades of work by the semantic web and biomedical ontologies communities, we have developed an approach for holistic matching and integration of ontologies from multiple domains to build KG to support AI-based personalized cancer therapy. The size, diversity and complexity of the underlying ontologies and overarching domain represented significant challenges that required evolving the current state of the art in ontology matching.

Acknowledgments. This work was supported by FCT through the LASIGE Research Unit (UIDB/00408/2020 and UIDP/00408/2020). It was also partially supported by the KATY project which has received funding from the European Union's Horizon 2020 research and innovation program under grant agreement No. 101017453.

References

1. Babalou, S., Grygorova, E., König-Ries, B.: $CoMerger$: a customizable online tool for building a consistent quality-assured merged ontology. In: Harth, A., et al. (eds.) ESWC 2020. LNCS, vol. 12124, pp. 19–24. Springer, Cham (2020). https://doi.org/10.1007/978-3-030-62327-2_4

2. Caldarola, E.G., Rinaldi, A.M.: An approach to ontology integration for ontology reuse. In: 2016 IEEE 17th International Conference on Information Reuse and Integration (IRI), pp. 384–393. IEEE (2016)

3. Chari, S., Gruen, D.M., Seneviratne, O., McGuinness, D.L.: Directions for explainable knowledge-enabled systems. arXiv preprint arXiv:2003.07523 (2020)

4. Chari, S., Gruen, D.M., Seneviratne, O., McGuinness, D.L.: Foundations of explainable knowledge-enabled systems. arXiv preprint arXiv:2003.07520 (2020)

5. Chatterjee, N., Kaushik, N., Gupta, D., Bhatia, R.: Ontology merging: a practical perspective. In: Satapathy, S.C., Joshi, A. (eds.) ICTIS 2017. SIST, vol. 84, pp. 136–145. Springer, Cham (2018). https://doi.org/10.1007/978-3-319-63645-0_15

6. Cruz, I.F., Stroe, C., Palmonari, M.: Interactive user feedback in ontology matching using signature vectors. In: ICDE 2012, pp. 1321–1324. IEEE (2012)

7. Euzenat, J., Shvaiko, P.: Ontology Matching, 2nd edn. Springer, Heidelberg (2013). https://doi.org/10.1007/978-3-642-38721-0

8. Faria, D., Pesquita, C., Mott, I., Martins, C., Couto, F.M., Cruz, I.F.: Tackling the challenges of matching biomedical ontologies. J. Biomed. Semant. **9**(1), 1–19 (2018)

9. Faria, D., Pesquita, C., Santos, E., Palmonari, M., Cruz, I.F., Couto, F.M.: The AgreementMakerLight ontology matching system. In: Meersman, R., et al. (eds.) OTM 2013. LNCS, vol. 8185, pp. 527–541. Springer, Heidelberg (2013). https://doi.org/10.1007/978-3-642-41030-7_38

10. Ferreira, J.D., Teixeira, D.C., Pesquita, C.: Biomedical ontologies: coverage, access and use. In: Wolkenhauer, O. (ed.) Systems Medicine Integrative, Qualitative and Computational Approaches, pp. 382–395. Academic Press, Else-

vier (2020). https://doi.org/10.1016/B978-0-12-801238-3.11664-2, http://www.sciencedirect.com/science/article/pii/B9780128012383116642

11. Gruetze, T., Böhm, C., Naumann, F.: Holistic and scalable ontology alignment for linked open data. LDOW **937**, 1–10 (2012)

12. Harrow, I., et al.: Matching disease and phenotype ontologies in the ontology alignment evaluation initiative. J. Biomed. Semant. **8**(1), 1–13 (2017)

13. Hertling, S., Paulheim, H.: Order matters: matching multiple knowledge graphs. arXiv preprint arXiv:2111.02239 (2021)

14. Jiménez-Ruiz, E.: Logmap family participation in the OAEI 2020. In: Proceedings of the 15th International Workshop on Ontology Matching (OM 2020), vol. 2788, pp. 201–203. CEUR-WS (2020)

15. Köhler, S.: Improving ontologies by automatic reasoning and evaluation of logical definitions. BMC Bioinf. **12**, 418 (2011)

16. Lecue, F.: On the role of knowledge graphs in explainable AI. Semantic Web **11**(1), 41–51 (2020)

17. Lima, B., Faria, D., Couto, F.M., Cruz, I.F., Pesquita, C.: Oaei 2020 results for aml and amlc. (2020)

18. Megdiche, I., Teste, O., Trojahn, C.: An extensible linear approach for holistic ontology matching. In: Groth, P., Simperl, E., Gray, A., Sabou, M., Krötzsch, M., Lecue, F., Flöck, F., Gil, Y. (eds.) ISWC 2016. LNCS, vol. 9981, pp. 393–410. Springer, Cham (2016). https://doi.org/10.1007/978-3-319-46523-4_24

19. Noy, N.F., Shah, N.H., Whetzel, P.L., et al.: Bioportal: ontologies and integrated data resources at the click of a mouse. Nucleic Acids Res. **37**(2), W170–W173 (2009)

20. Oliveira, D., Pesquita, C.: Improving the interoperability of biomedical ontologies with compound alignments. J. Biomed. Semant. **9**(1), 1–13 (2018)

21. Osman, I., Ben Yahia, S., Diallo, G.: Ontology integration: approaches and challenging issues. Inf. Fusion **71**, 38–63 (2021)

22. Otero-Cerdeira, L., Rodríguez-Martínez, F.J., Gómez-Rodríguez, A.: Ontology matching: a literature review. Expert Syst. Appl. **42**(2), 949–971 (2015)

23. Pesquita, C.: Towards semantic integration for explainable artificial intelligence in the biomedical domain. In: BIOSTEC 2021, vol. 5, pp. 747–753 (2020)

24. Pesquita, C., Faria, D., Santos, E., Couto, F.M.: To repair or not to repair: reconciling correctness and coherence in ontology reference alignments. In: Ontology Matching (2013)

25. Pesquita, C., Faria, D., Stroe, C., Santos, E., Cruz, I.F., Couto, F.M.: What's in a 'nym'? synonyms in biomedical ontology matching. In: Alani, H., et al. (eds.) ISWC 2013. LNCS, vol. 8218, pp. 526–541. Springer, Heidelberg (2013). https://doi.org/10.1007/978-3-642-41335-3_33

26. Pour, N., Algergawy, A., Amini, R., Faria, D., et al.: Results of the ontology alignment evaluation initiative 2020. In: OM 2020, vol. 2788, pp. 92–138. CEUR-WS (2020)

27. Rahm, E.: Towards large-scale schema and ontology matching. In: Schema Matching and Mapping, pp. 3–27. Springer, Heidelberg (2011). https://doi.org/10.1007/978-3-642-16518-4_1

28. Rahm, E.: The case for holistic data integration. In: Pokorný, J., Ivanović, M., Thalheim, B., Šaloun, P. (eds.) ADBIS 2016. LNCS, vol. 9809, pp. 11–27. Springer, Cham (2016). https://doi.org/10.1007/978-3-319-44039-2_2

29. Roussille, P., Megdiche, I., Teste, O., Trojahn, C.: Boosting holistic ontology matching: generating graph clique-based relaxed reference alignments for holistic evaluation. In: Faron Zucker, C., Ghidini, C., Napoli, A., Toussaint, Y. (eds.)

EKAW 2018. LNCS (LNAI), vol. 11313, pp. 355–369. Springer, Cham (2018). https://doi.org/10.1007/978-3-030-03667-6_23
30. Saleem, K., Bellahsene, Z., Hunt, E.: Porsche: performance oriented schema mediation. Inf. Syst. **33**(7–8), 637–657 (2008)
31. Silva, M.C., Faria, D., Pesquita, C.: Integrating knowledge graphs for explainable artificial intelligence in biomedicine? In: Ontology Matching Workshop at the International Semantic Web Conference (2021)
32. Stoilos, G., Geleta, D., Shamdasani, J., Khodadadi, M.: A novel approach and practical algorithms for ontology integration. In: Vrandečić, D., et al. (eds.) ISWC 2018. LNCS, vol. 11136, pp. 458–476. Springer, Cham (2018). https://doi.org/10. 1007/978-3-030-00671-6_27

Findsampo: A Linked Data Based Portal and Data Service for Analyzing and Disseminating Archaeological Object Finds

Heikki Rantala[1]([✉]) [iD], Esko Ikkala[1] [iD], Ville Rohiola[3] [iD], Mikko Koho[1,2] [iD],
Jouni Tuominen[1,2] [iD], Eljas Oksanen[2] [iD], Anna Wessman[4] [iD],
and Eero Hyvönen[1,2] [iD]

[1] Semantic Computing Research Group (SeCo), Aalto University, Espoo, Finland
{heikki.rantala,esko.ikkala,mikko.koho,jouni.tuominen,
eero.hyvonen}@aalto.fi
[2] HELDIG – Helsinki Centre for Digital Humanities, University of Helsinki,
Helsinki, Finland
{mikko.koho,jouni.tuominen,eljas.oksanen,eero.hyvonen}@helsinki.fi
[3] Finnish Heritage Agency, Helsinki, Finland
ville.rohiola@museovirasto.fi
[4] University Museum of Bergen, Bergen, Norway
anna.wessman@uib.no
http://seco.cs.aalto.fi, http://heldig.fi

Abstract. This paper presents the FINDSAMPO system for analyzing and disseminating archaeological object finds made by the public. The system is based on Linked Open Data (LOD), and consists of a web portal and an open data service. The underlying knowledge graph contains data of some 3000 archaeological object finds catalogued in the archaeological collection of the Finnish Heritage Agency (FHA) from 2015 to 2020. The portal and LOD service have been open to public use since May 2021.

1 Introduction

1.1 Web Services for Citizens and Researchers

The popularity of recreational metal detecting has grown rapidly in many countries such as in Finland during the last decade, creating a large amount of new archaeological data. This paper demonstrates how archaeological object finds made by the public can be analyzed using the Linked Open Data (LOD) based FINDSAMPO service [14,20]. FINDSAMPO research prototype has been created by the SuALT project[1] aiming to study and improve the reporting process and analysis of archaeological finds based on collaboration of the public, academic researchers, archaeologists, and the Finnish Heritage Agency (FHA) [6,22,29].

A demonstrator based on data of some 3000 archaeological object finds catalogued in the archaeological collection of FHA from 2015 to 2020 has been open

---

[1] SuALT project: https://blogs.helsinki.fi/sualt-project.

© The Author(s), under exclusive license to Springer Nature Switzerland AG 2022
P. Groth et al. (Eds.): ESWC 2022, LNCS 13261, pp. 478–494, 2022.
https://doi.org/10.1007/978-3-031-06981-9_28

to public use since May 2021, and has had over 3000 users during it's first six months. The FINDSAMPO demonstrator consists of a data service[2] and a semantic portal[3] with search functions and analytical tools. Figure 1 shows the front page of the FINDSAMPO portal with various perspectives to the data, and quick links to selected featured finds.

FINDSAMPO responds to a need for digital solutions to improve the management, accessibility and democratisation in cultural heritage management that stem from the recent popularity of recreational metal-detecting. In Finland, metal detecting is permitted but the Finnish Antiquities Act (295/1963) stipulates acts of law that must be followed also when metal detecting. There are also gerenal guidelines provided for recreational metal detectorists to prevent illegal acts and to protect cultural heritage. The Antiquities Act prohibits strictly metal detecting and especially digging at the ancient monuments and other archaeological sites. Certain areas are also protected by the Nature Conservaton Act. The Antiquities Act also requires that archaeological objects, including metal detected finds are expected to be at least 100 years old and do not have any known owner must be reported immediately to the FHA. The reporting of finds is guided to be done easily through FHA's electronic reporting service. In Finland, the FHA has the right to redeem archaeological finds to the national collections.[4] In the management process of the finds, the find information will also be entered into an electronic database that will feed the FINDSAMPO. Metal-detecting is therefore a form of crowd-sourcing information about the past. For the more serious (or avocational) metal-detectorists this activity is citizen science, where the citizen participates in the creation and discovery of new archaeological knowledge [30].

Metal-detecting in Finland has increased significantly in popularity in the last decade. The vastly increased amount of information generated challenges the heritage management. The larger context that this topic consequently links to is the pan-European need to develop an internationally operable and harmonised data infrastructure for using cultural heritage data from different countries in research.

In order to respond to the challenges in contemporary cultural heritage management, FINDSAMPO supports three overlapping stakeholder groups: 1) the public in analysing their finds and learning about archaeology, 2) cultural heritage professionals in analysing, managing, and publishing collection data, and 3) researchers in knowledge discovery using metal-detected citizen science data. In keeping with the ethos of open science and democratising access to cultural heritage the service has been designed to transfer knowledge from professionals to citizen scientists, and to provide a powerful set of digital tools for new knowledge discovery and creation among its users.

---

[2] https://www.ldf.fi/dataset/findsampo.
[3] https://loytosampo.fi/en.
[4] https://www.museovirasto.fi/uploads/Arkisto-ja-kokoelmapalvelut/Julkaisut/muinaisjaannokset-ja-metallinetsin-2017.pdf.

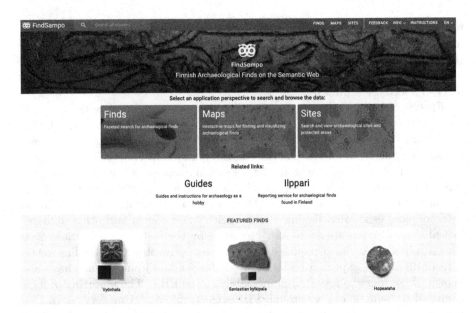

Fig. 1. Front page of the FINDSAMPO portal.

The archaeological finds included in the service have been recovered by the public mainly by metal-detecting and reported to the FHA for recording. The FINDSAMPO data constitutes an unprecedented reservoir of citizen science-generated archaeological heritage in Finland, equally accessible to researchers and to the broader public interested in heritage. The new archaeological object finds material it contains has proven to hold the potential for substantially adding to our understanding of the Finnish prehistorical and historical periods. Our goal is to allow the end users to improve themselves and also learn more about archaeology. In this way, archaeological data becomes more quickly and comprehensively available and accessible for research purposes and Digital Humanities [5].

This paper presents an overview of the service and the technical design principles and implementation of the FINDSAMPO Portal and it's underlying data service. FINDSAMPO Portal is yet another member in the "Sampo" series[5] of Linked Open Data services and semantic portals [12], based on a national Semantic Web infrastructure [11].

1.2   Related Work

Our work was motivated by the growing popularity of metal detecting in recent years. As a result, many countries are developing web services to collect, analyse, and study archaeological data.

---

[5] For a list of Sampo portals, see https://seco.cs.aalto.fi/applications/sampo/.

1. The largest of them, Portable Antiquities Scheme (PAS)[6], records[7] archaeological discoveries found by members of the public in England and Wales since 1997 [1].
2. Digital Metal Finds (DIME)[8] is an online platform for reporting metal detecting finds in Denmark [28].
3. Portable Antiquities of the Netherlands (PAN)[9] is an online portal in use in the Netherlands [26].
4. Metal-Detected Artefacts (MEDEA)[10] is an online portal developed in Flanders for metal detectors [2,28];
5. ILPPARI[11] is a service of the Finnish Heritage Agency (FHA) for reporting archaeological object finds found by the public in Finland. The service also includes tools for public to report archaeological sites and possible damages concurred to them [29].

Nomisma.org [23] is an example of an international project that concentrates on a specific type of objects, in this case coins. Nomisma.org uses a detailed model to describe various aspects of numismatic data using LOD. Numerous datasets have been published using the data model, and some of these datasets include also citizen finds.

FINDSAMPO is an application of the "Sampo model" [13], a collection of principles that have evolved gradually when creating a series of semantic portals[12]. The principles behind the Sampo model in use in FINDSAMPO have been explored and developed before in different contexts. For example, the notion of collaborative content creation by data linking is a fundamental idea behind the Linked Open Data Cloud movement[13] and has been developed also in various other settings, e.g., in ResearchSpace[14]. The idea of providing multiple analyses and visualizations to a set of filtered search results has been used in other portals, such as the ePistolarium[15] [21] for epistolary data, and using multiple perspectives have been studied as an approach in decision making [16]. Faceted search [24,25], also known as "view-based search" and "dynamic ontologies", is a well-known paradigm for explorative search and browsing [17] in computer science and information retrieval, based on S. R. Ranaganathan's original ideas of faceted classification in Libary Science in the 1930's. The two step filter-analyse usage model is used in prosopographical research [27] (without the faceted search component). The novelty of the Sampo model lies in consolidating several ideas

---

[6] PAS: https://finds.org.uk/database.
[7] 1.4 million finds have been reported by more than 14,000 citizens by now.
[8] DIME: https://www.metaldetektorfund.dk.
[9] PAN: https://portable-antiquities.nl.
[10] MEDEA: https://vondsten.be.
[11] ILPPARI: https://www.kyppi.fi/ilppari.
[12] This series is explained with references in https://seco.cs.aalto.fi/applications/sampo/.
[13] https://lod-cloud.net.
[14] https://www.researchspace.org.
[15] http://ckcc.huygens.knaw.nl.

and in operationalizing them for developing applications in Digital Humanities; something that the field of the Semantic Web seems to be missing as argued in [8].

1.3   Applying the Sampo Model on a Framework Level

The Sampo model principles can be used directly for creating semantic portals. However, it is also possible to apply them first to create an application domain specific framework and reuse it for developing different related application instances, which is arguably cost-efficient. Figure 2 illustrates the idea with FINDSAMPO and LetterSampo [9] frameworks as examples. The highest conceptual layer includes the Sampo model with its principles based on domain agnostic, logical Semantic Web standards of the W3C and Linked Data publishing principles. On the next, domain specific level, model level solutions and principles are applied to create a domain specific framework by using a domain specific data model that can be populated using domain specific vocabularies and ontologies (e.g., archaeological object types, archives involved, historical places, etc.). This layer includes also a domain specific template designed using the Sampo-UI framework [15] that can be copied and used as a starting point for creating application instances. The template tells, e.g., what thematic application perspectives, data-analysis tools, and ready-to-use UI components are available in this application domain. Finally, applications can be created by adding in specific datasets into the framework, by creating a Sampo-UI implementation of the portal interface, and by publishing the data in a Linked Data service with a SPARQL endpoint. In the figure, the LetterSampo framework has been used for two such applications corresponding to the epistolary datasets of CKCC (ca. 20 000 letters related to the Republic of Letters provided by the Huygens Institute [7,18], the Netherlands, and correspSearch (ca. 151 000 letter dataset [3,4] aggregated by the Berlin-Brandenburg Academy of Sciences [9]. In the case of FINDSAMPO, archaeological find collections from the Finnish Heritage Agency (FHA) are used in one instance and another one is based of the Portable Antiquities Schema (PAS) of the British Museum is being developed using the same framework[16] [19].

2   FindSampo Data Model and Data Service

2.1   Data Model

For the heritage agencies in Finland, and presumably around the world, a central problem are the limited available resources for information technology. We have therefore tried to use as simple model[17] for the data as possible. The data model is also constructed so that the faceted search for finds is efficient. The data

---

[16] https://seco.cs.aalto.fi/projects/diginuma/.

[17] The documentation of the data model, and other information, can be accessed at www.ldf.fi/dataset/findsampo.

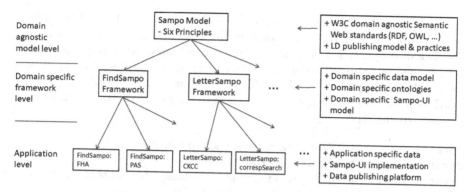

Fig. 2. Three conceptual layers for creating Sampo portals: Sampo Model, Sampo frameworks, and applications [9]. The idea is to re-use generic solutions of the model layer in domain specific frameworks and then frameworks for application instances in different domains.

points are attached to the Find entities, that are central to the data, with short property paths. We generally use a purpose build data model. In some cases we use certain relevant elements from well known standards, such as CIDOC Conceptual Reference Model (CRM)[18]. However these are mainly used as property names, and we do not implement, for example, full event structure of CIDOC CRM.

We created a FINDSAMPO Core ontology for representing the most relevant data relating to the object finds. Most important part of the data are the Find entities representing the object finds, and almost all the data is directly attached to those with either data type properties or object properties. The FINDSAMPO core properties include properties for object type, material, dating, and so on. We would expect any data relating to citizen science object finds would generally include these properties. In addition to the core properties we use specific properties to represent the data as it is in it's original source. These properties would be different for all different sources, and represent the data in the format used in that source. There can also be various properties for types of data that is not included as part of core properties. While the core properties aim to use ontologized object values, and standard data types, these source specific properties are only literal can can use various different data types.

Below is presented a simplified example, with selected properties, of a RDF representation of a single Find in Turtle format. Most of the properties here are either core properties, represented with prefix "findsampo-core" or properties specific to the FHA finds database, that are represented with the prefix "ltk-s". For example, in the Turtle notation below the property "ltk-s:length" denotes the original value for the object length in the FHA database as it is written there. The property "findsampo-core:length" on the other hand expresses the object length in standard decimal format. We also use SKOS prefLabel property

---

[18] https://www.cidoc-crm.org/.

for the main human readable label of the find, and we use Dublin Core source property to represent the source of data for the Find. Most properties attach the data directly to the Find, but with time spans and coordinate points we use a more complex representation where they are separate entities. Time spans are modelled with properties based on CIDOC CRM and for coordinates we use the W3C basic geographic vocabulary[19]. This is to make it easier to integrate with various existing tools.

```
@prefix crm: <http://erlangen-crm.org/current/> .
@prefix dct: <http://purl.org/dc/terms/> .
@prefix ltk-s: <http://ldf.fi/schema/findsampo/extended/ltk/> .
@prefix skos: <http://www.w3.org/2004/02/skos/core#> .
@prefix findsampo-core: <http://ldf.fi/schema/findsampo/core/> .
@prefix finds: <http://ldf.fi/findsampo/finds/> .


finds:km_39824-45 a findsampo-core:Find ;
    findsampo-core:find_site_coordinates
        <http://ldf.fi/findsampo/find_sites/find_site_of_39824-45> ;
    findsampo-core:has_creation_time_span
        [ crm:P82a_begin_of_the_begin "-0500-01-01"^^xsd:date ;
        crm:P82b_end_of_the_end "1300-12-31"^^xsd:date ] ;
    findsampo-core:identifier "KM39824:45" ;
    findsampo-core:length 56.0 ;
    findsampo-core:material
        <http://ldf.fi/findsampo/materials/p10> ;
    findsampo-core:object_type
        <http://ldf.fi/findsampo/object_types/hevosenkenkaesoljet> ;
    findsampo-core:period
        <http://ldf.fi/findsampo/periods/p17> ;
    findsampo-core:weight 35.0 ;
    findsampo-core:width 10.0 ;
    ltk-s:amount "1" ;
    ltk-s:find_name "Hevosenkenksolki" ;
    ltk-s:find_number "39824:45" ;
    ltk-s:length "56" ;
    dct:source "Museoviraston lyttietokanta" ;
    skos:prefLabel "Hevosenkenksolki KM39824:45" .
```

In the end, creating a simple model for the relevant properties of object finds is relatively easy. Internationally the central properties of the data seem to be mostly similar, and the data from other countries could be represent with this model as well. We have already created an initial conversion of PAS data to FINDSAMPO model to run tests with it using our system. This can be done with

---

[19] https://www.w3.org/2003/01/geo/.

relatively little effort. What would be a more difficult question is the harmonization of the various annotation ontologies used in the data.

2.2 Ontologies

Representing the properties of object finds requires specific ontologies[20]. These include ontologies for object type, materials, and periods. We have used two different kinds of ontologies to represent the data: "annotation ontologies" and "facet ontologies". Annotation ontologies are used to represent the concepts in the data in shared and machine understandable manner, while the facet ontologies are used to offer an easy to use hierarchy of concepts to the users. We have used the Finnish Ontology for Museum Domain and Applied Arts (MAO/TAO)[21] ontology as the basis of our annotation ontologies, but instead of using MAO identifiers directly, we have created new identifiers for the concepts, that then have an exact match concept in MAO ontology. We have also created mapping for the concepts to the Art and Architecture Thesaurus (AAT)[22]. We also added some semantic information to the concepts, such as machine readable dates for periods. The reasoning capabilities offered are limited, but make it possible to, for example, deduce the possible manufacture times of objects when given only a period.

MAO/TAO includes a hierarchy for the concepts, but this hierarchy can be unintuitive for users. For example "sword hilts" would not be under "swords" in MAO/TAO hierarchy, but instead under "parts", because sword hilt is not a sword. With domain experts we have created separate "facet ontologies" for the concepts that aim to make searching different types of finds easier for general public, and even for archaeologists. For example when a researcher would like to visualize find sites of iron age swords on a map, it isn't usually relevant if the find is a sword or only a part of a sword, and it is more convenient to simply select swords than having to know to select the sword and parts of swords. The facets ontologies can't be used to make certain logical deductions, but can be useful for quick analysis. It is obviously possible to classify archaeological objects in various different ways, and researches might want to use different classifications in different cases. One benefit of LOD is that it makes creating, using, and sharing various new hierarchy classifications easy.

It would have been possible to create the facet ontologies with some ontology editor. However, to make the creation easier, we opted to use a method where the hierarchies were created using a spread sheet and hierarchy was indicated using different column, so that the concepts on the first column is on the top level of the hierarchy, and a concept on the second column is a narrower concept of the first concept above it in the spreadsheet on the first column, and so on. We then had a purpose built Python script to convert these spreadsheets to RDF

---

[20] Note that we use the term "ontology" in a broad sense that does not make a clear division between a "vocabulary" and an "ontology".

[21] https://finto.fi/maotao/en/.

[22] https://www.getty.edu/research/tools/vocabularies/aat/.

format. Using such method meant that the archaeology expert creating the facet ontology did not have to learn to use the often quite complex ontology editors. On the other hand some issues became apparent only when the ontologies were actually put to use. An ontology editor would better show how the hierarchy actually operates and would also help to avoid some spelling errors.

We created mappings for the Finnish MAO terms to international ontologies, especially the Getty Art & Architecture Thesaurus[23] (AAT), to allow easy comparison of the Finnish find data to similar international data. A mapping to ATT allows also linking to other international vocabularies through AAT. For example we have used a mapping created by the Ariadne[24] project between the AAT and the Forum on Information Standards in Heritage (FISH) Archaeological Objects Thesaurus[25] to create mapping between the Finnish ontology and the FISH ontology that is used for example by the The Portable Antiquities Scheme (PAS) of the British Museum.

2.3   Data Conversion

Source data of FINDSAMPO is received in CSV format and is converted to RDF. The conversion pipeline consist of two main parts: the data conversion and the ontology conversion. Both processes use Python scripts mainly based on RDFLib[26] library, that convert CSV files to RDF. Figure 3 shows the basic steps in the conversion process. In addition to the Python libraries, we have not used any of the various existing tools for converting CSV data to RDF format, because we wanted to define the hierarchies for the ontologies in a specific format, but also because in our experience using the tools of Python, or some other programming language, directly is generally more convenient when dealing with imperfect data that needs cleaning.

First part of the pipeline is the ontology creation process, where the ontologies defined in CSV are converted to RDF format. The pipeline then runs an initial process that creates a simple RDF file with only literal values. Enriching process is then run which cleans up data and creates ontologized values for data based on the ontology definitions. After the data is updated, a triple store is automatically built with the updated data.

The data and ontology conversions are done with similar Python scripts for convenience, but the processes do not depended on each other and they could be done separately and in different ways. Ideally the existing ontology infrastructure would be so strong that we could use entirely ready made ontologies.

The most difficult part of the data conversion process was ontologization the the terms used in the data. The find database used by the FHA allows submitting data freely and no strict vocabulary had been used. It was relatively easy to ontologize the terms used for periods and materials in the FHA data, but

---

[23] https://www.getty.edu/research/tools/vocabularies/aat/.
[24] http://legacy.ariadne-infrastructure.eu/.
[25] http://www.heritage-standards.org.uk/fish-vocabularies/.
[26] https://rdflib.readthedocs.io/en/stable/.

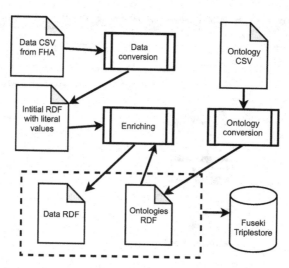

Fig. 3. Conversion pipeline

the terms for object types required more work. We mapped the object types of finds to MAO ontology using "object names" of each object. This was the most detailed description of the type of the objects available in structured format in the find database. An archaeology student was hired to manually do the mapping for object names, as well as to do the mapping from annotation ontology to AAT.

We run the whole conversion pipeline again when the source data is updated and the triple store is automatically build again with new data. The process is fully repeatable, and makes it easy to also fix possible errors in the data. In practice we expect to update the data only rarely. The LOD service[27] is run on the Linked Data Finland platform[28] [10], which is powered by a combination of the Fuseki SPARQL server[29] and a Varnish Cache web application accelerator[30] for routing URIs, content negotiation, and caching. Currently there are around 240 000 triples in the FINDSAMPO main graph.

3   FindSampo Semantic Portal

The FINDSAMPO DATA SERVICE includes currently over 3000 archaeological finds made by the public. The FINDSAMPO PORTAL queries this data service with SPARQL, and offers search, exploration, and analysis tools for DH researchers and hobbyists. The finds can be filtered using faceted search [24] with hierarchical facets based on ontologies, and then visualized using maps with external layers

---

[27] The service can be queried freely with SPARQL at http://ldf.fi/findsampo/sparql.
[28] https://ldf.fi.
[29] https://jena.apache.org/documentation/fuseki2/.
[30] https://varnish-cache.org.

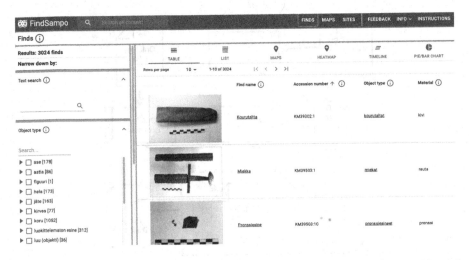

Fig. 4. The main search view of FINDSAMPO PORTAL showing finds as a paginated table on the right, and search facets on the left.

from the GIS services[31] of the FHA, various types of charts, and a timeline. On the front page, see Fig. 1, the user is presented with tree different perspectives: "Finds", "Maps", and "Sites".

The Finds perspective allows for searching and analyzing the archaeological object finds in the knowledge graph using facets and various visualizations. Maps perspective here is just a quick link for user for map visualizations of this perspective. Faceted search can be used to get the information of some specific find, and it can also be used to analyze and compare groups of finds. As default the individual finds are presented as a table as the default option on wider screens, or as a more mobile friendly list with mobile devices. The mobile friendly list option was created after the initial publication of the portal based on feedback from metal detectorists. Figure 4 shows one opened facet and results with pictures of individual finds when available. The various charts and timelines can be selected from tabs to visualize the relative distributions of the selected groups of finds. Currently the user can select visualizations from clustered map, heatmap, timeline, pie or bar charts, and line charts. In addition there is an option to download the data in CSV format.

The clustered map shows interactive markers on a map based on the find coordinates of each find. The clustering is made for performance reasons. The finds can be visualized using different base maps and map layers (selected in the box on the top right) including, e.g., street maps, satellite images, and a lidar-based elevation model.

Heatmap is a more research oriented tool that shows the data about the finds as a spatial heatmap. Figure 6 shows an example of a simple visualization for data-analysis, that can be made easily. The user has selected prehistory period

---

[31] https://kartta.museoverkko.fi/?lang=en.

Fig. 5. An entity landing page of a single find.

from the period facet and the heatmap tab to view the results. It is easy to see that the red area, that signify a lot of nearby finds, are concentrated in Häme and close to the city of Turku, and large areas of Finland have only a small number of finds. A researcher will have to determine if this tells something about the prehistory of Finland, or if this is related more to the popularity of metal-detecting hobby in certain areas.

The timeline tab can be used to visualize temporal spread of groups of finds in certain areas. Timeline component groups the finds by province in which they were found (y-axis), and by period (x-axis). The start and end years for the periods are retrieved from the period ontology developed with domain experts, instead of directly from the finds. Pie chart tab can be used to visualize distributions as pie or bar charts. These can be used to easily visualize, for example, the relative number of coins in finds from the medieval period.

Each individual object find has its own "home page" that contains detailed information about the find. Figure 5 shows an example of an entity landing page of a single find. This entity page of the find includes the detailed information of the find, a map showing the find coordinates, and recommended links to similar finds in Finland and abroad. The object types and periods have their own pages in the same way. The collect information such as the time span of a certain period and links to the related finds.

We have created links to PAS data as an example of connecting international data. This feature shows the possibilities and challenges that a linked data approach can have. The links to PAS as created through mapping FINDSAMPO object types to AAT vocabulary, that is then mapped the FISH vocabulary used by PAS. This means that there is one extra step that can cause various issues for the linking. In practice this can be seen in that in many cases a link is missing, or it can be less than optimal. For example an entity page of a sword find, as in Fig. 5, has a link to a certain object in PAS database that is determined to be similar, based on it's type "sword". Similarly entity page of object type of swords has a link to PAS database search for swords. In practice this can take user to a page of a sword pommel, as those are expressed in PAS data to be of type "sword". The accuracy of this kind of mapping is limited to the least accurate conceptualization.

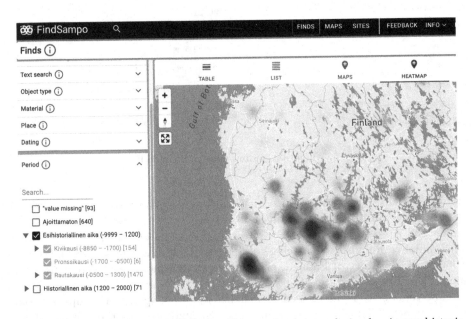

Fig. 6. An example of using FINDSAMPO PORTAL for data analysis: showing prehistoric finds as a heatmap.

The Sites perspective can be used to show finds made by the public and the registered archaeological sites of FHA. This data is received directly from FHA API. As an example, Fig. 7 shows finds (green markers, one of which is opened) and protected archaeological sites (red areas) along the Aura River in Turku, the former capital of Finland. A buffer zone of 200 m where metal detecting is not recommended is automatically calculated and shown around the sites with a dashed line. The maps can be used by researchers for analysis, and by hobbyists to get information on promising places to practice metal detecting as well as on protected sites where detecting should be avoided. This kind of mobile friendly map is particularly useful for metal-detectorists.

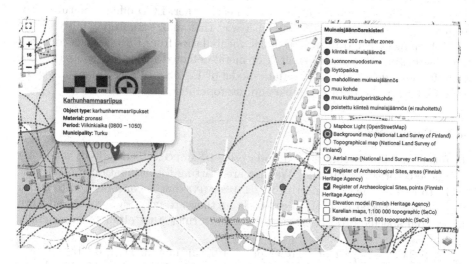

Fig. 7. Archaeological finds and protected sites along the Aura River in the City of Turku as shown in the FINDSAMPO portal.

The user interface of the portal is implemented with the Sampo-UI framework [15], and the source code is available on GitHub[32] with an open license. The performance of the portal is limited mainly by the underlying SPARQL endpoint. Faceted search can consume lot of computational resources, especially when using hierarchical facets. Counting the hit counts every time the facets are updated can be resource intensive. This is not an issue with the number of finds currently in the FINDSAMPO knowledge graph. However this would become issue with when using larger data with perhaps millions of finds. When adapting FINDSAMPO framework for larger number of finds, the hierarchical facets can be converted to flat ones, or perhaps counting the hit counts could be disabled.

4   Discussion

4.1   Contributions

The apps built into the FINDSAMPO offer a powerful set of tools for examining and analysing archaeological finds, and creating new knowledge and understanding of the past. As an archaeological cultural heritage service FINDSAMPO has

---

[32] https://github.com/SemanticComputing/findsampo-web-app.

been designed to organise, present and make widely available a complex form of crowd-sourced and heterogenous data.

Metal-detected public archaeology is an inherently international field. Find data cannot be viewed as restrictively national cultural heritage, and from a research and knowledge discovery perspective the various public finds archaeological databases are natural partners to each other. LOD offers a natural way of harmonizing international data in a way that makes interoperability possible. Mappings created to international vocabularies from FINDSAMPO concepts make it possible to create research of metal-detected finds that transcendents data from single countries. The data model of FINDSAMPO also offers a way to represent such find data in simple and interoperable way.

4.2 Future Work

In future we aim to continue to update FINDSAMPO with new finds made my metal-detectorists in Finland. We also are starting a continuation project that seeks to add a a new perspective to FINDSAMPO that is concentrated to coins. Coins are a special case of object finds as they are very numerous, and have many coin specific properties, such as mint or ruler, that other finds generally do not have.

FINDSAMPO is part of a larger pan-European movement in digital cultural heritage services. Major undertakings such as the EU-funded ARIADNEplus project[33] are presently developing data alignment methodology for combining diverse national archaeological databases. Research for transnational data services based on the FINDSAMPO framework is currently being taken forward by a new pilot project funded by the Helsinki Institute for Social Sciences and Humanities at the University of Helsinki, which seeks to integrate the PAS dataset within the FINDSAMPO framework as a test case of its international use-potential.

Acknowledgements. Our work was funded mostly by the Academy of Finland. The authors wish to acknowledge CSC - IT Center for Science, Finland, for providing computational resources.

References

1. Daubney, A., Nicholas, L.E.: Detecting heritage crime(s): what we know about illicit metal detecting in England and Wales. Int. J. Cult. Prop. **26**, 139–165 (2019). https://doi.org/10.1017/S0940739119000158
2. Deckers, P., et al.: MEDEA: crowd-sourcing the recording of metal-detected artefacts in Flanders (Belgium). Open Archaeol. **2**(1), 264–277 (2016). https://doi.org/10.1515/opar-2016-0019
3. Dumont, S.: correspSearch - connecting scholarly editions of letters. J. Text Encoding Initiat. (10) (2016). https://doi.org/10.4000/jtei.1742

---

[33] https://ariadne-infrastructure.eu/.

4. Dumont, S., Grabsch, S., Müller-Laackman, J.: correspSearch - connect scholarly editions of correspondence (2.0.0) [web service]. Berlin-Brandenburg Academy of Sciences and Humanities (2021). https://correspSearch.net

5. Gardiner, E., Musto, R.G.: The Digital Humanities: A Primer for Students and Scholars. Cambridge University Press, New York (2015)

6. Hassanzadeh, P.: FindSampo: A Citizen Science Platform for Archaeological Finds on the Semantic Web. Master's thesis, Aalto University, School of Science, Finland (2019). http://urn.fi/URN:NBN:fi:aalto-201912226669

7. van den Heuvel, C.: Mapping knowledge exchange in early modern europe: intellectual and technological geographies and network representations. Int. J. Human. Arts Comput. **9**(1), 95–114 (2015). https://doi.org/10.3366/ijhac.2015.0140

8. Hitzler, P.: A review of the semantic web field. Commun. ACM **64**(2), 76–83 (2021). https://doi.org/10.1145/3397512

9. Hyvönen, E., Leskinen, P., Tuominen, J.: Lettersampo - historical letters on the semantic web: a framework and its application to publishing and using epistolary data of the Republic of Letters (2022). https://seco.cs.aalto.fi/publications/2020/hyvonen-et-al-lettersampo-2020.pdf, submitted for peer review

10. Hyvönen, E., Tuominen, J., Alonen, M., Mäkelä, E.: Linked data Finland: a 7-star model and platform for publishing and re-using linked datasets. In: European Semantic Web Conference, vol. 8798, pp. 226–230 (2014). https://doi.org/10.1007/978-3-319-11955-7_24

11. Hyvönen, E.: Linked open data infrastructure for digital humanities in Finland. In: Digital Humanities in Nordic Countries, 5th Conference (DHN 2020), Proceedings. CEUR WS Proceedings, vol. 2612 (2020). http://ceur-ws.org/Vol-2612/

12. Hyvönen, E.: "Sampo" model and semantic portals for digital humanities on the semantic web. In: Digital Humanities in Nordic Countries, 5th Conference (DHN 2020), Proceedings. CEUR WS Proceedings, vol. 2612 (2020). http://ceur-ws.org/Vol-2612/

13. Hyvönen, E.: Digital humanities on the semantic web: Sampo model and portal series (2022). http://semantic-web-journal.org/content/digital-humanities-semantic-web-sampo-model-and-portal-series, submitted

14. Hyvönen, E., et al.: Citizen science archaeological finds on the semantic web: the FindSampo framework. Antiquity Rev. World Archaeol. **95**(382), e24 (2021). https://doi.org/10.15184/aqy.2021.87

15. Ikkala, E., Hyvönen, E., Rantala, H., Koho, M.: Sampo-UI: a full stack javascript framework for developing semantic portal user interfaces. Semant. Web Interoperabil. Usabil. Appl. **13**(1), 69–84 (2022). https://doi.org/10.3233/SW-210428

16. Linstone, H.A.: Multiple perspectives: concept, applications, and user guidelines. Syst. Pract. **2**(3), 307–331 (1989). https://doi.org/10.1007/BF01059977

17. Marchionini, G.: Exploratory search: from finding to understanding. Commun. ACM **49**(4), 41–46 (2006). https://doi.org/10.1145/1121949.1121979

18. van Miert, D.: What was the Republic of letters? a brief introduction to a long history (1417–2008). Groniek **204**(205), 269–287 (2016)

19. Oksanen, E., et al.: Digital humanities solutions for pan-European numismatic and archaeological heritage based on linked open data. In: Digital Humanities in Nordic and Baltic Countries conference (DHNB 2022) (2022). https://seco.cs.aalto.fi/publications/2022/oksanen-et-al-diginuma-dhnb-2022.pdf

20. Rantala, H., Ikkala, E., Koho, M., Tuominen, J., Rohiola, V., Hyvönen, E.: Using findsampo linked open data service and portal for spatio-temporal data analysis of archaeological finds in digital humanities. In: ISWC-Posters-Demos-Industry

2021 International Semantic Web Conference (ISWC) 2021: Posters, Demos, and Industry Tracks. CEUR Workshop Proceedings (2021). http://ceur-ws.org/Vol-2980/paper330.pdf

21. Ravenek, W., van den Heuvel, C., Gerritsen, G.: The ePistolarium: origins and techniques. In: van Hessen, A., Odijk, J. (eds.) CLARIN in the Low Countries, pp. 317–323. Ubiquity Press (2017). https://doi.org/10.5334/bbi

22. Thomas, S., et al.: SuALT: collaborative research infrastructure for archaeological finds and public engagement through linked open data. In: Digital Humanities in the Nordic Countries (DHN 2018), Book of Abtracts (2018). https://www2.helsinki.fi/sites/default/files/atoms/files/dhn2018-book-of-abstracts.pdf

23. Tolle, K., Wigg-Wolf, D.: Improving data quality by rules: a numismatic example. In: Digital Archaeologies, Material Worlds (Past and Present), Proceedings of the 45rd Annual Conference on Computer Applications and Quantitative Methods in Archaeology, pp. 193–201 (2020)

24. Tunkelang, D.: Faceted search. Synth. Lect. Inf. Conc. Retrieval Serv. 1(1), 1–80 (2009)

25. Tzitzikas, Y., Manolis, N., Papadakos, P.: Faceted exploration of RDF/S datasets: a survey. J. Intell. Inf. Syst. 48(2), 329–364 (2017)

26. Veen, V.: The Netherlands during the Napoleonic Era (1794–1815). Using detector finds to shed light on an under-researched period. In: Methods in Conflict Archaeology, 10th Fields of Conflict Conference, vol. 1, pp. 19–30. Mashantucket Pequot Museum & Research Center (2018)

27. Verboven, K., Carlier, M., Dumolyn, J.: A short manual to the art of prosopography. In: Prosopography Approaches and Applications: A handbook, pp. 35–70. Unit for Prosopographical Research (Linacre College) (2007)

28. Wessman, A., et al.: Citizen science in archaeology: developing a collaborative web service for archaeological finds in Finland. In: Jameson, J.H., Musteaţă, S. (eds.) Transforming Heritage Practice in the 21st Century. OWA, pp. 337–352. Springer, Cham (2019). https://doi.org/10.1007/978-3-030-14327-5_23

29. Wessman, A., et al.: A citizen science approach to archaeology: finnish archaeological finds recording linked open database (SuALT). In: Digital Humanities in the Nordic Countries 2019, pp. 469–478. CEUR WS Proceedings (2019)

30. Wessman, A.P.F., Thomas, S.E., Rohiola, V.: Digital archaeology and citizen science: introducing the goals of FindSampo and the SuALT project. SKAS 1, 2–17 (2019)

Author Index

Alam, Mirza Mohtashim 253
Alvarez-Coello, Daniel 406
An, Jingmin 39

Baader, Franz 130
Behrend, Andreas 253
Bento, Alexandre 289
Betz, Patrick 74
Bizer, Christian 113
Bobasheva, Anna 370
Both, Andreas 217
Breslin, John G. 93

Chávez-Feria, Serge 338
Chen, Jieying 56
Cocchi, Valerio 427
Curry, Edward 93

Daquino, Marilena 305
Delbecque, Stephanie 323
Demidova, Elena 353
Demir, Caglar 236
Deng, Chenglong 3
Dobriy, Daniil 21

Ehrich, Sofia Colette 387
Eltsova, Maria 217
Ettorre, Antonia 370

Faria, Daniel 461
Faron, Catherine 370
Ferranti, Nicolas 21
Frasincar, Flavius 183

Gänßinger, Merle 165
García-Castro, Raúl 338
Gashkov, Aleksandr 217
Giese, Martin 200
Gottschalk, Simon 353
Guan, Bei 3

Haller, Armin 21
Heindorf, Stefan 236

Hitzler, Pascal 323
Horrocks, Ian 427
Hyvönen, Eero 478

Ikkala, Esko 478

Janowicz, Krzysztof 323
Joshi, Unmesh 147
Jouanot, Fabrice 444

Kamburjan, Eduard 200
Kangyang, Yuxuan 3
Keil, Jan Martin 165
Khan, Md Tansen 253
Khan, Muhammad Jaleed 93
Klungre, Vidar Norstein 200
Koho, Mikko 478
Koopmann, Patrick 130
Kouagou, N'Dah Jean 236
Kriegel, Francesco 130
Kuculo, Tin 353

Laforest, Frédérique 289
Leemans, Inger 387
Lehmann, Jens 253
Le-Phuoc, Danh 406
Le-Tuan, Anh 406
Li, Guanyu 39
Lisena, Pasquale 387
Liu, Zilong 323

Ma, Yue 56
Mai, Gengchen 323
Marx, Lizzie 387
Médini, Lionel 289
Meilicke, Christian 74
Michel, Franck 370
Motik, Boris 427

Nayyeri, Mojtaba 253
Ngomo, Axel-Cyrille Ngonga 236
Nguyen-Duc, Manh 406
Nuradiansyah, Adrian 130

Oksanen, Eljas   478
Olivares, Jordi   427

Peñaloza, Rafael   56
Perevalov, Aleksandr   217
Peroni, Silvio   305
Persiani, Simone   305
Pesquita, Catia   461
Pietrasik, Marcin   270
Polleres, Axel   21
Poveda-Villalón, María   338
Primpeli, Anna   113

Rademaker, Mark   183
Rantala, Heikki   478
Reformat, Marek   270
Regalia, Blake   323
Rodríguez Méndez, Sergio J.   21
Rohiola, Ville   478
Rousset, Marie-Christine   444
Roy, Dylan   427

Sanchez-Ayte, Adam   444
Schneider, Patrik   406
Schwabe, Daniel   387
Shi, Meilin   323
Silva, Marta Contreiras   461

Singh, Kamal   289
Stuckenschmidt, Heiner   74

Teurlings, Tom   183
Troncy, Raphaël   387
Tullett, William   387
Tuominen, Jouni   478

Urbani, Jacopo   147

Vahdati, Sahar   253
van Erp, Marieke   387
van Lookeren Campagne, Roos   183
van Ommen, David   183

Wang, Yongji   3
Wenige, Lisa   253
Wessman, Anna   478
Wu, Bingchao   3

Xu, Wenjie   270

Yang, Hui   56

Zhang, Yujia   270
Zhu, Rui   323
Zou, Changlong   39

Printed in the United States
by Baker & Taylor Publisher Services